CARBON NANOTUBE–POLYMER COMPOSITES

CARBON NANOTUBE–POLYMER COMPOSITES
Manufacture, Properties, and Applications

BRIAN P. GRADY

Carbon Nanotube Technology Center (CaNTeC), the Institute for Applied Surfactant Research (IASR), and School of Chemical, Biological, and Materials Engineering University of Oklahoma, Norman, Oklahoma, USA

WILEY

A JOHN WILEY & SONS, INC., PUBLICATION

Published by John Wiley & Sons, Inc., Hoboken, New Jersey
Published simultaneously in Canada

Library of Congress Cataloging-in-Publication Data:

Grady, Brian P.
Carbon nanotube-polymer composites : manufacture, properties, and applications / Brian P. Grady.
p. cm.
Includes index.
ISBN 978-0-470-59641-8 (cloth)
1. Nanotubes. 2. Carbon composites. 3. Polymer composites. I. Title.
TA418.9.N35G73 2011
621.3815–dc22

2011006421

oBook ISBN: 9781118084380
ePDF ISBN: 9781118084366
ePub ISBN: 9781118084373

10 9 8 7 6 5 4 3 2 1

CONTENTS

PREFACE

I was first contacted by Jonathan Rose of Wiley in 2007 to write this book. I ran as fast as possible from this assignment because I had a vague idea of the amount of work involved! Fortunately, in 2009 Jonathan contacted a colleague of mine with a similar request and when he showed no interest, I decided I was ready to take up such a challenge. Although I certainly would not want to repeat the year gone by in terms of the time that I spent writing the book (now at least I have more than a vague idea of the amount of work involved!), I have to say that it is believed to be a "character-building experience." Hopefully, you, the reader, feel that the effort you spend reading the text was worth the time I spent writing it.

The order of the chapters in this book was logical, going from the fundamentals to the applied. A couple of comments are in order in terms of the nonobvious (at least in my opinion) decisions that were made. Chapter 2 on the properties of nanotubes was judged to be necessary while a similar chapter on the properties of polymers was not because most have at least some experience with the latter. Chapters 5 and 6 in particular could have been reversed without detriment in my opinion; in other words, covering mechanical properties prior to electrical properties was essentially an arbitrary decision. Separating thermal and electrical conductivity was another debatable decision, but I felt that separating the two was reasonable both because the applications are very different and because the underlying phenomena are quite different (e.g., the lack of a tunneling current in thermal conductivity).

In general, each chapter begins with a general introduction as does each subsection. My goal is to make the text simple enough for someone with a bachelor's education in science or engineering to understand. No previous polymer or carbon nanotube knowledge is assumed, so background concepts on subjects relevant to the topics important for this text are briefly described. Of course, most background topics are not covered in the detail that each subject deserves, so the text also specifically guides the reader, ideally to review articles or monographs, where the reader can go if more in-depth information is required.

I feel that the most useful part of this book are the tables. The book that I most tried to emulate in this regard was the one by Milton Rosen (*Surfactants and Interfacial Phenomena*, Wiley) in that 90% of the time I look in this book it is to examine one of its tables. This monograph contains a number of tables that are purported to be complete, see, for example, Table 2.1. The tables are complete until the end of 2009; any reference that occurred after this time frame is not included. The author is acutely aware of the number of papers that contain the keywords "polymer" and "nanotube" and it is impossible to know that all of the references from 2009 and prior are included, but the author has made an attempt to be as complete as possible.

However, the author would be surprised if each table did not contain significant omissions, and the fault is entirely mine. The author feels strongly that such tables are invaluable to the novice in the field, in helping to quickly identify papers that are of interest. For any additions (or in fact any changes) to this text, please report to the author at bpgrady@ou.edu; I will make a list of errata that will appear on my web page at http://coecs.ou.edu/Brian.P.Grady/index.html.

No preface would be complete without a list of people to thank. Jonathan Rose certainly deserves thanks, both for contacting me to write this book and for the significant amount of editing that I am sure this text required! Dan Resasco (University of Oklahoma) originally introduced me to nanotubes; if it weren't for Dan Resasco and his capabilities as a designer of catalysts, it is unlikely that I would have ever started working in carbon nanotubes in polymers. Dan also provided a figure for this manuscript. Micah Green (Texas Tech University) and Warren Ford (Oklahoma State University) had the thankless task of reading early versions of each chapter and providing detailed feedback. Any remaining errors are the sole responsibility of the author, and I apologize now for such errors. Alberto Striolo, Liu Shi, Rajesh Tummala, Dimitrios Argyris, and Lindsey Brinkmann all did computer work to provide figures for this book. My wife, Gina Grady, did some of the thankless tasks, such as correcting references. My family deserves the biggest praise, as they had to suffer through many weekends and nights when I was gone working on the "the book." So Gina, Ian, Nate, Owen, Luke, and Boomer, this book is for you.

Brian P. Grady

CHAPTER 1

INTRODUCTION

1.1 SIMILARITIES BETWEEN POLYMERS AND NANOTUBES

The introduction of synthetic polymers at the beginning of the twentieth century led to a revolution in the way people live their everyday lives. Polymers are found in homes, offices, and places of business all over the world; in fact, one could likely find synthetic polymers in every home or apartment in the world. From a molecular viewpoint, polymers have been around much longer than humans; many types of biomolecules including DNA, RNA, and proteins are polymers. Industrially, natural rubber has been used for various applications since the middle of the nineteenth century with the key technological step being the vulcanization of rubber first reported by Charles Goodyear in 1839. However, when Hermann Staudinger first proposed the existence of long-chain molecules just after World War I, many scientists in the established scientific community discounted that such molecules could possibly exist. Of course, Staudinger was correct; polymers are molecules of very high molecular weight. The per annum growth rate of polymer production from the time of Staudinger to 1975 was 15%; since 1975 the rate has slowed down to 8%. Even at this lower rate, the production of polymers far outstrips the per annum increase in the number of people on the planet indicating that people are using more and more polymers. The amount of polymers synthesized each year is approximately 50 pounds per person on the planet, and is more than twice that for people living in Western Europe or the United States. The author has a hard time even imagining a world without synthetic polymers (not to mention the fact he might be out of a job!).

One hundred years from now it might be said that carbon nanotubes had a similar life cycle to polymers, except that the key dates were shifted by about 100 years. The "birth" of carbon nanotubes is generally ascribed to the seminal publication by Iijima in 1991.[1] However carbon nanotubes have been around for a much longer time, just as polymers had been around long before Staudinger's hypothesis. A paper in Russian by L. V. Radushkevich and V. M. Lukyanovich[2] in 1952 first showed transmission electron microscopy images of carbon nanotubes. The first patent for something that, in hindsight, was clearly a carbon nanotube was issued to Hyperion Catalysis in 1987 (U.S. patent 4,663,230). Other examples available in the open literature prove the existence of carbon nanotubes long prior to 1991. In fact,

Carbon Nanotube–Polymer Composites: Manufacture, Properties, and Applications, First Edition.
Brian P. Grady.

Monthioux and Kuznetsov[3] provided a rather detailed answer to the question "Who should be given the credit for the discovery of carbon nanotubes?". It has also been stated that the confusion around who discovered carbon nanotubes has prevented the awarding of any Nobel Prize for their discovery.[4] Regardless, the explosion of interest in carbon nanotubes definitely dates from the Iijima's paper and, as Monthioux and Kuznetsov state, "the undoubted tremendous impact of the 1991 Iijima paper came from the right combination of favorable factors: a high quality paper, a top-rank journal read by all kinds of scientists, including those involved in basic research and fundamental physics, a boost received from its relation to the earlier worldwide research hit (fullerenes), and a fully mature scientific audience ready to surf on the 'nano' wave."

At the beginning of the twenty-first century, possible applications of nanotubes are being explored in medicine, electronics, energy, and polymers. Similar to polymers, nanotubes are attractive for a number of applications because of their unique architecture; nanotubes are just about the only material that has a diameter on the order of 1 nm, lengths on the order of microns, and are rigid. The fact that nanotubes are hollow is another unique property of nanotubes. Beyond the parallels in their history, carbon nanotubes and polymers are very much alike on a molecular basis as well. Consider Table 1.1 that lists the properties of an individual chain of the most common polymer, polyethylene, at a typical commercial length and an individual single-walled carbon nanotube at a typical commercial length.

The only significant difference in the values shown in Table 1.1 is for the persistence length. However, this stiffness difference means a great deal with respect to properties. As anyone who has worked with a carbon nanotube mat could tell you, a difference in chain stiffness means that the properties of a bulk polymer sample and a bulk nanotube sample are very different. Because of their flexibility, polymers can be melted and molded like other liquids; nanotubes cannot be melted and thus never form a liquid phase. Both nanotubes and polymers can be dispersed in liquids; however, upon drying a polymer can form a dense, nonporous film while nanotubes form porous films. There are other significant differences between the two materials not related to chain stiffness; for example, most polymers are thermal and electrical insulators, while nanotubes are thermal and electrical conductors. Nanotube electrical, mechanical, and thermal properties, as well as others, will be presented in Chapter 2.

TABLE 1.1 Comparison Between Polymers and Carbon Nanotubes

	Contour length[a]	Persistence length[b]	Diameter
Single-walled carbon nanotube	1 μm	50 μm; diameter dependent	1 nm
Polyethylene (MW = 100,000 g/mol)	0.9 μm	0.6 nm	0.5 nm

All values are representative.

[a]The contour length is the length of the molecule if it were stretched out and measured from end to end.

[b]The persistence length is a measure of chain flexibility. The longer the persistence length, the less flexible the chain (see Chapter 4).

In addition to history and size, a third important parallel between polymers and carbon nanotubes is the methods by which these two materials are synthesized. One type of very commercially important polymer, including polyethylene, is made using a solid catalytic particle and a gas that contains the reactive ingredient. The most commercially important synthetic method for carbon nanotubes also involves a solid catalytic particle and a gas that contains the reactive ingredient. One of the most important properties of polymers is the small per pound cost; carbon nanotubes definitely do not have this characteristic at present! The most significant component of the cost of a polymer is the cost of the reactive ingredient(s), that is, the monomer. In carbon nanotubes, the reactive ingredient cost is essentially a negligible portion of the total cost. In fact, the most common monomer used for carbon nanotubes, carbon monoxide, is actually less expensive than the most inexpensive polymer monomers. This difference in cost is due to one significant difference between the synthesis of polymers and that of carbon nanotubes. The yield, that is, pounds of product per pounds of catalyst, is on the order of 10,000,000 : 1 for polymers and at best 50 : 1 for carbon nanotubes. Not only is the catalyst cost substantial, but the carbon nanotube product must also be purified in order to remove the catalyst. In polymers, the catalyst is such a small part of the material that usually no purification step to remove catalyst is performed. Other methods can be used to make carbon nanotubes, but these methods are even more expensive and unsuitable for translation into processes of the size necessary for most commercial applications. Various synthetic methods to produce carbon nanotubes will be discussed in Chapter 2.

Synthetic polymers, with some notable exceptions such as silicone breast implants, certain plasticizers in poly(vinyl chloride), and bisphenol A in polycarbonate, are considered to have little or no effect on human health, and in fact have been used extensively in biomedical devices to improve human health. Nanotubes, because of their small size, will easily become airborne if special precautions are not taken. Since the effect of airborne nanotubes on human health is not understood (this issue is beyond the scope of this book), nanotubes must be handled carefully. This characteristic has a significant effect in the way in which nanotubes are processed when being combined with polymers, with the practical effect that the manufacturer of the tubes often must sell a product that is not simply nanotube powder.

1.2 ORGANIZATION OF THE BOOK

Polymers are typically categorized according to repeat unit structure. However, with a given repeat unit structure, polymers can be categorized with respect to chain architecture, that is, as linear, branched, and so on. The six-membered planar graphene ring serves as the repeat unit for all carbon nanotubes and all carbon nanotubes can be thought of as sheets of repeating graphene rings that have been rolled up into cylinders. Even so, carbon nanotubes can also be categorized structurally. Carbon nanotubes can be separated into one of the three types: multi-walled carbon nanotubes (MWCNTs), which consist of 3–30 concentric cylinders having an outside diameter generally from 5 to 20 nm; single-walled carbon nanotubes (SWCNTs), which consist of single cylinders having a diameter from 0.7 to 1.5 nm; and double-walled carbon nanotubes (DWCNTs), which consist of two concentric cylinders. The way in which the graphene sheets are rolled can also be different and this difference has a significant impact on properties. Rolling will be described more completely in Chapter 2; differences in rolling mean that over 50 types of nanotubes exist in many commercial SWCNTs with a much larger number in commercial MWCNTs. Like polymers, there is a broad distribution of tube lengths in a given sample. Hence, a single sample of carbon nanotubes is an extremely complicated mixture of many different products. A further complication rests in the fact that the efficacy of purification procedures to remove catalyst and reaction by-products can be very different. Hence, nominally identical materials from two different manufacturers are guaranteed to be different. A given manufacturer must also be very careful to ensure that batch-to-batch variations in nanotubes are not significant as well.

A number of companies are producing ton-size quantities of nanotubes for the polymer market. Predicting exactly which manufacturers will be making nanotubes at this scale or, for that matter, at a research scale at the time when the book is read would be an exercise in fortune telling. The interested reader can consult the online encyclopedia Wikipedia (http://en.wikipedia.org/wiki/Carbon_nanotube#List_of_Carbon_Nanotube_Suppliers) for an up-to-date complete (or almost complete) list of companies currently supplying nanotubes. Sending the reader to Wikipedia I don't believe is an exercise in fortune telling; Wikipedia will be around much longer than some of the companies currently listed as being manufacturers of carbon nanotubes!

Normally, single-walled tubes in particular are in an aggregated state in which the long axes of tubes are aligned in the same direction and the minimum distance between the tubes is roughly equal to the distance between the concentric cylinders in MWCNTs. This aggregated morphology is termed bundles. MWCNTs may also be bundled, but both MWCNTs and SWCNTs are also often found aggregated on a larger scale like a ball of yarn or string. Most applications, including those that involve polymers, would ideally use individually dispersed nanotubes. To reach this goal two issues must be addressed: deaggregation of nanotubes and the prevention of reaggregation during subsequent processing steps. As will become clear in Chapter 3, only perhaps in samples made by the most careful research laboratories where extraordinary efforts are used might there be polymers containing only individually dispersed tubes. So not only is it difficult to compare results between laboratories because different tubes are used, even in the case where the same tubes are used the dispersion is likely to be different and hence results are still difficult to compare.

Once the existence of carbon nanotubes was recognized by the scientific community, polymer composites were one of the most obvious applications. Carbon fibers, which have diameters 100–1000 times larger than nanotubes and are not hollow, have been used as fillers in polymer composites for over 40 years. Hence, the intellectual jump to use carbon nanotubes in composites was an easy one to make. A large number of companies are currently manufacturing products that contain carbon nanotubes in polymers or are considering products that contain carbon nanotubes in polymers. Partly because of health concerns, the production model for nanotubes in thermoplastics will likely be for a carbon nanotube manufacturer or partner to mix nanotubes with a given polymer at a much higher percentage than what will be used in the final product, i.e., make a masterbatch. Most economical processes for mixing nanotubes with thermoplastics involve high-shear mixing in a twin-screw extruder or the equivalent. The end user will then take the masterbatch and dilute it with polymer resin in order to make the final product. Some high-end uses will likely disperse nanotubes in a solvent to facilitate mixing. For thermosets with randomly organized nanotubes, it is likely that the resin will be sold with already dispersed nanotubes, and the final end user will perhaps dilute the nanotubes and then cure the product into its desired shape. As will become clear in Chapter 3, carbon nanotubes can be fashioned into yarns or fibers ideal for continuous composites. Even in this case, health concerns will likely force companies to sell prepreg, that is, a partially cured sheet of thermoset resin that contains nanotube fibers.

As stated in the previous paragraph, there are two broad classes of polymers: thermoplastics and thermosets. The difference between a thermoplastic and a thermoset is that the former consists of isolated chains (think of a ball of string that is cut into many pieces and the pieces are mixed together) while the latter is made up of interconnected chains (think of a net, although the tie points are not as regular as a net). Most, thermoplastics can be melted, that is, form a high-viscosity liquid, at elevated temperatures. Thermosets cannot be melted at any temperature without chemical degradation. Nanotubes are important for both kinds of polymers.

Polymer matrix nanocomposites, that is, mixtures of polymer and filler with the latter having at least one nanoscale dimension, have a number of interesting phenomena related to the fact that a large fraction of polymer is close to a solid interface. Although it has been understood for a great many years that interfaces, such as the amorphous–crystalline interface in a semicrystalline polymer, can affect the properties of polymers significantly, the proliferation of nanocomposites has popularized the fact that polymer properties near an interface are very different. Coupling that characteristic with high aspect ratio nanotubes means that phenomena are seen in nanotube composites that are seen nowhere else. One striking example is the unique polymer crystalline morphologies that can be achieved with carbon nanotubes. Chapter 4 describes how nanotubes affect and alter polymer physics, and these alterations contribute to some of the unique properties of carbon nanotube–polymer composites.

From a simple rule of mixtures, as well as more complicated models that will be described in Chapter 5, the promise of carbon nanotubes in making ultrastrong polymer composites is clear. At a nanotube volume percent of 10%, a common, inexpensive polymer such as polyethylene or polypropylene could be transformed into a polymer that has a stiffness and strength equivalent to a high-performance polymer such as Kevlar™. Alternatively, 10% nanotubes could be added to Kevlar™ or some other very high-modulus/high-strength polymer, to create a product having no counterpart in specific strength or stiffness. As will become clear, no such improvements have been seen at high loading levels. Carbon nanotubes added to polymers have led to marginal improvements in strength and stiffness, and the resulting composites have followed mixing rules to about 1% nanotube content; however, improvements in mechanical performance have not been large enough to use nanotubes instead of other less costly fillers. Some commercial success with respect to mechanical properties has been found by either replacing part of another filler or adding nanotubes as an additive to an already filled composite. Nanotubes that have been drawn into fibers or yarns and then incorporated into polymers can be loaded to significantly higher fractions with good enhancements in mechanical properties. The market is significantly smaller than the case where the nanotubes are mixed together with a polymer melt, but producing nanotube fibers and then adding a low-viscosity resin that is then cured holds some promise.

A property of which there is no question that the use of nanotubes is being driven by increases in performance is electrical conductivity. Because of the extremely high aspect ratio of a commercial sample of nanotubes, the amount of nanotubes required to achieve percolation, that is, the filler concentration where electrical conduction through a continuous conductive filler network begins, is much lower than that required for conventional fillers. In other words, in order to achieve a conductive material, a much smaller amount of nanotubes is required versus the standard conductive filler carbon black. Since the flexibility of polymers decreases with an increase in filler content, nanotube composites where the nanotubes are mixed in the polymer can have conductivity at higher flexibilities versus carbon black composites. Alternatively, pure mats can be made with carbon nanotubes to make transparent electrodes; these mats are ideal for polymeric substrates used in flexible electronics. Transparent electrodes might seem counterintuitive since carbon nanotubes are black, but such a thin mat can be used so that the absorption is actually quite small. Chapter 6 describes the fundamental science that underpins electrical applications.

When nanotubes were first considered as fillers for polymer composites, thermal conductivity was perhaps the property of most interest to industrial users. Heat management is an incredibly important area in the drive toward miniaturization and often times is the limiting factor for specific applications. Almost no composites made with discontinuous fillers, that is, fillers suitable for common processing operations such as injection molding or coatings, also have high thermal conductivities. If simple mixing rules were appropriate, then the thermal conductivity of a discontinuous carbon nanotube composite at 10% loading would approach the highest values of any polymer system filled with a continuous filler, and be one to two orders of magnitude higher than any discontinuous filled system. Unfortunately, thermal conductivity gains are orders of magnitude smaller than a mixing rule would predict for reasons described in Chapter 7. Still, without any improvements in efficiency, there are possible markets for nanotubes as supplementary thermal conductivity additives in systems filled with continuous carbon fibers.

Chapter 8 describes some applications for carbon nanotubes in or with polymers. Along with carbon fibers, the most commonly used fillers in polymers include carbon black, glass fibers, silica, talc, clays, alumina, and titanium dioxide. What exactly makes carbon nanotubes unique for polymer composite applications? Carbon nanotubes are unique morphologically because of their small diameter and relatively long length. Carbon nanotubes have very high electrical and thermal conductivities. Electrical, mechanical, and electromechanical applications will be highlighted in this chapter.

One bookkeeping note that must be mentioned: A fundamental description of composite characteristics requires that the volume fraction be specified, not the weight fraction. Of course, the weight fraction is much easier to measure. With most materials, there is a simple conversion between weight and volume fraction that involves the density. This conversion is also true for carbon nanotubes; however, the density depends on the diameter distribution of the tubes, with a low of approximately 1.30 g/cm^3 for large diameter single-walled tubes to an asymptotic high of 2.26 g/cm^3 (i.e., slightly less than the density) for multi-walled tubes with a large number of walls. In this text, the decision was made to use whatever was specified in a given paper, that is, weight or volume fraction, except for Chapter 6 where volume percent was used because the importance of modelling to quantifying changes in mechanical properties. For calculation purposes in that chapter, a value of 1.35 g/cm^3 was assumed as the density of SWCNTs and DWCNTs, and a value of 2.0 g/cm^3 was assumed for the density of all MWCNTs if values were reported in weight fractions.

1.3 WHY WRITE THIS BOOK?

Figure 1.1 shows the number of papers in the open literature, according to Web of Science, that have been published with the key words "carbon nanotube" in a given year along with the number of United States patents issued with the keyword "carbon nanotube." Shown on these graphs is the same for polypropylene to provide a basis for comparison. The number of papers and patents on nanotubes is increasing at a rate much faster than on a mature material such as polypropylene. Figure 1.2 shows that the

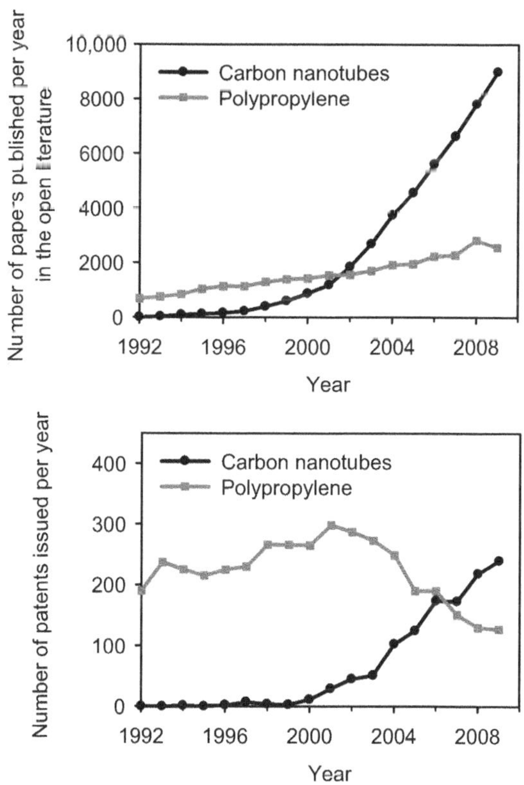

Figure 1.1 Number of papers and U.S. patents with carbon nanotube in either the title or abstract. Polypropylene is also shown to provide perspective.

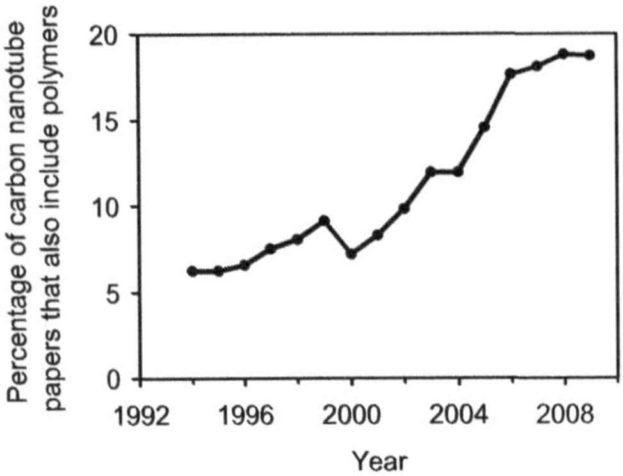

Figure 1.2 Number of papers with carbon nanotube and polymer in the title or abstract as a percentage of the total number of papers with carbon nanotube in the title or abstract.

percentage of carbon nanotube papers that also include the key word "polymer" is increasing. Clearly, there is an increasing acknowledgement, especially commercially, that polymers represent a key product area for nanotubes. Figure 1.3 shows how the percentage of papers with the key words "carbon nanotube" and "polymer" is being split between the two most significant categories of tubes, single-walled and multi-walled. This figure shows that the latter is increasing as a percentage of the whole, which is a direct result of these tubes being able to be produced cheaply as well as being able to be dispersed easily. These figures show the large number of papers that are available on this subject; a significant number of studies on carbon nanotubes and polymers are not referenced or described in the book. Instead of referencing all papers and patens on carbon nanotubes, the book communicates a fundamental understanding of what has been done in the carbon nanotube/polymer field and highlights those key studies that have had significant impact on this understanding.

However, unless otherwise noted, the author has tried to be comprehensive for all tables, that is, include ALL data that appears in refereed publications, although the

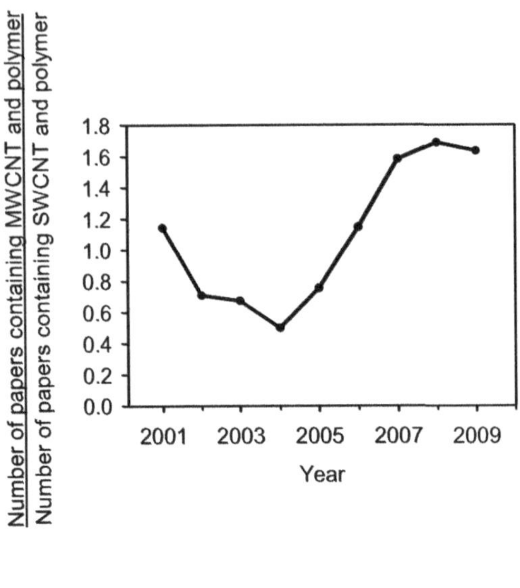

Figure 1.3 Ratio of papers containing multi-walled carbon nanotubes and polymers in the title or abstract to those containing single-walled carbon nanotubes and polymers in the title or abstract.

author did not attempt to translate the 10–20 articles that were in languages other than English, which might have data belonging in a table. Because of the scope of some tables, the author likely missed some studies and apologizes in advance to the authors of papers that were inadvertently not included. (I would appreciate missing references; feel free to report these to the author as described in the Preface.) This chapter was written during the year 2010. Hence, only those references that existed at the end of 2009 (including those that were published online in 2009, but have 2010 publication dates) are used in this text. This text describes completely what is known at the end of 2009, as well as what can reasonably be inferred.

One topic not covered in this text that represents an important field of polymers and nanotubes is biological applications of nanotubes. A number of different possible biological applications involve the interactions of biological macromolecules (DNA, RNA, proteins) with nanotubes, or, alternatively, the interaction of synthetic macromolecules with nanotubes designed for biological applications. The author felt that this field was too unrelated for the readers of this book. However, in some cases, the subject matter was relevant to nonbiological uses of carbon nanotube–polymer composites. In particular, interactions between polymers in solution and dispersed nanotubes described in Chapters 3 and 4 apply equally well to biological and synthetic polymers.

Besides giving the reader a fundamental understanding of carbon nanotubes and polymers, the other goal is to inform the reader of the challenges, in terms of both what we don't know and what needs to be done in order to increase the use of nanotubes in polymers. Exactly what will it take for carbon nanotubes to become as ubiquitous as polymers in our daily lives? These challenges are the reason so many are drawn to nanotubes. Nanotubes are fascinating materials that comprise one of the most exciting new areas of science and technology.

REFERENCES

1. Iijima, S. (1991). Helical microtubules of graphitic carbon, *Nature*, *354*, 56–58.
2. Lukyanovich, V. M., Radushkevich, L. V. (1953). O strukture ugleroda, obrazujucegosja pri termiceskom razlozenii okisi ugleroda na zeleznom kontakte, *Zh. Fiz. Khim.*, *26*, 88–95.
3. Monthioux, M., Kuznetsov, V. L. (2006). Who should be given the credit for the discovery of carbon nanotubes? *Carbon*, *44*, 1621–1623.
4. Harris, P. J. F. (2009). Carbon Nanotube Science Synthesis, Properties and Applications, Cambridge University Press, Cambridge.

CHAPTER 2

CARBON NANOTUBES

2.1 OVERVIEW

As every high school student knows, carbon is the basis of life on Earth. To ascribe only one reason why carbon is so important to living organisms would be foolish; however, the versatility of carbon certainly plays a critical role. A single carbon atom has the capability of forming one, two, three, or four chemical bonds, while the number of compounds that contain carbon is staggering. Four chemical bonds from a single carbon atom enable enantiomer formation, which is very important for, among other things, tacticity in polymers. Pure carbon can also form bonds in single-atom structures in two very different manners. In diamond, carbon forms four bonds with other carbon atoms. Electronically, this material is an insulator because all electrons are involved in single chemical bonds; that is, carbon is sp^3 hybridized. Graphene and fullerenes (see Figure 2.1) have carbon bonded to only three other carbon atoms. In this case, each carbon atom has on average an extra electron (sp^2 hybridization) that can act as a charge carrier, and hence the extra electron allows this form of carbon to be electrically conductive. The lowest energy geometric structure of these threefold bonded carbon atoms is a hexagon as shown in Figure 2.1; however, pentagons and heptagons are not uncommon. Diamond has an arrangement of atoms that is crystalline in three dimensions, while in the sp^2 hybridized materials the carbon atoms are planar or curved planar. Graphene sheets are planar (isolated graphene sheets often show kinks, etc.), while fullerenes have significant curvature.

Graphene is remarkable, having very high in-plane electrical and thermal conductivities, as well as very high strength. Graphite is a substance that consists of graphene sheets stacked on top of one another. Typical in-plane properties of graphite are a resistance of 2×10^6 S/m (about 1/30 the value of copper) and a thermal conductivity of about 200 W/(m K) (about half the value of copper); graphene has considerably higher values for both. The tensile (Young's) modulus of graphite is also quite high, with a measured value of 1.02 TPa.[1] Graphite is not a very strong material because the parallel planes tend to slip past one another; however, the tensile strength of an individual graphene sheet has been measured to be 130 GPa.[2] These values are approximately 5 and 200 times those of steel, respectively. Are the properties of a graphene sheet the limiting values for nanotubes? Surprisingly, as will be detailed in Section 2.4.1, the answer is probably yes for the modulus and no for the tensile

Carbon Nanotube–Polymer Composites: Manufacture, Properties, and Applications, First Edition. Brian P. Grady.

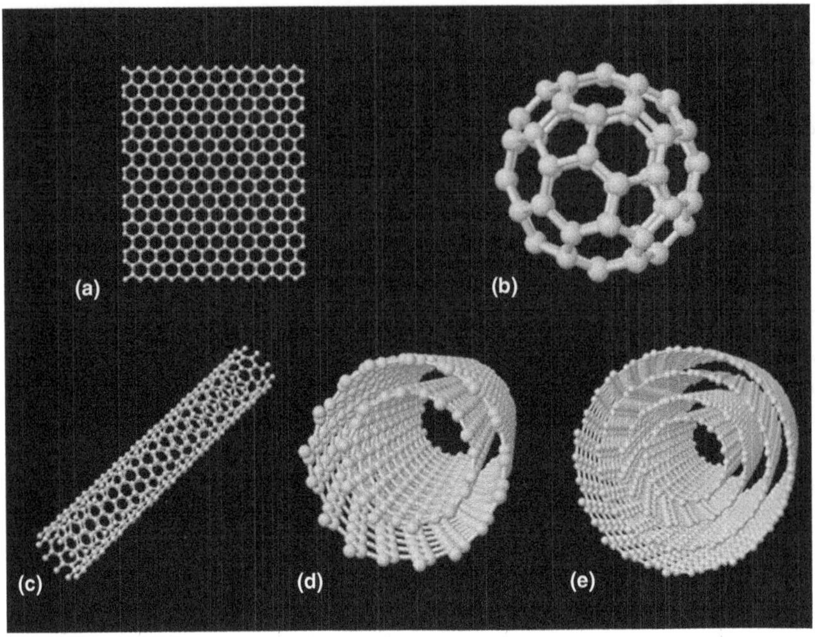

Figure 2.1 Various forms of sp^2 carbon. (a) Graphene sheet; (b) C60 fullerene; (c) SWCNT; (d) DWCNT; (e) MWCNT. Courtesy of Dimitrios Argyris and Alberto Striolo.

strength; strength values can actually be higher for nanotubes. Although nanotubes are correctly visualized as rolled sheets, the mechanism of nanotube formation is not the formation of individual graphene sheets followed by rolling and reaction. Tubes are grown as tubes. In fact, producing individual graphene sheets in large quantities requires a significant amount of effort.

Buckyballs and nanotubes (the latter were first termed buckytubes) are the two major categories of materials that comprise fullerenes. Fullerenes contain planar sp^2 carbon atoms that have been rolled into either a spheroidal or a tubular structure. The explosion of literature occurred first with buckyballs; the primary inventors of these materials, Rick Smalley, Robert Curl, and Harry Kroto, were awarded the Nobel Prize in Chemistry in 1996 because of their discovery of buckyballs roughly a decade earlier. Only a specific number of carbon atoms can be found in a buckyball; the two most common are C_{60} and C_{70}. The former is known to every school child as the structure found on a classic stitched soccer ball. (However recent designs, such as those used for the 2006 and 2010 World Cups, move away from a 32-panel stitched ball to a 14-panel thermally pressed ball!) In the C_{60} structure, all atoms are identical and hence the strain due to the curvature of the naturally planar graphene sheet is equally distributed among all carbon atoms. Figure 2.1 indicates that instead of the six-membered graphite ring found in graphene sheets, some of the rings consist of only five members, that is, pentagons instead of hexagons. The number of hexagons and pentagons is predictable according to Euler's law, which states $V + F - E = 2$, where V is the number of vertices, F is the number of faces, and E is the number of edges. For example, in C_{60}, $F=32$ (12 pentagons and 20 hexagons), $E=90$, and $V=60$. In fact, many synthetic strategies used to modify nanotubes were originally developed for their spheroidal counterparts. Although a Nobel Prize was awarded for the discovery of buckyballs, there are very few commercial applications of buckyballs. As evidence of the lack of interest, the number of issued patents is approximately a factor of 7 less than those issued for nanotubes as of the end of 2009; the number of papers is also approximately a factor of 7 less.

A single-walled carbon nanotube (SWCNT) consists of one tube, a double-walled carbon nanotube (DWCNT) consists of two concentric tubes, and a multi-walled carbon nanotube (MWCNT) consists of two or more concentric tubes. Double-walled carbon nanotubes have their own category because it is possible to synthesize these materials with a reasonably high degree of purity; in fact, multi-walled carbon nanotube samples have some small percentage of single-walled components and vice versa. Figure 2.1 illustrates the three types of tubes. There are essentially an infinite number of ways to roll a graphene sheet in order to form a single-walled carbon nanotube. On the lower end, the minimum diameter is set at the point where the bond strain of the graphene sheets from the preferred planar arrangement becomes too small. On the higher end, it becomes difficult to form SWCNTs of large diameter while preventing the formation of MWCNTs. The smallest diameter SWCNT experimentally verified[3] as well as the smallest inside diameter of a MWCNT[4] was a (3,3) nanotube, corresponding to a diameter of 0.4 nm. More typical diameters of SWCNTs are 0.7–2 nm. DWCNTs have typical diameters ranging from 0.7 to 4 nm. For both SWCNTs and DWCNTs, as-synthesized samples generally are not pure and have some contamination from multiwalled materials. Multiwalled nanotube samples always consist of nanotubes with varying number of concentric cylinders. The separation between the graphene layers is 0.34 nm, which is only slightly greater than 0.335 nm, which is the separation between graphene sheets in graphite; the curvature of the sheets causes the very slight increase in spacing. Defects having significantly larger spacings than 0.34 nm are not uncommonly found in transmission electron micrographs.

Considering a single-walled carbon nanotube, the way in which the graphene sheet is rolled will determine the chirality of the nanotube as shown in Figure 2.2. The chirality of the tube is given by its (n,m) designation. Two special chiralities deserve mention: $n=m$ and $m=0$. Both correspond to the case where the honeycomb lattices located 180° opposite to one another are always parallel. The first type is termed the armchair configuration while the second is termed the zigzag configuration. Strictly speaking, these two types are not chiral at all; that is, they do not have an enantiomeric pair. All other (n,m) tubes have a helical configuration of hexagons, which also means that these nanotubes are enantiomeric; that is, the mirror image of a given tube is not identical. The sharpness of the helical turns is a function of how far a (n,m) value is from either of the armchair or zigzag configurations. The diameter of a tube can be calculated from its (n,m) index as follows:

$$d=\frac{a_{\mathrm{C-C}}\sqrt{3(n^2+mn+m^2)}}{\pi} \tag{2.1}$$

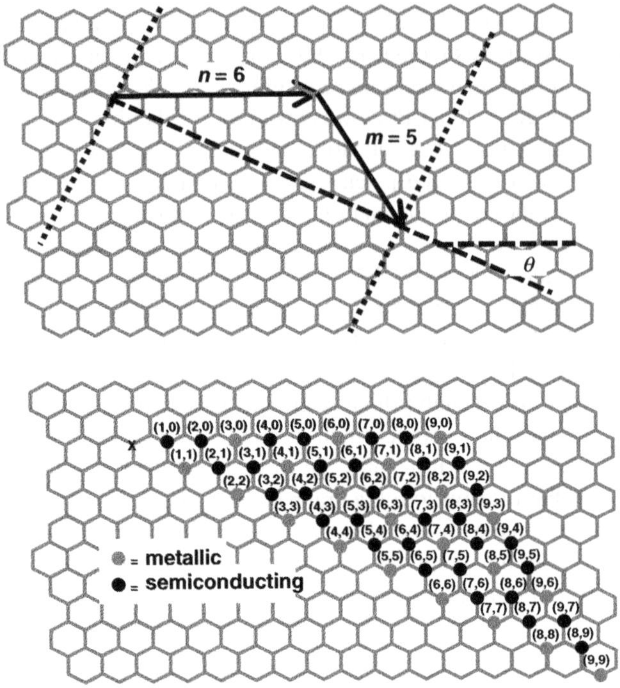

Figure 2.2 Schematic on how to determine the structure of a nanotube from its (*n*,*m*) designation (top). Shorthand showing (*n*,*m*) values on graphene sheet with designation of tube as either metallic or semiconducting (bottom).

where $a_{C=C}$ is the carbon=carbon bond length (1.42 Å). The orientation of the hexagons is characterized by the chiral angle, θ, that varies between 0° (for zigzag) and 30° (for armchair) and is given by

$$\theta = \tan^{-1}\left(\frac{\sqrt{3}m}{2n+m}\right) \tag{2.2}$$

For a double-walled or multiwalled tube, the outside tube determines the overall diameter; however, the same principles described here including Equations 2.1 and 2.2 apply.

Generally, concentric tubes in DWCNTs or MWCNTs do not have similar chiralities; that is, if the inner tube of a MWCNT is of a zigzag type, the rest of the tubes are generally not of the zigzag type. Chiralities cannot be random since the interplanar spacing is approximately constant; hence, there are a finite number of tubes that have the appropriate interplanar spacing. In other words, there must be certain relationships between the chirality of tubes in a concentric cylinder, although this is a very difficult question to confirm experimentally. In rare cases, as-synthesized single-walled nanotubes can switch in chirality; however, the diameter does not change greatly.[5,6] Defects in nanotubes are quite common; positive curvature is induced by pentagon formation and negative curvature by heptagon formation. Kinks are also quite common; a well-explored example involves the connection of (5,5) armchair tubes to (9,0) zigzag tubes, which involves a pentagonal ring on the outside of the elbow and a heptagonal ring on the inside of the elbow[7] although most experimentally observed kinks are much more complicated than this simple example. Of course, there is also the possibility of nanotube defects due to the presence of impurities during synthesis. In general, the number of defects for MWCNTs is larger than that for SWCNTs.

A related question to that of defects is whether the nanotube ends are open or closed. Nanotubes are almost always synthesized with closed ends. The structure of the ends requires some pentagons, just as buckyballs require pentagons in order to form closed structure. Since pentagons are less stable than hexagons, chemically it is possible to open the ends, and reclose them if desired.[8] In fact, freestanding cones (also termed nanohorns) have been formed. Hemispherical, almost flat, and conical ends have all been experimentally observed. For MWCNTs, asymmetric cones are the most common type of cap found on a MWCNT.

Most synthetic methods yield carbon nanotubes with lengths of $\sim$1 μm. Tubes grown from the surface of a plane can be much longer, and have reached centimeter lengths. Since these types of very long tubes cannot be grown in large quantities, there are few examples of ultralong tubes being added to a polymer. Length of nanotubes can be reduced via mechanical milling, sonication in solution, and so on; some high-shear dispersion methods can also reduce nanotube length. Tube length after mixing with a polymer is something that is not often studied or recorded because of the difficulty in making such measurements although such measurements are extremely useful.

It is useful at this point to compare carbon nanotubes to carbon fibers. Carbon fibers have been used in polymers as fillers for over 50 years. Carbon fibers can be made in a variety of different ways, most commonly from thermal treatment in a non-oxygen atmosphere of polyacrylonitrile fibers. The diameter of carbon fibers is on the order of 1 μm and the length is set by the length of precursor fiber; hence, essentially the length can be infinite. Carbon fibers also typically contain small pores because heating causes a reduction in density. The structure of carbon is locally similar to, but in long range very different from, any of the fullerene or graphene sheet structures described previously. The fundamental building block is still the hexagonal arrangement (threefold coordination); however, assuming little or no graphitization has taken place, the number of hexagons that lie in a plane is small and the planes are arranged at nonregular angles with respect to one another. Because of the formation mechanism, planes tend to be aligned in the fiber direction. Ungraphitized carbon fibers are termed amorphous carbon because of this misalignment; in general, the X-ray scattering peak corresponding to the interplanar spacing is not found. Carbon fibers can also have a small number of other atoms to make connections where two planes come together. Carbon black is another example of a very important amorphous carbon.

Through various thermal treatments, a precise description of which is beyond the scope of this book, the boundaries between the planes and pore size/structure can be altered in carbon fibers, which in turn causes a significant change in mechanical properties. Also, the arrangement of small planes can become more regular with thermal treatment (e.g., graphitization). Regardless, the mechanical properties of carbon fibers are significantly inferior to those of a graphene sheet. Typical values of modulus and tensile strength are 300 and 6 GPa for high-strength fibers, and 800 and 3 GPa for high-modulus carbon fibers. Comparing these data to those presented earlier for graphite clearly shows that modulus retention is more common than strength retention.

2.2 SYNTHESIS

There are three major procedures for the manufacture of carbon nanotubes: arc discharge, visible light vaporization, and chemical vapor deposition (CVD). Visible light methods actually consist of three separate methods: laser ablation where a pulsed laser is used, laser vaporization where a continuous laser is used, and solar vaporization where continuous multiwavelength light from a solar furnace is used. Other methods such as catalytic plastic pyrolysis,[9–11] flame synthesis,[12–17] and liquid hydrocarbon synthesis[18] will not be discussed because of their lesser importance compared to the major three methods. Arc discharge and visible light vaporization are very similar in the sense that both involve sublimating graphite into an inert gas at a pressure significantly less than atmospheric pressure and condensing the resulting vapor under a high temperature gradient. A catalyst may or may not be used; the purpose of the catalyst is to direct the growth of the nanotube toward a particular type. What differentiates the two processes is the method used for sublimating graphite: in arc discharge a plasma is used, while in visible light vaporization visible light is used. Instead of graphite, CVD uses gaseous hydrocarbon(s) as the carbon source, and generally temperatures are much lower. CVD requires the use of a catalyst. All three methods can be adapted to make either single-walled or multiwalled tubes; only a few examples have been published where DWCNTs are made with a method other than CVD.

A catalyst is required in order to synthesize SWCNTs by any of the three methods. For single-walled carbon nanotubes, the key step in the synthesis is the formation of a cap (i.e., something like half a fullerene), which occurs only on the surface of a catalytic species. This nucleation step is the key step in the synthesis, because this step determines whether a single-walled carbon nanotube or some other species will form. As anyone who has bought carbon nanotubes knows, single-walled nanotubes are much more expensive than multiwalled nanotubes. The cost of the catalyst is not why SWCNTs cost more; it is the fact that in the most economical process, which is CVD, the yield, that is, pounds of product/pounds of catalyst, is 10–100 for MWCNTs and approximately 0.1 for SWCNTs. In fact, the CVD process to make nanotubes differs from that used to make some polymers, e.g., high-density polyethylene (HDPE), only in the reaction temperature and chemical makeup of the catalyst. However, HDPE sells for ~\$1.00/lb because the yield is about 10,000,000:1, while MWCNTs sell for ~\$100/lb and SWCNTs for ~\$10,000/lb.

Applications involving polymers generally require a significant volume of nanotubes. Both arc discharge and visible light vaporization are inefficient from an energy perspective and are very difficult to adapt for production of large quantities of tubes. However, with respect to MWCNTs, the quality of tubes as measured by the number of defect sites is much lower than that for the CVD tubes. For electrical properties, quality as measured by the number of defects is a very important parameter. Combination methods, for example, arc discharge methods that use a nongraphitic feedstock with a catalyst, are also capable of producing carbon nanotubes; however, these methods have the same scale-up problems as the two major non-CVD methods. CVD is the only method that is capable at this time of providing enough nanotubes to be used in most polymer composite applications.

The growth mechanism of MWCNTs is specific to the type of process used, and will be described in the sections to follow. The fact that a metal catalyst must be present for the formation of SWCNTs should suggest to the reader that the mechanism of formation is likely the same for SWCNTs no matter what synthetic method is used, and it is likely that the reader is correct (there is still some disagreement about this issue). The metal catalyst for SWCNTs is in the form of a few atom cluster and serves as a nucleating site for the formation of the carbon nanotubes. A metal catalyst is required for SWCNTs because otherwise the open (growth) end of the tube will close; a metal catalyst is not required for MWCNTs because "lip–lip" interactions, that is, interactions between the open ends of the concentric cylinders, are able to stabilize the open structure and allow for further growth. The size and chemical composition of the metal cluster are critical in determining the type and chirality of the nanotube that forms.

The most well accepted model of nanotube clustering and growth for SWCNTs is termed the vapor–liquid–solid (VLS) model. In the first stage of growth, end caps, which can be envisioned as half a buckyball, form on a metal catalyst particle. In the arc discharge and visible light vaporization methods, metal atoms cocondense in clusters along with carbon atoms and solid diffusion processes coupled with further carbon atom deposition cause the formation of a cap, for example, something akin to half a buckyball. In the CVD method, a similar type of metal cluster condensation mechanism is possible; however, another route is to preform the metal cluster on a nonreactive catalyst support, such as a silica or alumina particle, and then have carbon atoms condense on the surface of the metal cluster to form the cap. The metal cluster is required to stabilize the end of the tube without a cap. Nanotube growth occurs from the insertion of individual carbon atoms at the base (e.g., non-cap end) of the tube, followed by growth of a ring comprised of hexagons. There are two possible mechanisms for the addition of carbon atoms to the growing structure. In the first, individual carbon atoms are present at the base of the cluster because the metal is supersaturated with dissolved carbon atoms and ejection of individual carbon atoms occurs at the surface. The metal cluster in this case can be either a liquid or a solid. In the second, the carbon atom does not diffuse into the surface of the solid metal cluster, instead carbon atom diffusion is along the surface until the growing nanotube is reached. Almost certainly, the rate-determining step is the insertion of the single carbon atom at the base of the structure stabilized by the metal catalyst particle; the hexagon growth to complete a ring of hexagons follows rapidly.

The diameter of the cluster plays a very important role in the diameter and characteristics of the SWCNT. In one situation, the cluster is small enough that the diameter of the nanotube is roughly the same size as the diameter of the cluster; so the end result is a single nanotube that has an encapsulated metal cluster at one end of the tube. In this case, the size of the cluster and the diameter of the nanotubes are related. In the second situation, the cluster is larger, and hence the nanotube is growing perpendicular from the surface and the result is the commonly observed nanotube "bundles" (see Chapter 3). In this case, the diameter of the tube and the size of the cluster are essentially uncorrelated. Even in the latter case, there appears to be a critical cluster size; that is, if the clusters are too large, SWCNTs will not form. The temperature is also critically important, since temperature will have a strong effect on the solubility of carbon in the particle as well as the probability of forming a tubular structure versus a carbon structure that simply coats the metal cluster.

For SWCNTs, the larger the n-value, the less the sp^2 bond distortion and the more stable the nanotube; however, it becomes more difficult to prevent MWCNT formation. Although armchair tubes are more thermodynamically favored, helical tubes, especially those with $(n,n-1)$ chiralities, are found in higher quantities than thermodynamically predicted because of the single carbon atom nature of the growth process. The insertion of a single carbon atom into the open end of a half-cap naturally leads to a helical type of tube; the process is similar to a screw dislocation from a crystal surface. The prevalence of $(n,n-1)$ chirality versus $(n,n-2)$ versus $(n, n-3)$ ... chirality is determined by the n-value; the larger the n-value, that is, the larger the diameter, the more common the latter species because the distortion of the preferred planar hexagon is less compared to the $(n,n-1)$ species.

The growth mechanism described here is very similar to that found for polymers that are made via heterogeneous catalytic growth, for example, Ziegler–Natta polymers. In both cases, metal catalyst initiates growth and individual monomers (for carbon nanotubes this would be the hexagonal ring) grow near the base of a catalyst. The termination mechanisms though are quite different. Polymers of this type generally terminate via some chemical means that has a significant amount of randomness, as evidenced by the molecular weight distribution. The length distribution of carbon nanotubes does not follow a distribution that is described well by some random process. The geometric effect of catalyst, that is, the fact that much longer nanotubes can be grown from a flat surface, indicates that a mechanical/geometric effect causes termination; that is, the nanotubes become too "heavy" to be supported and the reaction that favors deactivating the catalyst, by for example, coating it with an amorphous shell of carbon, becomes more favored. Of course, even in the absence of such forces this coating reaction can occur, which in turn limits the size of nanotubes that form.

Another theory for the growth mechanisms of SWCNTs will be briefly mentioned.[19] Essentially, this theory differs from the VLS model by the idea that carbon fragments (e.g., hexagons, etc.) grow in the vapor phase rather than individual carbon atoms being found dissolved in the solid and/or diffusing on the carbon surface. The primary evidence for this theory is that it is possible to control the growth of SWCNTs by careful control of the carbon nanotube "seed" required and the fact that metals that do not have significant carbon solubilities are able to produce carbon nanotubes.

2.2.1 Arc Discharge

In a typical arc discharge apparatus shown in Figure 2.3, two flat sheets of graphite, or more commonly the ends of graphite rods, are placed close to one another under reduced pressure. One of the graphite rods is fixed, while the other graphite rod is on a stage that allows the distance between the two ends to be controlled. The chamber is usually filled with either helium or argon (certain other gases, including mixtures, can be used as well) at a pressure around 500 Torr; a continuous flow of gas generally yields better product than simply filling the chamber and sealing the system. The potential between the rods is set at around 20 V, and the rods are slowly moved closer together until an arc occurs. The typical distance of the rods at this point is 1–3 mm, and the typical current that is achieved, depending on the size of the electrodes, pressure, and other experimental variables, is 50–120 A and the typical anode surface temperature is 4000–6000°C. An arc welder is normally sufficient to produce the energy and voltage required for this process. This temperature (2000°C and 3000°C) is high enough to sublimate carbon, that is, convert carbon from a solid to a gas without liquid formation. Because of the current, reduced pressure, and high temperature, a plasma forms. In order to produce the highest quality nanotubes, the plasma should be kept as stable as possible, which typically means keeping the current low. Distance between the graphite rods and the voltage are used to control the current. The anode (i.e., positive potential) loses material at a rate on the order of 1 mm/min, and, fairly quickly, the position of the adjustable stage must be changed in order to maintain the necessary distance between anode and cathode. The rate of nanotube synthesis is on the order of 50 mg/min, which is basically independent of the diameter of the rods. In order to improve performance, it is desirable to cool both the anode and the cathode. Variations in this process include using a liquid medium instead of gas, which removes the necessity of reduced pressures.

In the absence of metal catalyst, two products form, a soot on the reactor walls and a deposit on the cathode. The crumbly soot contains fullerenes, amorphous carbon, and some graphitic sheets, but no nanotubes. The deposit on the cathode consists of a hard outer shell of nanoparticles and MWCNTs, while the core contains about 2/3 MWCNTs and the rest graphitic nanoparticles. The inner diameters of the MWCNTs typically vary between 1 and 3 nm, while the outer diameters typically vary between 2 and 25 nm. Nanotube length is typically not more than 1 μm. These materials are almost always capped at both ends. When a liquid is used, nanotubes will form and drop to the bottom of the vessel, meaning that the process can be operated continuously.

A metal catalyst is typically introduced in this process by drilling a hole in the anode and filling it with a mixture of graphite and metal powder. There is rather a long list of metal particles that have been tried in this process; a complete list is beyond the scope of this text but the reader is referred to the various tomes on carbon nanotubes that all contain at least one dedicated chapter on nanotube synthesis. In this case, instead of two distinct areas where carbon species form, there are four. These are a soft, rubbery species on the walls, web-like material between the walls and the cathode, a hard cylindrical shell around the cathode, and a collaret that surrounds the hard cylindrical shell. The hard cylindrical shell contains MWCNTs, empty or filled

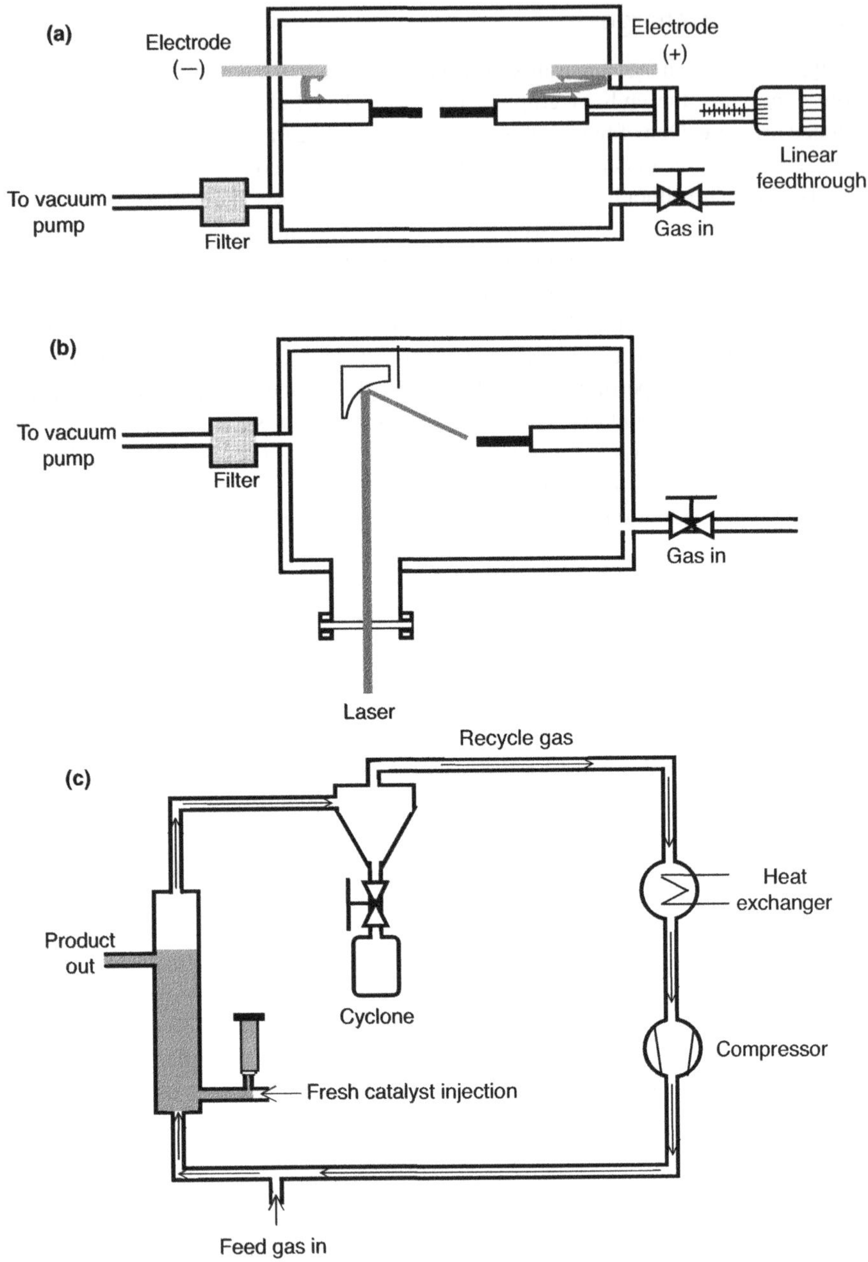

Figure 2.3 (a) Schematic of arc discharge apparatus. (b) Schematic of laser ablation apparatus. (c) Schematic of CVD apparatus.

graphitic nanoparticles, and round spherical metallic particles. If a catalyst that allows for the formation of SWCNTs is used, SWCNTs are found at largest concentration in the collaret. The spider web and walls have similar compositions, which in turn are similar to the compositions where no metals are used, except SWCNTs also found at both places. Several metal catalyst compositions produce SWCNTs, but the current standard widely used for SWCNT production is a Y:Ni mixture that has been shown to have a nanotube fraction of 90% SWCNT, with an average diameter of 1.2–1.4 nm.[20] DWCNTs can be produced using arc evaporation techniques if the proper catalyst and gas are used. Typically, mixed iron, cobalt, and nickel sulfides are used and a mixture of argon and hydrogen seems to give the most selectivity toward DWCNTs.[21] As with the catalytic methods for SWCNTs, not only DWCNTs are produced; both MWCNTs and SWCNTs are present.

The mechanism of SWCNT growth was described earlier and is thought to be no different from the growth mechanism for any procedure. A number of different theories have been advanced regarding the growth mechanism of MWCNTs in the arc and these can be divided into three types. In vapor phase theories, carbon atoms directly condense from the vapor (plasma) phase to form nanotubes. The electric field of the arc plays a critical role in forcing the growth to be along a particular local electric field linear gradient. Local variations in temperature also play an important role. Termination is the result of instabilities in the electric field gradient corresponding to the arc discharge.[22] A second theory involves the nucleation and growth of carbon nanotubes from the supercooled liquid centers of solid-shell carbon nanoparticles.[23] In the final model, carbon initially condenses on the cathode as primarily a two-atom carbon cluster, followed by conversion of these seeds into nanotube structures due to the high temperature of the arcing process.[24]

One advantage of arc discharge is the ease of setting up the system in a laboratory. The equipment required is inexpensive and can be easily operated to produce nanotubes. The number of defects on the nanotubes is also relatively low. One disadvantage is that the product consists of a significant amount of non-nanotube material; hence, the nanotubes must be purified (see Section 2.3). The amount of non-nanotube material is usually greater than 90%; that is, nanotubes are less than 10% of the product that is deposited on the cathode. Also, the amount of product produced is limited by the erosion rate of the target, and increasing the diameter of the target reduces the fraction of nanotubes.

2.2.2 Visible Light Vaporization

In this technique, a high-intensity light source is focused on a graphite block placed in a reduced pressure atmosphere that is maintained at a temperature around 1200°C. The process is shown schematically in Figure 2.3. Similar pressures (500 Torr) and inert gases (He or Ar) are used in the arc discharge and visible light vaporization processes. Laser or solar radiation converts the solid graphite into small vaporized particulates of carbon that will recombine into nanotubes provided a suitable temperature gradient is present. Gas flows gently through the system in order to both carry the particles and help provide the temperature gradient. The nanotubes are collected at cold finger that is placed downstream of the target. Using a pulsed laser requires significantly higher powers (100 kW/cm^2 compared to 0.01 kW/cm^2 for the continuous laser or solar methods). To concentrate enough energy for the production of nanotubes in the solar method, a solar furnace (plus daylight and a clear sky!) is required. As with the arc discharge technique, MWCNTs are collected if pure graphite is used, and SWCNTs are collected if the appropriate metal is used along with graphite. The same sort of preparation techniques for a metal-filled graphite that are used in arc discharge are used in visible light methods. In fact, to grow MWCNTs of reasonable length requires the use of a catalyst to force anisotropy into the system. The mechanism of growth for MWCNTs is thought to be similar to that for arc discharge methods.

Besides scalability, the primary disadvantage of this technique is the cost and difficulty of setting up the system to make nanotubes in this manner. The primary advantages are both the quality of nanotubes in terms of low defect density and their purity; about 50% of the material is nanotubes.

2.2.3 Chemical Vapor Deposition

CVD is by far the most studied of the methods used to make carbon nanotubes. As stated previously, the primary difference between CVD and arc discharge/visible light methods is that in CVD a variety of hydrocarbons can be used as the carbon source. To the polymer scientist (including the author!), this designation is confusing because CVD methods to produce nanotubes are often categorized as catalytic methods; however, as described previously, catalysts are used in both arc discharge and visible light methods. In fact, the acronym CCVD (catalytic chemical vapor deposition) is often used rather than CVD. All CVD methods require a catalyst because hydrocarbon feedstocks cannot be induced to form nanotubes without a catalyst. Because of the essentially infinite number of possibilities for reaction conditions, there is always hope that CVD could be used to produce a type of nanotube ideal for a given application.

CVD is essentially a thermal reaction whereby a transition metal catalyst, for example, iron, nickel, or cobalt, is used to lower the temperature required in order to "crack" a gaseous hydrocarbon feed into carbon and whatever else is in the material, typically hydrogen or oxygen. In other words, as opposed to graphite for the previous two methods, carbon atom formation is through the decomposition of carbon monoxide or some other carbon-containing species. Either a supported catalyst, where the metal particles sit on the surface of a nonreactive particle such as silica or alumina, or a floating catalyst, where metal particles are suspended in a gas, can be used. These processes are very similar to processes used to form vapor-grown carbon fibers, except that the metal clusters are much smaller for the manufacture of nanotubes. A very important class of floating catalyst used for SWCNT production is used in the HiPCO (high-pressure carbon monoxide) method. In this method, CO decomposition is catalyzed by Fe clusters generated *in situ* by decomposition of $Fe(CO)_5$ in continuously flowing CO at high pressure and elevated temperature. With supported catalysts, either fixed beds or fluidized beds have been used. A fluidized bed is one in which a gas is pumped through a bed of solid particulates at a velocity sufficient for the particulates to act as a fluid, whereas a fixed bed is one in which the particulates remain in the same place whether gas flow is present or not. The term "nonreacting" as regards the support is a bit misleading, since the support can help stabilize catalyst particle size and can be involved with electron processes that contribute to the reaction. Since fixed or fluidized beds are more commonly used because of the easier ability to scale-up, these are shown schematically in Figure 2.3. In fact, supported catalysts in fixed or fluidized beds are very commonly used to manufacture a wide variety of materials, including certain polymers.

The composition and, more particularly, the way in which the catalyst is prepared and its final morphology are closely guarded trade secrets. To prevent coalescence, refractory metals (tungsten, molybdenum, niobium, tantalum, and rhenium) would seem to be natural choices. However, nickel, cobalt, and iron have been found to be the most effective as CVD catalysts, whether being used for carbon fiber or carbon nanotube production. The key characteristic seems to be the ability of the material to solubilize sufficient amounts of carbon atoms. These metals are often mixed with refractory metals to reduce coalescence. There are two approaches to making the catalyst: one is to make the catalyst independent of the nanotube synthesis while the other is to make the catalyst *in situ* from an organometallic precursor or oxide solution/compound.

A wide variety of carbon-containing species have been used to produce nanotubes by CVD. The most common is carbon monoxide because of both cost and simplicity. Another critical feature is that the temperature required to crack CO is rather low. Low temperatures are important because the metal clusters must not coalesce; coalescence leads to carbon fibers instead of nanotubes. Other types of carbon-containing species include methane, acetylene, benzene, and alcohols. Generally, the carbon source is mixed with an inert gas (e.g., He or Ar) to allow for better reaction control.

The temperature of the reactor must be adjusted so as to allow for the cracking of the carbon source in question, but not too high so that pyrolytic carbon forms. In general, if a particular combination (catalyst, gas mixture) is found to be effective at producing nanotubes, at lower temperatures (say 700–800°C) MWCNTs are favored and at higher temperatures (say 850–950°C) SWCNTs are favored.

What are the characteristics that favor the formation of MWCNTs versus SWCNTs? Other than the lower temperature generalization described above, smaller metal cluster sizes favor the formation of SWCNTs. A very interesting recent experiment showed that the thickness of a thin catalytic film on a flat surface was critical to the formation of SWCNTs versus MWCNTs; thinner films promoted SWCNTs versus MWCNTs.[25] This result suggests that the effect of particulate size in determining the type of tube is related to the amount of solubilized carbon, as well as perhaps the size of the particle relative to the diameter of the tube. Again assuming that mechanisms on flat substrates are relevant to mechanisms for other types of geometries, in general, base growth has been found pretty much exclusively for SWCNTs, while tip growth (i.e., the catalyst particle is found on the top of the tubes, not the bottom) is found as the predominant mechanism for MWCNT growth. Whether a catalyst particle will stay on the surface or become part of the growing tube depends at least partly on the adhesion of the particle to the substrate as well as the abilities of the substrate and the metal cluster to stabilize an "open" end of a nanotube.

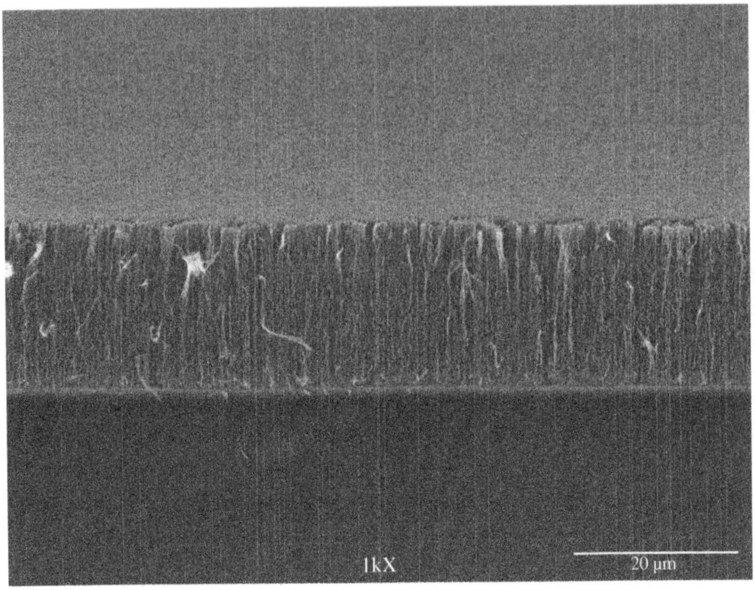

Figure 2.4 Micrograph showing nanotube forest. Courtesy of Dan Resasco.

One unique capability of the CVD process is the ability to use a support that has some characteristic geometry. The most obvious example of this is when flat substrates are used. Nanotube forests, an example of which is shown in Figure 2.4, consist of nanotubes that are grown perpendicular to a flat surface. Other than helping to answer mechanistic questions as explained in the previous paragraph, the unique structural feature of these nanotubes is that they can be made to be very long: 4 cm long SWCNTs[26] and 4 mm long MWCNTs.[27,28] These materials have been used to create yarns as described in Section 3.7 that in turn have been used similarly to the way continuous carbon fibers are used in thermosetting polymers. Other important possible uses of nanotube forests include as superhydrophobic surfaces and field emitters. Similarly, catalyst can be placed on a substrate in some pattern that allows for the growth of perpendicularly aligned nanotubes only in discrete places on the surface. Finally, catalyst can be patterned in such a manner so as to allow horizontal growth along a flat surface. Certainly, surface chemistry and surface characteristics also have a role in the ability to directionally grow nanotubes.

As stated previously, the main advantages of the CVD process are economics and scalability. It is certainly true at present and is expected to be true in the future that nearly all nanotubes that are made for use in polymers will be synthesized via a CVD process. Another advantage is that the amount of non-nanotube carbon material made can be a sufficiently low percentage (less than 1%) so as not to require removal. A significant disadvantage is that with current yields, the metal cluster must be removed because of its ability to cause undesirable side reactions in polymers; often the support is removed as well. Another disadvantage for MWCNTs is that the number of defects seems to be higher for CVD processes; for SWCNTs the number of defects seems to be the same as in the other processes used.

2.3 PURIFICATION

Purification can refer to the removal of carbon material, which is not nanotubes, or to the removal of non-carbon material, for example, metal or catalyst support. In addition, a great deal of effort is being expended to isolate a particular type of nanotube (e.g., either all metallic or all semiconducting), a particular chirality, or even a particular enantiomeric form of a certain chirality. Removal of non-nanotube carbon is only important for those applications that involve a small amount of material, because purification methods cannot economically be applied to large amounts of material. Removal of metal catalyst is critically important for polymers, because metal clusters are very deleterious to polymers because of their capability of catalyzing undesirable chain scission or cross-linking reactions. Catalyst support is generally chemically benign in a polymer; the only issue with nonremoval of support is the fact that it acts as a solid reinforcing agent that may not be desired depending on the amount of support. In practice, the level of catalyst support in MWCNTs is such that commercially it is generally not worth removing while in SWCNTs the support may be worth removing. Isolation of a particular type of tube is very important for certain biological and microelectronic applications and would be desirable for polymers. However, the difficulty, cost, and lack of scalability involved means that these methods are not relevant to polymers, and hence will not be discussed further; the interested reader is referred to Ref. 29 of this chapter.

Purification strategies for non-nanotube carbon are one of two types. The first involves purification via oxidation since non-nanotube carbon structures are more susceptible to oxidation. However, the difference between the oxidative susceptibility of nanotubes and some of the other carbon species found in the raw output from a nanotube reactor is not very high. Gas-phase, liquid-phase, and electrochemical methods have all been used. Because of the narrow operating window, these oxidative methods tend to reduce the length of tubes as well as leave functional groups on the surface that may not be desired. The other method involves physical separation that is possible due to the long length of nanotubes relative to the other carbon species. In this case, the nanotube-containing material must be dispersed in a liquid. One technique is simple filtration; nanotubes are going to remain on the retentate side while many impurities will pass through the filter. Another is flocculation; longer materials are more likely to flocculate; so the addition of a flocculating agent to the dispersion or centrifugation can be used to separate nanotubes from other carbon species.

Of more importance to purification schemes relevant to polymers is the removal of metal particles and catalyst support. These two materials are much easier to remove; that is, the purification methods have less effect on the nanotubes as well as retaining a higher fraction of nanotubes. These characteristics have a great deal of impact on the lower cost of CVD nanotubes; scalability is important but the relative ease of purification is also important. For use in a low-cost material such as polymers, purification is still a substantial component of the overall cost of nanotubes. This situation is very different from the situation with polymers, where in nearly all cases it is the monomer that is by far the most important component of polymer cost (monomer, e.g., gas, cost in nanotubes is negligible!).

To remove metal catalyst, assuming it is not encapsulated by a carbon species, the use of nitric or hydrochloric acid is preferred. Silica support can be removed by strong base or hydrofluoric acid. With MWCNTs, metal clusters and a significant fraction of non-nanotube carbon can be removed by vacuum annealing at high temperature (1600–3000°C), which also improves nanotube quality. If support is present, it must be removed prior to annealing at high temperatures. If necessary to remove non-nanotube carbon from SWCNTs, metal catalyst particles must be removed with acid prior to the more severe oxidation routines because tubes will be destroyed as well if metal catalyst is present.

As with the methods used to purify nanotubes, methods used to assess nanotube purity are often divided into methods to assess non-nanotube carbon content, methods to assess non-carbon content, and methods to assess chirality content. Thermogravimetry (TGA) in air is typically used to assess non-carbon content. Mass loss that occurs between roughly 300 and 1000°C can be attributed solely to carbon species; the remainder is either support or metal. In addition, mass loss between 300 and 1000°C as a function of temperature follows the general trend: non-nanotube carbon, nanotubes with a large number of defects, defect-free nanotubes; hence, this measurement can be used to semiquantitatively determine tube quality as well. Other more surface-sensitive spectroscopies, such as X-ray photoelectron spectroscopy and energy-dispersive spectroscopy, can be used to assess the amount of metals. Raman spectroscopy, which will be described more completely in Section 2.4.3, can be used to determine the amount of nanotube carbon relative to the non-nanotube carbon by integrated intensity of the band at $\sim$1340 cm^{-1} (D band) that is assigned to disordered graphite to the band at $\sim$1580 cm^{-1} (G band) that is assigned to nanotubes. Fluorescence spectroscopy can be used to assign (n,m) designations to semiconducting SWCNTs isolated in a solvent. A 3D plot with excitation wavelength on one axis, emission wavelength on the second axis, and intensity on the third axis can be used to assign each peak to a particular (n,m) type as described more completely in Section 2.4.3.

2.4 PROPERTIES

There are two difficulties in the measurement of individual carbon nanotube properties. First, because of their small size, measurements on individual nanotubes are difficult to make. Second, and more important, the values of the property in question depend on the tube type, number of shells (for MWCNTs), defect characteristics, and so on. With SWCNT samples containing tens of different types of tubes and MWCNTs containing thousands, it quickly becomes impractical to measure enough samples to achieve a true, representative average unless some assumptions are made, or unless the property can be measured reliably on a macroscopic sample containing a very large number number of tubes. An excellent example is density, which is critically important in calculating volume fraction from weight fractions in polymer composites (most theoretical descriptions of property alteration with filler depend on volume fraction of filler, not weight fraction). The density of a MWCNT with an infinite number of shells is almost that of graphite, which is 2.27 g/cm^3 at room temperature. As the number of shells drops, so does the density, with single-walled carbon nanotubes having a density between 1.3 and 1.4 g/cm^3, depending of course on the radius of the tube. The accurate measurement of nanotube density is essentially an impossible task, so a reasonable value is assumed in most situations.

2.4.1 Mechanical Properties

In terms of applications, the superior mechanical properties of carbon nanotubes are one of the primary reasons for considering the use of nanotubes in polymers. Given the aspect ratio of nanotubes and the uses of other high aspect ratio materials in polymers, it is obvious to first consider the tensile properties of nanotubes. A tensile test refers to a test where a sample is pulled in one direction; for a nanotube this test involves pulling a nanotube in the axial direction. The classical definition of the tensile modulus (E) is given by the following equation:

$$E = \frac{1}{V}\left(\frac{\partial^2 G}{\partial \varepsilon^2}\right)_{\varepsilon=0} \tag{2.3}$$

where G is the free energy of the system, ε is the axial strain (change in length/initial length), and V is the equilibrium volume of the system. Experimentally, this expression simplifies to

$$E = \left(\frac{\partial \sigma}{\partial \varepsilon}\right)_{\varepsilon=0} \tag{2.4}$$

where σ is the stress. Since σ has units of force/area, E also has such units since ε is dimensionless. For a single nanotube in either definition, there is a significant issue with respect to the thickness of the nanotube; what is the thickness of a nanotube that is one atom thick? Because of confusion regarding this rather simple issue, a table that reviews computational work in this area found values of the modulus between 0.5 and 5 TPa.[30] Using a consistent basis, for example, the adjacent layer thickness between adjacent walls in MWCNTs, that is, 0.34 nm, gives results that are close to one another. Theoretical values using this basis have varied between 1.26[31] and 0.97 TPa,[32] compared to the measured 1.02 TPa value for graphene.[1] As the diameter becomes smaller, the modulus falls because the additional bond strain due to curvature causes a reduction in the strength of the C=C sp^2 bond. Helicity will cause a decrease in modulus as well, also because of bond strain; in other words, armchair nanotubes will have a lower stiffness than a zigzag tube since all its C=C bonds are curved. The differences due to helicity and curvature are small, on the order of 1–5%, except for diameters below about 0.5 nm, where the modulus falls dramatically. Theoretical values of the MWCNTs are the average of the values for the individual nanotubes.

Experimental measurements generally confirm theoretically-calculated for both MWCNTs and SWCNTs. Again the proper definition of the wall thickness is a problem; most researchers assume a solid cylindrical structure for MWCNTs. Individual MWCNT moduli have been measured using TEM or AFM. For TEM, vibrational amplitudes upon heating of tubes anchored at one end can be used to estimate the modulus.[33] A number of different specific mechanisms can be used with AFM; these include pushing on a tube suspended across an opening but anchored on either side[34] and pulling on a tube that is anchored to both the AFM tip and a surface.[35] Arc discharge and visible light MWCNTs generally have a modulus that is comparable to that of graphene sheets, although the scatter in the data is substantial (between 0.27[35] and 1.8 TPa[33]). A rather detailed study using CVD-grown tubes yielded a modulus that is much lower (0.35 ± 0.11 TPa); both the low value and the high scatter can reasonably be taken as evidence that the number of defects on CVD tubes is higher.[36] Low values of modulus for CVD tubes have been found elsewhere as well.[34] Recent work with DWCNTs and nanotubes with only a slightly higher number of concentric cylinders indicates that CVD tubes can be manufactured with a modulus much closer to that of graphene sheets.[37] Studies on SWCNTs, most often carried out on ropes because of the difficulty of isolating an individual SWCNT, have measured values similar to or slightly larger than that of graphene (1.02 MPa), independent of the method used to synthesize the tubes.[38–40]

Higher strain properties, in both tension and compression, are of interest with respect to polymer composites. Nanotubes can buckle under load relatively easily, and will return to their unbuckled state when the stress is removed. A number of theoretical studies have examined the buckling mode in detail.[41–44] For tensile properties, measurement of individual tubes requires an AFM or similar device where a single tube can be attached at both ends and then pulled apart. For MWCNTs, Ruoff and coworkers were able to show that, in the case where the MWCNT was fixed to a substrate only via the outer tube, the outer tube broke and then the inner tubes were removed by a "sword-in-sheath" mechanism. The breaking strain ranged between 3% and 12%, and the strength ranged from 11 to 63 GPa.[35] The rather large variation was attributed to varying defects in the outer shell. A study on SWCNT ropes indicated a breaking stress in the same regime.[39] However, a study by Wagner and coworkers showed a much higher stress at break of 133 ± 73 GPa for MWCNTs[45] and a range from 10 to 300 GPa[46] using a similar technique as that described by Ruoff and coworkers; no sword-in-sheath mechanism was found, however. The difference in the two techniques was the method used to fix the nanotubes to a surface: Ruoff and coworkers used carbonaceous material while Wagner and coworkers used an epoxy.[47] The latter paper, as well as the response to a comment made on the latter paper, made the argument that the statistics of nanotube failure were best described by a Weibull distribution, so the use of a standard deviation is almost certainly inappropriate.

Measurements on MWCNTs assume that only the outside tube was broken, which micrographs appear to confirm but the resolution of the micrographs is not high enough for the statement to be definitive and hence reported tensile strengths should be recognized as upper limits. In particular, there must be some stress transfer to inner tubes for MWCNTs since in some cases the values presented in the previous paragraph fell outside the highest possible tensile strengths according to the Orowan–Polanyi equation. The Orowan–Polanyi equation states that the breaking stress is reached when the following is true:

$$\sigma = \sqrt{\frac{E\gamma}{a}} \tag{2.5}$$

where γ is the surface energy and a is the interplanar spacing. Using modulus and surface energy values for graphite and an interplanar spacing of 0.34 nm yields a value of 150 GPa. More sophisticated molecular simulations indicate that the maximum tensile strain is between 25% and 30%, while the failure stress is approximately 150 GPa.[48]

2.4.2 Electronic, Magnetic, and Thermal Properties

The electrical properties of nanotubes are extremely complicated, and in fact multiple books have been written essentially on this topic alone.[49,50] This section will give a very brief overview of the electronics of nanotubes with a focus on those properties that are relevant to nanotubes in polymers. So characteristics that are specific to, for example, field-effect transistors will not be discussed. Further, the author has decided to forgo a detailed discussion of band gap theory, which is necessary for the interested reader to have a detailed understanding of the electronic properties of nanotubes. Even a very cursory discussion of band gap theory would add many pages to this tome and is deemed to be outside the scope of this text. The interested reader is recommended to examine the references listed in this paragraph.

For graphite, it is reasonable to assume that conduction occurs only in the direction of the graphene sheets and not perpendicular to the sheets. Detailed and involved calculations for individual graphene sheets indicate an extremely high conductivity, higher than that for silver, which is in contrast to the measured conductivity of graphite that is roughly 30 times less than that for silver. Graphene also has a number of unique electronic characteristics, which have in large part driven a great deal of recent research in the area of using large area graphene sheets for various microelectronic applications.

The unique electronic properties of CNTs are caused by the confinement of electrons normal to the nanotube axis; that is, electrons cannot propagate normal to the long axis of the tube and can propagate only along the nanotube axis. The resulting one-dimensional conduction and valence bands effectively depend on the standing waves that arise around the circumference of the tube. Theoretical calculations of the electronic transport using band gap theory are generally carried out on long straight ribbons of nanotubes, e.g., "unzipped nanotubes." These calculations show that metallic conduction occurs when $n - m = 3q$, where q is an integer. This means that all armchair tubes are metallic, and of the remainder 1/3 are conducting and 2/3 are semiconducting. The term metallic indicates that no threshold energy (voltage) is required for the nanotube to be able to conduct electrons; in other words, the relationship between voltage and current is approximately linear starting at 0 V. The term semiconducting indicates that almost no transport of electrons will occur unless a given threshold voltage is reached (e.g., the band gap voltage); above that potential the relationship between voltage and current is approximately linear. As the tube diameter D increases, the band gap decreases with a $1/R$ dependence. The band gap for a 0.7 nm tube is about the same as silicon (1.1 eV) and drops by about a factor of 2 at a radius of about 1.6 nm. The effect of curvature is to make the metallic non-armchair tubes semiconducting; however, the size of the gap is so small that at room temperature these tubes are effectively metallic. Because of this effect, sometimes the terms large gap, tiny gap, and no gap are applied, rather than metallic and semiconducting. Finally, bundling of armchair tubes will cause the tubes to become semiconducting although the band gap is small; on the order of 0.1 eV.[51]

Conduction is ballistic in armchair nanotubes. The term ballistic indicates that the resistance does not scale with length; that is, the resistance is a certain value no matter how long the tube is. An alternative formulation of this statement is that there is no resistive heating as current passes through the material. However, that is not to say that there is no resistance; the resistance of metallic tubes is ~6.5 kΩ. The mobility of charge carriers, which is related to the speed at which devices work, is fast, about 10 times that of silicon. The current density is also very large, about three orders of magnitude larger than a metal such as aluminum or copper. Transport in semiconducting SWCNTs is much more complicated and appears to be diffusive, the same mechanism as most non-superconducting materials. However, electron mobilities of semiconducting tubes are also extremely high.

Similar to mechanical properties, measured conductivities of SWCNTs and MWCNTs are quite varied, certainly due to varying levels of defects as well as an unknown distribution of chiralities. Measurements on individual MWCNTs have shown both metallic and semiconducting behaviors, with conductivities between 2×10^7 and 8×10^5 S/m and a maximum band gap of 0.3 eV.[52,53] Values measured for bundles of SWCNTs were comparable to the lower value measured for MWCNTs.[54] Measurements of individual single-walled nanotubes gave a value of 32 kΩ for the resistance of a metallic single-walled nanotube ($l = 4$ mm, $d = 1.7$ nm);[55] the conductivity can be calculated as about 5×10^7 S/m. From a more practical perspective, nanotubes can be doped, typically with nitric acid or thionyl chloride, and the increase in conductivity is typically about a factor of 2 for a nanotube film.[56]

An interesting electrical property of carbon nanotubes is their ability to work as field emitters. Field emission is a property by which a material can be induced to eject electrons simply by putting a voltage difference between it and an object, that is, a sheet of metal or wire grid mesh. Carbon nanotubes are excellent field emitters because of their highly anisotropic nature and their small diameter. The ejection of electrons occurs at the tips of the nanotubes, where the nanotube axis is aligned perpendicular to the plane of the metal sheet or mesh. Although there has been a great deal of commercial interest in flat-screen displays using field emission, at present it appears that other technologies, in particular light-emitting diodes, will be the new technology that is used to replace liquid crystalline displays.

The magnetic properties of nanotubes will be briefly discussed primarily because of the role magnetic fields can play in aligning carbon nanotubes in a low-viscosity matrix (e.g., a thermoset resin prior to polymerization). The magnetic susceptibility is defined as the dimensionless proportionality constant between the applied magnetic field and the magnetization of the material, that is, the magnetic dipole moment per unit volume. The difference of this parameter in the direction along the tube axis versus that perpendicular to the tube axis is a measure of the ability of a magnetic field to align the tubes; a higher difference means more propensity to orient (note that magnetic susceptibility can be negative). The absolute value of the magnetic susceptibility of graphite is approximately an order of magnitude higher than most metals such as aluminum or silver, although still orders of magnitude lower than very strongly magnetic materials such as magnetite or iron. As with most weakly magnetic materials, graphite is diamagnetic; that is, a magnetic field is weakened by the graphite, and hence the susceptibility is negative. Further, the susceptibility is ~1.5 orders of magnitude higher when the magnetic field is perpendicular to the graphene plane as opposed to parallel to the plane. A simple-minded view of the susceptibility for nanotubes would suggest that ring currents (i.e., currents that travel around the circumference of the tubes) would dominate and hence the absolute value of the susceptibility would be much higher perpendicular to the tubes than parallel. In fact, studies have shown the exact opposite is true.[57,58] In fact, theoretical investigations have shown qualitatively different behaviors for metallic and semiconducting tubes. For metallic tubes, theoretical studies have indicated that the magnetic susceptibility parallel to the nanotube axis is positive while that perpendicular is negative. Theoretical studies have also indicated that both magnetic susceptibilities are negative for semiconducting tubes with the perpendicular value being larger (less negative) than the parallel value. An experimental measurement with individually isolated SWCNTs found that the difference between the parallel and perpendicular values was 1.4×10^{-5} (CGS units) that agreed with theoretical predictions.[59] Simple arguments from the behavior of graphene can be made that indicate that the measured susceptibility of SWCNTs should be roughly half that of the maximum graphene susceptibility, and measurements have indicated that this is true.[60] In fact, a magnetic field can be used to switch conduction in a SWCNT from metallic to semiconductive or vice versa; however, the magnetic fields required are significantly larger than are practical.

The thermal conductivity of multiwalled tubes has been measured as 3000[61] and 2000 W/(m K)[62] at room temperature and shows a maximum at about 320 K. Of particular interest for polymers, above about 320 K, the thermal conductivity of nanotubes drops by about 20% as the temperature changes by 50 K;[61] unfortunately, no data exist above this temperature but there is no reason to assume that the drop does not continue to extrapolate to higher temperatures.[63] Another study found the thermal conductivity at room temperature to be significantly lower, 650–830 W/(m K), for a single MWCNT.[64] Experimental studies on single-walled carbon nanotubes have given results varying from 2000 to 10,000 W/(m K) at room temperature,[63,65] with a maximum again around 320 K with a similar steep drop with temperature.[63] One particular issue with all of these studies is that various simulations have shown that the thermal conductivity should depend on nanotube length, increasing as nanotubes become longer.[66,67] Making measurements of thermal conductivity is very difficult, especially given the fact that simulations suggest nanotubes can interact with a substrate causing a reduction in the thermal conductivity.[68] Simulations suggest that the thermal conductivity does not depend on chirality[68,69] but has a significant dependence on diameter.[67,68]

Given their relative simplicity, experimental studies on the heat capacity of carbon nanotubes are surprisingly few and have been concentrated in the low-temperature range.[70–72] The reason for this is that the heat capacity has been found to be close to that of graphite for MWCNTs especially at room temperature and above,[71] and the heat capacity of graphite is well known. SWCNTs are a bit more complicated, but still the differences between SWCNT and graphite are small at temperatures relevant for polymers.[70,73] One study did find a significantly higher value for DWCNTs using a bulk measurement technique;[74] however, since this sample was a commercial sample, perhaps residual impurities caused the higher value.

2.4.3 Optical Properties

The existence of sp^2 carbons in graphite causes graphite to have a black color, meaning that all visible light is absorbed. However, this black can vary from a translucent black to an adsorbing black, depending on the area of single graphene sheets exposed on the surface. If the average area of a single graphene sheet on the surface is large enough, such as with highly ordered pyrolytic graphite or even pencil lead, there is a definite sheen that is a result of the conducting electrons found in graphite. If the average area exposed on the surface is smaller, such as with carbon black or charcoal, the color is a very deep black. Carbon nanotubes also have a black color; however, a sample of purely metallic tubes would likely appear much more like highly ordered pyrolytic graphite (the author is not aware of anyone who has produced a large enough sample of purely metallic tubes to confirm this statement!). Individual tubes are almost translucent; that is, the amount of absorption by a single tube is quite small because the diameters are small. In fact, thin sheets of nanotubes with extremely low overall absorption from the UV region to the near-infrared can be manufactured. The combination of electrical conduction and optical transparency has led to great interest of carbon nanotubes in the field of transparent electrodes; see Section 8.2.3.

The focus of the discussion of the optical properties in this section is on the characterization of nanotube chirality and/or dispersion, rather than applications. Nanotubes absorb all wavelengths of visible light (and actually all UV and near-infrared as well) to some degree; however, not all light is adsorbed equally. The adsorption of light by individual single-wall nanotubes at discrete wavelengths is driven by what are termed van Hove singularities, which are in turn electronic transitions between different energy states. In a solid material with one dimension that has a much higher characteristic length than the other two dimensions, electrons are located in discrete energy bands, and only certain transitions between those energy bands are allowed. These transitions are the source of peaks in absorption spectra, with absorption spectra being defined as the measure of absorption of light as a function of the wavelength or energy of the light. Band gaps in silicon and germanium, well-known materials with band gaps in the same energy range, cannot usually be probed using optical absorption because these are not usually made as one-dimensional materials. The ability to use optical methods to probe SWCNTs is a consequence of SWCNTs being direct band gap materials; that is, the minimum energy of the conduction band is directly above the maximum energy of the valence band in momentum space.

Peaks in optical absorption spectra are related to the band gap energies since both arise from the density of electronic energy states and primarily depend on the chirality of the nanotube, although the environment around the nanotube can alter the band gap energy. For a given diameter, the designations of the transitions from low energy to high energy are given by E_{11}^{S}, E_{22}^{S}, E_{11}^{M}, E_{33}^{S}, E_{44}^{S}, E_{22}^{M}, ... Subscripts represent electronic energy bands and superscripts represent semiconducting (S) or metallic (M). For each semiconducting transition E_{ii}, there is a split in energy; that is, absorption occurs at two wavelengths separated by a relatively small energy. The calculation of the location of the absorption maximum has been the focus of a great deal of theoretical effort, and because of the difficulty of calculation these transitions have been represented by a graph called a Kataura plot as shown in Figure 2.5. A relatively simple semiempirical expression for the location of each transition that was used to generate Figure 2.5 is given by the following equation:[75]

$$E_{ii} = \left(a\frac{p}{d}\left[1 + b\log\frac{c}{p/d}\right] + \frac{\beta_p \cos 3\theta}{d^2} \right) + \frac{\gamma}{d} \qquad (2.6)$$

where $a = 1.049\,\text{eV nm}$, $b = 0.456$, $c = 0.812\,\text{nm}^{-1}$, and $p = 1, 2, 3, 4, 5$ for E_{11}^{S}, E_{22}^{S}, E_{11}^{M}, E_{33}^{S}, E_{44}^{S}, E_{22}^{M}, ..., respectively. The β_p values for the two branches are (−0.07,0.05), (−0.19,0.14), (−0.19,none), (−0.42,0.42), and (−0.4,0.4) for E_{11}^{S}, E_{22}^{S}, E_{11}^{M}, E_{33}^{S}, and E_{44}^{S}, respectively, while $\gamma = 0.305\,\text{eV nm}$. The conversion between eV and wavelength is given by the following equation: wavelength (nm) = 1239.8/energy (eV). d and θ were defined in Equations 2.1 and 2.2, respectively. The term that contains γ applies only for $p > 3$.

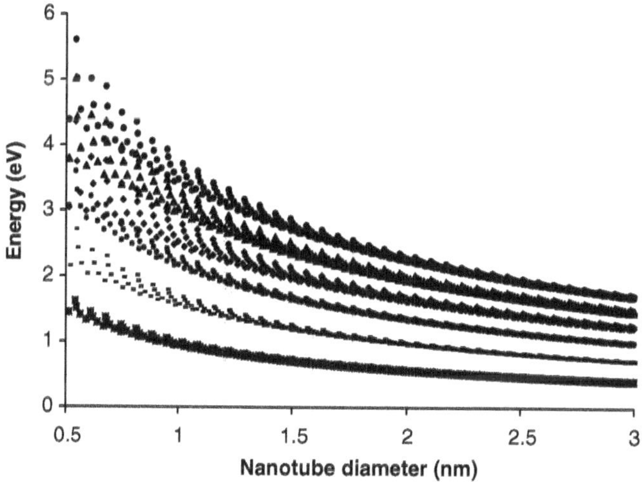

Figure 2.5 Kataura plot. From top to bottom: E_{11}^{S}, E_{22}^{S}, E_{11}^{M}, E_{33}^{S}, E_{44}^{S}, E_{22}^{M}. Courtesy of Lindsey Brinkmann.

The above discussion suggests that it is possible to determine the different types of nanotubes in a sample by the optical absorption spectra. However, the overlap between absorption energies coupled with the small effect of environment, including bundling, on the transition has meant that this technique has proven to be of little utility. As will be described more completely in Section 3.2.2 optical absorption has been used to qualitatively describe the dispersion; bundling generally causes a broadening of line widths and a slight shift to lower energies. With rather laborious purification procedures to isolate an individual type of nanotube (these procedures will not be discussed in this book because of their inapplicability to polymers), the narrowness of the energy transition for an individual tube can be used to build laboratory devices to absorb wavelengths of very precise values. If synthetic routes can be determined to make nanotubes of one particular type, these nanotubes would be used commercially in devices such as optical switches, among others.

Of significantly more use in identifying the specific types of nanotubes present in a given sample is fluorescence spectroscopy. The fundamental idea is that electrons are excited to a higher energy state, for example, E_{22}^{S}, using light of the necessary energy as predicted by Equation 2.6, then the higher energy electron relaxes through a nonradiative process to a lower energy excited state, for example, E_{11}^{S}, and finally a radiative relaxation to the lower energy state occurs, which in this example is the E_{11}^{S} to ground transition. Hence, the absorbed light photon has the energy associated with the ground to E_{22}^{S} transition, while the emitted light photon has the energy associated with the E_{11}^{S} to ground transition. This 3D plot (emission and adsorption on the x- and y-axis with intensity on the z-axis) allows for the semiquantitative identification of the SWCNT chiralities present in the mixture. The nanotubes must be debundled in order to determine chiralities present; typically, this is done by suspending the nanotubes in dilute solution usually with the aid of a dispersing agent. The intensity of zigzag or near-zigzag tubes is small or zero because the absorption coefficient of these tubes is quite small. Of course, only semiconducting tubes will be determined by fluorescence spectroscopy, since electrons in metallic tubes have no need to relax to a lower energy state. In fact, a qualitative measure of bundling can be made by measuring the intensity of luminescence since quenching will occur if metallic nanotubes are bundled with semiconducting tubes.

Raman spectroscopy measures the intensity as a function of the energy difference between light adsorbed and scattered. In most Raman experiments, laser light of one wavelength is used and the output intensity is measured as a function of energy or wavelength of the outgoing light. Normally, the scattered light is measured at an angle of either 90° or 180° relative to the direction of the incoming light. As with IR spectroscopy, the energy scale is normally reported in terms of wavenumber; the conversion is $1\,\text{eV} = 8065\,\text{cm}^{-1}$. This technique is also used extensively in polymer science to determine orientation, chemical identity, and, for thermosets, extent of reaction. As with polymers, the energy adsorption associated with the change in wavelength is due to vibration of molecular bonds in the carbon nanotubes.

There are typically five features in the Raman spectra of a single-walled carbon nanotube. Some of these features are also found in other types of carbon materials as will be described below.

1. Weak peaks between 1700 and 1800 cm^{-1} that are not useful in characterizing nanotubes.
2. A peak around 2500 and 2700 cm^{-1} that is the second-order harmonic of the D band and occurs at precisely twice the frequency of the D band (see point 4) and is labeled the G′ band. The G′ band is less sensitive to defects in an individual nanotube than the D band, and hence this band has been used to study some of the fundamental physics of electron and phonon dispersion in carbon nanotubes.[76]
3. A low-energy 100–400 cm^{-1} mode due to the radius of the tube expanding and contracting, which is termed the radial breathing mode (RBM). Except for very special types of MWCNTs that have extremely small diameter inner tubes, this feature occurs only in SWCNTs and DWCNTs, and the strength and position of this peak are a strong function of the excitation laser wavelength. The dependency on the excitation wavelength is due to the fact that the efficient absorption of energy from the laser is dependent on the nearness of the laser energy to an optical transition; the closer the energy, the more efficient the absorption. The ratio of the intensity of the Raman signal when the wavelength is at the resonance frequency versus when the wavelength is far away from the resonance frequency is on the order of 1000.

   The Raman wavenumber (ω_{RBM}) of the RBM is proportional to 1/diameter; the exact frequency depends on the environment of the nanotubes, for example, whether the tubes are bundled and, if not, then what solvent/dispersing agent is being used. Typically, semiempirical expressions of type $\omega_{RBM} = (A_1/d) + A_2$ are used to quantitatively relate the RBM frequency to the diameter. As shown in Table 8.2 of Ref. 77, the expressions differ by at most a couple of percent indicating that the state of the nanotubes has an effect, but not a large one, on the RBM. It should also be noted that once the constants A_1 and A_2 are determined for one type of nanotube under a particular set of conditions, it has been found that the same constants do a good job (differences at most 1–2 cm^{-1}) for all of the SWCNTs in the mixture.

   Unlike fluorescence spectroscopy, RBMs can be used to determine the presence of all (n,m) species in a sample of SWCNTs, not just the semiconducting species. The Raman intensity of an individual sample can be mapped between 100 and 400 cm^{-1} as a function of the energy of the incoming light. Because E_{ii} are known from Equation 2.6, the maximum intensity as a function of incoming laser energy and the maximum RBM frequency can be used to determine the species present in solution. The tubes must be individually separated for this procedure to be applied so as to eliminate complicating intertube effects. In fact, because bundling tends to red shift the excitation energy (i.e., the excitation energy is at lower energy) and blue shift ω_{RBM}, it is possible to use Raman spectroscopy as a measure of bundling in solution, which can be made semiquantitative with certain assumptions. This technique cannot be used to quantitatively determine tube amounts, because the maximum efficiency of the Raman intensity cannot be determined (i.e., some tubes are more efficient at producing a Raman scattering signal than others even when the incoming wavelength is optimized); however, armchair or chiral nanotubes with a large chiral angle have a higher probability for giving a strong signal.

4. A strong feature between 1250 and 1350 cm^{-1}, which is called the D band, with D representing disordered. The energy of this band shifts slightly as the energy of the incoming light changes; a reasonable value is ~5 cm^{-1} for every 100 nm change in the excitation wavelength. This feature arises from a sp^3 carbon atom that in turn arises from defects in the normal sp^2 bonding, and is found in all types of nanotubes as well as other types of graphitic materials. Fundamentally, this feature is due to a breaking of the symmetry associated with the hexagonal graphene arrangement. Hence, the relative intensity of this band to that of the G band (see point 5) is often used as a qualitative measure of nanotube purity. However, by its very nature, it is impossible to use this band to separate disorderness due to non-nanotube carbon or due to individual nanotube defects. Clearly, however, with a starting nanotube sample, this band is an excellent way to semi-quantitatively determine the amount of non-nanotube material, as well as determining the amount of functionalization as will be described in Section 2.5.

5. A strong peak, or series of peaks, that occurs between 1550 and 1600 cm^{-1} and is termed the G band. This peak is found in graphite at 1582 cm^{-1}. This absorption is due to the movement of carbon atoms in the plane of the graphene sheet in directions 180° relative to one another; in a nanotube these directions are in the tangential plane. For SWCNTs, this band is actually composed of six vibrational modes that in turn give six bands; however, these bands appear as two main bands, termed the G^+ and G^- bands that are roughly 20 cm^{-1} separated from one another. The wavenumber of the G^+ band does not depend on nanotube diameter and is due to vibrations in the same direction as the nanotube axis. The wavenumber of the vibrations of the G^- band does depend on diameter, but cannot be used to determine the (*n*,*m*) designation of a nanotube because the frequency does not depend on the chiral angle. A diameter dependence of $\omega_{G^-} = 1591 - (47.7/d^2)$ cm^{-1} with d in nm has been published.[78] The line shape of the G^- band can be used to determine if a tube is semiconducting or metallic; for semiconducting tubes the line shape is symmetric, while for metallic tubes the peak is asymmetric toward low wavenumbers. For MWCNTs, the position of this band, and its shape, is little, if any, different from that found in graphite.

2.5 CHEMISTRY

Modification of the structure of carbon nanotubes via a covalent bond has been a very voluminous area of research. Extensive studies with spherical fullerenes and to a lesser extent graphite allowed this area of research to advance fairly quickly, since it was a relatively simple matter to transfer approaches from these materials to nanotubes. The focus of this section will be on reactive chemical processes that have been used to improve the compatibility of nanotubes with polymers. The three types of nanotubes behave similarly with respect to covalent chemical reactions carried out on the nanotube surface, a process termed functionalization. Overall, the only significant difference is the reactivity of a given type of nanotube to various chemical synthesis procedures; the fundamental rule is that the higher the curvature or the more different the structure from the typical hexagon, the more reactive the site. So, spherical fullerenes are more reactive than nanotubes that in turn are more reactive than graphene. Further, the ends of nanotubes are more reactive than the sidewalls, and nanotubes synthesized with defects are more susceptible to functionalization than those that have no defects. Smaller diameter nanotubes will also be easier to functionalize, although the energy differences are such that defects and ends are typically more important factors. Finally, sidewall bonds neither perpendicular nor parallel to the tube axis can be functionalized more easily than those that are.

Direct covalent sidewall functionalization is associated with a change from sp^2 to sp^3 hybridization and hence a loss of conjugation. In the carbon nanotube literature, the term "noncovalent functionalization" is used to describe the process where moieties are attached to the nanotubes without covalent bond formation. In this author's opinion, this term is unfortunate and misleading since this process is no different from the more general term adsorption if the outside of the tube is considered (filling the inside of the tube is another matter entirely; however, this type of filling does not occur with polymers). Hence, the term "noncovalent functionalization will not be used in this tome, rather adsorption will be used." Of great interest to the biological community in particular is the fact that polymers adsorb to carbon nanotubes in dilute solution. In polymer physics terms, adsorption of a polymer to a nanotube alters the conformation of the polymer chain significantly. The adsorption of polymers on carbon nanotubes is unique only in that the highly curved surface as well as the small diameter could significantly alter the equilibrium chain configuration in a manner not otherwise found for an adsorbed polymer. This statement could have great practical implications, especially in biology. Pragmatically, determining chain conformation after adsorption is much simpler with nanotubes than with other surfaces. The effect of carbon nanotubes on polymer configuration will be discussed in detail in Chapter 4, and a full discussion of the details of the characteristics of polymer configuration after adsorption on nanotubes will be delayed until that chapter.

Returning to covalent bond formation, e.g., functionalization, the chemical modification of carbon nanotubes has been accomplished on both bundled tubes and tubes that were previously dispersed in solution, either with or without a dispersing agent. As might be expected, the efficiency of chemical modification is higher with individually dispersed tubes all other factors being equal. Further, chemical modification, if significant enough to promote sidewall functionalization, can lead to individually dispersed tubes. The effect of chemical modifications on dispersion and debundling will be delayed until Chapter 3, where these effects along with how adsorption can play a critical role will be discussed in detail.

The purpose of this section is *not* to give the reader a complete or thorough account of the various ways that nanotubes can be chemically functionalized. For a more complete treatment, the reader is urged to consult one or more of the many review chapters on this subject; two are listed in the References section. Figure 2.6 is an incomplete list of reactions that can be used to functionalize nanotubes but does give the reader some sense of the wide variety of methods that can be used. The discussion will be limited to those functionalization methods that are most important, and then a more complete discussion will follow describing the chemistry of polymers grafted to carbon nanotubes. Also, although a great deal of work has been published in the literature containing polymers mixed with functionalized tubes, the author is unaware of any large-volume commercial polymer/nanotube composite that contains functionalized tubes. However, companies are quite secretive about this aspect of their tubes and hence some commercial tubes are likely functionalized either purposefully or during purification procedures and this fact has not been released publicly.

2.5.1 Characterizing the Nature of Functionalization

Quantifying the level and location of functionalization is a nontrivial task. Ideally, the percentage of functionalized carbon atoms and whether the modifications are equally distributed along the axis of the carbon nanotube, and between different chiralities, would be the desired information. For certain modifications, the former can be determined quantitatively through the use of thermogravimetric measurements, while the latter two are difficult to determine. A percentage functionalization of between 1% and 10% of the nanotube carbon atoms is a reasonable value for a number of different techniques.

Raman spectroscopy is often used to semiquantitatively characterize the percentage of functionalized carbon atoms. An excellent review paper deals with this subject in detail;[79] the major conclusions will be summarized here. The first note is that the power of the laser must be set as not to cause the introduction of defects onto the nanotubes! Usually the increase in the D-band intensity, or the ratio of the D-band to G-band intensity, is used as a measure of functionalization. Correlation with independent measures of functionalization (e.g., thermogravimetric analysis or NMR) suggests that at low functionalization levels using either the D-band intensity or the D/G band ratio is probably reasonable as a measure of functionalization, but the approach breaks down at high functionalization levels. Whether the use of the G' band instead of the G band improves the situation is not clear. The use of RBM modes is even worse because of the fact that the RBM depends significantly on bundling and that the efficiency of absorption of light from the fixed wavelength source may change with functionalization. Charge transfer effects for nanotubes in solution must also be considered. Further, in a sample with more than one chirality (that is to say almost any real nanotube sample!), the effect of functionalization on the RBM modes may be different with all other things being maintained equal. In conclusion, Raman spectroscopy is often used to characterize semiquantitatively the amount of functionalization, but this technique is very open to misuse and the interested user is strongly suggested to consult the review chapter and other references therein.

Other methods used to characterize nanotube functionalization include thermgravimetry and temperature-programmed decomposition. Thermogravimetry (TGA)

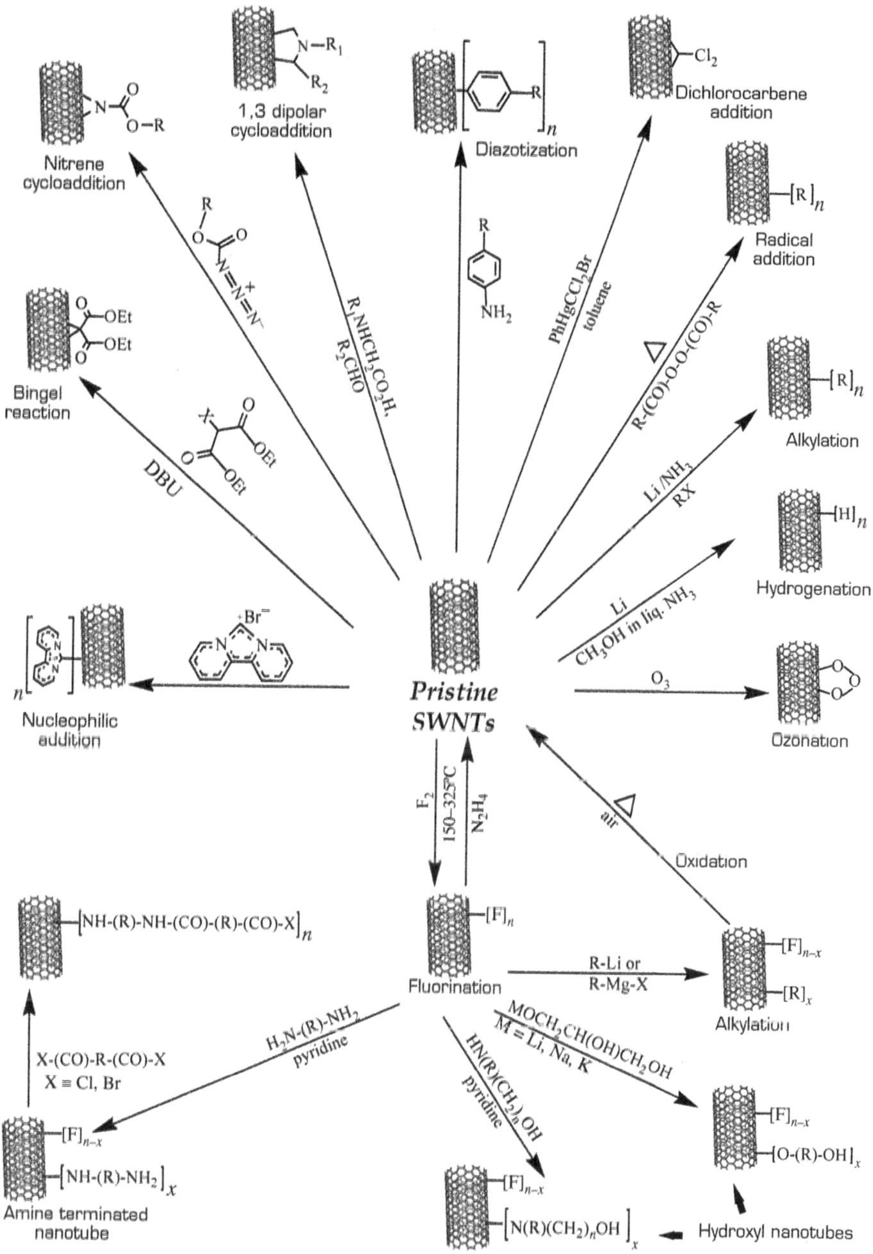

Figure 2.6 Schematic showing some of the synthesis routes available for functionalization of carbon nanotubes. Copyright 2004 Wiley-VCH Verlag GmbH & Co. Reproduced with permission from Ref. 84.

was briefly mentioned early in Section 2.3 as a method to quantify non-nanotube carbon; in TGA, a sample is heated at a constant temperature rate (i.e., 10 K/min) and the weight of the material is monitored with time. Since most covalent bonds involved with functionalization are weaker than the carbon=carbon bonds in nanotubes, this technique offers a way to quantify the amount of functionalization if the formula weight of the functionalizing unit is known and it is known that the nanotube itself does not change weight under the test conditions. Temperature-programming methods involve flowing a gas (typically either hydrogen or oxygen diluted in a carrier gas) over the tubes and heating the tubes at some temperature rate and monitoring the concentration of the gas with time; changes in the gas concentration are indicative of a reaction. Quantification of this method requires calibration with some other method.

Spectroscopic methods, either NMR or IR spectroscopy, are also used. Usually these techniques are used to assess the presence of a certain type of covalent bond between carbon on the nanotubes and another atom, or alternatively, the presence of characteristic features of the moiety itself after the sample has been functionalized and unreacted material eliminated from the functionalized sample. These methods are quantitative if the intensity of one bond at the energy of interest is known; this information is possible to determine most simply by calibration with TGA.

Before describing different chemical means for functionalizing nanotubes, it is first important to consider the effect of functionalization on the intrinsic properties of nanotubes. As surprising as this might seem, experimental investigations of how functionalization affects the mechanical, electrical, magnetic, and other properties of nanotubes are not very common. The reason for this is that it is difficult to be sure that two samples are identical (e.g., bundling is the same) in all aspects except for functionalization, so tedious single tube studies are required. The electrical conductivity of single-walled carbon nanotubes is very sensitive to sidewall functionalization, for example, drops of orders of magnitude in conductivity. The situation for multiwalled tubes is more complicated; since only the outer walls are affected, indirect evidence from polymer composites suggests that the conductivity does not change greatly with functionalization. Mechanical properties show a small drop at typical functionalization levels.[80,81] Computer simulations indicate that functionalization causes approximately a factor of 2 drop in thermal conductivity at typical 1–10% functionalization levels.[82,83]

2.5.2 Common Functionalization Chemistries

The addition of strong acids such as a mixture of nitric and sulfuric acids will cause the formation of various species on the surface of the nanotubes, primarily ketones, aldehydes, and carboxylic acids. Other strong oxidizing agents such as osmium tetraoxide, oxygen gas, and $KMnO_4/H_2SO_4$ can be used.[84] A very severe treatment consists of sonication in the presence of hot, strong nitric/sulfuric acid combination; a more mild treatment consists of using only hot nitric acid. A reduction in nanotube length usually accompanies strong acid treatments; although the more mild the treatment, the less the length reduction. Carboxylic acids are extremely useful functional groups for a variety of reactions; two very relevant to polymers are reactions with alcohols to make esters and reactions with primary amines to make amides. Alcohols and amines are terminating groups for a number of polymers including polyamides and polyesters. Acid-treated nanotubes are often reacted with some other reagent such as thionyl chloride to increase the reactivity. Acids or acid chlorides can be used in a wide variety of reactions that involve polymers; this route is the most important in covalently adding polymers to a carbon nanotube. The level of functionalization varies significantly with the strength of the acid, time, and temperature.

Another important reaction with the sidewalls of carbon nanotubes involves the reaction of substituted aryl diazonium salts, for example, O_2N-⟨◯⟩-N_2 in a reductive coupling reaction. The reaction forms an aryl to nanotube C=C bond at the carbon nanotube surface. The substituting group, typically a nitroxide (N_2O) in order to electrically stabilize the salt, can be used for further chemical reaction if desired, either directly or after reduction of the nitroxide. This reaction is extremely effective at functionalizing sidewalls and does not seem to lead to significant reductions in nanotube length. This chemistry is quite robust; organic media, water media, and superacids have been used; in addition, electrochemical processes have been used to make polymeric layers from these materials.[85] However, cases where this reaction route is used to graft a polymer on a nanotube are much fewer than the acid route.[86,87] Typical substitution levels vary from 1% to 10% for this process.[88] Fluorination is another route that is used to produce tubes that can be further modified to attach polymers; in this case, the fluorine is added most commonly via a high-temperature reaction with fluorine gas.[89] Substitution levels with this technique can be quite high, as much as 50%.[88]

Other reactions that are very obviously relevant to polymers include reduction with lithium metal in liquid ammonia, which is termed the Billups reaction; the lithiated nanotube can then add to vinyl monomers to initiate anionic polymerization. Electrochemical reactions have been used to make polymer-coated nanotubes. An electrochemical reaction occurs when a voltage is applied to a carbon nanotube electrode immersed in a solution that has a suitable reagent that can generate a radical species via electron transfer between the tube and the reagent. If the tubes are unfunctionalized, tubes will be coated with polymer; if tubes are functionalized appropriately, covalently bonded polymer can be present.

The next two sections will detail how different polymers can be attached via “grafting from” and “grafting to” reactions to nanotubes. In a “grafting to” reaction, the polymer is made in some other matter and then attached, often through reactive end groups, to the nanotubes. None of these types of reactions have been done to pristine nanotubes; the nanotubes themselves must be functionalized to enable “grafting to” attachment. In a “grafting from” reaction, the nanotubes are usually (but not always!) first functionalized in some manner that allows for polymerization of the monomer(s) from the functionalized nanotube. Modifying the substrate of a filler to be used in a thermoset so as to promote a chemical reaction is a very important component of many well-known composites, particularly those involving silica. Even in cases without explicit surface modification, for example, filled rubbers and carbon fiber epoxies, thermosets often have chemical reactions between the filler and the polymer, which are critical to the proper performance of the composites. The electrochemical polymerization scheme is an example of a "grafting from" approach. In both cases, most of the motivation for grafting polymers is to improve adhesion and dispersion between the nanotubes and the polymer. The effect of grafting and other tube functionalization strategies on the characteristics of polymer–nanotube interactions will be dealt with in the individual chapters on those particular areas.

2.5.3 Polymer Covalently Bonded to Nanotubes: "Grafting From"

The term "grafting from" implies that the reaction begins at the surface of the nanotube and proceeds from that point. In fact, most of the reactions that fall into the "grafting from" category cannot necessarily be considered as occurring in this manner. To understand why requires a bit of background on the way in which monomers react. Polymers form according to one of two general mechanisms: via addition or via step growth (and the terms used to describe these two types of polymerizations are not standard!). In the former, monomers add one at a time, e.g., monomers add to monomers to form dimers, monomers add to dimers to form trimers, monomers add to trimers to form tetramers, monomers add to tetramers to from pentamers, and so on. Common examples of mechanisms that fall into this category include the various forms of radical polymerizations (most common polymers such as polyethylene and polystyrene are made this way) as well as cationic and anionic polymerizations. In step-growth polymerizations, any two species can react, that is, dimers can add to dimers, dimers can add to trimers, trimers can add to trimers, octamers can add to dimers, and so on. Common examples of this mechanism are acids + alcohols to make polyesters and acids + amines to make polyamides.

With respect to nanotubes in the step-growth mechanism, functional groups are attached to the nanotubes that can react with the growing polymer when the latter is of any length. In other words, polymerization can start anywhere, either at the nanotube surface or in the monomer media. Statistically of course, the probability is that the two monomers will begin a polymerization rather than a functional group on a nanotube and a monomer. In this case, an individual carbon on a nanotube will only be directly bonded outward to one polymer chain; that is, the nanotube acts as an end group of the polymer chain. In the case of addition polymerization, again the growing polymer chain can add the nanotube when the former is at any length; however, in this case, from the perspective of the growing polymer the nanotube acts as a side group rather than a terminal group. Of course, in either case multiple polymer chains are attached to the same nanotube, so the nanotube has the function of a cross-linking agent from the perspective of the polymer.

There are some cases where implication of the term "grafting from" is strictly correct; that is, the reaction begins at the nanotube surface and proceeds outward from the nanotube. In these cases, the nanotube serves as the terminal group for one end of the growing polymer chain and no polymer exists that is not attached to a nanotube. This situation occurs when the group that is attached to the nanotube is the

TABLE 2.1 Monomers Grafted from Carbon Nanotubes to Form Polymer

Monomer	References	Monomer	References
Styrene	86,94,97,99–108	Isopropyl acrylamide	109
Methyl methacrylate	86,91,106,110–121	Various esters	122–131
Butyl acrylate	91,93,132,133	Styrene and acrylonitrile	134–136
Hydroxyethyl methacrylate	137–140	Methylstyrene	141
Butyl methacrylate	105,142–144	Glycopolymer monomer	145
Glycidyl methacrylate	146,147	Ethylene oxide	94,105,148–151
Other acrylates	152	Various amides	153–162
Styrene–maleic anhydride	92,116,163,164	Diphenyl amine	165,166
N-Vinyl carbazole	167	Various block copolymers	93,101,106,168
Water-soluble acids	169–177	Various urethanes	96,178–184
Vinyl pyridine	106	Epoxy	87,150,185–206
Various imides	161,207,208	Aniline	165,171,209–216
Pyrrole	217,218	Thiophene	219
Chlorinated propylene	220,221	Vinyl acetate	222
Dienes	223	*N*-Methyl-2-ethynlypyridinium triflate	224
Polysilsesquioxane	225,226	Cyanoester	227
Propylene	228–230	Maleic anhydride	231
Ethylene	95,232–238	Caprolactone	151
Lactide	239,240		

initiator, and is the only initiator present. Some specific examples include ring-opening metathesis polymerization,[90] atom-transfer radical polymerization,[91] reversible addition–fragmentation chain transfer polymerization,[92] and anionic polymerization.[93,94] Another case similar to the nanotube-initiated reaction is where the nanotubes are impregnated with a particulate catalyst, for example, a metallocene catalyst,[95] from which a polymer is grown. However, since the catalyst is not covalently bonded to the nanotube, and hence the tube is not functionalized this process will be considered in more detail in Chapter 3.

Addition polymerizations do occur where nanotubes do not have to be pre-functionalized in order to graft to a polymer. High-energy radiation such as microwaves[96] or gamma radiation[97] is able to cause the direct attachment of certain polymers to the nanotubes. In fact, the very common free radical initiator 2,2′-azoisobutyronitrile (AIBN) is capable of inducing breakage of the π bonds of carbon nanotubes,[98] which in turn allows "grafting from" the surface of the nanotubes during an otherwise normal free radical polymerization that occurs without active nanotube involvement.

Table 2.1 provides a list of monomers as of the end of 2009 that have been grafted from nanotubes. The only remarkable aspect of this list is that the number of thermoplastic monomers is quite high, even though commercially the use of nanotubes with thermoplastics is restricted to melt-mixed processing methods for economic reasons, which in turn eliminates "grafting from" approaches.

TABLE 2.2 Polymers Grafted to Carbon Nanotubes

Monomer	References	Monomer	References
Polystyrene	104,105,241–243	Polyurethane	244,245
Polymethyl methacrylate	246	Polyethylene oxide	105,244,247
Polybutyl methacrylate	105	Polyester	197,248
Polyoxyalkylene amine	249	Polycarbonate	250
Polyamide	251–256	Polyvinyl alcohol	257–259
Polypropylene	231,260,261	Polyamic acid	201
Linear low-density polyethylene	262,263	Polycaprolactone	264
Poly(*N*-vinyl carbazole)	265	Polyethylene-*co*-vinyl alcohol	266

2.5.4 Polymer Covalently Bonded to Nanotubes: "Grafting To"

As shown in Table 2.2, the number of polymers that have been attached to nanotubes via "grafting to" reactions is much smaller than that for "grafting from" reactions, even though these types of reactions can easily occur in melt-mixed thermoplastics if the proper chemistry is available. Of course, "grafting to" reactions are not sensible for thermosetting resins. The main barrier to "grafting to" reactions is not really the cost of the functionalization, but rather the paucity of thermoplastic polymers where such reactions could occur. This reality is reflected in the fact that in most thermoplastic composites with typical fillers such as glass, carbon fibers, and so on, there is no covalent bonding between the resin and the filler.

2.6 CHALLENGES

The key challenges are improving the selective syntheses of carbon nanotubes. The perfect nanotube would be of one type, that is, all semiconducting or all metallic, and preferably of all one chirality. Although the most significant impacts of this ability would rest in applications other than polymers, having all metallic nanotubes, for example, would significantly increase the measured conductivity in a polymer composite. Further, it might be possible to create composites with very little resistive heating and a very high charge carrying capability. Of lesser impact would be defect-free nanotubes; only marginal improvements would be expected in most composite properties. The ability to control length could be of great utility for polymer applications. Strictly from a property perspective in polymers, the longer the nanotubes, the better the situation (e.g., better properties, less material required for a given property, or both). Of course, longer nanotubes are going to be more difficult to mix with polymers, so the idea of length control becomes very appealing. A reasonable estimate of the critical entanglement length of an isolated SWCNT indicates that it is not unreasonable to consider the case where nanotubes could be made to be above the critical entanglement length. Work on nanotube forests that have the capability of making very long nanotubes is ongoing, and it will be very interesting to see what unique composites result with longer tubes.

To immediately increase use of nanotubes commercially, improving yield is perhaps the most important challenge. Hydrocarbon feedstocks used to make nanotubes via CVD methods are cheaper than most polymer monomers; if the yields could be increased to those achieved in catalyzed polymerizations, a number of problems related to both purification and cost would be eliminated. The amount of effort that has been spent on trying to improve yield has been significant; however, the inherent large number of possibilities offers the hope that perhaps there is a solution to this problem.

Improving the properties of composites through functionalization of nanotubes is an important area of work, although the author believes the marginal improvements that have in general been found have not justified the amount of work performed. Method of functionalization should also be described as a challenge; functionalization strategies that occur more or less simultaneously with nanotube synthesis, that is, in the gas phase of a CVD reactor, instead of after nanotube purification would be much more attractive economically.

A challenge is in the development of the procedures used to manufacture tubes in order to yield a more dispersible nanotube without significantly changing any other intrinsic nanotube properties. Such developments would perhaps include changes in both synthetic procedures and purification procedures. One interesting alternative that will be described fully in the next chapter is to make the nanotube and polymer more or less simultaneously using a catalytic approach for both.

REFERENCES

1. Blakslee, O. L. (1970). Elastic constants of compression-annealed pyrolytic graphite, *J. Appl. Phys.*, *41*, 3373–3383.
2. Lee, C., Wei, X. D., Kysar, J. W., Hone, J. (2008). Measurement of the elastic properties and intrinsic strength of monolayer graphene, *Science*, *321*, 385–388.
3. Wang, N., Tang, Z. K., Li, G. D., Chen, J. S. (2000). Single-walled 4 angstrom carbon nanotube arrays, *Nature*, *408*, 50–51.
4. Qin, L. C., Zhao, X. L., Hirahara, K., Miyamoto, Y., Ando, Y., Iijima, S. (2000). The smallest carbon nanotube, *Nature*, *408*, 50–50.
5. Doorn, S. K., O'Connell, M. J., Zheng, L. X., Zhu, Y. T., Huang, S. M., Liu, J. (2005). Raman spectral imaging of a carbon nanotube intramolecular junction, *Phys. Rev. Lett.*, *94*, 016802.
6. Wei, D. C., Liu, Y. Q. (2008). The intramolecular junctions of carbon nanotubes, *Adv. Mater.*, *20*, 2815–2841.
7. Dunlap, B. I. (1994). Relating carbon tubules, *Phys. Rev. B*, *49*, 5643–5650.
8. Zhou, O., Shimoda, H., Gao, B., Oh, S. J., Fleming, L., Yue, G. Z. (2002). Materials science of carbon nanotubes: fabrication, integration, and properties of macroscopic structures of carbon nanotubes, *Acc. Chem. Res.*, *35*, 1045–1053.
9. Chung, Y. H., Jou, S. (2005). Carbon nanotubes from catalytic pyrolysis of polypropylene, *Mater. Chem. Phys.*, *92*, 256–259.

10. Stamatin, I., Morozan, A., Dumitru, A., Ciupina, V., Prodan, G., Niewolski, J., Figiel, H. (2007). The synthesis of multi-walled carbon nanotubes (MWNTs) by catalytic pyrolysis of the phenol–formaldehyde resins, *Physica E*, *37*, 44–48.
11. Kong, Q. H., Zhang, J. H. (2007). Synthesis of straight and helical carbon nanotubes from catalytic pyrolysis of polyethylene, *Polym. Degrad. Stabil.*, *92*, 2005–2010.
12. Height, M. J., Howard, J. B., Tester, J. W. (2005). Flame synthesis of single-walled carbon nanotubes, *Proc. Combust. Inst.*, *30*, 2537–2543.
13. Hou, S. S., Chung, D. H., Lin, T. H. (2009). Flame synthesis of carbon nanotubes in a rotating counterflow, *J. Nanosci. Nanotechnol.*, *9*, 4826–4833.
14. Manciu, F. S., Camacho, J., Choudhuri, A. R. (2008). Flame synthesis of multi-walled carbon nanotubes using CH_4_H_2 fuel blends, *Fullerenes Nanotubes Carbon Nanostruct.*, *16*, 231–246.
15. Nakazawa, S., Yokomori, T., Mizomoto, M. (2005). Flame synthesis of carbon nanotubes in a wall stagnation flow, *Chem. Phys. Lett.*, *403*, 158–162.
16. Vander Wal, R. L., Hall, L., Berger, G. (2001). Flame synthesis of single- and multi-walled carbon nanotubes, *Abstr. Pap. Am. Chem. Soc.*, *222*, U473–U473.
17. Zhou, Q. L., Li, C. Z., Gu, F., Du, H. L. (2008). Flame synthesis of carbon nanotubes with high density on stainless steel mesh, *J. Alloys Compd.*, *463*, 317–322.
18. Zhang, Y. F., Gamo, M. N., Xiao, C. Y., Ando, T. (2002). Liquid phase synthesis of carbon nanotubes, *Physica B*, *323*, 293–295.
19. Harris, P. J. F. (2007). Solid state growth mechanisms for carbon nanotubes, *Carbon*, *45*, 229–239.
20. Journet, C., Maser, W. K., Bernier, P., Loiseau, A., de la Chapelle, M. L., Lefrant, S., Deniard, P., Lee, R., Fischer, J. E. (1997). Large-scale production of single-walled carbon nanotubes by the electric-arc technique, *Nature*, *388*, 756–758.
21. Li, Z. H., Wang, M., Yang, B., Xu, Y. B. (2007). The influence of different atmosphere gases on the growth and structure of double-walled carbon nanotubes, *Inorg. Mater.*, *43*, 475–479.
22. Lu, K. L., Lago, R. M., Chen, Y. K., Green, M. L. H., Harris, P. J. F., Tsang, S. C. (1996). Mechanical damage of carbon nanotubes by ultrasound, *Carbon*, *34*, 814–816.
23. de Heer, W. A., Poncharal, P., Berger, C., Gezo, J., Song, Z. M., Bettini, J., Ugarte, D. (2005). Liquid carbon, carbon-glass beads, and the crystallization of carbon nanotubes, *Science*, *307*, 907–910.
24. Harris, P. J. F., Tsang, S. C., Claridge, J. B., Green, M. L. H. (1994). High-resolution electron-microscopy studies of a microporous carbon produced by arc-evaporation, *J. Chem. Soc., Faraday Trans.*, *90*, 2799–2802.
25. Zhao, B., Futaba, D. N., Yasuda, S., Akoshima, M., Yamada, T., Hata, K. (2009). Exploring advantages of diverse carbon nanotube forests With tailored structures synthesized by supergrowth from engineered catalysts, *ACS Nano*, *3*, 108–114.
26. Zheng, L. X., O'Connell, M. J., Doorn, S. K., Liao, X. Z., Zhao, Y. H., Akhadov, E. A., Hoffbauer, M. A., Roop, B. J., Jia, Q. X., Dye, R. C., Peterson, D. E., Huang, S. M., Liu, J., Zhu, Y. T. (2004). Ultralong single-wall carbon nanotubes, *Nat. Mater.*, *3*, 673–676.
27. Musso, S., Porro, S., Rovere, M., Chiodoni, A., Taghaferro, A. (2007). Physical and mechanical properties of thick self-standing layers of multiwall carbon nanotubes, *Diam. Relat. Mater.*, *16*, 1174–1178.
28. Yun, Y. H., Shanov, V., Tu, Y., Subramaniam, S., Schulz, M. J. (2006). Growth mechanism of long aligned multiwall carbon nanotube arrays by water-assisted chemical vapor deposition, *J. Phys. Chem. B*, *110*, 23920–23925.
29. Hersam, M. C. (2008). Progress towards monodisperse single-walled carbon nanotubes, *Nat. Nanotechnol.*, *3*, 387–394.
30. Sears, A., Batra, R. C. (2004). Macroscopic properties of carbon nanotubes from molecular-mechanics simulations, *Phys. Rev. B*, *69*, 235406.
31. Hernandez, E., Goze, C., Bernier, P., Rubio, A. (1999). Elastic properties of single-wall nanotubes, *Appl. Phys. A*, *68*, 287–292.
32. Lu, J. P. (1997). Elastic properties of carbon nanotubes and nanoropes, *Phys. Rev. Lett.*, *79*, 1297–1300.
33. Treacy, M. M. J., Ebbesen, T. W., Gibson, J. M. (1996). Exceptionally high Young's modulus observed for individual carbon nanotubes, *Nature*, *381*, 678–680.

34. Salvetat, J. P., Kulik, A. J., Bonard, J. M., Briggs, G. A. D., Stockli, T., Metenier, K., Bonnamy, S., Beguin, F., Burnham, N. A., Forro, L. (1999). Elastic modulus of ordered and disordered multiwalled carbon nanotubes, *Adv. Mater.*, *11*, 161–165.
35. Yu, M. F., Lourie, O., Dyer, M. J., Moloni, K., Kelly, T. F., Ruoff, R. S. (2000). Strength and breaking mechanism of multiwalled carbon nanotubes under tensile load, *Science*, *287*, 637–640.
36. Guhados, G., Wan, W. K., Sun, X. L., Hutter, J. L. (2007). Simultaneous measurement of Young's and shear moduli of multiwalled carbon nanotubes using atomic force microscopy, *J. Appl. Phys.*, *101*, 033514.
37. Lukic, B., Seo, J. W., Bacsa, R. R., Delpeux, S., Beguin, F., Bister, G., Fonseca, A., Nagy, J. B., Kis, A., Jeney, S., Kulik, A. J., Forro, L. (2005). Catalytically grown carbon nanotubes of small diameter have a high Young's modulus, *Nano Lett.*, *5*, 2074–2077.
38. Salvetat, J. P., Briggs, G. A. D., Bonard, J. M., Bacsa, R. R., Kulik, A. J., Stockli, T., Burnham, N. A., Forro, L. (1999). Elastic and shear moduli of single-walled carbon nanotube ropes, *Phys. Rev. Lett.*, *82*, 944–947.
39. Yu, M. F., Files, B. S., Arepalli, S., Ruoff, R. S. (2000). Tensile loading of ropes of single wall carbon nanotubes and their mechanical properties, *Phys. Rev. Lett.*, *84*, 5552–5555.
40. Krishnan, A., Dujardin, E., Ebbesen, T. W., Yianilos, P. N., Treacy, M. M. J. (1998). Young's modulus of single-walled nanotubes, *Phys. Rev. B*, *58*, 14013–14019.
41. Kutana, A., Giapis, K. P. (2006). Transient deformation regime in bending of single-walled carbon nanotubes, *Phys. Rev. Lett.*, *97*, 245501.
42. Chang, T. C., Hou, J. (2006). Molecular dynamics simulations on buckling of multiwalled carbon nanotubes under bending, *J. Appl. Phys.*, *100*, 114327.
43. Kuang, Y. D., He, X. Q., Chen, C. Y., Li, G. Q. (2009). Buckling of functionalized single-walled nanotubes under axial compression, *Carbon*, *47*, 279–285.
44. Liew, K. M., Wong, C. H., He, X. Q., Tan, M. J., Meguid, S. A. (2004). Nanomechanics of single and multiwalled carbon nanotubes, *Phys. Rev. B*, *69*, 115429.
45. Barber, A. H., Cohen, S. R., Eitan, A., Schadler, L. S., Wagner, H. D. (2006). Fracture transitions at a carbon-nanotube/polymer interface, *Adv. Mater.*, *18*, 83–87.
46. Barber, A. H., Andrews, R., Schadler, L. S., Wagner, H. D. (2005). On the tensile strength distribution of multiwalled carbon nanotubes, *Appl. Phys. Lett.*, *87*, 203106.
47. Wagner, H. D., Barber, A. H., Andrews, R., Schadler, L. S. (2008). Response to "Comment on 'On the tensile strength distribution of multiwalled carbon nanotubes' [Appl. Phys. Lett. 92, 206101 (2008)]", *Appl. Phys. Lett.*, *92*, 206102.
48. Liew, K. M., He, X. Q., Wong, C. H. (2004). On the study of elastic and plastic properties of multiwalled carbon nanotubes under axial tension using molecular dynamics simulation, *Acta Mater.*, *52*, 2521–2527.
49. Javey, A., Kong, J., Eds. (2009). *Carbon Nanotube Electronics (Integrated Circuits and Systems)*, 1st ed., Springer, New York.
50. Leonard, F. (2008). *Physics of Carbon Nanotube Devices (Micro and Nano Technologies)*, William Andrew, Norwich, NY.
51. Ouyang, M., Huang, J. L., Cheung, C. L., Lieber, C. M. (2001). Energy gaps in "metallic" single-walled carbon nanotubes, *Science*, *292*, 702–705.
52. Dai, H. J., Wong, E. W., Lieber, C. M. (1996). Probing electrical transport in nanomaterials: conductivity of individual carbon nanotubes, *Science*, *272*, 523–526.
53. Ebbesen, T. W., Lezec, H. J., Hiura, H., Bennett, J. W., Ghaemi, H. F., Thio, T. (1996). Electrical conductivity of individual carbon nanotubes, *Nature*, *382*, 54–56.
54. Thess, A., Lee, R., Nikolaev, P., Dai, H. J., Petit, P., Robert, J., Xu, C. H., Lee, Y. H., Kim, S. G., Rinzler, A. G., Colbert, D. T., Scuseria, G. E., Tomanek, D., Fischer, J. E., Smalley, R. E. (1996). Crystalline ropes of metallic carbon nanotubes, *Science*, *273*, 483–487.
55. Mann, D., Javey, A., Kong, J., Wang, Q., Dai, H. J. (2003). Ballistic transport in metallic nanotubes with reliable Pd ohmic contacts, *Nano Lett.*, *3*, 1541–1544.
56. Jackson, R., Domercq, B., Jain, R., Kippelen, B., Graham, S. (2008). Stability of doped transparent carbon nanotube electrodes, *Adv. Funct. Mater.*, *18*, 2548–2554.
57. Tsui, F., Jin, L., Zhou, O. (2000). Anisotropic magnetic susceptibility of multiwalled carbon nanotubes, *Appl. Phys. Lett.*, *76*, 1452–1454.

58. Fujiwara, M., Oki, E., Hamada, M., Tanimoto, Y., Mukouda, I., Shimomura, Y. (2001). Magnetic orientation and magnetic properties of a single carbon nanotube, *J. Phys. Chem. A*, *105*, 4383–4386.
59. Zaric, S., Ostojic, G. N., Kono, J., Shaver, J., Moore, V. C., Hauge, R. H., Smalley, R. E., Xing, W. (2004). Estimation of magnetic susceptibility anisotropy of carbon nanotubes using magnetophotoluminescence, *Nano Lett.*, *4*, 2219–2221.
60. Heremans, J., Olk, C. H., Morelli, D. T. (1994). Magnetic susceptibility of carbon structures, *Phys. Rev. B*, *49*, 15122–15125.
61. Kim, P., Shi, L., Majumdar, A., McEuen, P. L. (2001). Thermal transport measurements of individual multiwalled nanotubes, *Phys. Rev. Lett.*, *87*, 215502.
62. Fujii, M., Zhang, X., Xie, H. Q., Ago, H., Takahashi, K., Ikuta, T., Abe, H., Shimizu, T. (2005). Measuring the thermal conductivity of a single carbon nanotube, *Phys. Rev. Lett.*, *95*, 065502.
63. Pop, E., Mann, D., Wang, Q., Goodson, K., Dai, H. J. (2006). Thermal conductance of an individual single-wall carbon nanotube above room temperature, *Nano Lett.*, *6*, 96–100.
64. Choi, T. Y., Poulikakos, D., Tharian, J., Sennhauser, U. (2005). Measurement of thermal conductivity of individual multiwalled carbon nanotubes by the 3-omega method, *Appl. Phys. Lett.*, *87*, 013108.
65. Yu, C. H., Shi, L., Yao, Z., Li, D. Y., Majumdar, A. (2005). Thermal conductance and thermopower of an individual single-wall carbon nanotube, *Nano Lett.*, *5*, 1842–1846.
66. Shiomi, J., Maruyama, S. (2008). Molecular dynamics of diffusive-ballistic heat conduction in single-walled carbon nanotubes, *Jpn. J. Appl. Phys.*, *47*, 2005–2009.
67. Wang, J. A., Wang, J. S. (2006). Carbon nanotube thermal transport: ballistic to diffusive, *Appl. Phys. Lett.*, *88*, 111909.
68. Savin, A. V., Hu, B. B., Kivshar, Y. S. (2009). Thermal conductivity of single-walled carbon nanotubes, *Phys. Rev. B*, *80*, 195423.
69. Zhang, G., Li, B. W. (2005). Thermal conductivity of nanotubes revisited: effects of chirality, isotope impurity, tube length, and temperature, *J. Chem. Phys.*, *123*, 114714.
70. Hone, J., Batlogg, B., Benes, Z., Johnson, A. T., Fischer, J. E. (2000). Quantized phonon spectrum of single-wall carbon nanotubes, *Science*, *289*, 1730–1733.
71. Mizel, A., Benedict, L. X., Cohen, M. L., Louie, S. G., Zettl, A., Budraa, N. K., Beyermann, W. P. (1999). Analysis of the low-temperature specific heat of multiwalled carbon nanotubes and carbon nanotube ropes, *Phys. Rev. B*, *60*, 3264–3270.
72. Yi, W., Lu, L., Zhang, D. L., Pan, Z. W., Xie, S. S. (1999). Linear specific heat of carbon nanotubes, *Phys. Rev. B*, *59*, R9015–R9018.
73. Hepplestone, S. P., Ciavarella, A. M., Janke, C., Srivastava, G. P. (2006). Size and temperature dependence of the specific heat capacity of carbon nanotubes, *Surf. Sci.*, *600*, 3633–3636.
74. Silva, G. G., Musumeci, A. W., Gomes, A. P., Liu, J. W., Waclawik, E. R., George, G. A., Frost, R. L., Pimenta, M. A. (2009). Characterization of commercial double-walled carbon nanotube material: composition, structure, and heat capacity, *J. Mater. Sci.*, *44*, 3498–3503.
75. Araujo, P. T., Doorn, S. K., Kilina, S., Tretiak, S., Einarsson, E., Maruyama, S., Chacham, H., Pimenta, M. A., Jorio, A. (2007). Third and fourth optical transitions in semiconducting carbon nanotubes, *Phys. Rev. Lett.*, *98*, 067401.
76. Dresselhaus, M. S., Dresselhaus, G., Saito, R., Jorio, A. (2005). Raman spectroscopy of carbon nanotubes, *Phys. Rep.*, *409*, 47–99.
77. Reich, S., Thomsen, C., Maultzsch, J. (2004). *Carbon Nanotubes: Basic Concepts and Physical Properties*, Wiley-VCH, New York.
78. Jorio, A., Souza, A. G., Dresselhaus, G., Dresselhaus, M. S., Swan, A. K., Unlu, M. S., Goldberg, B. B., Pimenta, M. A., Hafner, J. H., Lieber, C. M., Saito, R. (2002). G-band resonant Raman study of 62 isolated single-wall carbon nanotubes, *Phys. Rev. B*, *65*, 155412.
79. Graupner, R. (2007). Raman spectroscopy of covalently functionalized single-wall carbon nanotubes, *J. Raman Spectrosc.*, *38*, 673–683.
80. Garg, A., Sinnott, S. B. (1998). Effect of chemical functionalization on the mechanical properties of carbon nanotubes, *Chem. Phys. Lett.*, *295*, 273–278.
81. Kuang, Y. D., He, X. Q. (2009). Young's moduli of functionalized single-wall carbon nanotubes under tensile loading, *Compos. Sci. Technol.*, *69*, 169–175.
82. Shenogin, S., Bodapati, A., Xue, L., Ozisik, R., Keblinski, P. (2004). Effect of chemical functionalization on thermal transport of carbon nanotube composites, *Appl. Phys. Lett.*, *85*, 2229–2231.

83. Pan, R. Q., Xu, Z. J., Zhu, Z. Y., Wang, Z. X. (2007). Thermal conductivity of functionalized single-wall carbon nanotubes, *Nanotechnology*, *18*, 285704.
84. Banerjee, S., Hemraj-Benny, T., Wong, S. S. (2005). Covalent surface chemistry of single-walled carbon nanotubes, *Adv. Mater.*, *17*, 17–29.
85. Tasis, D., Tagmatarchis, N., Bianco, A., Prato, M. (2006). Chemistry of carbon nanotubes, *Chem. Rev.*, *106*, 1105–1136.
86. Matrab, T., Chancolon, J., L'Hermite, M. M., Rouzaud, J. N., Deniau, G., Boudou, J. P., Chehimi, M. M., Delamar, M. (2006). Atom transfer radical polymerization (ATRP) initiated by aryl diazonium salts: a new route for surface modification of multiwalled carbon nanotubes by tethered polymer chains, *Colloids Surf. A: Physicochem. Eng. Aspects*, *287*, 217–221.
87. Wang, S. R., Liang, R., Wang, B., Zhang, C. (2009). Covalent addition of diethyltoluenediamines onto carbon nanotubes for composite application, *Polym. Compos.*, *30*, 1050–1057.
88. Dyke, C. A., Tour, J. M. (2004). Covalent functionalization of single-walled carbon nanotubes for materials applications, *J. Phys. Chem. A*, *108*, 11151–11159.
89. Mickelson, E. T., Huffman, C. B., Rinzler, A. G., Smalley, R. E., Hauge, R. H., Margrave, J. L. (1998). Fluorination of single-wall carbon nanotubes, *Chem. Phys. Lett.*, *296*, 188–194.
90. Liu, Y. Q., Adronov, A. (2004). Preparation and utilization of catalyst-functionalized single-walled carbon nanotubes for ring-opening metathesis polymerization, *Macromolecules*, *37*, 4755–4760.
91. Yao, Z. L., Braidy, N., Botton, G. A., Adronov, A. (2003). Polymerization from the surface of single-walled carbon nanotubes: preparation and characterization of nanocomposites, *J. Am. Chem. Soc.*, *125*, 16015–16024.
92. Hong, C. Y., You, Y. Z., Pan, C. Y. (2006). A new approach to functionalize multi-walled carbon nanotubes by the use of functional polymers, *Polymer*, *47*, 4300–4309.
93. Chen, S. M., Chen, D. Y., Wu, G. Z. (2006). Grafting of poly(*t*BA) and P*t*BA-*b*-PMMA onto the surface of SWNTs using carbanions as the initiator, *Macromol. Rapid Commun.*, *27*, 882–887.
94. Sakellariou, G., Ji, H. N., Mays, J. W., Baskaran, D. (2008). Enhanced polymer grafting from multiwalled carbon nanotubes through living anionic surface-initiated polymerization, *Chem. Mater.*, *20*, 6217–6230.
95. Bonduel, D., Bredeau, S., Alexandre, M., Monteverde, F., Dubois, P. (2007). Supported metallocene catalysis as an efficient tool for the preparation of polyethylene/carbon nanotube nanocomposites: effect of the catalytic system on the coating morphology, *J. Mater. Chem.*, *17*, 2359–2366.
96. Chen, Y. H., Muthukumar, V. S., Wang, Y. B., Li, C., Krishnan, S. S., Sai, S. S. S., Venkataramaniah, K., Mitra, S. (2009). Microwave-assisted solid-state grafting of multi-walled carbon nanotubes on polyurethane for the synthesis of a composite with optical limiting properties, *J. Mater. Chem.*, *19*, 6568–6572.
97. Xu, H. X., Wang, X. B., Zhang, Y. F., Liu, S. Y. (2006). Single-step *in situ* preparation of polymer-grafted multi-walled carbon nanotube composites under Co-60 gamma-ray irradiation, *Chem. Mater.*, *18*, 2929–2934.
98. Jia, Z. J., Wang, Z. Y., Xu, C. L., Liang, J., Wei, B. Q., Wu, D. H., Zhu, S. W. (1999). Study on poly (methyl methacrylate)/carbon nanotube composites, *Mater. Sci. Eng. A*, *271*, 395–400.
99. Nayak, R. R., Lee, K. Y., Shanrnugharaj, A. M., Ryu, S. H. (2007). Synthesis and characterization of styrene grafted carbon nanotube and its polystyrene nanocomposite, *Eur. Polym. J.*, *43*, 4916–4923.
100. Liu, P. (2009). Facile graft polystyrene onto multi-walled carbon nanotubes via *in situ* thermo-induced radical polymerization, *J. Nanopart. Res.*, *11*, 1011–1016.
101. Zhao, X. D., Fan, X. H., Chen, X. F., Chai, C. P., Zhou, Q. F. (2006). Surface modification of multiwalled carbon nanotubes via nitroxide-mediated radical polymerization, *J. Polym. Sci. A*, *44*, 4656–4667.
102. Nayak, R. R., Shanmugharaj, A. M., Ryu, S. H. (2008). A novel route for polystyrene grafted single-walled carbon nanotubes and their characterization, *Macromol. Chem. Phys.*, *209*, 1137–1144.
103. Kong, H., Gao, C., Yan, D. Y. (2004). Functionalization of multiwalled carbon nanotubes by atom transfer radical polymerization and defunctionalization of the products, *Macromolecules*, *37*, 4022–4030.
104. Qin, S. H., Qin, D. Q., Ford, W. T., Resasco, D. E., Herrera, J. E. (2004). Functionalization of single-walled carbon nanotubes with polystyrene via grafting to and grafting from methods, *Macromolecules*, *37*, 752–757.

105. Zhang, Y., He, H. K., Gao, C. (2008). Clickable macroinitiator strategy to build amphiphilic polymer brushes on carbon nanotubes, *Macromolecules*, *41*, 9581–9594.
106. Wang, H. C., Li, Y., Yang, M. J. (2007). Sensors for organic vapor detection based on composites of carbon nonotubes functionalized with polymers, *Sens. Actuators B*, *124*, 360–367.
107. Viswanathan, G., Chakrapani, N., Yang, H. C., Wei, B. Q., Chung, H. S., Cho, K. W., Ryu, C. Y., Ajayan, P. M. (2003). Single-step *in situ* synthesis of polymer-grafted single-wall nanotube composites, *J. Am. Chem. Soc.*, *125*, 9258–9259.
108. Choi, H. J., Zhang, K., Lim, J. Y. (2007). Multi-walled carbon nanotube/polystyrene composites prepared by *in-situ* bulk sonochemical polymerization, *J. Nanosci. Nanotechnol.*, *7*, 3400–3403.
109. Hong, C. Y., You, Y. Z., Pan, C. Y. (2005). Synthesis of water-soluble multiwalled carbon nanotubes with grafted temperature-responsive shells by surface RAFT polymerization, *Chem. Mater.*, *17*, 2247–2254.
110. Vigolo, B., Mamane, V., Valsaque, F., Le, T. N. H., Thabit, J., Ghanbaja, J., Aranda, L., Fort, Y., McRae, E. (2009). Evidence of sidewall covalent functionalization of single-walled carbon nanotubes and its advantages for composite processing, *Carbon*, *47*, 411–419.
111. Liang, F., Beach, J. M., Kobashi, K., Sadana, A. K., Vega-Cantu, Y. I., Tour, J. M., Billups, W. E. (2006). *In situ* polymerization initiated by single-walled carbon nanotube salts, *Chem. Mater.*, *18*, 4764–4767.
112. Liu, M. H., Yang, Y. L., Zhu, T., Liu, Z. F. (2007). A general approach to chemical modification of single-walled carbon nanotubes with peroxy organic acids and its application in polymer grafting, *J. Phys. Chem. C*, *111*, 2379–2385.
113. Liu, M. H., Zhu, T., Li, Z. C., Liu, Z. F. (2009). One-step *in situ* synthesis of poly(methyl methacrylate)-grafted single-walled carbon nanotube composites, *J. Phys. Chem. C*, *113*, 9670–9675.
114. Baskaran, D., Dunlap, J. R., Mays, J. W., Bratcher, M. S. (2005). Grafting efficiency of hydroxy-terminated poly(methyl methacrylate) with multiwalled carbon nanotubes, *Macromol. Rapid Commun.*, *26*, 481–486.
115. Park, S. J., Cho, M. S., Lim, S. T., Choi, H. J., Jhon, M. S. (2003). Synthesis and dispersion characteristics of multi-walled carbon nanotube composites with poly(methyl methacrylate) prepared by *in-situ* bulk polymerization, *Macromol. Rapid Commun.*, *24*, 1070–1073.
116. Xu, G. Y., Wang, Y. S., Pang, W. M., Wu, W. T., Zhu, Q. R., Wang, P. H. (2007). Fabrication of multiwalled carbon nanotubes with polymer shells through surface RAFT polymerization, *Polym. Int.*, *56*, 847–852.
117. Kim, S. T., Lim, J. Y., Park, B. J., Choi, H. J. (2007). Dispersion-polymerized carbon nanotube/poly (methyl methacrylate) composite particles and their electrorheological characteristics, *Macromol. Chem. Phys.*, *208*, 514–519.
118. Zhou, Z., Wang, S. F., Lu, L., Zhang, Y., Zhang, Y. X. (2007). Preparation and rheological characterization of poly(methyl methacrylate)/functionalized multi-walled carbon nanotubes composites, *Compos. Sci. Technol.*, *67*, 1861–1869.
119. Yuen, S. M., Ma, C. C. M., Chiang, C. L., Chang, J. A., Huang, S. W., Chen, S. C., Chuang, C. Y., Yang, C. C., Wei, M. H. (2007). Silane-modified MWCNT/PMMA composites: preparation, electrical resistivity, thermal conductivity and thermal stability, *Compos. Part A*, *38*, 2527–2535.
120. Cui, L., Tarte, N. H., Woo, S. I. (2009). Synthesis and characterization of PMMA/MWNT nanocomposites prepared by *in situ* polymerization with $Ni(acac)_2$ catalyst, *Macromolecules*, *42*, 8649–8654.
121. Velasco-Santos, C., Martinez-Hernandez, A. L., Fisher, F. T., Ruoff, R., Castano, V. M. (2003). Improvement of thermal and mechanical properties of carbon nanotube composites through chemical functionalization, *Chem. Mater.*, *15*, 4470–4475.
122. Olalde, B., Aizpurua, J. M., Garcia, A., Bustero, I., Obieta, I., Jurado, M. J. (2008). Single-walled carbon nanotubes and multiwalled carbon nanotubes functionalized with poly(L-lactic acid): a comparative study, *J. Phys. Chem. C*, *112*, 10663–10667.
123. Castro, M., Lu, J. B., Bruzaud, S., Kumar, B., Feller, J. F. (2009). Carbon nanotubes/poly (ε-caprolactone) composite vapour sensors, *Carbon*, *47*, 1930–1942.
124. Ruelle, B., Peeterbroeck, S., Gouttebaron, R., Godfroid, T., Monteverde, F., Dauchot, J. P., Alexandre, M., Hecq, M., Dubois, P. (2007). Functionalization of carbon nanotubes by atomic

nitrogen formed in a microwave plasma Ar + N_2 and subsequent poly(ε-caprolactone) grafting, *J. Mater. Chem.*, *17*, 157–159.

125. Reza, S., Adeli, M., Astinchap, B., Kabiri, R. (2008). New nanocomposites containing metal nanoparticles, carbon nanotube and polymer, *J. Nanopart. Res.*, *10*, 1309–1318.
126. Li, J., He, W. D., Yang, L. P., Sun, X. L., Hua, Q. (2007). Preparation of multi-walled carbon nanotubes grafted with synthetic poly(L-lysine) through surface-initiated ring-opening polymerization, *Polymer*, *48*, 4352–4360.
127. Seyhan, A. T., Gojny, F. H., Tanoglu, M., Schulte, K. (2007). Rheological and dynamic-mechanical behavior of carbon nanotube/vinyl ester–polyester suspensions and their nanocomposites, *Eur. Polym. J.*, *43*, 2836–2847.
128. Seyhan, A. T., Gojny, F. H., Tanoglu, M., Schulte, K. (2007). Critical aspects related to processing of carbon nanotube/unsaturated thermoset polyester nanocomposites, *Eur. Polym. J.*, *43*, 374–379.
129. Seyhan, A. T., Tanoglu, M., Schulte, K. (2009). Tensile mechanical behavior and fracture toughness of MWCNT and DWCNT modified vinyl-ester/polyester hybrid nanocomposites produced by 3-roll milling, *Mater. Sci. Eng. A*, *523*, 85–92.
130. Jin, S. H., Yoon, K. H., Park, Y. B., Bang, D. S. (2008). Properties of surface-modified multiwalled carbon nanotube filled poly(ethylene terephthalate) composite films, *J. Appl. Polym. Sci.*, *107*, 1163–1168.
131. Mun, S. J., Jung, Y. M., Kim, J. C., Chang, J. H. (2008). Poly(ethylene terephthalate) nanocomposite fibers with functionalized multiwalled carbon nanotubes via *in-situ* polymerization, *J. Appl. Polym. Sci.*, *109*, 638–646.
132. Oh, S. B., Kim, H. L., Chang, J. H., Lee, Y. W., Han, J. H., An, S. S. A., Joo, S. W., Kim, H. K., Choi, I. S., Paik, H. J. (2008). Facile covalent attachment of well-defined poly(*t*-butyl acrylate) on carbon nanotubes via radical addition reaction, *J. Nanosci. Nanotechnol.*, *8*, 4598–4602.
133. Wu, W., Tsarevsky, N. V., Hudson, J. L., Tour, J. M., Matyjaszewski, K., Kowalewski, T. (2007). "Hairy" single-walled carbon nanotubes prepared by atom transfer radical polymerization, *Small*, *3*, 1803–1810.
134. Shi, J. H., Yang, B. X., Goh, S. H. (2009). Covalent functionalization of multiwalled carbon nanotubes with poly(styrene-*co*-acrylonitrile) by reactive melt blending, *Eur. Polym. J.*, *45*, 1002–1008.
135. Shanmugharaj, A. M., Bae, J. H., Nayak, R. R., Ryu, S. H. (2007). Preparation of poly(styrene-*co*-acrylonitrile)-grafted multiwalled carbon nanotubes via surface-initiated atom transfer radical polymerization, *J. Polym. Sci. A*, *45*, 460–470.
136. Yang, Y. K., Xie, X. L., Wu, J. G., Mai, Y. W. (2006). Synthesis and self-assembly of polystyrene-grafted multiwalled carbon nanotubes with a hairy-rod nanostructure, *J. Polym. Sci. A*, *44*, 3869–3881.
137. Kumar, N. A., Ganapathy, H. S., Kim, J. S., Jeong, Y. S., Jeong, Y. T. (2008). Preparation of poly 2-hydroxyethyl methacrylate functionalized carbon nanotubes as novel biomaterial nanocomposites, *Eur. Polym. J.*, *44*, 579–586.
138. Pei, X. W., Hao, J. C., Liu, W. M. (2007). Preparation and characterization of carbon nanotubes–polymer/Ag hybrid nanocomposites via surface RAFT polymerization, *J. Phys. Chem. C*, *111*, 2947–2952.
139. Ha, J. U., Kim, M., Lee, J., Choe, S., Cheong, I. W., Shim, S. E. (2006). A novel synthesis of polymer brush on multiwall carbon nanotubes bearing terminal monomeric unit, *J. Polym. Sci. A*, *44*, 6394–6401.
140. Kumar, N. A., Kim, S. H., Kim, J. T., Lim, K. T., Jeong, Y. T. (2008). Study on cluster formation of poly 2-hydroxyethyl methacrylate functionalized single-walled carbon nanotubes, *Surf. Rev. Lett.*, *15*, 689–697.
141. Jiang, S., Deng, J. P., Yang, W. T. (2008). Functionalization of multi-walled carbon nanotubes by thermo-grafting with alpha-methylstyrene-containing copolymers, *Macromol. Rapid Commun.*, *29*, 1521–1526.
142. Qin, S. H., Oin, D. Q., Ford, W. T., Resasco, D. E., Herrera, J. E. (2004). Polymer brushes on single-walled carbon nanotubes by atom transfer radical polymerization of *n*-butyl methacrylate, *J. Am. Chem. Soc.*, *126*, 170–176.

143. Shi, J. H., Yang, B. X., Pramoda, K. P., Goh, S. H. (2007). Enhancement of the mechanical performance of poly(vinyl chloride) using poly(*n*-butyl methacrylate)-grafted multi-walled carbon nanotubes, *Nanotechnology*, *18*, 375704.
144. Matrab, T., Save, M., Charleux, B., Pinson, J., Cabet-Deliry, E., Adenier, A., Chehimi, M. M., Delamar, M. (2007). Grafting densely-packed poly(*n*-butyl methacrylate) chains from an iron substrate by aryl diazonium surface-initiated ATRP: XPS monitoring, *Surf. Sci.*, *601*, 2357–2366.
145. Gao, C., Muthukrishnan, S., Li, W. W., Yuan, J. Y., Xu, Y. Y., Muller, A. H. E. (2007). Linear and hyperbranched glycopolymer-functionalized carbon nanotubes: synthesis, kinetics, and characterization, *Macromolecules*, *40*, 1803–1815.
146. Zheng, Y. P., Zhang, J. X., Xiaodong, Y., Chen, W. W., Wang, R. M. (2009). Effects of functionalized MWNTs with GMA on the properties of PMMA nanocomposites, *J. Appl. Polym. Sci.*, *112*, 1755–1761.
147. Ryu, J., Ramaraj, B., Yoon, K. R. (2009). Surface functionalization of multi-walled carbon nanotubes through surface-initiated atom transfer radical polymerization of glycidyl methacrylate, *Surf. Interface Anal.*, *41*, 303–309.
148. Han, J. H., Choi, J. H., Oh, S. B., Jang, J. H., An, S. S. A., Kim, H. K., Kim, B. G., Paik, H. J. (2006). Covalent attachment of poly(ethylene glycol) on multi-walled carbon nanotubes, *Compos. Interfaces*, *13*, 321–328.
149. Zhao, B., Hu, H., Yu, A. P., Perea, D., Haddon, R. C. (2005). Synthesis and characterization of water soluble single-walled carbon nanotube graft copolymers, *J. Am. Chem. Soc.*, *127*, 8197–8203.
150. Zhang, P., Henthorn, D. B. (2009). Fabrication of high-capacity biomolecular carriers from dispersible single-walled carbon nanotube–polymer composites, *Langmuir*, *25*, 12308–12314.
151. Priftis, D., Sakellariou, G., Hadjichristidis, N., Penott, E. K., Lorenzo, A. T., Muller, A. J. (2009). Surface modification of multiwalled carbon nanotubes with biocompatible polymers via ring opening and living anionic surface initiated polymerization. Kinetics and crystallization behavior, J. Polym. Sci. A, *47*, 4379–4390.
152. Xiao, Q., He, S. J., Liu, L. W., Guo, X. Z., Shi, K., Du, Z. J., Zhang, B. L. (2008). Coating of multiwalled carbon nanotubes with crosslinked silicon-containing polymer, *Compos. Sci. Technol.*, *68*, 321–328.
153. Yan, D. G., Yang, G. S. (2009). A novel approach of *in situ* grafting polyamide 6 to the surface of multi-walled carbon nanotubes, *Mater. Lett.*, *63*, 298–300.
154. Yan, D. G., Yang, G. S. (2009). Synthesis and properties of homogeneously dispersed polyamide 6/MWNTs nanocomposites via simultaneous *in situ* anionic ring-opening polymerization and compatibilization, *J. Appl. Polym. Sci.*, *112*, 3620–3626.
155. Gao, J. B., Itkis, M. E., Yu, A. P., Bekyarova, E., Zhao, B., Haddon, R. C. (2005). Continuous spinning of a single-walled carbon nanotube–nylon composite fiber, *J. Am. Chem. Soc.*, *127*, 3847–3854.
156. Gao, J. B., Zhao, B., Itkis, M. E., Bekyarova, E., Hu, H., Kranak, V., Yu, A. P., Haddon, R. C. (2006). Chemical engineering of the single-walled carbon nanotube–nylon 6 interface, *J. Am. Chem. Soc.*, *128*, 7492–7496.
157. Ge, J. J., Zhang, D., Li, Q., Hou, H. Q., Graham, M. J., Dai, L. M., Harris, F. W., Cheng, S. Z. D. (2005). Multiwalled carbon nanotubes with chemically grafted polyetherimides, *J. Am. Chem. Soc.*, *127*, 9984–9985.
158. Kim, H. S., Park, B. H., Yoon, J. S., Jin, H. J. (2007). Nylon 610/functionalized multiwalled carbon nanotubes composites by *in situ* interfacial polymerization, *Mater. Lett.*, *61*, 2251–2254.
159. Qu, L. W., Veca, L. M., Lin, Y., Kitaygorodskiy, A., Chen, B. L., McCall, A. M., Connell, J. W., Sun, Y. P. (2005). Soluble nylon-functionalized carbon nanotubes from anionic ring-opening polymerization from nanotube surface, *Macromolecules*, *38*, 10328–10331.
160. Yuen, S. M., Ma, C. C. M., Lin, Y. Y., Kuan, H. C. (2007). Preparation, morphology and properties of acid and amine modified multiwalled carbon nanotube/polyimide composite, *Compos. Sci. Technol.*, *67*, 2564–2573.
161. Yuen, S. M., Ma, C. C. M., Chiang, C. L., Lin, Y. Y., Teng, C. C. (2007). Preparation and morphological, electrical, and mechanical properties of polyimide-grafted MWCNT/polyimide composite, *J. Polym. Sci. A*, *45*, 3349–3358.
162. Saeed, K., Park, S. Y. (2007). Preparation of multiwalled carbon nanotube/nylon-6 nanocomposites by *in situ* polymerization, *J. Appl. Polym. Sci.*, *106*, 3729–3735.

163. Wang, G. J., Qu, Z. H., Liu, L., Shi, Q., Guo, H. L. (2008). Study of SMA graft modified MWNT/PVC composite materials, *Mater. Sci. Eng. A*, *472*, 136–139.
164. Choi, J. H., Oh, S. B., Chang, J. H., Kim, I., Ha, C. S., Kim, B. G., Han, J. H., Joo, S. W., Kim, G. H., Paik, H. J. (2005). Graft polymerization of styrene from single-walled carbon nanotube using atom transfer radical polymerization, *Polym. Bull.*, *55*, 173–179.
165. Santhosh, P., Manesh, K. M., Lee, K. P., Gopalan, A. Y. (2006). Enhanced electrocatalysis for the reduction of hydrogen peroxide at new multiwall carbon nanotube grafted polydiphenylamine modified electrode, *Electroanalysis*, *18*, 894–903.
166. Lee, K. P., Gopalan, A. Y., Kim, K. S., Santhosh, P. (2007). Synthesis and characterization of processable multi-walled carbon nanotubes–sulfonated polydiphenylamine graft copolymers, *J. Nanosci. Nanotechnol.*, *7*, 3386–3393.
167. Wu, H. X., Qiu, X. Q., Cai, R. F., Qian, S. X. (2007). Poly(*N*-vinyl carbazole)-grafted multiwalled carbon nanotubes: synthesis via direct free radical reaction and optical limiting properties, *Appl. Surf. Sci.*, *253*, 5122–5128.
168. Kong, H., Gao, C., Yan, D. Y. (2004). Constructing amphiphilic polymer brushes on the convex surfaces of multi-walled carbon nanotubes by *in situ* atom transfer radical polymerization, *J. Mater. Chem.*, *14*, 1401–1405.
169. De Falco, A., Fascio, M. L., Lamanna, M. E., Corcuera, M. A., Mondragon, I., Rubiolo, G. H., D'Accorso, N. B., Goyanes, S. (2009). Thermal treatment of the carbon nanotubes and their functionalization with styrene, *Physica B*, *404*, 2780–2783.
170. Zhao, B., Hu, H., Haddon, R. C. (2004). Synthesis and properties of a water-soluble single-walled carbon nanotube–poly(*m*-aminobenzene sulfonic acid) graft copolymer, *Adv. Funct. Mater.*, *14*, 71–76.
171. Kim, D. K., Oh, K. W., Kim, S. H. (2009). Synthesis of conducting composite of polyaniline and multi wall carbon nanotube grafted with sulfonated polystyrene, *Mol. Cryst. Liq. Cryst.*, *510*, 51–59.
172. Adeli, M., Bahari, A., Hekmatara, H. (2008). Carbon nanotube-*graft*-poly(citric acid) nanocomposites, *Nano*, *3*, 37–44.
173. Li, Y., Wang, H. C., Cao, X. H., Yuan, M. Y., Yang, M. J. (2008). A composite of polyelectrolyte-grafted multi-walled carbon nanotubes and *in situ* polymerized polyaniline for the detection of low concentration triethylamine vapor, *Nanotechnology*, *19*, 015503.
174. Lee, K. P., Gopalan, A. I., Komathi, S. (2009). Direct electrochemistry of cytochrome *c* and biosensing for hydrogen peroxide on polyaniline grafted multi-walled carbon nanotube electrode, *Sens. Actuators B*, *141*, 518–525.
175. You, Y. Z., Hong, C. Y., Pan, C. Y. (2006). Directly growing ionic polymers on multi-walled carbon nanotubes via surface RAFT polymerization, *Nanotechnology*, *17*, 2350–2354.
176. Kong, H., Luo, P., Gao, C., Yan, D. (2005). Polyelectrolyte-functionalized multiwalled carbon nanotubes: preparation, characterization and layer-by-layer self-assembly, *Polymer*, *46*, 2472–2485.
177. Datsyuk, V., Billon, L., Guerret-Piecourt, C., Dagreou, S., Passade-Boupatt, N., Bourrigaud, S., Guerret, O., Couvreur, L. (2007). *In situ* nitroxide-mediated polymerized poly(acrylic acid) as a stabilizer/compatibilizer carbon nanotube/polymer composites, *J. Nanomater.* 74769.
178. Yang, Y., Wang, X., Liu, L., Xie, X., Yang, Z., Li, R. K. Y., Mai, Y. W. (2007). Structure and photoresponsive behaviors of multiwalled carbon nanotubes grafted by polyurethanes containing azobenzene side chains, *J. Phys. Chem. C*, *111*, 11231–11239.
179. Song, H. J., Zhang, Z. Z., Men, X. H. (2007). Surface-modified carbon nanotubes and the effect of their addition on the tribological behavior of a polyurethane coating, *Eur. Polym. J.*, *43*, 4092–4102.
180. Zhang, C., Guo, S. Z., Wang, W. Z., Liu, T. X. (2009). Fluorinated multi-walled carbon nanotubes and fabrication of polyurethane composite films with hydrophobic surface, *Acta Chim. Sin.*, *67*, 1001–1007.
181. Wang, X., Zhang, C., Liu, Y. X., Li, C. J., Du, Z. J., Li, H. Q. (2007). Synthesis and characterization of polyurethane-grafted multi-walled carbon nanotubes, *Chem. J. Chin. Univ.*, *28*, 366–370.
182. Xia, H. S., Song, M. (2006). Preparation and characterisation of polyurethane grafted single-walled carbon nanotubes and derived polyurethane nanocomposites, *J. Mater. Chem.*, *16*, 1843–1851.
183. Chen, X. H., Chen, X. J., Lin, M., Zhong, W. B., Chen, Z. H. (2007). Functionalized multi-walled carbon nanotubes prepared by *in situ* polycondensation of polyurethane, *Macromol. Chem. Phys.*, *208*, 964–972.

184. Chen, X. H., Wang, J. F., Zou, J. G., Wu, X. L., Chen, X. J., Xue, F. (2009). Mechanical and thermal properties of functionalized multiwalled carbon nanotubes and multiwalled carbon nanotube–polyurethane composites, *J. Appl. Polym. Sci.*, *114*, 3407–3413.
185. Zhang, A. B., Liu, W., Li, M., Zheng, Y. P. (2009). Wear performance of tetrazine modified multiwalled carbon nanotube/epoxy composites, *J. Reinf. Plast. Compos.*, *28*, 2405–2412.
186. Yang, K., Gu, M. Y. (2009). The effects of triethylenetetramine grafting of multi-walled carbon nanotubes on its dispersion, filler–matrix interfacial interaction and the thermal properties of epoxy nanocomposites, *Polym. Eng. Sci.*, *49*, 2158–2167.
187. Ma, P. C., Kim, J. K., Tang, B. Z. (2007). Effects of silane functionalization on the properties of carbon nanotube/epoxy nanocomposites, *Compos. Sci. Technol.*, *67*, 2965–2972.
188. Yang, K., Gu, M. Y., Guo, Y. P., Pan, X. F., Mu, G. H. (2009). Effects of carbon nanotube functionalization on the mechanical and thermal properties of epoxy composites, *Carbon*, *47*, 1723–1737.
189. Tseng, C. H., Wang, C. C., Chen, C. Y. (2007). Functionalizing carbon nanotubes by plasma modification for the preparation of covalent-integrated epoxy composites, *Chem. Mater.*, *19*, 308–315.
190. Yuen, S. M., Ma, C. C. M., Chuang, C. Y., Hsiao, Y. H., Chiang, C. L., Yu, A. D. (2008). Preparation, morphology, mechanical and electrical properties of TiO_2 coated multiwalled carbon nanotube/epoxy composites, *Compos. Part A*, *39*, 119–125.
191. Chen, W., Lu, H. B., Nutt, S. R. (2008). The influence of functionalized MWCNT reinforcement on the thermomechanical properties and morphology of epoxy nanocomposites, *Compos. Sci. Technol.*, *68*, 2535–2542.
192. Wang, S., Liang, R., Wang, B., Zhang, C. (2008). Reinforcing polymer composites with epoxide-grafted carbon nanotubes, *Nanotechnology*, *19*, 085710.
193. Wang, S. R., Liang, Z. Y., Liu, T., Wang, B., Zhang, C. (2006). Effective amino-functionalization of carbon nanotubes for reinforcing epoxy polymer composites, *Nanotechnology*, *17*, 1551–1557.
194. Che, J. F., Yuan, W., Jiang, G. H., Dai, J., Lim, S. Y., Chan-Park, M. B. (2009). Epoxy composite fibers reinforced with aligned single-walled carbon nanotubes functionalized with generation 0-2 dendritic poly(amidoamine), *Chem. Mater.*, *21*, 1471–1479.
195. Baudot, C., Tan, C. M. (2009). Solubility, dispersion and bonding of functionalised carbon nanotubes in epoxy resins, *Int. J. Nanotechnol.*, *6*, 618–627.
196. Wang, J. G., Fang, Z. P., Gu, A. J., Xu, L. H., Liu, F. (2006). Effect of amino-functionalization of multi-walled carbon nanotubes on the dispersion with epoxy resin matrix, *J. Appl. Polym. Sci.*, *100*, 97–104.
197. Zheng, Y. P., Zhang, J. X., Yu, P. Y., Liu, L. L., Gao, Y. (2009). Influence of hyperbranched polyamine-ester modified carbon nanotubes on properties of epoxy nanocomposites, *J. Compos. Mater.*, *43*, 2771–2783.
198. LeMieux, M. C., Julthongpiput, D., Bergman, K. N., Cuong, P. D., Ahn, H. S., Lin, Y. H., Tsukruk, V. V. (2004). Ultrathin binary grafted polymer layers with switchable morphology, *Langmuir*, *20*, 10046–10054.
199. Prolongo, S. G., Gude, M. R., Urena, A. (2009). Synthesis and characterisation of epoxy resins reinforced with carbon nanotubes and nanofibers, *J. Nanosci. Nanotechnol.*, *9*, 6181–6187.
200. Chen, W., Shen, H. B., Auad, M. L., Huang, C. Z., Nutt, S. (2009). Basalt fiber-epoxy laminates with functionalized multi-walled carbon nanotubes, *Compos. Part A*, *40*, 1082–1089.
201. Yuen, S. M., Ma, C. C. M., Chiang, C. L., Teng, C. C. (2008). Morphology and properties of aminosilane grafted MWCNT/polyimide nanocomposites, *J. Nanomater.* 786405.
202. Yuen, S. M., Ma, C. C. M., Chiang, C. L. (2008). Silane grafted MWCNT/polyimide composites: preparation, morphological and electrical properties, *Compos. Sci. Technol.*, *68*, 2842–2848.
203. Yuen, S. M., Ma, C. C. M., Chiang, C. L., Teng, C. C., Yu, Y. H. (2008). Poly(vinyltriethoxysilane) modified MWCNT/polyimide nanocomposites: preparation, morphological, mechanical, and electrical properties, *J. Polym. Sci. A*, *46*, 803–816.
204. Zou, W., Du, Z. J., Liu, Y. X., Yang, X., Li, H. Q., Zhang, C. (2008). Functionalization of MWNTs using polyacryloyl chloride and the properties of CNT-epoxy matrix nanocomposites, *Compos. Sci. Technol.*, *68*, 3259–3264.
205. Yang, K., Gu, M. Y., Jin, Y. P. (2008). Cure behavior and thermal stability analysis of multiwalled carbon nanotube/epoxy resin nanocomposites, *J. Appl. Polym. Sci.*, *110*, 2980–2988.

206. Zhou, W., Wang, B., Zheng, Y. P., Zhu, Y. C., Wang, J. J., Qi, N. (2008). Effect of surface decoration of CNTs on the interfacial interaction and microstructure of epoxy/MWNT nanocomposites, *ChemPhysChem*, *9*, 1046–1052.
207. Zhang, Q. H., Li, J., Zhao, X., Chen, D. J. (2009). Preparation and characterization of alkylated carbon nanotube/polyimide nanocomposites, *Polym. Int.*, *58*, 557–563.
208. Chou, W. J., Wang, C. C., Chen, C. Y. (2008). Characteristics of polyimide-based nanocomposites containing plasma-modified multi-walled carbon nanotubes, *Compos. Sci. Technol.*, *68*, 2208–2213.
209. Santhosh, P., Manesh, K. M., Gopalan, A., Lee, K. P. (2006). Fabrication of a new polyaniline grafted multi-wall carbon nanotube modified electrode and its application for electrochemical detection of hydrogen peroxide, *Anal. Chim. Acta*, *575*, 32–38.
210. Manesh, K. M., Santhosh, P., Gopalana, A. L., Lee, K. P. (2006). Electrocatalytic dioxygen reduction at glassy carbon electrode modified with polyaniline grafted multiwall carbon nanotube film, *Electroanalysis*, *18*, 1564–1571.
211. Zhang, L. J., Peng, H., Kilmartin, P. A., Soeller, C., Travas-Sejdic, J. (2007). Polymeric acid doped polyaniline nanotubes for oligonucleotide sensors, *Electroanalysis*, *19*, 870–875.
212. Xu, J., Yao, P., Wang, Y. W., He, F., Wu, Y. (2009). Synthesis and characterization of HCl doped polyaniline grafted multi-walled carbon nanotubes core-shell nano-composite, *J. Mater. Sci.*, *20*, 517–527.
213. Yao, P., Xu, J., Wang, Y., Zhu, C. (2009). Preparation and characterization of soluble and DBSA doped polyaniline grafted multi-walled carbon nanotubes nano-composite, *J. Mater. Sci.*, *20*, 891–898.
214. Xu, J., Yao, P., Li, X., He, F. (2008). Synthesis and characterization of water-soluble and conducting sulfonated polyaniline/*para*-phenylenediamine-functionalized multi-walled carbon nanotubes nano-composite, *Mater. Sci. Eng. B*, *151*, 210–219.
215. Reddy, K. R., Sin, B. C., Ryu, K. S., Kim, J. C., Chung, H., Lee, Y. (2009). Conducting polymer functionalized multi-walled carbon nanotubes with noble metal nanoparticles: synthesis, morphological characteristics and electrical properties, *Synth. Met.*, *159*, 595–603.
216. Baibarac, M., Baltog, I., Godon, C., Lefrant, S., Chauvet, O. (2004). Covalent functionalization of single-walled carbon nanotubes by aniline electrochemical polymerization, *Carbon*, *42*, 3143–3152.
217. Cosnier, S., Holzinger, M. (2008). Design of carbon nanotube–polymer frameworks by electropolymerization of SWCNT-pyrrole derivatives, *Electrochim. Acta*, *53*, 3948–3954.
218. Wu, T. M., Chang, H. L., Lin, Y. W. (2009). Synthesis and characterization of conductive polypyrrole/multi-walled carbon nanotubes composites with improved solubility and conductivity, *Compos. Sci. Technol.*, *69*, 639–644.
219. Kumar, N., Kim, S., Cho, B., Lim, K., Jeong, Y. (2009). Surface functionalization of multiwalled carbon nanotubes with poly(3,4-propylenedioxythiophene) and preparation of its random copolymers: new hybrid materials, *Colloid Polym. Sci.*, *287*, 97–102.
220. Blake, R., Gun'ko, Y. K., Coleman, J., Cadek, M., Fonseca, A., Nagy, J. B., Blau, W. J. (2004). A generic organometallic approach toward ultra-strong carbon nanotube polymer composites, *J. Am. Chem. Soc.*, *126*, 10226–10227.
221. Blake, R., Coleman, J. N., Byrne, M. T., McCarthy, J. E., Perova, T. S., Blau, W. J., Fonseca, A., Nagy, J. B., Gun'ko, Y. K. (2006). Reinforcement of poly(vinyl chloride) and polystyrene using chlorinated polypropylene grafted carbon nanotubes, *J. Mater. Chem.*, *16*, 4206–4213.
222. Vandervorst, P., Lei, C. H., Lin, Y., Dupont, O., Dalton, A. B., Sun, Y. P., Keddie, J. L. (2006). The fine dispersion of functionalized carbon nanotubes in acrylic latex coatings, *Prog. Org. Coat.*, *57*, 91–97.
223. Barroso-Bujans, F., Verdejo, R., Perez-Cabero, M., Agouram, S., Rodriguez-Ramos, I., Guerrero-Ruiz, A., Lopez-Manchado, M. A. (2009). Effects of functionalized carbon nanotubes in peroxide crosslinking of diene elastomers, *Eur. Polym. J.*, *45*, 1017–1023.
224. Ku, B. C., Kim, D. K., Lee, J. S., Blumstein, A., Kumar, J., Samuelson, L. A. (2009). Synthesis and properties of water soluble single-walled carbon nanotube graft ionic polyacetylene nanocomposites, *Polym. Compos.*, *30*, 1817–1824.
225. Yuen, S. M., Ma, C. C. M., Teng, C. C., Wu, H. H., Kuan, H. C., Chiang, C. L. (2008). Molecular motion, morphology, and thermal properties of multiwall carbon nanotube/polysilsesquioxane composite, *J. Polym. Sci. Polym. Phys.*, *46*, 472–482.

226. Yuen, S. M., Ma, C. C. M. (2008). Morphological, electrical, and mechanical properties of multiwall carbon nanotube/polysilsesquioxane composite, *J. Appl. Polym. Sci.*, *109*, 2000–2007.
227. Wang, J. H., Liang, G. Z., Yan, H. X., He, S. B. (2009). Mechanical and thermal properties of functionalized multiwalled carbon nanotubes/cyanate ester composite, *Polym. Eng. Sci.*, *49*, 680–684.
228. Wiemann, K., Kaminsky, W., Gojny, F. H., Schulte, K. (2005). Synthesis and properties of syndiotactic poly(propylene)/carbon nanofiber and nanotube composites prepared by *in situ* polymerization with metallocene/MAO catalysts, *Macromol. Chem. Phys.*, *206*, 1472–1478.
229. Funck, A., Kaminsky, W. (2007). Polypropylene carbon nanotube composites by *in situ* polymerization, *Compos. Sci. Technol.*, *67*, 906–915.
230. Kaminsky, W., Funck, A., Klinke, C. (2008). *In-situ* polymerization of olefins on nanoparticles or fibers by metallocene catalysts, *Top. Catal.*, *48*, 84–90.
231. Li, W. H., Chen, X. H., Li, S. N., Xu, L. S., Yang, Z. (2007). Synthesis of polypropylene wrapped carbon nanotubes composite via *in situ* graft method with maleic anhydride, *Mater. Sci. Technol.*, *23*, 1181–1185.
232. Trujillo, M., Arnal, M. L., Muller, A. J., Laredo, E., Bredeau, S., Bonduel, D., Dubois, P. (2007). Thermal and morphological characterization of nanocomposites prepared by *in-situ* polymerization of high-density polyethylene on carbon nanotubes, *Macromolecules*, *40*, 6268–6276.
233. Trujillo, M., Arnal, M. L., Mueller, A. J., Bredeau, S., Bonduel, D., Dubois, P., Hamley, I. W., Castelletto, V. (2008). Thermal fractionation and isothermal crystallization of polyethylene nanocomposites prepared by *in situ* polymerization, *Macromolecules*, *41*, 2087–2095.
234. Bonduel, D., Mainil, M. L., Alexandre, M., Monteverde, F., Dubois, P. (2005). Supported coordination polymerization: a unique way to potent polyolefin carbon nanotube nanocomposites, *Chem. Commun.* 781–783.
235. Vega, J. F., Martinez-Salazar, J., Trujillo, M., Arnal, M. L., Muller, A. J., Bredeau, S., Dubois, P. (2009). Rheology, processing, tensile properties, and crystallization of polyethylene/carbon nanotube nanocomposites, *Macromolecules*, *42*, 4719–4727.
236. Kim, J., Hong, S. M., Kwak, S., Seo, Y. (2009). Physical properties of nanocomposites prepared by *in situ* polymerization of high-density polyethylene on multiwalled carbon nanotubes, *Phys. Chem. Chem. Phys.*, *11*, 10851–10859.
237. McDaniel, N. D., McDaniel, M. P., Balzano, L., Resasco, D. E. (2009). Silica supported single-walled carbon nanotubes as a modifier in polyethylene composites, *J. Appl. Polym. Sci.*, *111*, 589–601.
238. Pötschke, P., Pegel, S., Claes, M., Bonduel, D. (2008). A novel strategy to incorporate carbon nanotubes into thermoplastic matrices, *Macromol. Rapid Commun.*, *29*, 244–251.
239. Kim, H. S., Park, B. H., Yoon, J. S., Jin, H. J. (2007). Thermal and electrical properties of poly(L-lactide)-graft-multiwalled carbon nanotube composites, *Eur. Polym. J.*, *43*, 1729–1735.
240. Kim, H. S., Chae, Y. S., Park, B. H., Yoon, J. S., Kang, M., Jin, H. J. (2008). Thermal and electrical conductivity of poly(L-lactide)/multiwalled carbon nanotube nanocomposites, *Curr. Appl. Phys.*, *8*, 803–806.
241. Xie, L., Xu, F., Qiu, F., Lu, H. B., Yang, Y. L. (2007). Single-walled carbon nanotubes functionalized with high bonding density of polymer layers and enhanced mechanical properties of composites, *Macromolecules*, *40*, 3296–3305.
242. Hill, D. E., Lin, Y., Rao, A. M., Allard, L. F., Sun, Y. P. (2002). Functionalization of carbon nanotubes with polystyrene, *Macromolecules*, *35*, 9466–9471.
243. Xiong, H., Gao, Y., Li, H. M. (2007). Non-isothermal crystallization kinetics of syndiotactic polystyrene–polystyrene functionalized SWNTs nanocomposites, *Express Polym. Lett.*, *1*, 416–426.
244. Deng, J. N., Zhang, X. Q., Wang, K., Zou, H., Zhang, Q., Fu, Q. (2007). Synthesis and properties of poly(ether urethane) membranes filled with isophorone diisocyanate-grafted carbon nanotubes, *J. Membr. Sci.*, *288*, 261–267.
245. Buffa, F., Abraham, G. A., Grady, B. P., Resasco, D. (2007). Effect of nanotube functionalization on the properties of single-walled carbon nanotube/polyurethane composites, *J. Polym. Sci. Polym. Phys.*, *45*, 490–501.
246. Guo, G. Q., Yang, D., Wang, C. C., Yang, S. (2006). "Fishing" polymer brushes on single-walled carbon nanotubes by *in-situ* free radical polymerization in a poor solvent, *Macromolecules*, *39*, 9035–9040.

247. Sun, Y. F., Wu, F., Deng, X. Y., Xiong, D. M., Zhao, B., Wu, M. H., Jiao, Z. (2008). Effect of chemical modification on functionalization of carbon nanotubes by poly(ethylene glycol), *Chin. J. Inorg. Chem.*, *24*, 98–104.

248. d'Arlas, B. F., Goyanes, S., Rubiolo, G. H., Mondragon, I., Corcuera, M. A., Eceiza, A. (2009). Surface modification of multiwalled carbon nanotubes via esterification using a biodegradable polyol, *J. Nanosci. Nanotechnol.*, *9*, 6064–6071.

249. Liao, S. H., Yen, C. Y., Hung, C. H., Weng, C. C., Tsai, M. C., Lin, Y. F., Ma, C. C. M., Pan, C., Su, A. (2008). One-step functionalization of carbon nanotubes by free-radical modification for the preparation of nanocomposite bipolar plates in polymer electrolyte membrane fuel cells, *J. Mater. Chem.*, *18*, 3993–4002.

250. Mormann, W., Lu, Y., Zou, X. N., Berger, R. (2008). Modification and grafting of multi-walled carbon nanotubes with bisphenol-A-polycarbonate, *Macromol. Chem. Phys.*, *209*, 2113–2121.

251. Feng, Q. P., Xie, X. M., Liu, Y. T., Zhao, W., Gao, Y. F. (2007). Synthesis of hyperbranched aromatic polyamide–imide and its grafting onto multiwalled carbon nanotubes, *J. Appl. Polym. Sci.*, *106*, 2413–2421.

252. Li, J., Fang, Z. P., Tong, L. F., Gu, A. J., Uu, F. (2007). Improving dispersion of multiwalled carbon nanotubes in polyamide 6 composites through amino-functionalization, *J. Appl. Polym. Sci.*, *106*, 2898–2906.

253. Li, J., Fang, Z. P., Zhu, Y., Tong, L. F., Gu, A. J., Uu, F. (2007). Isothermal crystallization kinetics and melting behavior of multiwalled carbon nanotubes/polyamide-6 composites, *J. Appl. Polym. Sci.*, *105*, 3531–3542.

254. Chen, G. X., Kim, H. S., Park, B. H., Yoon, J. S. (2006). Multi-walled carbon nanotubes reinforced nylon 6 composites, *Polymer*, *47*, 4760–4767.

255. Meng, H., Sui, G. X., Fang, P. F., Yang, R. (2008). Effects of acid- and diamine-modified MWNTs on the mechanical properties and crystallization behavior of polyamide 6, *Polymer*, *49*, 610–620.

256. Moniruzzaman, M., Du, F. M., Romero, N., Winey, K. I. (2006). Increased flexural modulus and strength in SWNT/epoxy composites by a new fabrication method, *Polymer*, *47*, 293–298.

257. Paiva, M. C., Zhou, B., Fernando, K. A. S., Lin, Y., Lopes, P. E., Pennington, W. T., Kennedy, J. M., Sun, Y. P. (2005). Physical and mechanical characterization of nanocomposites with carbon nanotubes functionalized with the matrix polymer, *Compos. Interfaces*, *12*, 757–768.

258. Paiva, M. C., Zhou, B., Fernando, K. A. S., Lin, Y., Kennedy, J. M., Sun, Y. P. (2004). Mechanical and morphological characterization of polymer–carbon nanocomposites from functionalized carbon nanotubes, *Carbon*, *42*, 2849–2854.

259. Lin, Y., Zhou, B., Fernando, K. A. S., Liu, P., Allard, L. F., Sun, Y. P. (2003). Polymeric carbon nanocomposites from carbon nanotubes functionalized with matrix polymer, *Macromolecules*, *36*, 7199–7204.

260. Li, W. H., Chen, X. H., Yang, Z., Xu, L. S. (2009). Structure and properties of polypropylene-wrapped carbon nanotubes composite, *J. Appl. Polym. Sci.*, *113*, 3809–3814.

261. Causin, V., Yang, B. X., Marega, C., Goh, S. H., Marigo, A. (2009). Nucleation, structure and lamellar morphology of isotactic polypropylene filled with polypropylene-grafted multiwalled carbon nanotubes, *Eur. Polym. J.*, *45*, 2155–2163.

262. Causin, V., Yang, B. X., Marega, C., Goh, S. H., Marigo, A. (2008). Structure–property relationship in polyethylene reinforced by polyethylene-grafted multi-walled carbon nanotubes, *J. Nanosci. Nanotechnol.*, *8*, 1790–1796.

263. Yang, B. X., Pramoda, K. P., Xu, G. Q., Goh, S. H. (2007). Mechanical reinforcement of polyethylene using polyethylene-grafted multiwalled carbon nanotubes, *Adv. Funct. Mater.*, *17*, 2062–2069.

264. Jana, R. N., Im, C., Bhunia, H. (2009). Effect of multiwalled carbon nanotubes on crystallization behavior of poly(ε-caprolactone)diol, *J. Thermoplast. Compos. Mater.*, *22*, 531–546.

265. Wang, W., Lin, Y., Sun, Y. P. (2005). Poly(*N*-vinyl carbazole)-functionalized single-walled carbon nanotubes: synthesis, characterization, and nanocomposite thin films, *Polymer*, *46*, 8634–8640.

266. Fernando, K. A. S., Lin, Y., Zhou, B., Grah, M., Joseph, R., Allard, L. F., Sun, Y. P. (2005). Poly (ethylene-*co*-vinyl alcohol) functionalized single-walled carbon nanotubes and related nanocomposites, *J. Nanosci. Nanotechnol.*, *5*, 1050–1054.

CHAPTER 3

DISPERSION, ORIENTATION, AND LENGTHS OF CARBON NANOTUBES IN POLYMERS

3.1 OVERVIEW

The term "dispersion" in this text refers to the spatial distribution of different volume elements of a nanotube relative to the volume elements of all other nanotubes in the system. Each nanotube unit, either MWCNT, DWCNT, or SWCNT, is characterized as a single species when discussing the word dispersion, even though the former two are in fact more than one molecule. Dispersion is logically characterized on two length scales: a scale on the order of the diameter of the nanotubes, that is, a nanometer length scale; and a scale on the order of the length of nanotubes, that is, a micron length scale.

On the nanometer length scale, nanotubes have a tendency to self-assemble in the same manner as uncooked spaghetti, with their long axis parallel to one another. Individual nanotubes are arranged in a hexagonal pattern inside this self-assembled structure, termed a bundle, as schematically illustrated in Figure 3.1. Nanotubes are separated by a distance in the bundle that is at its closest similar to the distance found between graphene sheets. Because nanotubes, especially SWCNTs, have some flexibility, an individual nanotube can be a part of more than one bundle. This leads to a classical "fringed micelle" type structure, of which a further description is given in Section 4.3.1. To quantitatively characterize the dispersion of a sample of nanotubes requires a complete description of the bundle size distribution, for example, what fraction of nanotubes are in a bundle having one nanotube, what fraction in a bundle having two nanotubes, what fraction in a bundle having three nanotubes, and so on. Of course, this description is not alone sufficient to characterize dispersion on a nanometer length scale, but even this measure cannot be determined currently as described more fully in Section 3.2.

On a micron length scale (or sometimes slightly larger), nanotubes are grouped together in what is best described as appearing like a ball of yarn, with many individual pieces interlocked with one another to form the particle. Sometimes

Carbon Nanotube–Polymer Composites: Manufacture, Properties, and Applications, First Edition.
Brian P. Grady.

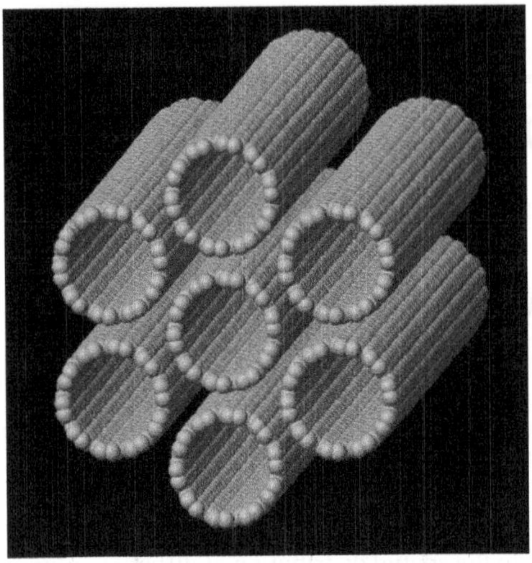

Figure 3.1 Bundling of carbon nanotubes into hexagonal arrangements. SWCNTs are shown.

instead of a ball of yarn, the tubes will appear as a more fibrous structure; see Figure 3.2 for representative micrographs of different types of these particles. So, on a macroscopic length scale, nanotubes appear as a powder with a density far less than the density of the individual tubes. The density of these powders, that is, the density of the particles, has an extremely large effect on the dispersibility of the tubes; lower density powders tend to disperse easily. Without dispersing the tubes over both length scales, that is, the micron length scale and the nanoscale length scale, nearly all if not all of the important advantages of having nanotubes combined with polymers will be negated. How much dispersion/debundling must occur is far from a settled issue, and likely depends critically on the application of interest. At this point, the common strategy is to maximize dispersion and debundling via processing within some constraints (such as no reduction in length, cost, etc.), and then control, if possible, the amount of reaggregation/rebundling that occurs during further processing steps. Control does not necessarily mean maintain maximum dispersion or debundling; electrical conductivity in composites can be increased if some reaggregation occurs.[1]

Dispersion has more aspects than reducing the size or eliminating the powder particles or bundling/debundling. Either insufficient macroscopic mixing or macroscopic demixing can lead to regions of high or low average nanotube concentration on large length scales. As an example of insufficient macroscopic mixing, nanotubes can be isolated at the edges of particles after powder mixing and compression molding. In the case of macroscopic demixing, because of large energetic driving forces nanotubes can self-organize into locally anisotropic fringed micelle structures over large length scales. Both types of nonuniformity in micron-scale spatial distribution can be extremely valuable with respect to increasing electrical conductivity, for example. Based on mostly anecdotal evidence, good debundling on the nanoscale and some separation on the microscale is a good combination for many applications.

Figure 3.2 SEM micrographs of various MWCNT powders from commercial manufacturers at two different magnifications. Identical letters represent identical samples. Further details concerning each material can be found in Ref. 301. Copyright Elsevier Ltd. Reproduced with permission from Ref. 301.

Because nanotubes are anisotropic objects, orientation is an important component of dispersion. For example, nanotubes can be bundled so that individual nanotube orientations within a bundle are identical, but on the average nanotubes are unoriented because of randomness in bundle domain orientation. Since nanotubes have essentially the same geometric shape as a typical polymer (as stated in Chapter 1, only the flexibility is different), stress fields that have the tendency to align polymers in a certain manner will have the same tendency with nanotubes. In fact, because of their stiffness, nanotubes have a much higher probability of being aligned than polymers in most situations. Since nearly all products made from thermoplastics have residual polymer orientation, articles made with thermoplastic nanotube composites will also have residual nanotube orientation, likely at a higher level than in the pure polymer.

The same mathematical concepts used in quantifying polymer orientation are useful in characterizing nanotube orientation. In both polymers and nanotubes, the most common approach to quantitatively characterize orientation is to express orientation in terms of the orientation of the long axis of the chain or tube. A very common orientation field is uniaxial orientation, that is, orientation where the angle between the long axis of the polymer or nanotube and some reference direction is not random, but is random in all other orthogonal directions. Fiber spinning operations yield uniaxial orientated polymers and nanotubes, and to a first approximation many injection molding operations yield uniaxially oriented chains and tubes. In uniaxial orientation, only 90° of a possible 360° rotation needs to be considered because of symmetry. Legendre polynomials are the typical mathematical construct used to quantify uniaxial orientation, and knowledge of the numerical value of each polynomial enables the calculation of the full orientation distribution function. The second-order term has been given the name "orientation function" or alternatively the "Herman's orientation parameter" or "order parameter" because of its importance and is shown below.

$$\text{Orientation function} = \frac{3\langle\cos^2\theta\rangle - 1}{2} \tag{3.1}$$

where $\langle\,\rangle$ represents the average value and θ is the angle between the reference axis and the long axis of the nanotube or polymer. A value of 1 for the orientation function indicates perfect orientation along the reference axis, a value of 0 indicates random orientation, and a value of $-1/2$ indicates perfect orientation perpendicular to the reference axis. The reference axis is almost always chosen to be the extrusion or flow direction. More complicated stress fields are also found in polymers, in particular biaxial stress fields that arise in blow molding operations, but their mathematical description is beyond the scope of this book. Further, the author is unaware of any other quantitative experimental measure of orientation of nanotubes in polymers more complicated than uniaxial at this time (although certainly the possibility exists to make such measurements). Methods used to calculate orientation for nanotubes as well as dispersion will be described in Section 3.2.

The distribution of nanotube lengths is also an important parameter. Just as in polymers, the distribution of lengths in a nanotube sample is initially determined by the synthetic procedures. The fundamental question addressed in this chapter will be whether this length is reduced by the procedures used to disperse nanotubes in polymers. As with polymers, number and weight average lengths could be defined. In polymers, the reason for using number and weight averages instead of the more typical number average and standard deviation is that polymer molecular weights tend to be highly asymmetric with long tails at high molecular weights and this two-parameter measure provides some representation of that asymmetry. Without processing procedures that reduce nanotube length, the distribution of nanotube lengths tends to be more narrow than in the typical polymer. Hence, the use of a one-parameter number average length is typical. However, the length distribution reported by one group of authors suggests that the distribution is asymmetric with a higher length tail since a log-normal distribution was a better description of the distribution.[2] The effect of the various dispersion methods on the average nanotube length will be discussed in this chapter as well.

In the previous paragraph, the term "length" was used to signify the contour length, that is, the length if the nanotube were straightened. The end-to-end length, that is, the length from one end of the tube to another, is not the contour length since nanotubes can be curved. Persistence length is a quantitative measure of chain or tube flexibility; the lower the persistence length, the more flexible the chain. For example, the persistence length is key for determining at what chain length polymers will entangle sufficiently so as to behave elastically. Microscopy has been most effective with nanotubes in measuring persistence lengths[3–6] because the typical scattering methods used for polymers are not able to access the low angular range to enable measurement of extremely long persistence lengths. The most satisfying measurement of persistence length was one that showed that the persistence length was well described by an empirical formula: persistence length $= (63 \times d^3)\,\mu\text{m}$, where d is the nanotube diameter in nanometers.[3] These authors argue that the persistence length of MWCNTs should be orders of magnitude larger; however, similar visualization techniques yielded a value that was in the hundreds of nanometer range[5] while ultrasmall-angle scattering was used to determine, via a model fit to scattering angle, a smaller persistence length of 80 nm in polyamide 6.[7] The clear discrepancy between these values is explained by the authors of the latter paper by the presence of static bend points, i.e. defects that cause a significant lessening of the persistence length.

Strictly speaking, if nanotubes are single molecules, then the term solubilization should be used rather than dispersion in some situations. What is the difference between "dispersion" and "solubilization"? A recent review paper drew a distinction between dispersed surfactant-stabilized individual nanotubes and solubilized individual nanotubes without surfactant stabilization.[8] To define the difference between these terms in terms of the presence or absence of an adsorbed layer seems arbitrary; if the surfactant molecules were covalently bonded would the term solubilization be appropriate? A recent paper makes the argument that it is possible to solubilize pristine (e.g., unfunctionalized) nanotubes under certain conditions[9] and defines solubilization as corresponding to the case where the free energy of mixing is negative. This definition for solubilization is perfectly correct; however, it is not very satisfying because experimental measurements of free energy are notoriously difficult. A more satisfying definition that is also consistent with the free energy definition is to define solubilization as when two nanotubes will spontaneously separate no matter what their starting center-to-center distance is. True solubilization requires that all nanotubes be isolated at some small but finite concentration; in fact, these authors[9] find that a plateau exists in the systems where solubility is claimed, which corresponds roughly to 20% isolated tubes as measured by AFM. It should be noted that the claim of solubility made for NMP in the original paper was softened considerably in a review paper written by the same author.[10] Another, more recent publication[11] measures the phase diagram of solubilized nanotubes in superacids of chlorosulfonic acid and shows that a thermodynamic model of athermal mixing for rods is able to describe the observed phase diagram quite well. True athermal mixing would of course mean a negative free energy of mixing since the entropy will increase at very low concentrations. More importantly, a truly isotropic phase consisting of only individual tubes was found. In this case, the term "solubilization" is the correct one to use.

Frankly, most authors do not explicitly note the difference between the terms solubilization and dispersion. Regardless of the correct term, in this book the term dispersion will be used when referring to carbon nanotubes, regardless of whether solubilization would be strictly correct. This choice avoids the problem of ascertaining whether a particular sample consists of nanotubes that are either (a) solubilized or (b) dispersed or (c) portions of the sample are solubilized, and portions are dispersed. The term dispersion is used with respect to polymers when groups of polymer molecules are present in a liquid, and solubilization when individual polymer molecules are present in a liquid. Although the two definitions are not consistent, systems where a dispersed polymer in a liquid is in equilibrium with a significant amount of dissolved polymer are not common, while such coexistence could be common in nanotubes mixed with liquids. Coexistence is possible not only because the concentration is such that some nanotubes (or some parts of nanotubes) are in bundles and some are not, but also because different nanotube chiralities or different functionalization levels could lead to some tubes in a sample being soluble, and others not. The author feels comfortable that the reader will not be confused by this inconsistency.

Figure 3.3 compiles the six basic methods (dispersion–reaction, dissolution–dispersion–precipitation, dispersion–dispersion–evaporation, melt mixing, no fluid mixing, impregnation/infusion) used to disperse nanotubes in a polymer, with significant variants of the methods raising the total to 12. Impregnation/infusion refers to where a pure nanotube sample is produced and then polymer or monomer is allowed to infuse into the sample. The key difference of impregnation/infusion processes is that the spatial distribution of nanotubes is already in its final form, while the other five methods in Figure 3.3 all suppose some nanotube–polymer mixing processes. This chapter will discuss each method in a roughly left to right sequence according to Figure 3.3.

The methods listed in Figure 3.3 are not complete because three important considerations are not addressed. The first is whether a dispersing agent is used; is a molecule(s) added that adsorbs on the surface of the tubes? Typical dispersing agents are small molecules that increase the compatibility, that is, reduce the interfacial tension, between the nanotubes' surface and a liquid via adsorption of the molecule on the surface of the nanotube. Since unmodified carbon nanotubes are hydrophobic, in general the more polar the solvent, the more the need for a dispersing agent. The second consideration is whether tubes are functionalized, that is, a molecule has been covalently bonded to the nanotubes. Functionalization can have two impacts from a dispersion perspective. First, the functionalizing molecule can reduce the interfacial tension with the liquid and/or polymer. Second, functionalization can make tubes easier to disperse because tube–tube attractive forces in a bundle (at optimal

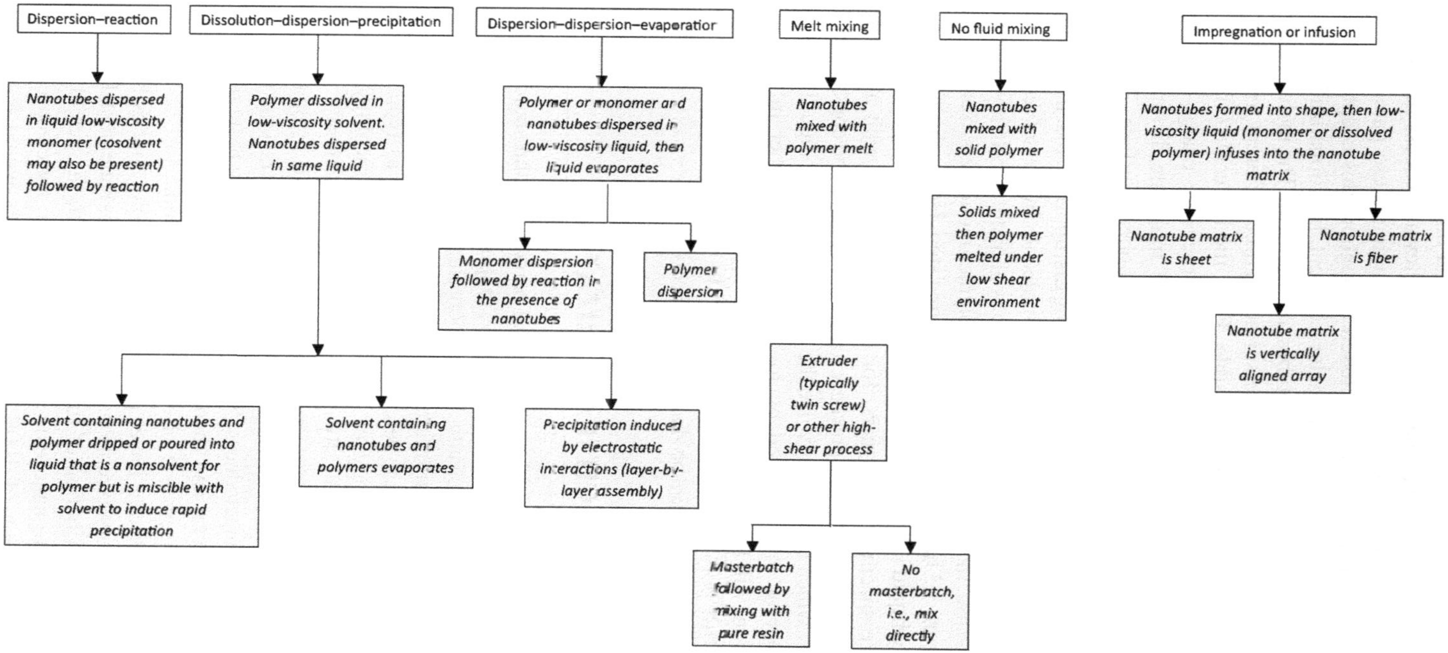

Figure 3.3 Schematic of various ways in which nanotubes are mixed with polymers to manufacture composites.

separation distance) can be reduced. The third and final consideration is the mixing procedure itself. For example, there are many relevant process parameters in twin-screw extrusion or in ultrasonication, and changing these parameters can affect dispersion. These factors will be discussed throughout this chapter.

Many requirements of a particular covalent functionalization or adsorbing molecule to effectively disperse nanotubes are identical. The molecule must be able to effectively adsorb or react with the surface. Three different mechanisms can be operative; combinations are also possible. First, the molecule effectively disperses the tubes by introducing a charge, which leads to charge–charge repulsion in polar (high dielectric constant) solvents. Second, the molecule disrupts the nanotube–nanotube packing by modifying the smooth graphene surface. Third, a molecule protrudes from the surface of the nanotubes, which causes stabilization via steric repulsion. Steric repulsion differs from packing disruption in that the driving force for stabilization is solubility of the protruding molecule in the solvent.

The chapter will first consider dispersions of nanotubes in low-viscosity liquids in a general sense, that is, not restricted to polymer monomers. Some of the methods shown in Figure 3.3 have this procedure as a first step. How a dispersed nanotube in a low-viscosity liquid is mixed or made into a polymer will be the next topic. Methods that are more specific to polymers, for example, mixing with a polymer melt or mixing nanotube and polymer powders together, will then be considered. Impregnation methods, which typically involve the use of a nanotube–liquid mixture to form a nanotube fiber or mat followed by low-viscosity monomer infusion and finally polymerization, will be discussed last. First though, before the methods used to mix nanotubes with polymers can be described, the way in which the dispersion of nanotubes is quantified will be discussed in Section 3.2.

3.2 DISPERSION CHARACTERIZATION

Measurement of the electrical or rheological percolation threshold has been used to qualitatively determine dispersion quality in polymers. The term "percolation" refers to long-range connectivity, so the percolation threshold for a filled system is defined as the filler concentration where the filler network first exhibits long-range connectivity. The rheological percolation threshold,[12,13] as typically measured by torsional oscillatory rheology or tensional oscillatory rheology,[14] and electrical percolation threshold, as typically measured by DC conductivity,[15] are by far the two most common percolation threshold measurements used. In other words, distinct changes in the electrical or rheological properties occur when the filler network first shows long-range connectedness. Section 5.2.2 contains a more detailed discussion of percolation in general and rheological percolation specifically, and Section 6.2 contains a detailed discussion of electrical percolation.

In a general sense, a lower percolation threshold implies better debundling. However, there are many exceptions and cautions to this statement. For systems where the micron-scale distribution of tubes is not uniform, the percolation threshold is essentially unrelated to nanoscale dispersion. Further, for electrical conductivity at least, perfect dispersion of 100% isolated nanotubes would lead to a very high percolation threshold depending on nanotube alignment. Hence, practically there is some characteristic dispersion that does not likely involve 100% debundling that corresponds to the lowest electrical percolation threshold. Inhomogeneous macro-scale dispersion in particular can lead to very low percolation thresholds. Especially at low percolation thresholds (below 1%), it is possible for a lower percolation threshold material to have significantly worse dispersion than a higher percolation threshold material when electrical conductivity is used. Since rheological percolation in a polymer involves both polymers and nanotubes,[16] it is not clear exactly whether the lowest rheological percolation threshold would correspond to perfect dispersion. The length of the tubes, which has nothing per se to do with dispersion, will also affect both percolation thresholds. Overall though, using oscillatory rheology to characterize dispersion is less ambiguous than using electrical conduction. A disadvantage of using the percolation threshold to characterize dispersion is that measurements are laborious since samples of varying concentration must be made and measured to determine percolation threshold. Overall, if the percolation threshold is below about 1%, then the nanoscale dispersion is probably pretty good; trying to directly compare two values and draw conclusions about the level of dispersion is done on a regular basis and is fine in some, but definitely not all, cases.

Other methods of characterizing dispersion in polymers are more direct than using percolation. Dispersion characterization tools can be thought of as one of the two types: tools based on spectroscopy or tools based on microscopy. Four types of microscopy have been used to characterize dispersion, transmission electron microscopy (TEM), scanning electron microscopy (SEM), atomic force microscopy (AFM), and optical microscopy, including confocal microscopy. Spectroscopic methods include Raman spectroscopy, UV–Vis spectroscopy, and scattering methods. All these methods can be applied to both low-viscosity liquids and polymers; the latter will obviously be the focus although the former will be briefly mentioned. Other methods that can only be applied to nanotubes dispersed in a low-viscosity liquid will be briefly discussed as well. The author strongly feels that the microscopy methods have the most potential for characterizing dispersion, but these are also difficult to apply. Although the techniques themselves are not likely to change substantially, the way in which microscopy techniques are applied to nanotubes and polymers might change; this book is already significantly out of date with respect to microscopic methods to characterize nanotube dispersion! Spectroscopic methods are relatively more mature, and being out of date is not expected to be nearly as much of an issue.

3.2.1 Microscopy

TEM measures the spatial arrangement of objects via the absorption of electrons, in a similar manner as X-rays used in medicine, except that electrons are used in TEM. If the components of a spatially inhomogeneous sample absorb electrons differently, then the three-dimensional inhomogeneous distribution appears as a two-dimensional inhomogeneous grey-scale image, again in the same manner as a medical X-ray image. Carbon nanotubes generally are more able to absorb electrons than liquids or polymers, although the difference between absorptions, termed the contrast, is typically not large. Hence, TEM can be used to probe dispersion. However, there are two other significant issues with nanotube samples. The first is that the nanotubes must be in a matrix that is able to withstand vacuum conditions, which means liquids are not suitable unless imaging occurs under cryogenic conditions. The second issue is that a very thin sample must be used in order to minimize overlap effects. If a sample is too thick, nanotubes at different places in the thickness direction may appear to overlap even though they are widely separated in the thickness direction. With polymers, samples are frozen and then cut into very thin slices in order to produce samples thin enough to minimize overlap effects. TEM has the ability to resolve individual nanotubes, even single-walled carbon nanotubes,[11] although making a composite sample thin enough coupled with the rather low contrast is a problem that makes imaging difficult. Staining to increase contrast has not been used presumably because currently no staining agents have been found that significantly improve the contrast.

As might be expected, TEM is most often used to characterize dispersions in solid polymers rather than liquids. A complete listing of papers that have used TEM to examine dispersion in nanotube–polymer samples numbers in the many hundreds. A representative micrograph is shown in Figure 3.4.[17] As the micrograph indicates, clusters can be identified, and semiquantitative analyses are possible. Arguably, nanotube lengths can be determined as well, although cutting of the tubes during sample preparation must be considered. An example of a detailed examination of TEM images to better quantify the state of dispersion is given in Ref. 18. A correlation function was determined from a similar micrograph as that shown in Figure 3.4 and qualitatively related to the number of clusters. However, to the author's knowledge, no researchers have at this time used TEM to present a quantitative relation to the more satisfying average number of tubes per cluster or fraction of tube volume found in a cluster. A measure of orientation was also determined;[18] although not explicitly related to the more common orientation function or orientation distribution function, such a measure could have been made. These measures allowed the authors to make the conclusion that electrical conductivity decreased as either orientation increased or clustering decreased.

SEM is also used to look at bundling characteristics of nanotubes. SEM is similar to TEM in the sense that electrons impact a sample, but in this case the image is collected in reflection rather than transmission. Hence, usually SEM is able to image only surface topology. In its most typical use, SEM is used to image a fracture surface of nanotube–polymer composite, where the composite is frozen to guarantee brittle fracture. Two items are typically of interest in such a micrograph. The first is the manner in which the nanotubes are found with respect to the fracture surface. For maximum or close to maximum reinforcement, nanotubes would be fractured at the fracture surface. Instead, generally, some nanotubes stick out of the fracture surface while holes are found as well. Both indicate nanotube pullout from the fracture surface and usually are taken to be evidence of nonbreaking of the nanotubes. A polymer sheath covering the nanotubes is taken as evidence of good polymer–nanotube adhesion, since pullout caused the breakage of polymer chains rather than sliding of the nanotube through the matrix. These qualitative measures of matrix adhesion are very useful because producing two identical nanotube

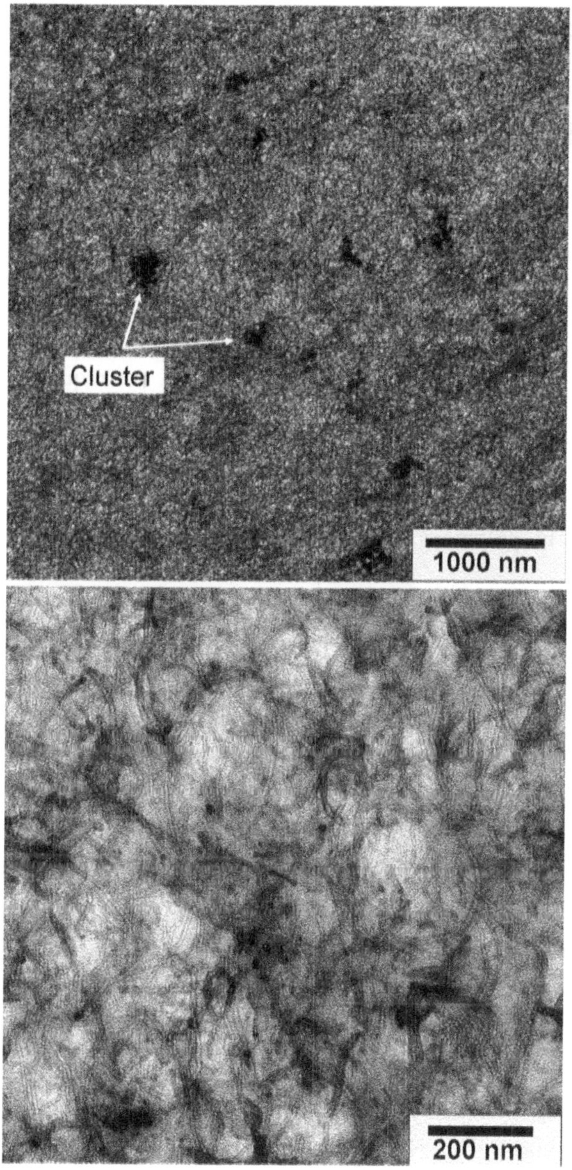

Figure 3.4 Representative TEMs of a well-dispersed MWCNT sample in a polymer. In this particular case, the polymer is polyamide 6. Copyright Elsevier Ltd. Reproduced with permission from Ref. 17.

composites that differ only in matrix adhesion is impossible since inevitably bundling characteristics also change. One common example where SEM is used in this qualitative manner to highlight differences in nanotube–polymer adhesion is in functionalization studies.[19,20]

The resolution of SEM is typically on the order of a couple of nanometers, too large to distinguish individual SWCNTs and DWCNTs from bundles containing a small number of tubes. This resolution should allow quantification of the average bundle size for MWCNTs. However, in systems where the bundle size is small enough and hence the properties are good enough to warrant such a laborious effort, the MWCNTs are generally much larger. The author is not aware of any effort to resolve whether a large diameter feature is due to bundles or polymer-coated tubes. Further, the fact that a MWCNT sample consists of a distribution of diameters also makes such a determination difficult, except in the cases of large bundles.

As stated previously, normally SEM images surface topology. To image electrically insulating materials including most polymers, SEM surfaces are usually coated with a conductive layer, typically gold or carbon. This coating does reduce the resolution of SEM. Environmental SEMs, which still are performed in an evacuated environment but at much higher pressures than normal SEMs, do not require a conductive topcoat for nonconductive samples. Further, nanotube composites can reach sufficient conductivity at higher nanotube contents so that a coating is not required even in standard SEM. Although an advantage due to improved resolution, the lack of a coating does not significantly change the ability or capabilities of SEM in cases relevant to nanotubes in polymers, except as detailed in the following paragraph.

Because nanotubes are electrically conductive and most polymers are insulating, SEM can be used to image the interior of a sample and not just the surface through a technique termed charge contrast imaging. The contrast is due to the charging ability of nanotubes relative to the insulating matrix. However, this technique seems to require that the sample be sufficiently above the percolation threshold to be useful.[21] This technique was first used on SWCNT samples, and one interesting result was that some tubes were highly curved (circular structures corresponding to radii less than 50 nm), although no quantitative measures such as persistence length were attempted.[22] In terms of resolution, a single 1 nm tube will not appear as a 1 nm image in the system due to charging of the nearby polymer; the actual apparent diameter is not known although an order of magnitude increase is probably expected. Unfortunately as the last statement implies, the exact resolution may be dependent on the polymer in question. The other question is the depth of vertical probing. In one study,[23] the authors found that the imaging depth was limited to 50 nm, which is consistent with the current understanding of the escape depth of a secondary electron. In a more recent paper,[24] the authors claim to image to approximately 1000 nm, which is certainly deep enough to consider this a bulk measurement. The micrograph from that paper is shown in Figure 3.5. The authors calculate an average bundle diameter; however, the bundle diameter is much larger than is likely possible, indicative of the resolution problems mentioned earlier. Unfortunately, a limiting lower bundle size is not found, which in turn could be used to determine the inherent resolution. Although average length was not measured in any of these studies, determination of average bundle length is possible as well. One-parameter statistical measures of the bundle size distribution are also found in this paper, although the measures are different from those used in the TEM study discussed earlier. This difference in approach in the author's opinion highlights the need to present dispersion characterization parameters in terms of average number of tubes per bundle or average diameter.

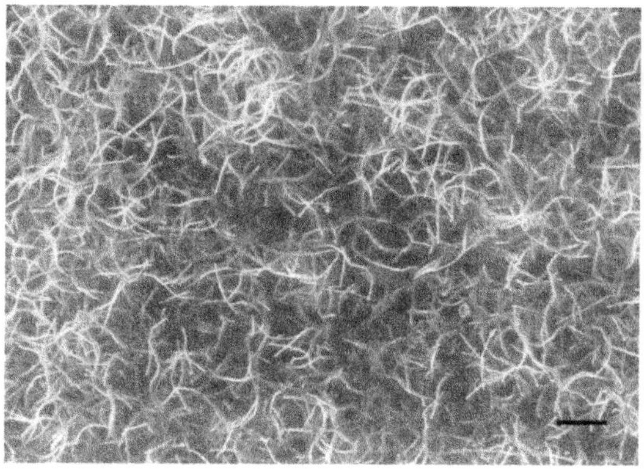

Figure 3.5 Scanning electron micrograph (scale bar = 2 μm) measured under high accelerating voltages that enables the signals from as deep as 1000 nm in the sample according to the author. The spaghetti-like features are SWCNTs. It is not clear whether individually dispersed tubes are present in this sample. Copyright IOPscience. Reproduced with permission from Ref. 24.

AFM measures the topology of a surface using a very thin mechanical probe, and is capable of imaging topology on surfaces suspended in liquids. The horizontal resolution of AFM is on a par with SEM but AFM cannot image surfaces with rapid changes in height. Consequently, AFM is difficult to use on surfaces created by brittle fracture, although there are some examples in the literature.[21] AFM is truly a three-dimensional probe of topology, while SEM is three dimensional only if multiple images are taken at different sample tilt angles. A common use of AFM to measure dispersion of nanotubes in a liquid is to place a drop of liquid containing the nanotubes on a flat surface, and then evaporate the liquid and measure average bundle size and, if more effort is taken, the distribution of bundle sizes.[25,26] This method is also the most common to measure average nanotube length. SEM or TEM can also be used for this same measurement, but AFM is easier because of the higher flexibility in the substrate afforded, although TEM has significantly higher resolution. Of course, the assumption is that evaporation does not affect dispersion characteristics, which may not be true in some cases.

The standard way in which AFM is used to characterize dispersion in nanotube–polymer composites is to image the surface and use the characteristics of the surface dispersion to infer the characteristics of the bulk dispersion. Although much less common, it is possible to examine nanotube orientation in the same manner.[27] The difference in modulus between a polymer and a nanotube can be used in AFM imaging to distinguish between the two even in cases where the topology is smooth. Further, use of an electrically charged tip allows for the imaging of nanotubes not right at the surface, although the depth probed is very limited.[28] One of the more interesting sidelights is that nanotubes make excellent AFM tips because of their small diameter and large stiffness, and a number of commercial firms offer AFM tips made with carbon nanotubes.

Optical microscopy relies on the contrast between the dark-colored nanotubes and the approximately optically clear matrix. The image provided by such a micrograph can be considered to represent the bulk dispersion. Resolution is poor compared to the previous methods, so instead of imaging individual nanotubes or even individual bundles, optical microscopy images dispersion on a minimum of

~1 μm length scale.[29] The thickness of the sample coupled with the nanotube loading is an issue since this measurement is made in transmission with a high-intensity multiwavelength bulb. For proper measurements to be made, samples must be of uniform thickness and at least mm^2 in area. Unless samples are extremely poorly dispersed, nanotube levels are typically in the fraction of percent range in order to obtain usable micrographs. Rather difficult filtering procedures can be used on optical micrographs, which in turn allows for the calculation of the entire orientation distribution function for the bundles being imaged.[30] Fluorescence microscopy can be used to image individual SWCNTs by using the inherent fluorescence of an isolated tube.[3] A related method, that is not a transmission method so that problems associated with sample thickness are not an issue, is laser confocal microscopy.[31] Laser confocal microscopy allows for depth profiling, so that a true three-dimensional image can be generated.[32] The resolution of this method, ~0.1 μm, is much finer than in optical microscopy. An example of the image generated by confocal microscopy is shown in Figure 3.6. Statistical measures have been used to quantify the distribution.

3.2.2 Spectroscopy

As described in Chapter 2, nanotubes are Raman active. Using a Raman microscope, the intensity of Raman scattering can be used as a measure of nanotube concentration in a given volume element with a resolution of approximately 1 μm^2.[15] Raman measurements are often made in reflection; the depth of penetration is such that the measurement is a bulk measurement. However, using intensity of Raman scattering to quantify nanotube concentration is extremely risky, since some intensities are a strong function of whether tubes are bundled or not. A number of papers have described how Raman shifts in intensity at a given wavelength, or a shift in the wavelength where the maximum scattered intensity occurs, can be used to semi-quantitatively characterize dispersion in solution.[33–35] The use of this with respect to solid polymers is very limited,[36] because changes in strain on the nanotubes as well as interactions of the nanotube with the polymer can cause similar changes in Raman spectra.

Raman spectroscopy also has the ability to quantitatively characterize the orientation of the nanotubes via the use of polarized radiation. The most common approach is to have the incoming and scattered radiation polarization directions parallel to one another. In fact, both the second- and fourth-order Legendre polynomials can be quantified using Raman spectroscopy, and with the assumption of maximum entropy the entire orientation distribution function can be determined with only the second- and fourth-order polynomials.[37] At least three different relative orientations of the polarization axis with the sample axis must be measured in order to determine the second- and fourth-order moments. Nanotubes are much more amenable than polymers to orientation measurements with Raman spectroscopy, since in polymers knowledge is required of the angle between the molecular vibration responsible for a given Raman feature and the polymer chain axis. In nanotubes, all Raman-active vibrations can be assumed to be active only in the direction of the nanotube axis. In other words, there would be no Raman activity if nanotubes are

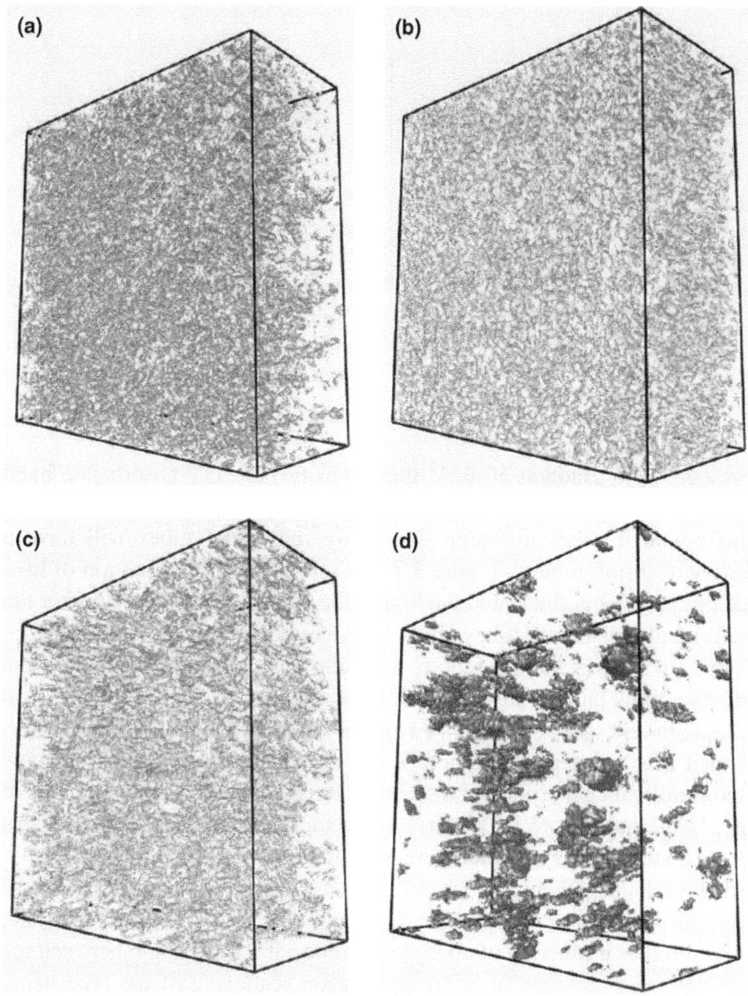

Figure 3.6 Four different images showing SWCNTs dispersed in PMMA at 0.5 wt% tube fraction. The difference in the images is due to the method of dispersion. Copyright Elsevier Ltd. Reproduced with permission from Ref. 32.

perfectly orientated and the polarization directions are 90° different from the orientation direction. Given the importance of orientation, the number of studies that have examined orientation in this manner is actually rather limited.[37,38] A more common approach is to measure the intensity when the sample axis is parallel to the two parallel polarization directions and to measure the intensity when the sample axis is perpendicular. The ratio of these intensities can be used to determine the orientation function, that is, the second-order Legendre polynomial.[39,40]

Related to the complete bundle size distribution would be the fraction of tubes that are in bundles versus those that are not. In papers by Regev and coworkers,[41–43]

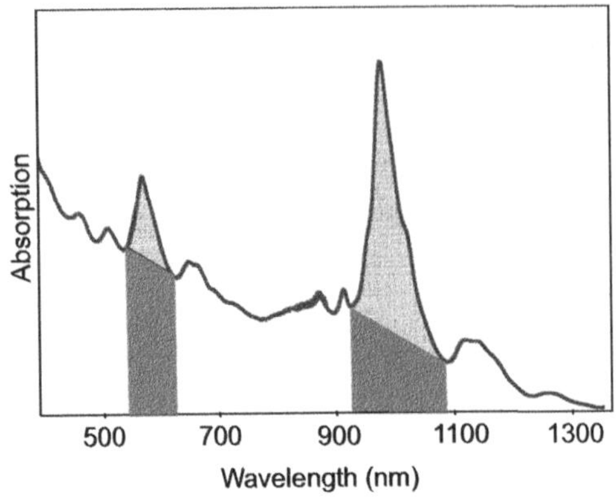

Figure 3.7 A schematic representation of how the resonance ratio is determined. The resonance ratio would be the ratio of the lightly shaded area to the dark gray shaded area. One or more than one peak can be used in this determination.

Tan and Resasco,[44] and Haddon et al.,[45] the intensity of a UV absorbance band (or multiple bands) can be used to monitor the relative concentration of individually dispersed single-walled tubes in water since only individual tubes will have a relatively narrow absorption band. Figure 3.7 is a schematic representation of how the resonance ratio can be defined, and the higher the resonance ratio, the higher the fraction of individually dispersed tubes. Since the incremental effect on the nonresonant background by adding another tube to a bundle is much less than that on the resonant band by adding an individual tube, it is many times perfectly reasonable to consider the peak height as a measure of exfoliation. Other groups[14,32] have qualitatively extended this concept to a nanotube–polymer composite; however, many polymers have significant absorption in the UV region and hence this technique is of limited utility. Of perhaps more importance is the fact that it is not clear how this technique treats a single tube of which part is bundled and part is isolated.

Scattering methods, either neutrons or X-rays, have the capability of measuring both nanoscale dispersion and orientation. In contrast to Raman scattering where intensity is measured as a function of energy difference at a fixed angle between the incoming and outgoing photons, in X-ray or neutron scattering of the type being discussed here there is no energy difference between incoming and outgoing photons, and intensity is measured as a function of scattering angle. Because of the characteristic size of nanotubes and their bundles, scattering angles being probed are in the small-angle regime, that is, less than 5°. The scattering pattern of a perfectly dispersed sample of nanotubes will have slope of -1 in a log-log plot of scattering intensity versus q at certain angles when q and scattering intensity both have the same units of inverse length. Bragg's law in Equation 3.2 relates the magnitude of the scattering vector (q) to the distance (d) being probed with q defined in terms of the wavelength λ of the scattering radiation and scattering angle (2θ).

$$d = \frac{2\pi}{q}, \quad q = \frac{4\pi \sin\theta}{\lambda} \tag{3.2}$$

The high-angle limit of the -1 dependence will be determined by the diameter of the nanotubes, that is, the larger the diameter, the smaller the angle where the deviation from −1 will occur. The low-angle limit, in the case of perfect dispersion, will be determined by the average flexibility of the tubes, or the average length of a tube if all tubes are perfectly straight. Practically, because the angles achieved are usually not small enough to be affected by flexibility or length, a transition away from the −1 slope at low q indicates bundling, and the length scale corresponding to the deviation from the −1 dependence is a measurement of some average run length of an individually dispersed tube prior to bundling. In many cases, including to the author's knowledge all polymer systems, the value of the slope is typically between −2 and −3 indicating that dispersion is far from perfect. Some authors argue that it is possible to use the magnitude of the deviation away from −1 as a measure of the nanoscale dispersion.[46,47]

Scattering methods definitely work best on liquids for a number of reasons. Although an amorphous polymer should in theory have a flat scattering profile, it is a well-known (but unpublicized) fact that most amorphous polymers do not because of long-range density fluctuations that are difficult to eliminate. A semicrystalline polymer will have significant scattering due to crystalline–amorphous scattering; since nanotubes usually affect the crystalline superstructure of a polymer (see Section 4.2), this component cannot be appropriately subtracted. Neutrons are better to use than X-rays because the intensity of scattering relative to background sources is typically much larger. In the angular regime where a slope of −1 would be found with perfect dispersion, the slope is typically −2 to −3 as mentioned earlier. At very low angles, the slope transitions to a different value, and this value has been interpreted as being caused by micron-scale inhomogeneity in the carbon nanotube spatial distribution.[46,48] Other interpretations for the characteristic slope in this regime have been given representative of the ambiguity of such measurements as applied to nanotube dispersion.[7,49]

The use of neutron or X-ray scattering methods to measure nanotube orientation in polymers has been semiquantitative. In the case of perfect uniaxial orientation, the scattering due to nanotubes in the direction perpendicular to the orientation direction will be a maximum, while that in the parallel direction will be zero over the angular range that can be probed. The more perfect the distribution along the reference axis, the more narrow the width of the peak at a given scattering angle in the perpendicular direction. This peak width has been used to characterize the relative orientation;[50] this type of procedure has been mathematically formalized to produce a quantity analogous to the orientation function.[51] However, the meaning of that function is not the same as the orientation function defined previously, because the off-axis scattering intensity does not vary in a simple manner as does the off-axis Raman signal.

Light scattering can also be used to characterize nanotube dispersion, although the wavelength of light is such that the micron length scale is imaged. As with X-ray and neutron scattering, intensity as a function of scattering angle is measured where the incoming and outgoing light have the same wavelength. Light scattering has much the same issues as optical microscopy; hence, light scattering has only been used in a couple of cases to examine dispersion since optical microscopy gives a much easier-to-interpret image and is simpler to set up and perform. Light scattering can provide a value for the average macroscale cluster size.[48]

Light scattering can also be used to measure nanotube orientation, which has been used to some advantage. However, the situation is much more complicated than with X-rays or neutrons. Individually dispersed carbon nanotubes as well as small bundles will not contribute to the signal perpendicular to the nanotube axis, because the wavelength of light is much larger than that of the diameter. However, in this case nanotube length is not longer than the wavelength of light, so there is some small, but finite, contribution of the signal from the length. As the bundle size grows, bundles can now contribute to the signal. Further, nonhomogeneous variations in nanotube concentration on the micron scale could very well have some preferred orientation, and this will also contribute to the anisotropy of the signal. Using polarized incoming and outgoing radiation, the location of the intensity peak maximum relative to the reference axis is a measure of orientation; with no orientation, the value will be 45° and will decrease as the orientation becomes more perfect.[30] Using unpolarized radiation allows for the calculation of the various moments of the Legendre polynomials;[52] however, the meaning of those moments is not clear as it is in Raman spectroscopy.

Other techniques, not applicable to solid polymers, have also been used to measure dispersion. Fluorescence spectroscopy (or photoluminescence) is reported as a sensitive measure of SWCNT dispersion in solution, since one metallic tube in a SWCNT bundle negates the fluorescence associated with the band gap of the semiconducting SWCNT species. In other words, fluorescence, like UV absorption, gives a measure proportional to the number of isolated tubes in solution. Because fluorescence is very sensitive to any adsorbed species on the surface, comparing the efficiency of debundling via fluorescence should be confirmed via the use of a second technique. Fluorescence in general cannot be used in most polymers because of interference. Size exclusion chromatography has also been used to separate bundles by size in solution, which in turn allows for a size distribution characterization.[53] Since separation is on the basis of hydrodynamic volume, differing lengths will confound the measurement. In order to convert the size distribution to a bundle size requires a calibration that will be difficult if not impossible to achieve. Finally, as mentioned previously, rheological methods can also be used to monitor dispersion in low-viscosity liquids and polymers.[54–57] Normally, a maximum in viscosity is interpreted as the best dispersion in a Newtonian fluid;[54] however, in one case a minimum in viscosity was interpreted as the best dispersion.[56] The latter is not intuitive and is explained by the authors as aggregates interfering with one another, that is, accidental jamming, and the authors noted erratic rheological properties in support of this hypothesis. However, the viscosity will also decrease with nanotube length and the authors might have been measuring tube breakage. A general procedure to separate the two characteristics, dispersion and average length, via rheology has not been presented at the time of the writing of this book. Although microscopy is normally used to characterize nanotube length, colloidal hydrodynamic fractionation has been used to characterize nanotube length in a solvent after sonication.[58] Dynamic light scattering can also be used to determine nanotube length assuming that nanotubes behave as rigid rods.[59]

3.3 METHODS TO DISPERSE NANOTUBES INTO LOW-VISCOSITY LIQUIDS, INCLUDING MONOMERS

There are four different protocols that involve the dispersion of nanotubes into a low-viscosity liquid in order to eventually achieve a polymer–nanotube mixture: (1) dispersion of nanotubes into a monomer followed by polymerization; (2) dispersion of nanotubes into a liquid that also contains dissolved polymer followed by evaporation of the liquid; (3) dispersion of nanotubes into a liquid that also contains dispersed polymer or monomer followed by polymerization (if necessary) followed by evaporation of the liquid; (4) dispersion of nanotubes into monomer followed by infusion of monomer/nanotube followed by polymerization, for example, impregnation methods. In methods 1 and 4, the monomer may also contain a solvent to improve dispersion, which has to be evaporated either before or during polymerization. Hence, a separate section that discusses dispersions of nanotubes into low-viscosity liquids is required because of its importance to nanotubes in polymers. This section also has particular relevance for most other applications as well, since good dispersion in both places in this sentence is critically important for most applications and the use of low-viscosity liquids is the best method to achieve debundling.

Dispersion of carbon nanotubes into low-viscosity liquids is best viewed as being comprised of two different components. The first component is a kinetic process and consists of reducing the average bundle size, that is, debundling, and eliminating large yarn-like particulate aggregates. Except possibly for the aforementioned superacids, simply placing nanotubes into a solvent either with or without a dispersing agent will not yield a dispersed nanotube system without significant energy input in terms of either high-velocity mixing or ultrasonics. In other words, this process is an activated process having a substantial energy barrier that must be overcome. During this stage, the energetics between the nanotube surface and the solvent are essentially irrelevant The second part is the prevention of reagglomeration, which is where the energetics of the nanotube surface and the solvent are critically important. An interesting corollary to this analysis is that a original sample and a sample that was well dispersed and allowed to reagglomerate will have very different characteristics, which is the fundamental principle implicitly used in many impregnation methods.

Some estimates of the energy barrier are possible. In a calculation by Girifalco, the van der Waals interaction is estimated for 1 nm diameter tube to be 0.36 eV/A at the graphite separation distance,[60] which is equivalent to an energy of 5.8×10^{-16} J to separate two tubes that are 1 μm long. A calculation based on classical mechanics can be used to estimate the energy required for separation for MWCNTs. The energy is given by

$$\text{Energy of separation} = \frac{ALd^{1/2}H^{-3/2}}{24} \tag{3.3}$$

where A is the Hamaker constant ($\sim 2 \times 10^{-19}$ J for nanotubes in a low-permittivity medium[61]), L is the length, d is the diameter, and H is the separation distance at the point of closest approach. With the same parameters as used previously for H and L, the energy required to separate two 10 nm diameter MWCNTs has a value roughly 1/3 of that for the SWCNTs.

Centrifugation is often used to improve the apparent dispersion. After a given procedure to disperse the nanotubes, the material is centrifuged and the material that settles at the bottom is discarded. The amount of material that is discarded is often used as a quick quantitative method to characterize dispersion. Because centrifugation characteristics are typically not controlled from laboratory to laboratory, it is a very difficult matter to compare results from different laboratories, similar to the difficulties encountered in sonication as will be described below. Of course, centrifugation does not really improve the dispersion; centrifugation just eliminates material that cannot be dispersed so that the material does not appear in subsequent processing steps.

An interesting method to compare the quality of all the different types of dispersing media for SWCNTs (presumably from an identical source, although that issue was not entirely clear) was presented in a recent review paper by Coleman[10] and is represented in Figure 3.8. From a practical perspective, the most efficient dispersing medium will be the one where a large absolute number of nanotubes are debundled into individual tubes. For a perfect medium, that number will increase linearly with concentration (see dotted line in the graph) but for an imperfect one that value shows a maximum with concentration. The concentration where that maximum occurs was used as the measure of dispersion efficiency. The graph clearly shows that functionalization is most effective at dispersing tubes, and that solvents vary widely in their ability to disperse tubes (CHP represents cyclohexyl pyrrolidone, while GBL represents γ-butrylactone). The best dispersing biomolecules and surfactants were roughly equivalent.

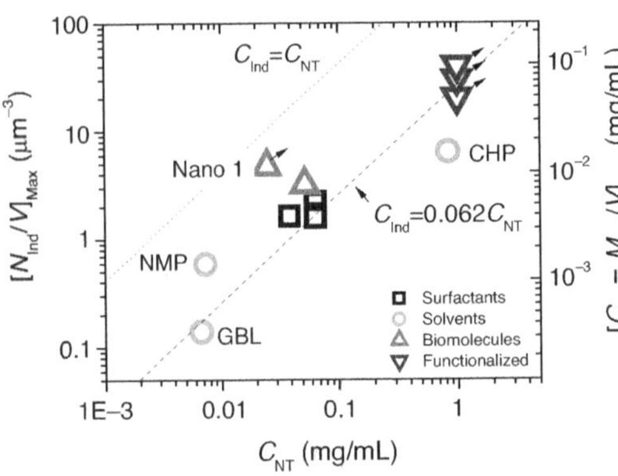

Figure 3.8 Graph showing the efficiency of dispersion for SWCNT dispersion for various types of liquids with and without dispersants. Efficiency is measured by the concentration where a maximum occurs when the number of individual nanotubes per unit volume is plotted versus nanotube concentration. The former is determined via AFM. For details regarding the source of data as well as more specifics, please consult the original reference. Copyright Wiley-VCH Verlag GmbH & Co. KGaA. Reproduced with permission from Ref. 10.

This section is divided into three separate components. The first describes in some detail the mixing processes used, that is, sonication (also termed ultrasonication) and high-shear mixers. Water is the so-called "universal" solvent, but for nanotubes water is actually a very poor dispersing liquid. The relatively nonpolar graphene ring has a relatively unfavorable energetic interaction with the polar water, and either a dispersing agent or functionalized tubes are required. The second section provides a discussion of nanotube dispersions in water. The third section includes organic solvents, including most monomers, and superacids; these low-viscosity liquids do not necessarily require dispersing agents although in certain cases these are required. In this latter case, functionalization does often improve the dispersion, so functionalization will be discussed as well.

3.3.1 Mixing Protocols: Sonication and High-Shear Mixing

The amount of energy required to reduce the average size of a bundle is not insignificant; the energies given previously (10^{-16} J) are approximately $10{,}000k_bT$. The use of low-viscosity liquids means that the forces generated by mixing are not as substantial as those that can be generated by higher viscosity materials such as polymer melts. To overcome this force limitation, ultrasonic mixing is often used. Sonication works on the principle of generating alternating low- and high-pressure waves at a frequency corresponding to the frequency of sound waves. The low-pressure waves lead to the formation of bubbles that have an internal pressure less than atmospheric pressure, e.g. vacuum bubbles, while the high-pressure waves cause violent collapse of the bubbles. The collapse of the bubbles in turn generates the high forces associated with ultrasonic processing. The phenomenon of the generation and collapse of bubbles is termed cavitation. There are two typical devices used for sonication, sonicating baths and sonicating probes. The latter are also termed horn sonicators. The latter generally produce much higher forces because the energy is localized to a very small area and hence ultrasonic probes are generally preferred for the separation of nanotubes.

In the case of a probe, the mechanism for producing high- and low-pressure waves is by oscillating the tip of the probe along its long axis. Two of the three of amplitude, frequency, and power must be specified; in general, the frequency is fixed, the amplitude is specified, and the power is measured. The power required for a given amplitude and frequency, as well as the efficiency of cavitation generation, depends on solution viscosity. To account for varying conditions, a feedback control mechanism is typically operative that forces the power to be constant. Typically, horns work at much lower frequency (~25 kHz), which means that the bubble size is smaller than that in the case of baths (~50–75 kHz). Baths create an essentially uniform operating zone, while horns create a conical zone that occurs immediately below the tip of the device. Because the tip diameter is typically of the order of 1 cm, forces are very localized. The important parameters that control the efficiency of debundling are the ultrasonic amplitude and frequency, and the presence and amount of dissolved gases. For horns, the location of the ultrasound source relative to the geometry of the container is also critically important. Sonication is often viewed as more of an art than a science because of the difficulty of control of these parameters from laboratory to laboratory. Surprisingly, solvent viscosity does not seem to be an important factor in the ability of the sonicator to create good mixing. Local temperatures higher than 5000 K are possible in sonication; sonication also creates an overall temperature rise that usually requires some sort of cooling.

From a dispersion perspective, the first consideration is the magnitude of the shear forces involved. Calculations performed elsewhere suggest that the shear rate in an implosion process (e.g., collapse of a bubble) can reach as high as $10^7\ s^{-1}$,[62] which is much higher than the few thousand s^{-1} shear rate possible in an extruder or a high-shear mixer. A very simple calculation not reproduced here indicates that the mechanical forces involved in sonication are in the tens of gigapascal range.[63] This simple analysis demonstrates why sonication is very effective in debundling nanotubes; the reader should remember that the tensile strength of nanotubes is in the tens of gigapascal range as well indicating that sonication has the capability of breaking nanotubes. This calculation agrees with observation; it is well known that sonication can reduce the size of carbon nanotubes.[64,65] For MWCNTs, there is also a "peeling" of layers, that is, tubes become thinner as well as shorter.[64] The fact that both temperature and mechanical forces are involved requires a determination of whether the reduction in size is due to chemical or mechanical effects, although obviously the cause will depend on the chemical identity of the solvent. If chemical effects are dominant, then the expectation would be that the length would be reduced over time until the length becomes almost zero. If mechanical forces dominate, then the nanotube length will reach some constant limiting value over time. In water at approximately neutral pH, the length reaches a constant limiting value, indicating that mechanical forces dominate.[54,59,66] The same result was found for dispersions in toluene.[2] However, interestingly, different rates of length reduction have been seen; in one case a cubic dependence with time was found,[2] in another case a square dependence.[66]

In systems where a dispersant is necessary to maintain a stable dispersion, for example, water, a key parameter becomes the time required for the dispersant to diffuse to the surface of the nanotube relative to the characteristic time that the tubes are separated. Taking the latter as the inverse of the shear rate, for example, 10^{-7} s, this suggests that diffusion rates must be of this order in order to form a stable dispersion. It is well known that macromolecules are much less effective in sonication processes than small-molecule surfactants, and the time required for diffusion of 100 nm is in the 10^{-7} s time frame for small molecules but not for polymers, suggesting that this approximate analysis is reasonable. Another important consideration is the polydispersity of the sample with respect to diameter. With SWCNTs, specific synthetic processes that produce a more narrow SWCNT diameter distribution are generally more difficult to separate and maintain separation.

A fundamental important question is whether it is possible to completely debundle tubes without any tube breakage using sonication. Because of the difficulty in controlling the parameters that affect the efficiency of sonication, there is no simple answer to this question. A related question is whether there is an optimal time for sonication. Maximum debundling seems to occur prior to maximum length reduction, at least according to one study.[54] Most laboratories do not concern themselves greatly with the question of the optimal sonication time because it is difficult to characterize dispersion and nanotube length. Generally, laboratories will set the sonication time and parameters by some simple criterion, such as the minimum time required to minimize the amount of material that can be separated by centrifugation.

Shear mixing with low viscosity liquids at very high shear rates is generally only successful at dispersing MWCNTs; the forces involved are not large enough to significantly reduce the average bundle size for SWCNTs. Typical forces involved in high-shear mixing are in the tens of kilopascal range, which is typically not large enough to cause mechanical breakage of tubes although contact with solid surfaces involved in the mixing can cause breakage. The advantage of a high-shear mixing process vs. sonication is that parameters that affect the mixing process are much easier to control. The mechanics of screw extruders are very different from the type of mixers that are the focus here, and a discussion of this type of dispersion will be delayed until Section 3.5. High-shear mixers include injector nozzles, which have been shown to cause severe tube breakage under these circumstances.[2] A rather detailed study of a Couette system with a high-viscosity poly(dimethyl siloxane) liquid used the point at which the viscosity was constant to set the minimum time required for mixing and found that this minimum time was linearly dependent on the nanotube concentration.[56] A quantitative or semiquantitative measure of tube dispersion was not given, however. A high-speed vibration mill can also be used to disperse nanotubes in a low-viscosity solvent; this mixer works on the principle of putting nanotubes and a solvent in a vessel that contains balls and oscillating the vessel at 10–100 Hz. This procedure is able to suspend nanotubes well,[67] although one would expect significant breaking of the tubes. Calendering, which involves two rollers with a small (1–100 μm) gap between them, has also been used to debundle nanotubes.[68] Calculation of the actual shear rate experienced is not a simple problem, because smaller gaps coupled with faster and more different velocities of the two rolls increase the shear rate. Further, samples with the viscosity of water are unsuitable for calendaring; uncured epoxy resins that have a thousand-fold higher viscosity of water are appropriate for this approach. A qualitative comparison to horn sonication using electron microscopy for the dispersion of DWCNTs has been presented,[69] and these researchers found that calendaring was superior in terms of smaller and less numerous microscale aggregates. A more detailed study using electron and atomic force microscopy indicated the presence of both individual and bundled nanotubes indicating that it is possible to use calendaring to achieve significant debundling, at least in multiwalled nanotubes.[21] The author is unaware of any studies with more quantitative nanoscale characterization.

3.3.2 Dispersions of Nanotubes in Water

Water has many advantages as a dispersing liquid for carbon nanotubes. Water is inexpensive, nontoxic, nonflammable, and is easily removed via evaporation. These advantages mean that water is a very important dispersing liquid for many uses of carbon nanotubes. Even though most synthetic polymers are not water soluble, there has been a great deal of work with nanotube dispersions in water that also involve synthetic polymers. The importance of water-dispersible polymers is growing irrespective of nanotubes, for the same reasons that water is preferred as a dispersing liquid for carbon nanotubes. Hence, mixtures of polymers and nanotubes both dispersed in water are growing. Water-soluble polymers also play a role in the importance of water dispersions of nanotubes, with particular importance for biological systems. As stated previously, discussion of interactions between biological moieties and nanotubes is beyond the scope of this book, although interactions of water-soluble macromolecules with nanotubes in solution are certainly not and will be covered in this section.

As stated previously, water and nanotubes do not have a favorable energetic interaction, and without either a dispersing agent or functionalization nanotube dispersions in water are not stable. Dispersing agents will be considered first. Surfactants are the most common dispersing agents used to disperse nanotubes in water because surfactants have a propensity to accumulate at surfaces. Surfactants are water-soluble small molecules with a hydrophobic tail and a hydrophilic head group; a good schematic representation is an eraser (hydrophilic part) on the end of a pencil (hydrophobic part). The active ingredient in the common-use definition of the word "soap" is a surfactant. Head groups can be ionic, typically quaternary amines [$-N(CH_3)_3^+$] if cationic, typically sulfates [$-SO_4^-$] or carboxylates [$-CO_2^-$] if anionic, and typically ethylene oxide $(-CH_2CH_2O-)_n$ if nonionic. The pencil part is a hydrophobic chain; examples include saturated hydrocarbons, $(-CH_2CH_2-)_n$, and propylene oxide, $(-CH_2CH_2CH_2O-)_n$, although there are many other types of common hydrocarbon chains. Some common surfactant structures are shown in Figure 3.9. Ionic surfactants have hydrophobic chain lengths typically between 8 and 18 carbon atoms because shorter chains do not have surfactant properties, while longer chains are not water soluble. Nonionic surfactants can have much longer hydrophobic lengths, because the hydrophilic part can

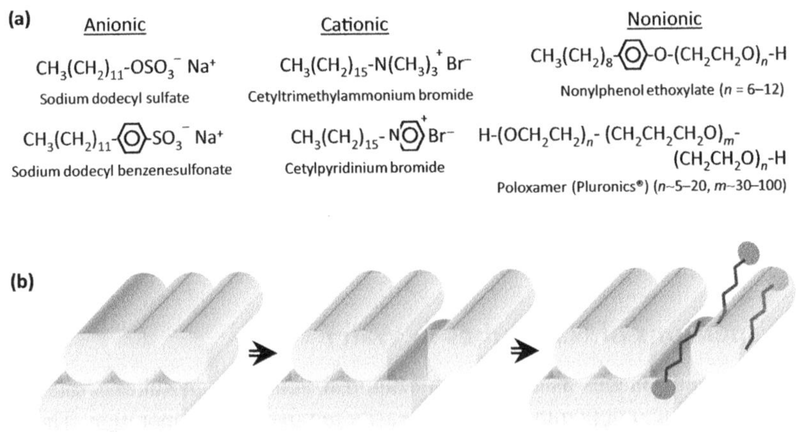

Figure 3.9 (a) Structures of some common surfactants. (b) Schematic representing mechanism of how nanotubes are debundled with surfactants; initially the tubes are bundled, then a tube is removed from the bundle, and finally a surfactant dispersant adsorbs to the bundle.

also be much longer. Other structures such as those having multiple charged groups on one molecule, a combination of water solubility from both nonionic and anionic parts, or multiple tails are possible. Surfactants have the ability to disperse nanotubes in water because the hydrophobic tail is able to energetically interact with the nanotube favorably while the hydrophilic head is able to interact with the water. Using the same pencil–eraser analogy, it should be noted that the diameter of the pencil and the eraser can be different, and that a stiff pencil is not accurate; the stiffness of the hydrophobic part of the molecule is typically more reasonably approximated by a flexible wire. For ionic surfactants, an eraser is a reasonable representation; nonionic surfactants also typically have more flexible units than an eraser would suggest.

The ability of a surfactant to act as a good dispersing agent for nanotubes depends on the speed at which the surfactant diffuses into the interstitial sites when nanotubes are temporarily debundled due to high-shear forces, the energetic interaction of the surfactant with the nanotube surface that includes surface–surfactant interactions as well as surfactant–surfactant interactions. This process is shown schematically in Figure 3.9b. In one study, cationic and anionic surfactants of identical hydrophobic chain length and type were compared and the conclusion is that the nanotubes were better dispersed by the anionic rather than the cationic surfactant.[70] The matching of chain lengths means that the diffusion constants are more or less identical, which allows for a direct comparison of the energetics of the interaction. This result is very interesting, since in two cases pristine nanotubes have been measured to have a negative surface charge that would possibly allow head down adsorption for cationic surfactants.[71,72] This result indicates that a bilayer adsorption mechanism (e.g., head-down adsorption at the nanotube surface) is likely not active for any dispersion of surfactants with nanotubes, at least at relatively neutral pH values, because otherwise better performance would be expected from the cationic molecule. Exactly why the anionic surfactants performed better is not entirely understood.

One structural question that has clearly been answered is the effect of surfactant structure on the ability to disperse surfactants. Three prominent studies used fluorescence[73] and resonance ratio[44] and both[74] to examine a wide number of surfactants for their ability to disperse single-walled carbon nanotubes in water. Both studies found that the more aromatic the character of the surfactant, the better the surfactant was as a dispersant. The obvious logical conclusion is that the aromatic group is able to better interact with the nanotube. Cholate or oxycholate surfactants deserve special mention because these surfactants are consistently the best or among the best at dispersing carbon nanotubes. This molecule has three six-membered rings and one five-member ring, although none of the rings are unsaturated. A carboxylic acid group is the anion; the rings have three hydroxyl groups that allow the surfactant to be soluble in water. Clearly, these rings are able to very effectively interact with the surface of the nanotubes. Along similar lines, two nonionic surfactants that differed only in the fact that one had unsaturation in the hydrophobic tail were compared and the one that contained the unsaturation was better able to disperse the nanotubes.[71] A study that ran counter to these statements used AFM of dried solutions to measure the ability of surfactants to debundle nanotubes and found that the surfactants that did not contain unsaturation did a better job in terms of dispersing the nanotubes, and correlated this measurement to measured zeta potential.[75] However, it is very possible that reaggregation dynamics were faster than evaporation times so that the result did not actually present a true picture of the ability of a surfactant to maintain a dispersion at typical surfactant concentrations. Finally, the use of aromatic surfactants with multiple aromatic moieties is capable of selectively adsorbing on only nanotubes of certain helicities.[76]

The amount of surfactant required to disperse carbon nanotubes should be presented in terms of an adsorption isotherm; where the x-axis is surfactant concentration in solution (surfactant added − surfactant adsorbed) and the y-axis is the amount of surfactant adsorbed per unit area of surface or per unit weight. For nanotubes, clearly the latter is preferred since determining the surface area is an impossible task. The saturation adsorption ratio of surfactant on SWCNTs is about 0.004 mol/g of nanotube for an octylphenol ethoxylate,[77] which is close to the reported values for sodium dodecylbenzene sulfonate[78] and sodium dodecyl sulfate.[79] Normally, the amount of adsorbed surfactant is constant after the surfactant begins to form micelles; because of the inherent nonequilibrium process involved in sonication, the amount of adsorbed surfactant may not be constant after the surfactant forms micelles although the number of studies that have looked at this issue is few.

The structure of adsorbed surfactants at the nanotube surface is very interesting; different possibilities are shown in Figure 3.10. On flat graphene, it is well established that alkyl tail surfactants form an effectively irreversibly adsorbed flat

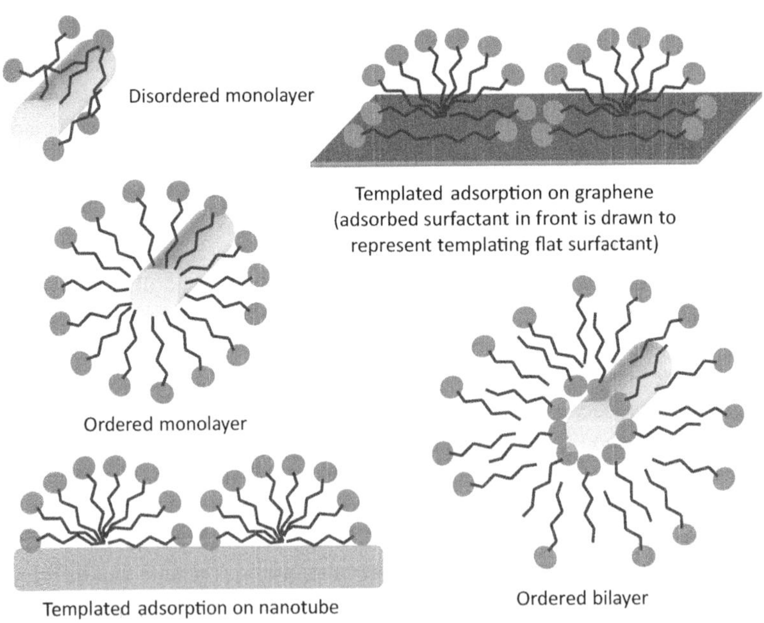

Figure 3.10 Qualitative representation of possible structures of surfactants on carbon nanotubes as well as a schematic representation of adsorption on flat graphene. The dimensions are drawn to represent a single-walled carbon nanotube.

layer at the surface, that is, surfactants lie flat on the graphene surface to maximize the interactions between the all-*trans* configured alkyl chain and the unsaturated carbon hexagons, which in turn templates further adsorption of half-cylindrical aggregate.[80] The preponderance of evidence, some of which is given in the previous paragraph, indicates tail-down adsorption consistent with adsorption on flat graphene. Based on the fact that adsorption is significantly stronger for surfactants that do not consist of alkyl chains, it is almost certain that templated adsorption does not occur because the curvature of nanotubes reduces or eliminates the interaction between the alkyl chain and the hexagons. Further, no evidence has been presented in the literature of an essentially irreversible adsorption although no experimental studies to this author's knowledge have investigated the reversibility of adsorption on nanotubes. An unanswered question is at what diameter does this type of essentially irreversible flat adsorption become possible again, although very recent molecular dynamic simulations have begun to explore this issue.[81] On flat surfaces, it is well established that at sufficient concentrations (starting between 10% and 50% of the maximum adsorbed amount) the critical driving force for adsorption is tail–tail interactions within the adsorbed layer. In other words, the adsorbed layer is relatively ordered. The geometry of the adsorbed structure depends largely on the area of the head group relative to the cross-sectional area of the tail; spherical, cylindrical, and flat layers are possible. Even qualitative representations such as Figure 3.10 suggest significant problems with a relatively ordered arrangement; the tail–tail interaction that tends to drive surfactant adsorption is difficult to achieve on a curved substrate. Given the observation that cholate surfactants are typically among the best dispersing agents for nanotubes provides indirect evidence that a disordered structure with significant laying down of the surfactant on the surface of the nanotube (see schematic in Figure 3.10) is likely the structure. Both molecular simulations[82,83] and neutron scattering experiments[84] suggest that a disordered structure is likely preferred on SWCNTs. A very interesting result suggests that nanotubes are likely to adsorb at higher local concentration at nanotube crossings.[83] However, the paucity of systems investigated certainly leaves open the possibility that more structured arrangements might be possible for some surfactant/nanotube combination.

Surfactants are not the only molecules that are able to disperse carbon nanotubes; in fact, a recent study indicates that this class of molecules is not even the best for SWCNTs when sonication is used to suspend nanotubes.[74] Oligonucleotides, which are molecules that contain phosphate groups, a five-carbon ring sugar, and aromatic amine bases repeated from 5 to 30 times, were found to be as good as the best surfactant, an oxycholate, at dispersing carbon nanotubes. These types of materials are able to "wrap" around a single-walled tube in a helical fashion, which is a polymer conformation unique to small diameter nanotubes. Details of this wrapping mechanism, and polymer adsorption in general, will be discussed more fully in Section 4.2.1. Another study, using high-speed vibration milling, showed that the dispersibility depended strongly on the number of phosphate groups and the type of base in the oligonucleotide.[85]

All three stabilization mechanisms, charge–charge repulsion, packing disruption, and steric repulsion, can be operative with polymer adsorption. As an example of the latter mechanism, one very common approach is to use a block copolymer, where one block adsorbs to the nanotube and the other block is water soluble.[54,86,87] Steric stabilization mechanisms can be very temperature sensitive partly because the conformation of a polymer chain can be temperature sensitive.[88] Wrapping, which works on a packing reduction mechanism, is not limited to biopolymers[89] or even polymers dissolved in water since polymers in other solvents have been shown to interact with nanotubes in this fashion.[90,91] The stability of nanotubes suspended in such a manner has been found to be quite high. Further, depending on the polymer, it is possible to separate tubes based on their ability to be wrapped by a certain polymer.[92] A significant advantage of polymer adsorption is that it is possible to dry the nanotubes and then resuspend them without high shear or sonication,[93,94] a procedure that is not possible with surfactants.

Covalent functionalization is also an effective strategy for producing nanotubes that are easily dispersible in water, without the need for a dispersing agent. Some of the functional groups that have been used to improve water dispersibility include hydroxyl[95] and carboxylic acids.[96] Grafted water-soluble small molecules or polymers have also been used to great effect to improve dispersibility in polymers. The list of these molecules is far too involved to completely reproduce here, but some prominent examples include various organic acids,[97] glucosamine,[98] proteins[99] and peptides,[100] amine-containing dendrimers,[101] other aminopolymers,[102] poly(ethylene glycol),[103] poly(styrene sulfonate),[104] and poly(vinyl alcohol).[105] Functionalization is an excellent mechanism for improving the dispersibility of nanotubes; however, as made clear earlier many of the intrinsic properties of nanotubes are severely affected.

3.3.3 Dispersions of Nanotubes in Other Solvents

The significant difference between some organic solvents and water is that it is possible to disperse nanotubes without functionalization or the use of a dispersant. A classical study published in 2000 tested a wide variety of solvents for their ability to disperse single-walled carbon nanotubes.[106] Some of the solvents that were found to disperse SWCNTs well include *N*-methyl pyrrolidone (NMP), cyclopentanone, α-caprolactone, and dimethyl formamide (DMF). In general, these solvents are good Lewis bases without hydrogen donors. Further studies in this area have not significantly changed these conclusions. A great deal of work has been carried out on NMP in particular because of its ability to do an excellent job at debundling,[107] which led to likely incorrect claims[10] that this solvent is truly able to solubilize tubes.[9] It should be noted that generally the concentrations achievable with pure solvent dispersions tend to be much less than those that are possible with dispersant-assisted dispersions.

The use of small molecules to assist dispersion in organic solvents is much less common than in water. Surfactants have many of the same properties in organic solvents as in water; of importance to nanotubes is that if the energetics are appropriate, surfactants can still adsorb to carbon nanotubes and play the same role as they do in water in enhancing dispersion of nanotubes in a low-viscosity liquid.[108–110] Of course, the structure of the adsorbed layer can be different from that in water although detailed studies of this question have not been undertaken. Other small-molecule dispersants are also not common; one of the rather few examples includes trifluoroacetic acid, which has been shown to improve dispersion in solvents such as DMF, dichloromethane, and tetrahydrofuran[111] and oligomers of phenylenevinylene to help dispersion in chloroform.[112]

Although the use of small-molecule dispersants has not been nearly as widespread in organic solvents as in water, the use of polymers to assist in the dispersion of nanotubes in organic solvents is quite widespread. The logic of using polymeric dispersants is usually not to assist with initial dispersion, since the diffusion constant of polymers is typically too long to be effective in a kinetic sense. Polymeric dispersants are often used to allow for the removal of the solvent and produce a dry sample of nanotubes that can be easily redispersed in a liquid or a polymer melt. The use of dry nanotube samples is critically important for the practical application of nanotubes since manufacturers would much prefer to use dry ingredients.

One particular class of polymeric dispersants for organic solvents has proved to be of significant commercial importance. Dispersants that contain a rigid conjugated backbone of poly(*p*-phenylene ethynylene)s are used commercially by Zyvex (www.zyvexpro.com) to disperse nanotubes in organic solvents.[114] The basic structure of this molecule is 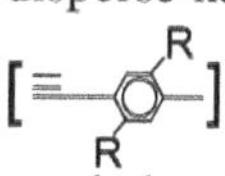 where the R groups are often alkyl chains of 10–20-carbon length. Although the choices of R and the end groups of the polymer affect the dispersibility in the solvent, these groups are chosen commercially on the basis of the polymer that the nanotubes are going to be mixed with to make a composite. Further, it is possible to choose R groups so that the material is water soluble,[115] although the focus commercially has been on thermoset structural materials. Because of interactions of the conjugated backbones with the nanotube, these materials cannot be washed off after addition.[114] The original paper also claimed[114] that the materials are too stiff to wrap the nanotubes although recent studies where the R groups are chosen to make the poly(*p*-phenylene ethynylene)s water soluble do find significant wrapping via TEM and AFM studies.[116]

Functionalization is a very common strategy to improve dispersion in organic solvents. As with water, the number of functional groups/small molecules/polymers that have been attached to nanotubes to promote dispersion in organic solvents is far too lengthy to list. Some commercial nanotubes sold in large quantities are likely functionalized to promote better dispersion in organic polymers, but the author is unaware of any of these modifications and it is entirely possible that these modifications have remained trade secrets. Certainly, functionalized nanotubes modified with molecules designed to promote better dispersion in organic media can be purchased at present in laboratory-scale quantities.

Superacids are very effective dispersion solvents for carbon nanotubes.[11,117] Superacids are acids that have acidity greater than 100% sulfuric acid with some examples being oleum (100% sulfuric acid with added SO_3), trifluoromethanesulfonic acid, and chlorosulfonic acid. These acids are able to promote true solubilization of nanotubes via protonation of the nanotube, which in turn allows the free energy of debundling to be larger than that of bundling at small, but finite, concentrations. A key practical issue is that superacids are extremely hydroscopic, and even small amounts of water will cause nanotube aggregation.[118] Because polymers will degrade in superacids, from a polymer perspective the only use of superacids is with respect to infusion processes. Given the difficulty of working with these materials, it is not clear that superacids offer any substantial advantages over good dispersing solvents such as DMF or NMP. However, the use of these materials to help understand the phase behavior of nanotubes has been invaluable.

3.4 POLYMER–NANOTUBE DISPERSIONS: SOLUTION METHODS

Three different methods can be used to disperse nanotubes in polymers that also involve low-viscosity liquids: (1) nanotubes are dispersed in a low-viscosity monomer followed by polymerization (in some cases, the nanotubes are dispersed in a low-viscosity liquid and then added to the monomer), (2) nanotubes are dispersed in a liquid that also includes dissolved polymer followed by removal of the solvent either by evaporation or by dilution of the solvent into a second miscible liquid that is also a nonsolvent for the polymer, and (3) dispersed nanotubes are added to dispersed monomer, followed by polymerization; or added to dispersed polymer followed by evaporation of the liquid. The only example of this latter scheme is where the liquid is water.

3.4.1 Dispersion–Reaction

The process is simple: choose a monomer(s), disperse nanotubes in the monomer using one of the procedures described in Section 3.3, and polymerize the monomer. Of course, the final dispersion is the key to making good material. Both thermoplastic[119–123] and thermoset[124–127] materials have been manufactured using this approach, with the latter more common. As of the writing of this chapter, dispersing nanotubes in a thermosetting monomer and then reacting the monomer to make a final, finished product is one of the only two methods used commercially to make nanotube-containing polymer parts. Materials used in this type of procedure include common thermosets such as epoxy,[127–130] vinyl ester,[131–133] and polyimides,[134–136] as well as a host of less common thermosets. As with any thermoset, the material must be formed into its final shape prior to reaction. Most commonly, those thermoplastics that have liquid monomers such as polystyrene[137,138] and various acrylates[19,139,140] are used in the dispersion–reaction method since solvents that dissolve monomers are typically undesirable organics. Exceptions where organic solvents have been used to facilitate the polymerization of other thermoplastics include electrically conducting polymers[141,142] and liquid crystalline polymers[143,144] because these thermoplastics cannot be processed as a melt without degradation. The number of papers describing this method is in hundreds, if not thousands.

Although not typical, dispersants have been used to assist debundling of the nanotubes in the monomer.[110] If the dispersant remains with the solid polymer, then the dispersant will usually have a deleterious effect on mechanical properties, and can also have negative effects on properties such as chemical resistance. Another approach is to first debundle nanotubes in a liquid that is able to better disperse tubes, and then add this mixture of nanotubes and liquid to the monomer. Of course, the liquid and liquid monomer must be miscible. This approach can lead to improved performance versus the case where the nanotubes are added directly to the monomer.[145] Not only must the solvent evaporate, but the solvent must also not interfere with the polymerization reaction or else the properties of the composite can be significantly degraded.[146]

Functionalization[147,148] and polymer adsorption[149] have been used to assist dispersion in dispersion–reaction systems. One study compared functionalization versus polymer adsorption versus no treatment in an epoxy, and not surprisingly found functionalization the most effective and no treatment the least effective in improving dispersion.[150] Strictly speaking, improving the dispersion is not the purpose of these treatments, but rather improvement in properties, usually mechanical properties, is the goal. However, as will become clear in Chapter 5, improvements in dispersion are often correlated with improvements in mechanical properties. The effect of various treatments to improve composite properties will be discussed at length in Chapters 5–7, but briefly functionalization has the advantages of usually giving better dispersion and adhesion with the disadvantages of increasing cost and possibly affecting some desired properties of the nanotubes, in particular electrical conductivity for SWCNTs. Practically, assigning changes in mechanical properties to either dispersion or adhesion is very difficult to do since the two are usually highly correlated.

Polymerization is initiated in dispersion-reaction methods by some combination of an increase in temperature, the addition of a small amount of a chemical that initiates the reaction, or the mixing of two monomers. Initiations with ultraviolet light are normally common elsewhere, but nanotubes are UV absorbing, so this strategy is usually ineffective. Although nanotubes cannot serve as the initiators of polymerizations directly, two indirect methods where nanotubes initiate polymerization do exist. An interesting recent development, which is possible with nanotubes and only a few other materials, has been the use of microwave radiation to induce polymerization.[151,152] Nanotubes are strongly microwave absorbing causing a local increase in temperature that is capable of initiating many chemical reactions on the surface.[153,154] Hence, it becomes possible to initiate polymerization at the nanotube surface, rather than uniformly throughout the matrix. The issue with this procedure is control: it becomes difficult to control the reaction because of the extremely high temperatures that can result due to the very strong absorption of microwave radiation by nanotubes.[155] This technique is new enough that it is not clear whether the resultant product is any different from that produced using more typical initiation strategies.

A second indirect method is directly impregnating polymerization catalysts on the tubes enabling growth of polymer from the surface of the tube.[156–159] Further processing, for example, injection or compression molding, usually must take place because the material is not a continuous film. This method is different from functionalization methods where the final product is a solution of grafted-from polymer-coated nanotubes because in the catalyst case the polymer is not necessarily directly covalently bonded to the tube.[160–162] The upper limit of the amount of polymer that can be grown seems to be around 95%, so further reduction would have to come through compounding with another material. As a composite at 5% nanotube content, the dispersion of the nanotubes seems to be quite good as indicated qualitatively by TEM micrographs;[163] however, a more quantitative determination of the dispersion compared to other methods still needs to be determined.

For electrically conducting polymers, an alternative dispersion–reaction mechanism is electrochemical polymerization. In a typical electrochemical polymerization, a conducting surface (electrode) has an electrical potential relative to the solution that contains soluble monomer, and the monomer and/or oligomer is drawn to the surface because it is oppositely charged to the electrode in that solution. A chemical initiator is also present in the solution; the purpose of the potential is to increase the concentration of monomer at the electrode so that most of the polymerization takes place there (although polymerization in solution does also occur). With carbon nanotubes, two approaches are possible: nanotubes can be predeposited on the surface of an electrode followed by subsequent polymerization[164,165] or both the monomer and the nanotubes can be dissolved/dispersed in solution, and simultaneously drawing both the electrodes is also possible.[166,167] Nanotube coated with electrically conducting polymer[168,169] is generally the morphology of the materials manufactured electrochemically, although other dispersion–reaction schemes with electrically conducting polymers often give this sort of morphology as well.[170,171] It should also be noted that the sometimes strong oxidizing agents used in polymerization of monomers to electrically conducting polymers can cause nanotube breakage in both the electrochemical and chemical cases[172] as well as covalent grafting of the monomer to the nanotube.[173]

3.4.2 Dissolution–Dispersion–Precipitation

In this case, the polymer is dissolved in a solvent that is the same solvent used to disperse the nanotubes. Dispersion of the nanotubes is performed separately from dissolution of the polymer because the typical method of sonication would likely cause molecular weight degradation of the polymer. Also, the rise in viscosity that would occur upon addition of the polymer reduces the effectiveness of debundling relative to typically undesirable tube shortening. Hence, in the most typical method, dissolution is done separately from dispersion and the two solutions are then mixed. In a laboratory, dissolution–dispersion–precipitation is an excellent dispersion method since very good dispersions of nanotubes are possible in low-viscosity solvents. From a commercial perspective, an economic process with water-soluble polymers is possible. However, unless biological applications are of interest, water-soluble polymers are not suitable for most envisioned applications with carbon nanotube–polymer composites. For polymers soluble in organic solvents, the large volume of solvent required to disperse the nanotubes and dissolve the polymer makes this method commercially unsuitable in most instances. Exceptions are found for polymers that cannot be melt processed and where the cost is justified by the unique properties of the polymer, for example, a category that includes some liquid crystalline polymers and ultrahigh molecular weight polyethylene. Regardless, nearly all thermoplastic polymers, including difficult to dissolve materials such as polyethylene and polypropylene, have had nanotube composites prepared with dissolution–dispersion–precipitation. The number of papers describing this method to disperse nanotubes into polymers is in many hundreds, if not thousands.

An important issue in these systems is the manner in which the solvent is removed; in particular, how fast does solvent removal take place? This question is important because nanotubes will often reaggregate on the nanoscale[9,174] as the nanotube concentration increases in a low-viscosity liquid. One approach to remove solvent quickly is to rapidly add the solvent–polymer–nanotube solution to an excess of liquid that is miscible with the solvent, but for which the polymer does not dissolve and the nanotubes are not dispersed.[50,175] In this coagulation process, some further processing method is usually required to make continuous films. A generally slower method, but one that can directly result in a continuous film or fiber, is to evaporate the solvent;[176–178] continuous films or fibers can still result via solvent evaporation even if higher temperatures are used.[179,180] Surprisingly, the number of papers that have directly contrasted different methods of solvent removal with respect to characterizing dispersion is quite small. A recent study showed that evaporation gave a smaller percolation threshold than coagulation.[181] This seeming anomaly is likely due to rebundling/reaggregation during evaporation causing a lower percolation threshold with an inferiorly dispersed material.

Two fiber-producing dispersion–dissolution methods should be mentioned. Gel spinning, a technique that involves coagulation after extrusion of the polymer–nanotube–solvent mixture through a small opening (i.e., the solvent–polymer–nanotube fiber extrudate is passed through a second liquid), is able to produce fibers directly.[40,182,183] The solvent-swollen fibers are typically wound on a drum to make a continuous long fiber. The wet fibers can be handled and stretched when immersed in a liquid (not necessarily the coagulation liquid), often to as much as 40 times their original length, prior to final evaporation of the solvent.[184,185] Electrospinning is another way to take nanotubes dispersed in a solution containing polymer and make very thin fibers of polymers and nanotube;[186,187] the characteristic diameter of electrospun fibers is between 0.1 and 1 μm, while gel-spun fibers typically have diameters between 1 and 100 μm. In electrospinning, pressure is not used to cause the solvent–polymer–nanotube mixture to exit the small opening, rather a ~10 kV voltage difference between the metal opening (typically a needle) and the collector is applied that forces liquid to exit (a small pressure force may also be applied). Because electrospun fibers have very small diameters, solvent evaporation tends to be rapid.

A related method is to mix two separate solutions using the same low-viscosity liquid as originally developed by Poulin and coworkers.[188–190] One liquid contains nanotubes dispersed with surfactant (or possibly some other small-molecule dispersing agent, although the author is not aware of any example other than surfactant) and the other contains dissolved polymer. Injecting small quantities of the nanotube-containing solution into the polymer-containing solution causes the polymer to replace surfactant at the surface and the nanotubes will agglomerate into a nanotube–polymer complex. This process is possible because the desorption rate of surfactant from the tubes is typically quite fast. The production of fibers has been the most common use of this approach; the nanotube solution is extruded through a small opening and the polymer-containing solution is flowing. It is not entirely clear why significant amounts of polymer are entrapped with the nanotubes; likely the fact the solution is flowing plays a role.

The layer-by-layer (LBL) method uses an electrostatic driving force to cause successive adsorption of some combination of oppositely charged polyelectrolytes and/or colloids. The general procedure is to dip a solid such as a glass slide into a solution containing species of one charge and hence coat with one of the layers, then dip into a second solution with the oppositely charged species that will overcoat the first layer because of electrostatic attraction, then dip in the first solution to overcoat the second layer, then dip in the second solution to overcoat the second coat of the first layer, and so on. The material can be dried and then removed from the surface and in some cases a freestanding film results. A large number of layers can be build quickly and easily since adsorption tends to be a very fast (minutes) process. Because unfunctionalized nanotubes are only weakly charged, the general procedure is to first mix the nanotubes with a positively or negatively charged polyelectrolyte and use this polyelectrolyte/nanotube combination as one of the oppositely charged species in the adsorption process.[191,192] Alternatively, a small-molecule salt that has aromatic rings that promote adsorption can be used to generate positively or negatively charged nanotubes.[193] Of course, functionalized tubes with a certain charge because of the functionalization can also be used, eliminating the need for a polyelectrolyte dissolved with the dispersed nanotube.[194,195] In fact, with nanotubes oppositely charged layers are not required; in one example, the poly(styrene sulfonate)-wrapped nanotubes have a negative charge while poly(vinyl alcohol) served as the "positively" charged layer.[196] An LBL process can yield aligned nanotubes via the application of high-speed air across the surface during drying[197] or via the application of magnetic fields.[198] An LBL approach can be used with nanotubes serving as the surface for which polymer adsorption occurs via successive addition of oppositely charged polyelectrolytes. In this way, layers of adsorbed polymer are built up on the surface of the tubes.[199] A somewhat similar, but still different method, is to swell a thin sheet of a polymer in solvent and then allow nanotubes to diffuse into the material, and then remove the film and evaporate the solvent.[200]

One significant problem with the dissolution–dispersion–precipitation method is the fact that no environment-friendly solvent exists that is able to both dissolve most polymers and disperse carbon nanotubes well without the use of a dispersing agent. Supercritical carbon dioxide, or carbon dioxide mixed with water or alcohols, can dissolve a number of polymers, and these solvents are considered to be environment friendly. Evaporation can be very quick by simply releasing the pressure. Using a mixture of organic solvent and supercritical carbon dioxide, interesting microstructures of various polymers on carbon nanotubes have been produced.[201–203]

3.4.3 Dispersion–Dispersion–Evaporation

As opposed to the previous method, polymer is dispersed rather than dissolved; in other words, polymer molecules are aggregated together instead of being individually solubilized. The use of organic liquids is eliminated since water is used as the dispersion liquid for the polymer and nanotubes (although in theory water would not have to be used). Because of environmental considerations, manufacturers are developing water-dispersible polymers regularly for the coatings market; even non-polar materials such as polypropylene are being marketed commercially in a water-dispersible form.[204] Mixtures of water-dispersible nanotubes and polymer could be used to produce mixtures suitable for direct application as coatings, although rheological alteration by nanotubes would likely interfere with performance. One disadvantage of this process is that the dispersing agent necessary for the nanotubes will be present, unless this agent is volatile. Another disadvantage is that some water-dispersible polymers have inferior properties compared to non-water-dispersible counterparts. The number of papers describing this method to disperse nanotubes into polymers is in many tens, and might reach 100. The author expects this method to gain in importance during the coming years.

Two approaches are possible: reacting the monomer in a dispersion polymerization in the presence of nanotubes, or mixing a solution of the dispersed polymer with a solution of the dispersed nanotubes. A potential problem with the first approach is that the dispersing agent can migrate from the nanotubes to the reacting micelles, causing nanotube precipitation.[14] Regardless, there have been examples where this approach has been used to make composite materials with good dispersions.[205,206] The second approach has been more typically used. The most common example involves mixing the dispersed nanotubes with polymer colloids, that is, a polymer dispersed into small (<10 μm) spheres and most often stabilized via charge repulsion. A number of polymers have been used including poly(methyl methacrylate),[207] poly(vinyl acetate),[208] polystyrene,[209] polyacrylonitrile,[210] and polyurethane.[211] Variations include spraying polymer pellets or powder with water-dispersed nanotubes.[212,213] Another variation is to form a water-based emulsion with the water containing suspended nanotubes and the suspended oil droplets containing dissolved polymer, and then evaporating the lower boiling organic liquid.[214]

Surfactant-stabilized carbon nanotubes have served as stabilizers for emulsions, that is, oil or polymer droplets in water.[215] Surfactants or polymers normally serve as stabilizers for emulsions; the purpose of these molecules is to prevent coalescence of individual droplets. Emulsions stabilized with dispersed solids rather than dissolved molecules are termed Pickering emulsions. Because the dispersed nanotube solution was used directly without removal of the excess dispersing surfactant, the proper view, which was clearly stated in the paper, was that the combination of the surfactant and the nanotubes was stabilizing the emulsion. The authors did note that the surfactant alone was not an effective stabilizing moiety.[215] Another type of emulsion that can be stabilized with nanotubes is termed high internal phase emulsion;[216,217] in these emulsions, oil is the continuous phase but water is present in high amounts. The resulting product, because the nanotubes are located at the surfaces of the foam cells, is conductive at a very low volume fraction of nanotubes.

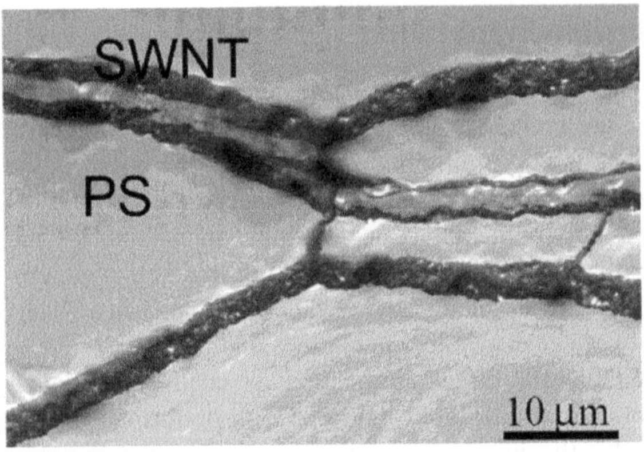

Figure 3.11 Scanning electron micrograph showing nanotubes at the edges of original polymer particles after coating the surface of the polymer and then compression molding under conditions that do not allow for complete interdiffusion of polymers and nanotubes. Copyright Elsevier Ltd. Reproduced with permission from Ref. 220.

In order to analyze the resulting polymer after mixing the dispersed polymer and the nanotubes, the sample must be dried. Often, further processing is required to make a continuous film without significant porosity, unlike the dissolution–dispersion–precipitation method where nonporous continuous films can be usually easily directly cast from solution. The percolation threshold, when samples are processed in such a way as to lead to a uniform microscale dispersion in a one-phase sample, can be quite low[209,218] indicating that the nanotubes are relatively well dispersed on the nanoscale. The relatively uniform microscale dispersion means that diffusion of the polymer chains into the nanotubes and diffusion of the nanotubes are significant enough at typical polymer melt viscosities so that the tubes can migrate away from one another. Films with nanotubes concentrated at the dispersed polymer-dispersed polymer interface can also be produced using this method. Such films can be cast directly from this mixture, as first demonstrated by Grunlan et al.[208,219] However, these cast films had a significant void level since the conditions were not adjusted to cause sufficient flow of the polymer. Films without voids having nonuniform microscale distributions can be made if the viscosity of the polymer is controlled so that the sample becomes uniform without causing complete inter-diffusion of polymer.[220,221] After drying, nanotubes are isolated at the edges of the polymer particles as shown in Figure 3.11. Because of the nonuniform microscale dispersion of nanotubes, the percolation threshold tends to be quite low and the maximum conductivity quite high. The mechanical properties of these films with nanotubes aggregated at the polymer particle interface are likely inferior to those films where the polymer flows out fully to produce a uniform distribution of nanotubes on the microscale; however, a study to address this question has not been published to the author's knowledge.

3.5 POLYMER–NANOTUBE DISPERSIONS: MELT MIXING

Melt mixing consists of melting the polymer and mixing with nanotubes in a high-shear process. High-shear forces are responsible for producing well-dispersed nanotubes on both the micro- and nanoscale. In general, the amount of nanoscale dispersion resulting from melt mixing is inferior to that resulting from solution methods. The reason for the generally inferior dispersion is the fact that small molecules are generally not available to help exfoliate the tubes via diffusion, and, compared to sonication, the forces involved are orders of magnitude smaller, with shear rates on the order of $10^3\ s^{-1}$ instead of $10^7\ s^{-1}$. Still, because of its commercial importance, the number of papers describing this method to disperse nanotubes into polymers is in many hundreds, if not thousands. The vast majority of papers involve multiwalled carbon nanotubes since single-walled carbon nanotubes do not disperse very well using forces generated by shearing a high-viscosity polymer. For thermoplastic polymers, melt mixing is the most desirable way to combine nanotubes and polymers for most applications. A great deal of the reason for the increase in the percentage of papers published with MWCNTs versus SWCNTs over the past 10 years shown in Figure 1.3 is because MWCNTs can be dispersed adequately in a twin-screw extruder while SWCNTs cannot.

A twin-screw extruder has generally been used to mix polymers and nanotubes in order to generate the high forces for maximum dispersion. The corotating mode is generally used as opposed to counterrotating one because of the higher dispersive mixing of the former. However, a very recent paper suggests that the mode of mixing (counter versus co) has no effect on the dispersion achieved.[222] Small conical twin screws have been very popular because of the small amount of material required and because such devices are equipped with recycle capability so as to control the residence time independent of shear rate.[223,224] As expected, higher shear rates tend to provide better dispersion, to the point where masterbatch manufacture with a high-viscosity resin followed by a reduction in concentration in the lower viscosity resin that has the desired properties seems to improve dispersion (as measured by the percolation threshold) as opposed to simply mixing with the low-viscosity resin.[225] This was the result found for polycarbonate; a more recent study with polypropylene found the exactly opposite behavior; that is, masterbatch formation with a lower viscosity resin followed by dilution with a higher viscosity resin led to a lower percolation threshold that the authors attributed to better wetting by the lower viscosity resin.[226] One conclusion from these results is that for high surface energy resins, high-viscosity resins for masterbatches are better, while low surface energy resins are best served with lower viscosity resins. Overall the best choice of masterbatch resin likely depends both on chemical identity as well as tube characteristics especially of the type shown in Figure 3.2. A recent study on various configurations and speeds of a twin-screw extruder indicated that mixing sections, as opposed to kneading sections, improved dispersibility.[227] In that paper, the size and number of black aggregates in optical micrographs were used to assess dispersion. Also, as expected based on experience with other fillers, the addition of a small amount of polymer containing highly adhesive functional groups, for example, maleic anhydride, to a polyolefin–nanotube mix improves dispersion[228,229] although the effect may be reduced at high nanotube loadings.[230] Another study found that extensional flows were much more effective at dispersing nanotubes than shear flows.[231]

With the prevalent use of ultrasound to disperse nanotubes in low-viscosity liquids, an obvious modification to standard twin-screw extrusion is to use ultrasound to assist in the dispersion of nanotubes. Overall, the results from the addition of ultrasound on the outlet of a twin-screw extruder showed minimal, if any, improvements in dispersion as measured by the various percolation or mechanical property measurements.[232] Perhaps the most interesting thing about this study was that a significant reduction in tube breakage occurred, which was attributed to a reduction of pressure at the outlet.

One concept that is in theory possible with a number of different dispersion protocols but is most obviously adopted for melt mixing is the use of phase-separated polymers to control the location of nanotubes within a bulk polymer.[17,233] The concept of "double percolation" becomes important, that is, the idea that the nanotubes are percolating in one phase of a two-phase material as the one phase is also percolating in the second phase. In order for double percolation, the nanotubes must be localized to one component of the two-component blend and one component must be continuous. In fact, such localization has been achieved in a number of different systems with resulting low percolation thresholds.[17,234,235] Nanotubes might also be able to serve as morphology stabilizers in polymer blends, in particular if reactive groups are attached. As mentioned previously, nanotubes can act as surface stabilizers in water-based Pickering emulsions; one paper claims that nanotubes might be able to act as morphology stabilizers in polymer–polymer blends.[236]

Maintenance of a nonprecipitating solution after the application of shear was a very important consideration in dispersion of nanotubes in low-viscosity liquids (this concept will be further explored in Section 5.2); in polymers, the inherent viscosity is much higher providing a significantly higher kinetic barrier to reaggregation. However, as recent work has showed, reaggregation in viscous polymers is critically important.[1,224,237–241] Reaggregation has been measured via both rheology and electrical conductivity; effects on other properties are at this time unknown. Because reaggregation is a highly kinetically dependent phenomenon, the ability to control dispersion is complicated dramatically. Further, this phenomenon is not limited to nanotubes; carbon nanofibers[241] and carbon black show such behavior as well. Reaggregation is very disquieting in the sense that seemingly small changes in processing could have dramatic impacts on observed behavior.

3.6 POLYMER–NANOTUBE DISPERSIONS: NO FLUID MIXING

Without a fluid to disperse the tubes, the ability to debundle tubes is extremely limited. The total number of papers that have used no fluids during mixing is less than 50. Nonetheless, examples exist where a nanotube powder is dry blended with a polymer powder and the mixture is processed in a low-shear environment. The obvious materials for such a procedure are polymers that cannot be processed as melts, such as ultrahigh molecular weight polyethylene.[242] Nanoscale dispersion would not be expected to be very good, and there is no evidence to suggest that this is not the case. One study showed substantially lower dispersion quality as opposed to the dispersion–dispersion–evaporation method.[243] However, percolation thresholds obtained by powder mixing protocols can be quite low, because the nanotubes were localized to the interface between the powder particles.[244]

Using a high-shear process when mixing the two solid materials together might, on the other hand, yield good dispersions. A variety of high-shear powder–powder mixing methods have been used including ball milling,[245,246] pan milling,[247,248] and solid-state shear pulverization.[249,250] The polymer is then usually treated in a normal melt-mixing process, for example, twin-screw extrusion. The dispersion that results after these methods, at least in some cases, seems to be better than what was found without the high-shear environment in the solid state, although these high-shear solid mixing processes tend to shorten nanotube length.

3.7 POLYMER–NANOTUBE DISPERSIONS: IMPREGNATION/INFUSION

Previous methods all had in common significant movement of the center of masses of nanotubes during the mixing process. Another set of processes involves the formation of nanotubes into their final shape, and then adding low-viscosity monomer followed by polymerization, although a solution of polymer can also be used in some cases. Neat polymer is generally not suitable for this process because flow will not be sufficient to prevent significant voids in the final product. However, two papers, one with high-density polyethylene[251] and the other with poly(ether ether ketone),[252] were able to infuse melted polymer into buckypaper using a compression molding approach, although in neither case was the void content reported. The number of papers on composites made via these types of processes is around 50. The author believes that the number of papers in this area will grow rapidly in the future. There are three common shapes for the final form of the nanotubes: nanotube fibers, nanotube sheets, and nanotube forests. A separate subsection is devoted to each of these forms, and a brief description of how to manufacture the various forms is given as well. In addition, nanotubes can be grown or deposited on already existing fibers or fiber mats followed by resin infusion; a separate subsection is devoted to this process as well.

3.7.1 Nanotube Fiber–Polymer Composites

A number of methods are used to make nanotube fibers. Nanotubes can be synthesized in a CVD process performed in such a manner so as to enable capture of the nanotube fibers on a spindle. The starting material in this process is ethanol containing ferrocene and thiophene; using a hydrogen carrier gas, it is possible to form nanotubes in a hot zone at temperatures between 1100 and 1200°C; either SWCNTs or MWCNTs can be formed depending on the conditions and composition of the ingredients. The density of the fiber is about 0.01 g/cm^3 compared to the actual nanotube density of about 1.3 g/cm^3, indicating that the fibers contain a great deal of air. After a post-reaction densification procedure, which involves running the fiber through an acetone vapor stream followed by surface tension-driven densification caused by evaporation, fibers with a density of about 1.0 g/cm^3, strengths of 9 GPa, and stiffnesses of 350 GPa have been reported.[253] These values are significantly larger than the typical values for a high-strength polymer fiber such as Kevlar® (about 4 and 130 GPa, respectively) and are the highest ever reported for nanotubes.

However, the tensile strengths quoted throughout this section are easily misconstrued as will be explained more completely in Section 5.3.2.

Another approach uses nanotubes grown on a flat surface, that is, vertically oriented nanotubes or nanotube forests (Section 2.2.3). Through a simple mechanical technique, for example, using tweezers to grip a portion of the tubes and then pulling, fibers can be produced.[254,255] The width of the nanotube forest can be used to set the diameter of the fiber, and twisting during pulling can improve mechanical properties. The highest strengths and moduli reported via drawing from nanotube forests are 1.9 and 330 GPa, respectively.[255] The fibers produced by this method are very long. A similar technique that uses nanotube dispersions in a low-viscosity liquid instead of vertically grown carbon nanotubes has also been used to produce fibers.[256]

The third method used to produce carbon nanotube fibers involves spinning the fibers from solution. This process is essentially identical to that used to produce fibers for difficult to process polymers such as ultrahigh molecular weight polyethylene and liquid crystalline polymers such as poly(*p*-phenylene benzobisoxazole) (PPTA; i.e., Kevlar®) and poly(*p*-phenylene-2,6-benzobisoxazole) (PBO; i.e., Zylon®). The process is also similar to the gel spinning techniques described in Section 3.4.2 except that no polymer is used in the solution. The tubes are dispersed in a liquid and then extruded through a nozzle and the solvent is evaporated or the fiber is immersed in a miscible liquid, that is, a coagulation process, or the two are done in series. Solvents that have been used to form fibers in this manner include superacids,[257] water,[258] and ethylene glycol.[259] Controlling the rheological properties of the nanotube–liquid mixture in order to produce continuous fibers is a key requirement. The stiffness of fibers produced in this manner is roughly equivalent to that of Kevlar®, while the strength is typically an order of magnitude or more lower than that of Kevlar®.

The fourth and final method used to produce fibers cannot readily be adapted to make continuous (long) fibers, but has been used to make the highest strength and modulus fibers reported. A nanotube film is made by direct synthesis onto a quartz tube in a chemical vapor deposition process.[260] Fibers are formed by taking the films and twisting them into fibers; typical lengths were on the order of 4–8 cm. As with the directly formed fibers in the gas, these fibers also need a densification process with acetone to achieve high strengths; strengths as high 850 MPa and modulus as high as 18 GPa were reported.[261]

For most of the applications of high-strength fibers, for example, hard body armor, the high-strength fiber is encased in a polymer, almost always a low-viscosity thermoset such as an epoxy. The fibers may be weaved into a mat prior to encapsulation or may be laid down parallel to one another. Weaving of neat nanotube fibers into mats has been reported in the literature.[262] A concern during weaving is not to damage the fibers. The resin must have low viscosity in order to properly infiltrate into the fibers and between the fibers. In addition, wetting of the fibers is a large concern; if the nanotubes are not properly wet by the resin, then voids will result and composite properties will be compromised. A solution containing dissolved polymer can be used to infiltrate into the fiber mat to help with wetting issues.

Exceedingly poor quality fibers (tensile strength = 0.001 GPa; tensile modulus = 0.1 GPa) made via drawing from nanotube forests were infiltrated with methyl methacrylate and there was a marked improvement in properties.[263] Using much higher quality fibers (tensile strength = 0.55 GPa; tensile modulus = 22 GPa) with an epoxy gave results after infiltration that were roughly equivalent to a rule of mixtures at a nanotube level of about 25% by volume, similar to the results obtained for an unsized carbon fiber. The authors point out that in these carbon nanotube fibers there is a great deal of void space, making proper determination of the pure fiber properties difficult. More common fiber strength and modulus values that normalize to weight rather than cross-sectional area are more informative for the case of nanotube fibers.[264]

3.7.2 Nanotube Sheet–Polymer Composites

A nanotube sheet has preferential in-plane orientation of nanotubes. As described in more detail in Section 8.2.3, thin sheets of carbon nanotubes, often termed bucky-paper, infiltrated with polymers have possible commercial applications as transparent electrodes.

The most common method used to make nanotube sheets is to simply filter a nanotube solution through a medium that can trap nanotubes. The thickness and void content of the sheet is controlled by nanotube concentration, the volume of liquid that is filtered, and the dispersion conditions. The bundle size distribution depends on dispersion conditions as well. Nanotubes in the film show curvature and are randomly oriented. Very thin sheets can be made by dissolving the media[265] because the forces required to lift the nanotube membrane from the substrate are the limiting factor in terms of the minimum thickness that can be manufactured. Epoxy monomer has been infused into a thin sheet of SWCNT film and then cured, but acetone was required to reduce the viscosity in order to properly infuse the sheet.[266,267] A preformed polymer, poly(vinyl pyridine), in solution has been used to infiltrate nanotube sheets; the effect of molecular weight was tested with the higher molecular weight material giving better improvements in reinforcing capability.[268] Other polymers in solution have been used as well.[269,270] Laminates of sheets with thermoplastics and high-viscosity epoxies, where the elimination of voids will be a challenge, have been constructed using vacuum bagging or compression processes common in the composite industry.[251,271] The use of electrically conducting polymers, which are polymerized on the sheets directly, represents an obvious way to make highly conducting sheets of material,[272] although transparency will be lessened. Electrical conductivities of composites using buckypapers can be significantly higher than 100 S/m, which is much higher than that found for randomly distributed systems that result from the procedure described in Section 3.4.1.

Langmuir–Blodgett (LB) film formation is another process used in the formation of nanotube films. The LB process involves the formation of a film at a liquid–air interface followed by lifting off of the film. Moving surface boundaries that are used to compress the film on the liquid–air surface increase the nanotube surface concentration and eventually induce and/or increase orientation of the tubes. Horizontal lifting of the film is done by positioning a solid substrate parallel to the air–liquid interface and lifting the film off carefully. More standard LB procedure is vertical dipping; that is, the substrate surface is normal to liquid–air surface and dipped into the liquid, which is the same orientation and procedure used in the LBL process. However, the LB and the LBL processes are different; in the LBL process material suspended in bulk comprises the majority of the film, while in the LB process material at the liquid–air interface comprises the majority of the film. In both vertical and horizontal dipping, the film must be removed from the substrate to achieve a freestanding nanotube film. The thickness of the film can be controlled via multiple dips, just as in the LBL process, although typically more time between dips is used in the LB process to allow diffusion to the air–liquid interface. Colloidal solids as well as surface-active species such as surfactants are routinely made into films using the LB process.

Both horizontal and vertically dipping were used in studies reported by one group,[273,274] while in another only horizontal dipping yielded a cohesive film.[275] In all these cases, modified tubes were used; using a solvent other than water enabled the use of unfunctionalized tubes.[276] In the latter case, the organic liquid dichloroethane containing the nanotubes was spread on a water surface and, upon the evaporation of the organic liquid, a LB film was vertically lifted from the surface after compression. An aromatic polymer was also present in the organic solution, and presumably was part of the LB film as well. A similar process to the LB method utilizes the addition of a few drops of a less dense miscible liquid that will form a surface layer on another bulk liquid, and the surface liquid has a greater affinity for nanotubes than the bulk liquid. Nanotubes that are initially dispersed in the bulk liquid will be drawn to the surface under these conditions. The particulars are the solvents being water and ethanol, and DWCNTs were used in the process. This process deserves special mention because of the high modulus, 12.2 GPa, and high strength, 0.75 GPa, measured for the films. The density was also quite high at 0.8 g/cm^3.[277] None of the LB films has been infiltrated with polymer at this time.

Other miscellaneous methods have been used to produce all nanotube films including the LBL process using negatively and positively charged functionalized nanotubes.[278] In one case, simply drying a water dispersion of tubes led to film formation on a substrate that was repellant to nanotubes.[279] Producing buckypaper with orientation in one direction instead of in a planar direction can be achieved via the use of magnetic fields during filtration. The buckypapers were infiltrated with polycarbonate to yield final CNT concentrations in the 40–60 wt% range. The modulus and tensile strength for the buckypaper were about 2 GPa and 6.5 MPa and increased by a factor of ~2.5 with the addition of polycarbonate, and the electrical conductivity dropped by about a factor of 3 to about 100 S/cm.[270] A unique method, which is more like an aerogel rather than a sheet, involves removing the polymer from a nanotube composite after a procedure such as melt mixing or dissolution–dispersion–evaporation, and then infusing this material with a thermosetting polymer.[280] A CVD process similar to that described earlier to manufacture nanotube fibers directly in the reactor can be used to manufacture sheets. A substrate is used and the thickness of the films can be controlled by controlling the substrate position and deposition time. The films are freestanding and have sufficient mechanical integrity to be handled.[281,282] Laminated composites were manufactured by compression molding poly(ether ether ketone) with a SWCNT CVD film.[252] Although small increases in properties occurred, no attempt was made to quantify void content of the films.

Nanotube forests can be used directly to make nanotube sheets. By using a flexible, somewhat adhesive microporous membrane and pushing down starting on one end yields flat buckypaper with a uniaxial nanotube orientation in the pushing direction in the plane parallel to the substrate, and the film can then be lifted off the membrane with the possible assistance of a liquid. After infiltration, this method has the capability of yielding a composite with very long tubes, since vertically grown tubes can be quite long as discussed in Section 2.2.3. The density of this method is quite large compared to other buckypapers, about 0.6 g/cm^3.[283] Another method uses the fiber method as described in the previous section for nanotube forests but with a wider drawing implement than tweezers to yield a film with uniaxial oriented nanotubes in a plane. Infusion of numerous plies of this drawn sheet after surface tension-driven densification with a dimethyl siloxane monomer was used to form a composite with a negative Poisson's ratio.[284]

3.7.3 Nanotube Forests–Polymer Composites

Nanotube forests on a substrate can be infused directly with a low-viscosity resin and then cured, yielding a product with nanotubes oriented in the thickness direction. Although macroscopically geometrically identical, nanotube sheets have fibers oriented in the sheet plane while these materials have nanotubes oriented perpendicular to the sheet plane. Removal of nanotubes from the supporting substrate prior to infiltration, if feasible, eliminates the issue of removal after infiltration and curing. As with the other methods, good wetting is the key to achieving a product with no voids, and in one study three different epoxies with three different viscosities were used and in all cases the lack of void formation indicated that the resin properly infused the nanotube forest.[285] Modulus enhancement in the direction of nanotube alignment as measured by nanocompression tests was over 200%.[286] Biaxial surface tension-driven densification of the forests prior to resin infusion has been used to increase the volume fraction of tubes to 20%.[287] The rule of mixtures with respect to modulus is off by about a factor of 10 after densification and infusion, which the authors attribute to the imperfections in nanotube orientation, and the electrical conductivity is between 0.1 and 1 S/cm.[288] Electrically conducting polymer has also been introduced into nanotube forests via monomer vapor deposition followed by polymerization.[289]

3.7.4 Nanotubes on Already Existing Fibers

Nanotubes can be grown or deposited on already existing fibers, followed by resin infusion. Essentially, this procedure adds an extra step to the already used production of continuous thermoset composites. In a simplified description, with the extra step in italics: fibers are weaved into mats, mats are placed in a mold, *nanotubes are then deposited on the mats*, and then the resin is added and the material is cured. One deposition process involves simply spraying the fibers with a solution containing nanotubes followed by evaporation;[290] a more subtle approach uses electric fields to deposit nanotubes on the surface of the fibers. [291] Another approach is to disperse nanotubes into the infusing resin, either with or without a dispersing solvent, and then infuse the resin/nanotube mixture into the fiber mat. This procedure has the disadvantage of creating a nonuniform distribution of nanotubes due to filtering by the fibers located on the outside of the mat.[292] Finally, the fiber can be drawn through a resin that contains dispersed nanotubes and conditions adjusted so that nanotubes become part of the coated fiber.[293] More resin is then added to the coated fibers, which is then cured.

Direct deposition of a nanotube catalyst on a unique substrate for use in a polymer was already detailed in Section 3.4.1, where the catalyst for nanotube growth and catalyst for polymer growth were deposited on the same support. Growth of nanotubes on an already existing fiber, most commonly carbon fiber, is another case where nanotubes are put in their final location prior to infusion of resin.[294–297] Although certainly more difficult than infiltrating the nanotubes into the fiber directly, growth of the nanotubes on the fibers has the following advantages: more uniform concentration of nanotubes spatially; nanotube adhesion may be improved; and nanotube orientation will be perpendicular to the fiber direction. The latter is of great importance if the properties in the thickness direction, that is, electrical or thermal conductivity, are of interest. However, the high temperatures used to grow CNTs typically lead to undesirable fiber damage. The use of alumina fibers eliminates this concern, although alumina fibers are not nearly as common as carbon fibers.[298]

3.8 CHALLENGES

One obvious challenge is the use of better and more consistent dispersion metrics. Average bundle size, or something related to that, is a good nanoscopic metric to use. The use of microscopy to assess microscopic dispersion is appropriate, but needs to be more widespread. The use of percolation threshold as a dispersion metric, although widespread, is not optimal. A simple-to-use indication of dispersion at the nanoscopic level in particular would be useful.

As stated in Chapter 2, the development of easily dispersible tubes is a challenge. The author does not feel that there are likely to be new dispersion methods available, and hence only through changing the nanotubes are significant improvements in dispersion likely. A better understanding of the interrelationships between nanoscopic and microscopic dispersion with properties is necessary. The increasing recognition that reaggregation occurs in a high-viscosity melt polymer complicates the pursuit of this understanding dramatically. An unanswered question is whether nanotubes that are easily dispersible in one medium via one procedure, say water via sonication and surfactant addition, are also easily dispersible in another, say a polymer via melt mixing. The number of studies that have examined this question is few,[299,300] although certainly anecdotal evidence suggests that nanotube qualitative dispersibility is independent of the media being investigated.

Infiltration/infusion processes have been not nearly as well studied as their importance suggests. The reason for this delay is the difficulty of making sheets and fibers with consistent high strength over large length scales. However significant advances in this area are occurring. The number of studies of these types of processes is increasing, driven by the advances of making good sheets and fibers. Of likely lesser impact, but still important, is the ability to understand how to use dispersed nanotubes to make good coatings.

REFERENCES

1. Alig, I., Lellinger, D., Engel, M., Skipa, T., Pötschke, P. (2008). Destruction and formation of a conductive carbon nanotube network in polymer melts: in-line experiments, *Polymer*, *49*, 1902–1909.
2. Hilding, J., Grulke, E. A., Zhang, Z. G., Lockwood, F. (2003). Dispersion of carbon nanotubes in liquids, *J. Dispers. Sci. Technol.*, *24*, 1–41.
3. Fakhri, N., Tsyboulski, D. A., Cognet, L., Weisman, R. B., Pasquali, M. (2009). Diameter-dependent bending dynamics of single-walled carbon nanotubes in liquids, *Proc. Natl. Acad. Sci. USA*, *106*, 14219–14223.
4. Duggal, R., Pasquali, M. (2006). Dynamics of individual single-walled carbon nanotubes in water by real-time visualization, *Phys. Rev. Lett.*, *96*, 246104.
5. Lee, H. S., Yun, C. H. (2008). Translational and rotational diffusions of multiwalled carbon nanotubes with static bending, *J. Phys. Chem. C*, *112*, 10653–10658.
6. Lee, H. S., Yun, C. H., Kim, H. M., Lee, C. J. (2007). Persistence length of multiwalled carbon nanotubes with static bending, *J. Phys. Chem. C*, *111*, 18882–18887.
7. Zhao, C. G., Hu, G. J., Justice, R., Schaefer, D. W., Zhang, S. M., Yang, M. S., Han, C. C. (2005). Synthesis and characterization of multi-walled carbon nanotubes reinforced polyamide 6 via *in situ* polymerization, *Polymer*, *46*, 5125–5132.
8. Kharisov, B. I., Kharissova, O. V., Gutierrez, H. L., Mendez, U. O. (2009). Recent advances on the soluble carbon nanotubes, *Ind. Eng. Chem. Res.*, *48*, 572–590.
9. Bergin, S. D., Nicolosi, V., Streich, P. V., Giordani, S., Sun, Z. Y., Windle, A. H., Ryan, P., Niraj, N. P. P., Wang, Z. T. T., Carpenter, L., Blau, W. J., Boland, J. J., Hamilton, J. P., Coleman, J. N. (2008). Towards solutions of single-walled carbon nanotubes in common solvents, *Adv. Mater.*, *20*, 1876–1881.
10. Coleman, J. N. (2009). Liquid-phase exfoliation of nanotubes and graphene, *Adv. Funct. Mater.*, *19*, 3680–3695.
11. Davis, V. A., Parra Vasquez, A. N. G., Green, M. J., Rai, P. K., Behabtu, N., Prieto, V., Booker, R. D., Schmidt, J., Kesselman, E., Zhou, W., Fan, H., Adams, W. W., Hauge, R. H., Fischer, J. E., Cohen, Y., Talmon, Y., Smalley, R. E., Pasquali, M. (2009). True solutions of single-walled carbon nanotubes for assembly into macroscopic materials, *Nat. Nanotechnol.*, *4*, 830–834.
12. Pötschke, P., Fornes, T. D., Paul, D. R. (2002). Rheological behavior of multiwalled carbon nanotube/polycarbonate composites, *Polymer*, *43*, 3247–3255.
13. Mitchell, C. A., Bahr, J. L., Arepalli, S., Tour, J. M., Krishnamoorti, R. (2002). Dispersion of functionalized carbon nanotubes in polystyrene, *Macromolecules*, *35*, 8825–8830.
14. Ha, M. L. P., Grady, B. P., Lolli, G., Resasco, D. E., Ford, W. T. (2007). Composites of single-walled carbon nanotubes and styrene–isoprene copolymer lattices, *Macromol. Chem. Phys.*, *208*, 446–456.
15. Du, F. M., Scogna, R. C., Zhou, W., Brand, S., Fischer, J. E., Winey, K. I. (2004). Nanotube networks in polymer nanocomposites: rheology and electrical conductivity, *Macromolecules*, *37*, 9048–9055.
16. Pötschke, P., Abdel-Goad, M., Alig, I., Dudkin, S., Lellinger, D. (2004). Rheological and dielectrical characterization of melt mixed polycarbonate–multiwalled carbon nanotube composites, *Polymer*, *45*, 8863–8870.
17. Meincke, O., Kaempfer, D., Weickmann, H., Friedrich, C., Vathauer, M., Warth, H. (2004). Mechanical properties and electrical conductivity of carbon-nanotube filled polyamide-6 and its blends with acrylonitrile/butadiene/styrene, *Polymer*, *45*, 739–748.
18. Pegel, S., Pötschke, P., Villmow, T., Stoyan, D., Heinrich, G. (2009). Spatial statistics of carbon nanotube polymer composites, *Polymer*, *50*, 2123–2132.
19. Blond, D., Barron, V., Ruether, M., Ryan, K. P., Nicolosi, V., Blau, W. J., Coleman, J. N. (2006). Enhancement of modulus, strength, and toughness in poly(methyl methacrylate)-based composites by the incorporation of poly(methyl methacrylate)-functionalized nanotubes, *Adv. Funct. Mater.*, *16*, 1608–1614.
20. Qiu, J. J., Zhang, C., Wang, B., Liang, R. (2007). Carbon nanotube integrated multifunctional multiscale composites, *Nanotechnology*, *18*, 275708.
21. Chang, L., Friedrich, K., Ye, L., Toro, P. (2009). Evaluation and visualization of the percolating networks in multi-wall carbon nanotube/epoxy composites, *J. Mater. Sci.*, *44*, 4003–4012.

22. Loos, J., Alexeev, A., Grossiord, N., Koning, C. E., Regev, O. (2005). Visualization of single-wall carbon nanotube (SWNT) networks in conductive polystyrene nanocomposites by charge contrast imaging, *Ultramicroscopy*, *104*, 160–167.
23. Kovacs, J. Z., Andresen, K., Pauls, J. R., Garcia, C. P., Schossig, M., Schulte, K., Bauhofer, W. (2007). Analyzing the quality of carbon nanotube dispersions in polymers using scanning electron microscopy, *Carbon*, *45*, 1279–1288.
24. Lillehei, P. T., Kim, J. W., Gibbons, L. J., Park, C. (2009). A quantitative assessment of carbon nanotube dispersion in polymer matrices, *Nanotechnology*, *20*, 325708.
25. Tchoul, M. N., Ford, W. T., Lolli, G., Resasco, D. E., Arepalli, S. (2007). Effect of mild nitric acid oxidation on dispersability, size, and structure of single-walled carbon nanotubes, *Chem. Mater.*, *19*, 5765–5772.
26. Lee, J. Y., Kim, J. S., An, K. H., Lee, K., Kim, D. Y., Bae, D. J., Lee, Y. H. (2005). Electrophoretic and dynamic light scattering in evaluating dispersion and size distribution of single-walled carbon nanotubes, *J. Nanosci. Nanotechnol.*, *5*, 1045–1049.
27. Yonemura, H., Yamamoto, Y., Yamada, S., Fujiwara, Y., Tanimoto, Y. (2008). Magnetic orientation of single-walled carbon nanotubes or their composites using polymer wrapping, *Sci. Technol. Adv. Mater.*, *9*, 024213.
28. Phang, I. Y., Liu, T. X., Zhang, W. D., Schonherr, H., Vancso, G. J. (2007). Probing buried carbon nanotubes within polymer–nanotube composite matrices by atomic force microscopy, *Eur. Polym. J.*, *43*, 4136–4142.
29. Inam, F., Peijs, T. (2006). Transmission light microscopy of carbon nanotubes–epoxy nanocomposites involving different dispersion methods, *Adv. Compos. Lett.*, *15*, 7–13.
30. Hobbie, E. K., Wang, H., Kim, H., Han, C. C., Grulke, E. A., Obrzut, J. (2003). Optical measurements of structure and orientation in sheared carbon-nanotube suspensions, *Rev. Sci. Instrum.*, *74*, 1244–1250.
31. Bellayer, S., Gilman, J. W., Eidelman, N., Bourbigot, S., Flambard, X., Fox, D. M., De Long, H. C., Trulove, P. C. (2005). Preparation of homogeneously dispersed multiwalled carbon nanotube/polystyrene nanocomposites via melt extrusion using trialkyl imidazolium compatibilizer, *Adv. Funct. Mater.*, *15*, 910–916.
32. Kashiwagi, T., Fagan, J., Douglas, J. F., Yamamoto, K., Heckert, A. N., Leigh, S. D., Obrzut, J., Du, F. M., Lin-Gibson, S., Mu, M. F., Winey, K. I., Haggenmueller, R. (2007). Relationship between dispersion metric and properties of PMMA/SWNT nanocomposites, *Polymer*, *48*, 4855–4866.
33. Heller, D. A., Barone, P. W., Swanson, J. P., Mayrhofer, R. M., Strano, M. S. (2004). Using Raman spectroscopy to elucidate the aggregation state of single-walled carbon nanotubes, *J. Phys. Chem. B*, *108*, 6905–6909.
34. Yoon, D., Choi, J. B., Han, C. S., Kim, Y. J., Baik, S. (2008). The quantitative characterization of the dispersion state of single-walled carbon nanotubes using Raman spectroscopy and atomic force microscopy, *Carbon*, *46*, 1530–1534.
35. Fagan, J. A., Landi, B. J., Mandelbaum, I., Simpson, J. R., Bajpai, V., Bauer, B. J., Migler, K., Walker, A. R. H., Raffaelle, R., Hobbie, E. K. (2006). Comparative measures of single-wall carbon nanotube dispersion, *J. Phys. Chem. B*, *110*, 23801–23805.
36. Karachevtsev, V. A., Glamazda, A. Y., Dettlaff-Weglikowska, U., Leontiev, V. S., Mateichenko, P. V., Roth, S., Rao, A. M. (2006). Spectroscopic and SEM studies of SWNTs: polymer solutions and films, *Carbon*, *44*, 1292–1297.
37. Perez, R., Banda, S., Ounaies, Z. (2008). Determination of the orientation distribution function in aligned single wall nanotube polymer nanocomposites by polarized Raman spectroscopy, *J. Appl. Phys.*, *103*, 074302.
38. Liu, T., Kumar, S. (2003). Quantitative characterization of SWNT orientation by polarized Raman spectroscopy, *Chem. Phys. Lett.*, *378*, 257–262.
39. Zhou, W., Vavro, J., Guthy, C., Winey, K. I., Fischer, J. E., Ericson, L. M., Ramesh, S., Saini, R., Davis, V. A., Kittrell, C., Pasquali, M., Hauge, R. H., Smalley, R. E. (2004). Single wall carbon nanotube fibers extruded from super-acid suspensions: preferred orientation, electrical, and thermal transport, *J. Appl. Phys.*, *95*, 649–655.
40. Minus, M. L., Chae, H. G., Kumar, S. (2009). Interfacial crystallization in gel-spun poly(vinyl alcohol)/single-wall carbon nanotube composite fibers, *Macromol. Chem. Phys.*, *210*, 1799–1808.

41. Grossiord, N., Loos, J., Meuldijk, J., Regev, O., Miltner, H. E., Van Mele, B., Koning, C. E. (2007). Conductive carbon-nanotube/polymer composites: spectroscopic monitoring of the exfoliation process in water, *Compos. Sci. Technol.*, *67*, 778–782.
42. Grossiord, N., Regev, O., Loos, J., Meuldijk, J., Koning, C. E. (2005). Time-dependent study of the exfoliation process of carbon nanotubes in aqueous dispersions by using UV–visible spectroscopy, *Anal. Chem.*, *77*, 5135–5139.
43. Edri, E., Regev, O. (2009). "Shaken, not stable": dispersion mechanism and dynamics of protein-dispersed nanotubes studied via spectroscopy, *Langmuir*, *25*, 10459–10465.
44. Tan, Y. Q., Resasco, D. E. (2005). Dispersion of single-walled carbon nanotubes of narrow diameter distribution, *J. Phys. Chem. B*, *109*, 14454–14460.
45. Haddon, R. C., Sippel, J., Rinzler, A. G., Papadimitrakopoulos, F. (2004). Purification and separation of carbon nanotubes, *MRS Bull.*, *29*, 252–259.
46. Chatterjee, T., Jackson, A., Krishnamoorti, R. (2008). Hierarchical structure of carbon nanotube networks, *J. Am. Chem. Soc.*, *130*, 6934–6935.
47. Zhou, W., Islam, M. F., Wang, H., Ho, D. L., Yodh, A. G., Winey, K. I., Fischer, J. E. (2004). Small angle neutron scattering from single-wall carbon nanotube suspensions: evidence for isolated rigid rods and rod networks, *Chem. Phys. Lett.*, *384*, 185–189.
48. Brown, J. M., Anderson, D. P., Justice, R. S., Lafdi, K., Belfor, M., Strong, K. L., Schaefer, D. W. (2005). Hierarchical morphology of carbon single-walled nanotubes during sonication in an aliphatic diamine, *Polymer*, *46*, 10854–10865.
49. Justice, R. S., Wang, D. H., Tan, L. S., Schaefer, D. W. (2007). Simplified tube form factor for analysis of small-angle scattering data from carbon nanotube filled systems, *J. Appl. Crystallogr.*, *40*, S88–S92.
50. Du, F. M., Fischer, J. E., Winey, K. I. (2003). Coagulation method for preparing single-walled carbon nanotube/poly(methyl methacrylate) composites and their modulus, electrical conductivity, and thermal stability, *J. Polym. Sci. Polym. Phys.*, *41*, 3333–3338.
51. Pujari, S., Rahatekar, S. S., Gilman, J. W., Koziol, K. K., Windle, A. H., Burghardt, W. R. (2009). Orientation dynamics in multiwalled carbon nanotube dispersions under shear flow, *J. Chem. Phys.*, *130*, 214903.
52. Hobbie, E. K. (2004). Optical anisotropy of nanotube suspensions, *J. Chem. Phys.*, *121*, 1029–1037.
53. Bauer, B. J., Becker, M. L., Bajpai, V., Fagan, J. A., Hobbie, E. K., Migler, K., Guttman, C. M., Blair, W. R. (2007). Measurement of single-wall nanotube dispersion by size exclusion chromatography, *J. Phys. Chem. C*, *111*, 17914–17918.
54. Cotiuga, I., Picchioni, F., Agarwal, U. S., Wouters, D., Loos, J., Lemstra, P. J. (2006). Block-copolymer-assisted solubilization of carbon nanotubes and exfoliation monitoring through viscosity, *Macromol. Rapid Commun.*, *27*, 1073–1078.
55. Battisti, A., Skordos, A. A., Partridge, I. K. (2009). Monitoring dispersion of carbon nanotubes in a thermosetting polyester resin, *Compos. Sci. Technol.*, *69*, 1516–1520.
56. Huang, Y. Y., Ahir, S. V., Terentjev, E. M. (2006). Dispersion rheology of carbon nanotubes in a polymer matrix, *Phys. Rev. B*, *73*, 125422.
57. Choi, H. J., Zhang, K., Park, S. Y., Lee, B. Y. (2009). Effect of preparation method of multi-walled carbon nanotube/poly(methyl methacrylate) nanocomposite on its characteristics, *J. Nanosci. Nanotechnol.*, *9*, 6089–6095.
58. Park, H. J., Noh, H., Park, M., Park, K., Chang, J. Y., Lee, H. (2008). Length characterization of multi-walled CNT by capillary hydrodynamic fractionation (CHDF), *J. Nanosci. Nanotechnol.*, *8*, 5220–5223.
59. Badaire, S., Poulin, P., Maugey, M., Zakri, C. (2004). *In situ* measurements of nanotube dimensions in suspensions by depolarized dynamic light scattering, *Langmuir*, *20*, 10367–10370.
60. Girifalco, L. A., Hodak, M., Lee, R. S. (2000). Carbon nanotubes, buckyballs, ropes, and a universal graphitic potential, *Phys. Rev. B*, *62*, 13104–13110.
61. Shvartzman-Cohen, R., Nativ-Roth, E., Baskaran, E., Levi-Kalisman, Y., Szleifer, I., Yerushalmi-Rozen, R. (2004). Selective dispersion of single-walled carbon nanotubes in the presence of polymers: the role of molecular and colloidal length scales, *J. Am. Chem. Soc.*, *126*, 14850–14857.
62. Nguyen, T. Q., Liang, O. Z., Kausch, H. H. (1997). Kinetics of ultrasonic and transient elongational flow degradation: a comparative study, *Polymer*, *38*, 3783–3793.

63. Ahir, S. V., Huang, Y. Y., Terentjev, E. M. (2008). Polymers with aligned carbon nanotubes: active composite materials, *Polymer, 49*, 3841–3854.
64. Lu, K. L., Lago, R. M., Chen, Y. K., Green, M. L. H., Harris, P. J. F., Tsang, S. C. (1996). Mechanical damage of carbon nanotubes by ultrasound, *Carbon, 34*, 814–816.
65. Kuijpers, M. W. A., Iedema, P. D., Kemmere, M. F., Keurentjes, J. T. F. (2004). The mechanism of cavitation-induced polymer scission; experimental and computational verification, *Polymer, 45*, 6461–6467.
66. Hennrich, F., Krupke, R., Arnold, K., Stutz, J. A. R., Lebedkin, S., Koch, T., Schimmel, T., Kappes, M. M. (2007). The mechanism of cavitation-induced scission of single-walled carbon nanotubes, *J. Phys. Chem. B, 111*, 1932–1937.
67. Ikeda, A., Hayashi, K., Konishi, T., Kikuchi, J. (2004). Solubilization and debundling of purified single-walled carbon nanotubes using solubilizing agents in an aqueous solution by high-speed vibration milling technique, *Chem. Commun.* 1334–1335.
68. Thostenson, E. T., Chou, T. W. (2006). Processing–structure–multi-functional property relationship in carbon nanotube/epoxy composites, *Carbon, 44*, 3022–3029.
69. Gojny, F. H., Wichmann, M. H. G., Kopke, U., Fiedler, B., Schulte, K. (2004). Carbon nanotube-reinforced epoxy-composites: enhanced stiffness and fracture toughness at low nanotube content, *Compos. Sci. Technol., 64*, 2363–2371.
70. Chatterjee, T., Yurekli, K., Hadjiev, V. G., Krishnamoorti, R. (2005). Single-walled carbon nanotube dispersions in poly(ethylene oxide), *Adv. Funct. Mater., 15*, 1832–1838.
71. Vaisman, L., Marom, G., Wagner, H. D. (2006). Dispersions of surface-modified carbon nanotubes in water-soluble and water-insoluble polymers, *Adv. Funct. Mater., 16*, 357–363.
72. Jiang, L. Q., Gao, L., Sun, J. (2003). Production of aqueous colloidal dispersions of carbon nanotubes, *J. Colloid Interface Sci., 260*, 89–94.
73. Moore, V. C., Strano, M. S., Haroz, E. H., Hauge, R. H., Smalley, R. E., Schmidt, J., Talmon, Y. (2003). Individually suspended single-walled carbon nanotubes in various surfactants, *Nano Lett., 3*, 1379–1382.
74. Haggenmueller, R., Rahatekar, S. S., Fagan, J. A., Chun, J. H., Becker, M. L., Naik, R. R., Krauss, T., Carlson, L., Kadla, J. F., Trulove, P. C., Fox, D. F., DeLong, H. C., Fang, Z. C., Kelley, S. O., Gilman, J. W. (2008). Comparison of the quality of aqueous dispersions of single wall carbon nanotubes using surfactants and biomolecules, *Langmuir, 24*, 5070–5078.
75. Sun, Z., Nicolosi, V., Rickard, D., Bergin, S. D., Aherne, D., Coleman, J. N. (2008). Quantitative evaluation of surfactant-stabilized single-walled carbon nanotubes: dispersion quality and its correlation with zeta potential, *J. Phys. Chem. C, 112*, 10692–10699.
76. Marquis, R., Greco, C., Sadokierska, I., Lebedkin, S., Kappes, M. M., Michel, T., Alvarez, L., Sauvajol, J. L., Meunier, S., Mioskowski, C. (2008). Supramolecular discrimination of carbon nanotubes according to their helicity, *Nano Lett., 8*, 1830–1835.
77. Wang, H., Zhou, W., Ho, D. L., Winey, K. I., Fischer, J. E., Glinka, C. J., Hobbie, E. K. (2004). Dispersing single-walled carbon nanotubes with surfactants: a small angle neutron scattering study, *Nano Lett., 4*, 1789–1793.
78. Matarredona, O., Rhoads, H., Li, Z. R., Harwell, J. H., Balzano, L., Resasco, D. E. (2003). Dispersion of single-walled carbon nanotubes in aqueous solutions of the anionic surfactant NaDDBS, *J. Phys. Chem. B, 107*, 13357–13367.
79. Grossiord, N., van der Schoot, P., Meuldijk, J., Koning, C. E. (2007). Determination of the surface coverage of exfoliated carbon nanotubes by surfactant molecules in aqueous solution, *Langmuir, 23*, 3646–3653.
80. Kiraly, Z., Findenegg, G. H. (2005). Pulsed-flow microcalorimetric study of the template-monolayer region of nonionic. Surfactants adsorbed at the graphite/water interface, *Langmuir, 21*, 5047–5054.
81. Tummala, N. R., Striolo, A. (2009). Curvature effects on the adsorption of aqueous sodium-dodecyl-sulfate surfactants on carbonaceous substrates: structural features and counterion dynamics, *Phys. Rev. E, 80*, 021408.
82. Tummala, N. R., Striolo, A. (2009). SDS surfactants on carbon nanotubes: aggregate morphology, *ACS Nano, 3*, 595–602.
83. Angelikopoulos, P., Bock, H. (2008). Directed self-assembly of surfactants in carbon nanotube materials, *J. Phys. Chem. B, 112*, 13793–13801.

84. Yurekli, K., Mitchell, C. A., Krishnamoorti, R. (2004). Small-angle neutron scattering from surfactant-assisted aqueous dispersions of carbon nanotubes, *J. Am. Chem. Soc.*, *126*, 9902–9903.
85. Ikeda, A., Hamano, T., Hayashi, K., Kikuchi, J. (2006). Water-solubilization of nucleotides-coated single-walled carbon nanotubes using a high-speed vibration milling technique, *Org. Lett.*, *8*, 1153–1156.
86. Mountrichas, G., Tagmatarchis, N., Pispas, S. (2007). Synthesis and solution behavior of carbon nanotubes decorated with amphiphilic block polyelectrolytes, *J. Phys. Chem. B*, *111*, 8369–8372.
87. Wang, Z. M., Liu, Q. C., Zhu, H., Liu, H. F., Chen, Y. M., Yang, M. S. (2007). Dispersing multi-walled carbon nanotubes with water-soluble block copolymers and their use as supports for metal nanoparticles, *Carbon*, *45*, 285–292.
88. Yang, Y., Grulke, E. A., Zhang, Z. G., Wu, G. F. (2007). Temperature effects on the rheological properties of carbon nanotube-in-oil dispersions, *Colloids Surf. A*, *298*, 216–224.
89. O'Connell, M. J., Boul, P., Ericson, L. M., Huffman, C., Wang, Y. H., Haroz, E., Kuper, C., Tour, J., Ausman, K. D., Smalley, R. E. (2001). Reversible water-solubilization of single-walled carbon nanotubes by polymer wrapping, *Chem. Phys. Lett.*, *342*, 265–271.
90. Shin, J. Y., Kim, C., Geckeler, K. E. (2009). Single-walled carbon nanotube–polystyrene nanocomposites: dispersing nanotubes in organic media, *Polym. Int.*, *58*, 579–583.
91. Hasan, T., Scardaci, V., Tan, P. H., Rozhin, A. G., Milne, W. I., Ferrari, A. C. (2008). Dispersibility and stability improvement of unfunctionalized nanotubes in amide solvents by polymer wrapping, *Physica E*, *40*, 2414–2418.
92. Chou, S. G., Ribeiro, H. B., Barros, E. B., Santos, A. P., Nezich, D., Samsonidze, G. G., Fantini, C., Pimenta, M. A., Jorio, A., Plentz, F., Dresselhaus, M. S., Dresselhaus, G., Saito, R., Zheng, M., Onoa, G. B., Semke, E. D., Swan, A. K., Unlu, M. S., Goldberg, B. B. (2004). Optical characterization of DNA-wrapped carbon nanotube hybrids, *Chem. Phys. Lett.*, *397*, 296–301.
93. Glamazda, A. Y., Dettlaff-Weglikowska, U., Leontiev, V. S., Mateichenko, P. V., Karachevtsev, V. A. (2006). Raman spectroscopy and SEM study of SWNTs in aqueous solution and films with surfactant or polymer surroundings, *Fullerenes Nanotubes Carbon Nanostruct.*, *14*, 221–225.
94. Yamamoto, T., Motoyanagi, J., Murakami, Y., Miyauchi, Y., Maruyama, S., Kato, M. (2009). Surfactant-stabilized single-walled carbon nanotubes using triphenylene derivatives remain individually dispersion in both liquid and dried solid states, *Appl. Phys. Express*, *2*, 055501.
95. Chen, L. F., Xie, H. Q., Li, Y., Yu, W. (2009). Carbon nanotubes with hydrophilic surfaces produced by a wet-mechanochemical reaction with potassium hydroxide using ethanol as solvent, *Mater. Lett.*, *63*, 45–47.
96. Lee, G. W., Kim, J., Yoon, J., Bae, J. S., Shin, B. C., Kim, I. S., Oh, W., Ree, M. (2008). Structural characterization of carboxylated multi-walled carbon nanotubes, *Thin Solid Films*, *516*, 5781–5784.
97. Peng, H. Q., Alemany, L. B., Margrave, J. L., Khabashesku, V. N. (2003). Sidewall carboxylic acid functionalization of single-walled carbon nanotubes, *J. Am. Chem. Soc.*, *125*, 15174–15182.
98. Pompeo, F., Resasco, D. E. (2002). Water solubilization of single-walled carbon nanotubes by functionalization with glucosamine, *Nano Lett.*, *2*, 369–373.
99. Huang, W. J., Taylor, S., Fu, K. F., Lin, Y., Zhang, D. H., Hanks, T. W., Rao, A. M., Sun, Y. P. (2002). Attaching proteins to carbon nanotubes via diimide-activated amidation, *Nano Lett.*, *2*, 311–314.
100. Pantarotto, D., Partidos, C. D., Graff, R., Hoebeke, J., Briand, J. P., Prato, M., Bianco, A. (2003). Synthesis, structural characterization, and immunological properties of carbon nanotubes functionalized with peptides, *J. Am. Chem. Soc.*, *125*, 6160–6164.
101. Zeng, Y. L., Huang, Y. F., Jiang, J. H., Zhang, X. B., Tang, C. R., Shen, G. L., Yu, R. Q. (2007). Functionalization of multi-walled carbon nanotubes with poly(amidoamine) dendrimer for mediator-free glucose biosensor, *Electrochem. Commun.*, *9*, 185–190.
102. Lin, Y., Rao, A. M., Sadanadan, B., Kenik, E. A., Sun, Y. P. (2002). Functionalizing multiple-walled carbon nanotubes with aminopolymers, *J. Phys. Chem. B*, *106*, 1294–1298.
103. Amiran, J., Nicolosi, V., Bergin, S. D., Khan, U., Lyons, P. E., Coleman, J. N. (2008). High quality dispersions of functionalized single walled nanotubes at high concentration, *J. Phys. Chem. C*, *112*, 3519–3524.
104. Zhao, X. D., Fan, X. H., Chen, X. F., Chai, C. P., Zhou, Q. F. (2006). Surface modification of multiwalled carbon nanotubes via nitroxide-mediated radical polymerization, *J. Polym. Sci. A*, *44*, 4656–4667.

105. Lin, Y., Zhou, B., Fernando, K. A. S., Liu, P., Allard, L. F., Sun, Y. P. (2003). Polymeric carbon nanocomposites from carbon nanotubes functionalized with matrix polymer, *Macromolecules*, *36*, 7199–7204.
106. Ausman, K. D., Piner, R., Lourie, O., Ruoff, R. S., Korobov, M. (2000). Organic solvent dispersions of single-walled carbon nanotubes: toward solutions of pristine nanotubes, *J. Phys. Chem. B*, *104*, 8911–8915.
107. Giordani, S., Bergin, S. D., Nicolosi, V., Lebedkin, S., Kappes, M. M., Blau, W. J., Coleman, J. N. (2006). Debundling of single-walled nanotubes by dilution: observation of large populations of individual nanotubes in amide solvent dispersions, *J. Phys. Chem. B*, *110*, 15708–15718.
108. Wu, B., Bai, L., Gong, Q. M., Liang, J. (2009). Effect of non-ionic surfactants on the dispersion of multiwalled carbon nanotubes at high loading in ethanol, *Acta Physico-Chim. Sin.*, *25*, 1065–1069.
109. Cho, J., Daniel, I. M., Dikin, D. A. (2008). Effects of block copolymer dispersant and nanotube length on reinforcement of carbon/epoxy composites, *Compos. Part A*, *39*, 1844–1850.
110. Geng, Y., Liu, M. Y., Li, J., Shi, X. M., Kim, J. K. (2008). Effects of surfactant treatment on mechanical and electrical properties of CNT/epoxy nanocomposites, *Compos. Part A*, *39*, 1876–1883.
111. Chen, H., Muthuraman, H., Stokes, P., Zou, J. H., Liu, X., Wang, J. H., Huo, Q., Khondaker, S. I., Zhai, L. (2007). Dispersion of carbon nanotubes and polymer nanocomposite fabrication using trifluoroacetic acid as a co-solvent, *Nanotechnology*, *18*, 415606.
112. Kimura, M., Miki, N., Adachi, N., Tatewaki, Y., Ohta, K., Shirai, H. (2009). Organization of single-walled carbon nanotubes wrapped with liquid-crystalline pi-conjugated oligomers, *J. Mater. Chem.*, *19*, 1086–1092.
113. Chen, J., Ramasubramaniam, R., Xue, C., Liu, H. (2006). A versatile, molecular engineering approach to simultaneously enhanced, multifunctional carbon-nanotube–polymer composites, *Adv. Funct. Mater.*, *16*, 114–119.
114. Chen, J., Liu, H. Y., Weimer, W. A., Halls, M. D., Waldeck, D. H., Walker, G. C. (2002). Noncovalent engineering of carbon nanotube surfaces by rigid, functional conjugated polymers, *J. Am. Chem. Soc.*, *124*, 9034–9035.
115. Mao, J., Liu, Q., Lv, X., Liu, Z. F., Huang, Y., Ma, Y. F., Chen, Y. S., Yin, S. G. (2007). A water-soluble hybrid material of single-walled carbon nanotubes with an amphiphilic poly(phenyleneethynylene): preparation, characterization, and photovoltaic properties, *J. Nanosci. Nanotechnol.*, *7*, 2709–2718.
116. Kang, Y. K., Lee, O. S., Deria, P., Kim, S. H., Park, T. H., Bonnell, D. A., Saven, J. G., Therien, M. J. (2009). Helical wrapping of single-walled carbon nanotubes by water soluble poly(*p*-phenyleneethynylene), *Nano Lett.*, *9*, 1414–1418.
117. Sreekumar, T. V., Liu, T., Kumar, S., Ericson, L. M., Hauge, R. H., Smalley, R. E. (2003). Single-wall carbon nanotube films, *Chem. Mater.*, *15*, 175–178.
118. Davis, V. A., Ericson, L. M., Parra-Vasquez, A. N. G., Fan, H., Wang, Y. H., Prieto, V., Longoria, J. A., Ramesh, S., Saini, R. K., Kittrell, C., Billups, W. E., Adams, W. W., Hauge, R. H., Smalley, R. E., Pasquali, M. (2004). Phase behavior and rheology of SWNTs in superacids, *Macromolecules*, *37*, 154–160.
119. Jia, Z. J., Wang, Z. Y., Xu, C. L., Liang, J., Wei, B. Q., Wu, D. H., Zhu, S. W. (1999). Study on poly (methyl methacrylate)/carbon nanotube composites, *Mater. Sci. Eng. A*, *271*, 395–400.
120. Yao, Z. L., Braidy, N., Botton, G. A., Adronov, A. (2003). Polymerization from the surface of single-walled carbon nanotubes: preparation and characterization of nanocomposites, *J. Am. Chem. Soc.*, *125*, 16015–16024.
121. Jeong, J. Y., Lee, H. J., Kang, S. W., Tan, L. S., Baek, J. B. (2008). Nylon 610/functionalized multiwalled carbon nanotube composite prepared from *in-situ* interfacial polymerization, *J. Polym. Sci. A*, *46*, 6041–6050.
122. Kang, M., Myung, S. J., Jin, H. J. (2006). Nylon 610 and carbon nanotube composite by *in situ* interfacial polymerization, *Polymer*, *47*, 3961–3966.
123. Haggenmueller, R., Du, F. M., Fischer, J. E., Winey, K. I. (2006). Interfacial *in situ* polymerization of single wall carbon nanotube/nylon 6,6 nanocomposites, *Polymer*, *47*, 2381–2388.
124. Schadler, L. S., Giannaris, S. C., Ajayan, P. M. (1998). Load transfer in carbon nanotube epoxy composites, *Appl. Phys. Lett.*, *73*, 3842–3844.

125. Zhu, J., Peng, H. Q., Rodriguez-Macias, F., Margrave, J. L., Khabashesku, V. N., Imam, A. M., Lozano, K., Barrera, E. V. (2004). Reinforcing epoxy polymer composites through covalent integration of functionalized nanotubes, *Adv. Funct. Mater.*, *14*, 643–648.
126. Seyhan, A. T., Gojny, F. H., Tanoglu, M., Schulte, K. (2007). Rheological and dynamic-mechanical behavior of carbon nanotube/vinyl ester–polyester suspensions and their nanocomposites, *Eur. Polym. J.*, *43*, 2836–2847.
127. Gojny, F. H., Wichmann, M. H. G., Fiedler, B., Kinloch, I. A., Bauhofer, W., Windle, A. H., Schulte, K. (2006). Evaluation and identification of electrical and thermal conduction mechanisms in carbon nanotube/epoxy composites, *Polymer*, *47*, 2036–2045.
128. Bryning, M. B., Islam, M. F., Kikkawa, J. M., Yodh, A. G. (2005). Very low conductivity threshold in bulk isotropic single-walled carbon nanotube–epoxy composites, *Adv. Mater.*, *17*, 1186–1191.
129. Kim, Y. J., Shin, T. S., Choi, H. D., Kwon, J. H., Chung, Y. C., Yoon, H. G. (2005). Electrical conductivity of chemically modified multiwalled carbon nanotube/epoxy composites, *Carbon*, *43*, 23–30.
130. Pumera, M., Merkoci, A., Alegret, S. (2006). Carbon nanotube–epoxy composites for electrochemical sensing, *Sens. Actuators B*, *113*, 617–622.
131. Gryshchuk, O., Karger-Kocsis, J., Thomann, R., Konya, Z., Kiricsi, I. (2006). Multiwall carbon nanotube modified vinylester and vinylester-based hybrid resins, *Compos. Part A*, *37*, 1252–1259.
132. Thostenson, E. T., Ziaee, S., Chou, T. W. (2009). Processing and electrical properties of carbon nanotube/vinyl ester nanocomposites, *Compos. Sci. Technol.*, *69*, 801–804.
133. Grujicic, M., Sun, Y. P., Koudela, K. L. (2007). The effect of covalent functionalization of carbon nanotube reinforcements on the atomic-level mechanical properties of poly-vinyl-ester-epoxy, *Appl. Surf. Sci.*, *253*, 3009–3021.
134. Park, C., Ounaies, Z., Watson, K. A., Crooks, R. E., Smith, J., Lowther, S. E., Connell, J. W., Siochi, E. J., Harrison, J. S., Clair, T. L. S. (2002). Dispersion of single wall carbon nanotubes by *in situ* polymerization under sonication, *Chem. Phys. Lett.*, *364*, 303–308.
135. Wise, K. E., Park, C., Siochi, E. J., Harrison, J. S. (2004). Stable dispersion of single wall carbon nanotubes in polyimide: the role of noncovalent interactions, *Chem. Phys. Lett.*, *391*, 207–211.
136. Zhu, B. K., Xie, S. H., Xu, Z. K., Xu, Y. Y. (2006). Preparation and properties of the polyimide/multi-walled carbon nanotubes (MWNTs) nanocomposites, *Compos. Sci. Technol.*, *66*, 548–554.
137. Choi, H. J., Zhang, K., Lim, J. Y. (2007). Multi-walled carbon nanotube/polystyrene composites prepared by *in-situ* bulk sonochemical polymerization, *J. Nanosci. Nanotechnol.*, *7*, 3400–3403.
138. Liu, P. (2009). Facile graft polystyrene onto multi-walled carbon nanotubes via *in situ* thermo-induced radical polymerization, *J. Nanopart. Res.*, *11*, 1011–1016.
139. Kwon, S. M., Kim, H. S., Myung, S. J., Jin, H. J. (2008). Poly(methyl methacrylate)/multiwalled carbon nanotube microspheres fabricated via *in-situ* dispersion polymerization, *J. Polym. Sci. Polym. Phys.*, *46*, 182–189.
140. Xia, H. S., Qiu, G. H., Wang, Q. (2006). Polymer/carbon nanotube composite emulsion prepared through ultrasonically assisted *in situ* emulsion polymerization, *J. Appl. Polym. Sci.*, *100*, 3123–3130.
141. Wu, T. M., Lin, Y. W. (2006). Doped polyaniline/multi-walled carbon nanotube composites: preparation, characterization and properties, *Polymer*, *47*, 3576–3582.
142. Konyushenko, E. N., Stejskal, J., Trchova, M., Hradil, J., Kovarova, J., Prokes, J., Cieslar, M., Hwang, J. Y., Chen, K. H., Sapurina, I. (2006). Multi-wall carbon nanotubes coated with polyaniline, *Polymer*, *47*, 5715–5723.
143. Bliznyuk, V. N., Singamaneni, S., Sanford, R. L., Chiappetta, D., Crooker, B., Shibaev, P. V. (2006). Matrix mediated alignment of single wall carbon nanotubes in polymer composite films, *Polymer*, *47*, 3915–3921.
144. Lafuente, E., Pinol, M., Oriol, L., Munoz, E., Benito, A. M., Maser, W. K., Dalton, A. B., Serrano, J. L., Martinez, M. T. (2006). Polyazomethine/carbon nanotube composites, *Mater. Sci. Eng. C*, *26*, 1198–1201.
145. Song, Y. S., Youn, J. R. (2005). Influence of dispersion states of carbon nanotubes on physical properties of epoxy nanocomposites, *Carbon*, *43*, 1378–1385.
146. Lau, K. T., Lu, M., Lam, C. K., Cheung, H. Y., Sheng, F. L., Li, H. L. (2005). Thermal and mechanical properties of single-walled carbon nanotube bundle-reinforced epoxy nanocomposites: the role of solvent for nanotube dispersion, *Compos. Sci. Technol.*, *65*, 719–725.

147. Ma, P. C., Kim, J. K., Tang, B. Z. (2007). Effects of silane functionalization on the properties of carbon nanotube/epoxy nanocomposites, *Compos. Sci. Technol.*, *67*, 2965–2972.
148. Gojny, F. H., Schulte, K. (2004). Functionalisation effect on the thermo-mechanical behaviour of multi-wall carbon nanotube/epoxy-composites, *Compos. Sci. Technol.*, *64*, 2303–2308.
149. Liu, X. Q., Chan-Park, M. B. (2009). Facile way To disperse single-walled carbon nanotubes using a noncovalent method and their reinforcing effect in poly(methyl methacrylate) composites, *J. Appl. Polym. Sci.*, *114*, 3414–3419.
150. Liu, L., Etika, K. C., Liao, K. S., Hess, L. A., Bergbreiter, D. E., Grunlan, J. C. (2009). Comparison of covalently and noncovalently functionalized carbon nanotubes in epoxy, *Macromol. Rapid Commun.*, *30*, 627–632.
151. Chowdhury, S. R., Chen, Y., Wang, Y., Mitra, S. (2009). Microwave-induced rapid nanocomposite synthesis using dispersed single-wall carbon nanotubes as the nuclei, *J. Mater. Sci.*, *44*, 1245–1250.
152. Yuan, J. M., Fan, Z. F., Chen, X. H. C., X. H., Wu, Z. J., He, L. P. (2009). Preparation of polystyrene–multiwalled carbon nanotube composites with individual-dispersed nanotubes and strong interfacial adhesion, *Polymer*, *50*, 3285–3291.
153. Wang, Y. B., Iqbal, Z., Mitra, S. (2005). Microwave-induced rapid chemical functionalization of single-walled carbon nanotubes, *Carbon*, *43*, 1015–1020.
154. Tian, R., Wang, X. B., Xu, Y., Li, S. Q., Wan, L., Li, M. J., Cheng, J. (2009). Microwave-assisted functionalization of single-walled carbon nanotubes with 3-chloropropene, *J. Nanopart. Res.*, *11*, 1201–1208.
155. Higginbotham, A. L., Moloney, P. G., Waid, M. C., Duque, J. G., Kittrell, C., Schmidt, H. K., Stephenson, J. J., Arepalli, S., Yowell, L. L., Tour, J. M. (2008). Carbon nanotube composite curing through absorption of microwave radiation, *Compos. Sci. Technol.*, *68*, 3087–3092.
156. Bonduel, D., Mainil, M. L., Alexandre, M., Monteverde, F., Dubois, P. (2005). Supported coordination polymerization: a unique way to potent polyolefin carbon nanotube nanocomposites, *Chem. Commun.* 781–783.
157. Trujillo, M., Arnal, M. L., Mueller, A. J., Bredeau, S., Bonduel, D., Dubois, P., Hamley, I. W., Castelletto, V. (2008). Thermal fractionation and isothermal crystallization of polyethylene nanocomposites prepared by *in situ* polymerization, *Macromolecules*, *41*, 2087–2095.
158. Bonduel, D., Bredeau, S., Alexandre, M., Monteverde, F., Dubois, P. (2007). Supported metallocene catalysis as an efficient tool for the preparation of polyethylene/carbon nanotube nanocomposites: effect of the catalytic system on the coating morphology, *J. Mater. Chem.*, *17*, 2359–2366.
159. Bredeau, S., Peeterbroeck, S., Bonduel, D., Alexandre, M., Dubois, P. (2008). From carbon nanotube coatings to high-performance polymer nanocomposites, *Polym. Int.*, *57*, 547–553.
160. Shieh, Y. T., Liu, G. L., Hwang, K. C., Chen, C. C. (2005). Crystallization, melting and morphology of PEO in PEO/MWNT-*g*-PMMA blends, *Polymer*, *46*, 10945–10951.
161. Fragneaud, B., Masenelli-Varlot, K., Gonzalez-Montiel, A., Terrones, M., Cavaille, J. Y. (2008). Mechanical behavior of polystyrene grafted carbon nanotubes/polystyrene nanocomposites, *Compos. Sci. Technol.*, *68*, 3265–3271.
162. Hwang, G. L., Shieh, Y. T., Hwang, K. C. (2004). Efficient load transfer to polymer-grafted multiwalled carbon nanotubes in polymer composites, *Adv. Funct. Mater.*, *14*, 487–491.
163. Trujillo, M., Arnal, M. L., Muller, A. J., Laredo, E., Bredeau, S., Bonduel, D., Dubois, P. (2007). Thermal and morphological characterization of nanocomposites prepared by *in-situ* polymerization of high-density polyethylene on carbon nanotubes, *Macromolecules*, *40*, 6268–6276.
164. Ferrer-Anglada, N., Kaempgen, M., Skakalova, V., Dettlaf-Weglikowska, U., Roth, S. (2004). Synthesis and characterization of carbon nanotube–conducting polymer thin films, *Diam. Relat. Mater.*, *13*, 256–260.
165. Kaempgen, M., Roth, S. (2006). Transparent and flexible carbon nanotube/polyaniline pH sensors, *J. Electroanal. Chem.*, *586*, 72–76.
166. Wu, M. Q., Snook, G. A., Gupta, V., Shaffer, M., Fray, D. J., Chen, G. Z. (2005). Electrochemical fabrication and capacitance of composite films of carbon nanotubes and polyaniline, *J. Mater. Chem.*, *15*, 2297–2303.
167. Branzoi, V., Pilan, L., Branzoi, F. (2009). Nanocomposite films obtained by electrochemical codeposition of conducting polymers and carbon nanotubes, *Electroanalysis*, *21*, 557–562.

168. Wei, D., Kvarnstrom, C., Lindfors, T., Ivaska, A. (2007). Electrochemical functionalization of single walled carbon nanotubes with polyaniline in ionic liquids, *Electrochem. Commun.*, *9*, 206–210.
169. Hughes, M., Chen, G. Z., Shaffer, M. S. P., Fray, D. J., Windle, A. H. (2004). Controlling the nanostructure of electrochemically grown nanoporous composites of carbon nanotubes and conducting polymers, *Compos. Sci. Technol.*, *64*, 2325–2331.
170. Canobre, S. C., Almeida, D. A. L., Fonseca, C. P., Neves, S. (2009). Synthesis and characterization of hybrid composites based on carbon nanotubes, *Electrochim. Acta*, *54*, 6383–6388.
171. Gopalan, A. Y., Lee, K. P., Santhosh, P., Kim, K. S., Nho, Y. C. (2007). Different types of molecular interactions in carbon nanotube/conducting polymer composites: a close analysis, *Compos. Sci. Technol.*, *67*, 900–905.
172. Baibarac, M., Baltog, I., Lefrant, S., Mevellec, J. Y., Chauvet, O. (2003). Polyaniline and carbon nanotubes based composites containing whole units and fragments of nanotubes, *Chem. Mater.*, *15*, 4149–4156.
173. Baibarac, M., Baltog, I., Godon, C., Lefrant, S., Chauvet, O. (2004). Covalent functionalization of single-walled carbon nanotubes by aniline electrochemical polymerization, *Carbon*, *42*, 3143–3152.
174. Martin, C. A., Sandler, J. K. W., Shaffer, M. S. P., Schwarz, M. K., Bauhofer, W., Schulte, K., Windle, A. H. (2004). Formation of percolating networks in multi-wall carbon-nanotube–epoxy composites, *Compos. Sci. Technol.*, *64*, 2309–2316.
175. Ramanathan, T., Liu, H., Brinson, L. C. (2005). Functionalized SWNT/polymer nanocomposites for dramatic property improvement, *J. Polym. Sci. Polym. Phys.*, *43*, 2269–2279.
176. Pham, J. Q., Mitchell, C. A., Bahr, J. L., Tour, J. M., Krishanamoorti, R., Green, P. F. (2003). Glass transition of polymer/single-walled carbon nanotube composite films, *J. Polym. Sci. Polym. Phys.*, *41*, 3339–3345.
177. Grady, B. P., Pompeo, F., Shambaugh, R. L., Resasco, D. E. (2002). Nucleation of polypropylene crystallization by single-walled carbon nanotubes, *J. Phys. Chem. B*, *106*, 5852–5858.
178. Liu, J., Liu, T., Kumar, S. (2005). Effect of solvent solubility parameter on SWNT dispersion in PMMA, *Polymer*, *46*, 3419–3424.
179. Chang, T. E., Kisliuk, A., Rhodes, S. M., Brittain, W. J., Sokolov, A. P. (2006). Conductivity and mechanical properties of well-dispersed single-wall carbon nanotube/polystyrene composite, *Polymer*, *47*, 7740–7746.
180. Singh, S., Pei, Y. Q., Miller, R., Sundararajan, P. R. (2003). Long-range, entangled carbon nanotube networks in polycarbonate, *Adv. Funct. Mater.*, *13*, 868–872.
181. Tchoul, M. N., Ford, W. T., Ha, M. L. P., Chavez-Sumarriva, I., Grady, B. P., Lolli, G. L., Resasco, D. E., Arepalli, S. (2008). Composites of single walled carbon nanotubes and polystyrene: preparation and electrical conductivity, *Chem. Mater.*, *20*, 3120–3126.
182. Zhang, X. F., Liu, T., Sreekumar, T. V., Kumar, S., Hu, X. D., Smith, K. (2004). Gel spinning of PVA/SWNT composite fiber, *Polymer*, *45*, 8801–8807.
183. Ruan, S. L., Gao, P., Yu, T. X. (2006). Ultra-strong gel-spun UHMWPE fibers reinforced using multiwalled carbon nanotubes, *Polymer*, *47*, 1604–1611.
184. Razal, J. M., Coleman, J. N., Munoz, E., Lund, B., Gogotsi, Y., Ye, H., Collins, S., Dalton, A. B., Baughman, R. H. (2007). Arbitrarily shaped fiber assemblies from spun carbon nanotube gel fibers, *Adv. Funct. Mater.*, *17*, 2918–2924.
185. Chae, H. G., Minus, M. L., Rasheed, A., Kumar, S. (2007). Stabilization and carbonization of gel spun polyacrylonitrile/single wall carbon nanotube composite fibers, *Polymer*, *48*, 3781–3789.
186. Saeed, K., Park, S. Y., Lee, H. J., Baek, J. B., Huh, W. S. (2006). Preparation of electrospun nanofibers of carbon nanotube/polycaprolactone nanocomposite, *Polymer*, *47*, 8019–8025.
187. Naebe, M., Lin, T., Staiger, M. P., Dai, L. M., Wang, X. G. (2008). Electrospun single-walled carbon nanotube/polyvinyl alcohol composite nanofibers: structure–property relationships, *Nanotechnology*, *19*, 305702.
188. Vigolo, B., Penicaud, A., Coulon, C., Sauder, C., Pailler, R., Journet, C., Bernier, P., Poulin, P. (2000). Macroscopic fibers and ribbons of oriented carbon nanotubes, *Science*, *290*, 1331–1334.
189. Vigolo, B., Poulin, P., Lucas, M., Launois, P., Bernier, P. (2002). Improved structure and properties of single-wall carbon nanotube spun fibers, *Appl. Phys. Lett.*, *81*, 1210–1212.
190. Poulin, P., Vigolo, B., Launois, P. (2002). Films and fibers of oriented single wall nanotubes, *Carbon*, *40*, 1741–1749.

191. Sugikawa, K., Numata, M., Kaneko, K., Sada, K., Shinkai, S. (2008). Alternate layer-by-layer adsorption of single- and double-walled carbon nanotubes wrapped by functionalized beta-1,3-glucan polysaccharides, *Langmuir, 24*, 13270–13275.
192. Gheith, M. K., Sinani, V. A., Wicksted, J. P., Matts, R. L., Kotov, N. A. (2005). Single-walled carbon nanotube polyelectrolyte multilayers and freestanding films as a biocompatible platform for neuroprosthetic implants, *Adv. Mater., 17*, 2663–2670.
193. Paloniemi, H., Lukkarinen, M., Aaritalo, T., Areva, S., Leiro, J., Heinonen, M., Haapakka, K., Lukkari, J. (2006). Layer-by-layer electrostatic self-assembly of single-wall carbon nanotube polyelectrolytes, *Langmuir*, 22, 74–83.
194. Shi, J. H., Chen, Z. Y., Qin, Y. J., Guo, Z. X. (2008). Multiwalled carbon nanotube microspheres from layer-by-layer assembly and calcination, *J. Phys. Chem. C, 112*, 11617–11622.
195. Kim, B. S., Kim, B., Suh, K. D. (2008). Electrorheological properties of carbon nanotube/polyelectrolyte self-assembled polystyrene particles by layer-by-layer assembly, *J. Polym. Sci. A, 46*, 1058–1065.
196. Shim, B. S., Zhu, J., Jan, E., Critchley, K., Ho, S. S., Podsiadlo, P., Sun, K., Kotov, N. A. (2009). Multiparameter structural optimization of single-walled carbon nanotube composites: toward record strength, stiffness, and toughness, *ACS Nano, 3*, 1711–1722.
197. Shim, B. S., Kotov, N. A. (2005). Single-walled carbon nanotube combing during layer-by-layer assembly: from random adsorption to aligned composites, *Langmuir, 21*, 9381–9385.
198. Tian, Y., Park, J. G., Cheng, Q. F., Liang, Z. Y., Zhang, C., Wang, B. (2009). The fabrication of single-walled carbon nanotube/polyelectrolyte multilayer composites by layer-by-layer assembly and magnetic field assisted alignment, *Nanotechnology, 20*, 335601.
199. Moya, S. E., Ilie, A., Bendall, J. S., Hernandez-Lopez, J. L., Ruiz-Garcia, J., Huck, W. T. S. (2007). Assembly of polyelectrolytes on CNTs by Van der Waals interactions and fabrication of LBL polyelectrolyte/CNT composites, *Macromol. Chem. Phys., 208*, 603–608.
200. O'Connor, I., De, S., Coleman, J. N., Gun'ko, Y. K. (2009). Development of transparent, conducting composites by surface infiltration of nanotubes into commercial polymer films, *Carbon, 47*, 1983–1988.
201. Zhang, F., Zhang, H., Zhang, Z. W., Chen, Z. M., Xu, Q. (2008). Modification of carbon nanotubes: water-soluble polymers nanocrystal wrapping to periodic patterning with assistance of supercritical CO_2, *Macromolecules, 41*, 4519–4523.
202. Yue, B. H., Wang, Y. B., Huang, C. Y., Pfeffer, R., Iqbal, Z. (2007). Polymeric nanocomposites of functionalized carbon nanotubes synthesized in supercritical CO_2, *J. Nanosci. Nanotechnol., 7*, 994–1000.
203. Yue, J., Xu, Q., Zhang, Z. W., Chen, Z. M. (2007). Periodic patterning on carbon nanotubes: supercritical CO_2-induced polyethylene epitaxy, *Macromolecules, 40*, 8821–8826.
204. Lu, K. B., Grossiord, N., Koning, C. E., Miltner, H. E., van Mele, B., Loos, J. (2008). Carbon nanotube/isotactic polypropylene composites prepared by latex technology: morphology analysis of CNT-induced nucleation, *Macromolecules, 41*, 8081–8085.
205. Xia, H. S., Wang, Q., Qiu, G. H. (2003). Polymer-encapsulated carbon nanotubes prepared through ultrasonically initiated *in situ* emulsion polymerization, *Chem. Mater., 15*, 3879–3886.
206. Ham, H. T., Choi, Y. S., Jeong, N., Chung, I. J. (2005). Singlewall carbon nanotubes covered with polypyrrole nanoparticles by the miniemulsion polymerization, *Polymer, 46*, 6308–6315.
207. Zhang, K., Lim, J. Y., Park, B. J., Jin, H. J., Choi, H. J. (2009). Carboxylic acid functionalized multi-walled carbon nanotube-adsorption onto poly(methyl methacrylate) microspheres, *J. Nanosci. Nanotechnol., 9*, 1058–1061.
208. Grunlan, J. C., Mehrabi, A. R., Bannon, M. V., Bahr, J. L. (2004). Water-based single-walled-nanotube-filled polymer composite with an exceptionally low percolation threshold, *Adv. Mater., 16*, 150–153.
209. Regev, O., ElKati, P. N. B., Loos, J., Koning, C. E. (2004). Preparation of conductive nanotube–polymer composites using latex technology, *Adv. Mater., 16*, 248–251.
210. Han, S. J., Kim, B., Suh, K. D. (2007). Electrical properties of a composite film of poly(acrylonitrile) nanoparticles coated with carbon nanotubes, *Macromol. Chem. Phys., 208*, 377–383.
211. Zhao, C. X., Zhang, W. D., Sun, D. C. (2009). Preparation and mechanical properties of waterborne polyurethane/carbon nanotube composites, *Polym. Compos., 30*, 649–654.

212. Zhang, Q. H., Lippits, D. R., Rastogi, S. (2006). Dispersion and rheological aspects of SWNTs in ultrahigh molecular weight polyethylene, *Macromolecules, 39*, 658–666.

213. Mierczynska, A., Friedrich, J., Maneck, H. E., Boiteux, G., Jeszka, J. K. (2004). Segregated network polymer/carbon nanotubes composites, *Cent. Eur. J. Chem., 2*, 363–370.

214. Jung, R., Park, W. I., Kwon, S. M., Kim, H. S., Jin, H. J. (2008). Location-selective incorporation of multiwalled carbon nanotubes in polycarbonate microspheres, *Polymer, 49*, 2071–2076.

215. Hobbie, E. K., Fagan, J. A., Obrzut, J., Hudson, S. D. (2009). Microscale polymer–nanotube composites, *ACS Appl. Mater. Interfaces, 1*, 1561–1566.

216. Hermant, M. C., Verhulst, M., Kyrylyuk, A. V., Klumperman, B., Koning, C. E. (2009). The incorporation of single-walled carbon nanotubes into polymerized high internal phase emulsions to create conductive foams with a low percolation threshold, *Compos. Sci. Technol., 69*, 656–662.

217. Hermant, M. C., Klumperman, B., Koning, C. E. (2009). Conductive pickering-poly(high internal phase emulsion) composite foams prepared with low loadings of single-walled carbon nanotubes, *Chem. Commun.* 2738–2740.

218. Grossiord, N., Miltner, H. E., Loos, J., Meuldijk, J., Van Mele, B., Koning, C. E. (2007). On the crucial role of wetting in the preparation of conductive polystyrene–carbon nanotube composites, *Chem. Mater., 19*, 3787–3792.

219. Grunlan, J. C., Kim, Y. S., Ziaee, S., Wei, X., Abdel-Magid, B., Tao, K. (2006). Thermal and mechanical behavior of carbon-nanotube-filled latex, *Macromol. Mater. Eng., 291*, 1035–1043.

220. Mu, M. F., Walker, A. M., Torkelson, J. M., Winey, K. I. (2008). Cellular structures of carbon nanotubes in a polymer matrix improve properties relative to composites with dispersed nanotubes, *Polymer, 49*, 1332–1337.

221. Grossiord, N., Kivit, P. J. J., Loos, J., Meuldijk, J., Kyrylyuk, A. V., van der Schoot, P., Koning, C. E. (2008). On the influence of the processing conditions on the performance of electrically conductive carbon nanotube/polymer nanocomposites, *Polymer, 49*, 2866–2872.

222. Kim, S., Drzal, L. T. (2009). Comparison of exfoliated graphite nanoplatelets (xGnP) and CNTs for reinforcement of EVA nanocomposites fabricated by solution compounding method and three screw rotating systems, *J. Adhes. Sci. Technol., 23*, 1623–1638.

223. Lin, B., Sundararaj, U., Potschke, P. (2006). Melt mixing of polycarbonate with multi-walled carbon nanotubes in miniature mixers, *Macromol. Mater. Eng., 291*, 227–238.

224. Kasaliwal, G., Godel, A., Pötschke, P. (2009). Influence of processing conditions in small-scale melt mixing and compression molding on the resistivity and morphology of polycarbonate–MWNT composites, *J. Appl. Polym. Sci., 112*, 3494–3509.

225. Pötschke, P., Bhattacharyya, A. R., Janke, A., Pegel, S., Leonhardt, A., Taschner, C., Ritschel, M., Roth, S., Hornbostel, B., Cech, J. (2005). Melt mixing as method to disperse carbon nanotubes into thermoplastic polymers, *Fullerenes Nanotubes Carbon Nanostruct., 13*, 211–224.

226. Micusik, M., Omastova, M., Krupa, I., Prokes, J., Pissis, P., Logakis, E., Pandis, C., Pötschke, P., Pionteck, J. (2009). A comparative study on the electrical and mechanical behaviour of multi-walled carbon nanotube composites prepared by diluting a masterbatch with various types of polypropylenes, *J. Appl. Polym. Sci., 113*, 2536–2551.

227. Villmow, T., Pötschke, P., Pegel, S., Haussler, L., Kretzschmar, B. (2008). Influence of twin-screw extrusion conditions on the dispersion of multi-walled carbon nanotubes in a poly(lactic acid) matrix, *Polymer, 49*, 3500–3509.

228. Prashantha, K., Soulestin, J., Lacrampe, M. F., Claes, M., Dupin, G., Krawczak, P. (2008). Multi-walled carbon nanotube filled polypropylene nanocomposites based on masterbatch route: improvement of dispersion and mechanical properties through PP-*g*-MA addition, *Express Polym. Lett., 2*, 735–745.

229. Wu, D. F., Sun, Y. R., Zhang, M. (2009). Kinetics study on melt compounding of carbon nanotube/polypropylene nanocomposites, *J. Polym. Sci. Polym. Phys., 47*, 608–618.

230. Lee, S. H., Kim, M. W., Kim, S. H., Youn, J. R. (2008). Rheological and electrical properties of polypropylene/MWCNT composites prepared with MWCNT masterbatch chips, *Eur. Polym. J., 44*, 1620–1630.

231. Hong, J. S., Kim, C. (2007). Extension-induced dispersion of multi-walled carbon nanotube in non-Newtonian fluid, *J. Rheol., 51*, 833–850.

232. Isayev, A. I., Kumar, R., Lewis, T. M. (2009). Ultrasound assisted twin screw extrusion of polymer–nanocomposites containing carbon nanotubes, *Polymer, 50*, 250–260.
233. Pötschke, P., Bhattacharyya, A. R., Janke, A. (2004). Carbon nanotube-filled polycarbonate composites produced by melt mixing and their use in blends with polyethylene, *Carbon, 42*, 965–969.
234. Goldel, A., Kasaliwal, G., Pötschke, P. (2009). Selective localization and migration of multiwalled carbon nanotubes in blends of polycarbonate and poly(styrene–acrylonitrile), *Macromol. Rapid Commun., 30*, 423–429.
235. Bose, S., Bhattacharyya, A. R., Khare, R. A., Kulkarni, A. R., Patro, T. U., Sivaraman, P. (2008). Tuning the dispersion of multiwall carbon nanotubes in co-continuous polymer blends: a generic approach, *Nanotechnology, 19*, 335704.
236. Khare, R. A., Bhattacharyya, A. R., Kulkarni, A. R., Sarp, M., Biswas, A. (2008). Influence of multiwall carbon nanotubes on morphology and electrical conductivity of PP/ABS blends, *J. Polym. Sci. Polym. Phys., 46*, 2286–2295.
237. Alig, I., Skipa, T., Engel, M., Lellinger, D., Pegel, S., Pötschke, P. (2007). Electrical conductivity recovery in carbon nanotube–polymer composites after transient shear, *Phys. Status Solidi B, 244*, 4223–4226.
238. Alig, I., Lellinger, D., Dudkin, S. M., Pötschke, P. (2007). Conductivity spectroscopy on melt processed polypropylene–multiwalled carbon nanotube composites: recovery after shear and crystallization, *Polymer, 48*, 1020–1029.
239. Alig, I., Skipa, T., Lellinger, D., Pötschke, P. (2008). Destruction and formation of a carbon nanotube network in polymer melts: rheology and conductivity spectroscopy, *Polymer, 49*, 3524–3532.
240. Lellinger, D., Xu, D. H., Ohneiser, A., Skipa, T., Alig, I. (2008). Influence of the injection moulding conditions on the in-line measured electrical conductivity of polymer–carbon nanotube composites, *Phys. Status Solidi B, 245*, 2268–2271.
241. Cipriano, B. H., Kota, A. K., Gershon, A. L., Laskowski, C. J., Kashiwagi, T., Bruck, H. A., Raghavan, S. R. (2008). Conductivity enhancement of carbon nanotube and nanofiber-based polymer nanocomposites by melt annealing, *Polymer, 49*, 4846–4851.
242. Grady, B. P., Arthur, D. J., Ferguson, J. (2009). Single-walled carbon nanotube/ultrahigh-molecular-weight polyethylene composites with percolation at low nanotube contents, *Polym. Eng. Sci., 49*, 2440–2446.
243. Mierczynska, A., Mayne-L'Hermite, M., Boiteux, G., Jeszka, J. K. (2007). Electrical and mechanical properties of carbon nanotube/ultrahigh-molecular-weight polyethylene composites prepared by a filler prelocalization method, *J. Appl. Polym. Sci., 105*, 158–168.
244. Lisunova, M. O., Mamunya, Y. P., Lebovka, N. I., Melezhyk, A. V. (2007). Percolation behaviour of ultrahigh molecular weight polyethylene/multi-walled carbon nanotubes composites, *Eur. Polym. J., 43*, 949–958.
245. Gorrasi, G., Sarno, M., Di Bartolomeo, A., Sannino, D., Ciambelli, P., Vittoria, V. (2007). Incorporation of carbon nanotubes into polyethylene by high energy ball milling: morphology and physical properties, *J. Polym. Sci. Polym. Phys., 45*, 597–606.
246. Olmos, D., Dominguez, C., Castrillo, P. D., Gonzalez-Benito, J. (2009). Crystallization and final morphology of HDPE: effect of the high energy ball milling and the presence of TiO_2 nanoparticles, *Polymer, 50*, 1732–1742.
247. Xia, H. S., Wang, Q., Li, K. S., Hu, G. H. (2004). Preparation of polypropylene/carbon nanotube composite powder with a solid-state mechanochemical pulverization process, *J. Appl. Polym. Sci., 93*, 378–386.
248. Shao, W. G., Wang, Q., Wang, F., Chen, Y. H. (2006). The cutting of multi-walled carbon nanotubes and their strong interfacial interaction with polyamide 6 in the solid state, *Carbon, 44*, 2708–2714.
249. Masuda, J., Torkelson, J. M. (2008). Dispersion and major property enhancements in polymer/multiwall carbon nanotube nanocomposites via solid-state shear pulverization followed by melt mixing, *Macromolecules, 41*, 5974–5977.
250. Pujari, S., Ramanathan, T., Kasimatis, K., Masuda, J., Andrews, R., Torkelson, J. M., Brinson, L. C., Burghardt, W. R. (2009). Preparation and characterization of multiwalled carbon nanotube dispersions in polypropylene: melt mixing versus solid-state shear pulverization, *J. Polym. Sci. Polym. Phys., 47*, 1426–1436.

251. Nasibulin, A. G., Ollikainen, A., Anisimov, A. S., Brown, D. P., Pikhitsa, P. V., Holopainen, S., Penttila, J. S., Helisto, P., Ruokolainen, J., Choi, M., Kauppinen, E. I. (2008). Integration of single-walled carbon nanotubes into polymer films by thermo-compression, *Chem. Eng. J.*, *136*, 409–413.
252. Song, L., Zhang, H., Zhang, Z., Xie, S. S. (2007). Processing and performance improvements of SWNT paper reinforced PEEK nanocomposites, *Compos. Part A*, *38*, 388–392.
253. Koziol, K., Vilatela, J., Moisala, A., Motta, M., Cunniff, P., Sennett, M., Windle, A. (2007). High-performance carbon nanotube fiber, *Science*, *318*, 1892–1895.
254. Zhang, S., Zhu, L., Minus, M. L., Chae, H. G., Jagannathan, S., Wong, C. P., Kowalik, J., Roberson, L. B., Kumar, S. (2008). Solid-state spun fibers and yarns from 1-mm long carbon nanotube forests synthesized by water-assisted chemical vapor deposition, *J. Mater. Sci.*, *43*, 4356–4362.
255. Zhang, X. F., Li, Q. W., Tu, Y., Li, Y. A., Coulter, J. Y., Zheng, L. X., Zhao, Y. H., Jia, Q. X., Peterson, D. E., Zhu, Y. T. (2007). Strong carbon-nanotube fibers spun from long carbon-nanotube arrays, *Small*, *3*, 244–248.
256. Jang, E. Y., Kang, T. J., Im, H., Baek, S. J., Kim, S., Jeong, D. H., Park, Y. W., Kim, Y. H. (2009). Macroscopic single-walled-carbon-nanotube fiber self-assembled by dip-coating method, *Adv. Mater.*, *21*, 4357–4361.
257. Ericson, L. M., Fan, H., Peng, H. Q., Davis, V. A., Zhou, W., Sulpizio, J., Wang, Y. H., Booker, R., Vavro, J., Guthy, C., Parra-Vasquez, A. N. G., Kim, M. J., Ramesh, S., Saini, R. K., Kittrell, C., Lavin, G., Schmidt, H., Adams, W. W., Billups, W. E., Pasquali, M., Hwang, W. F., Hauge, R. H., Fischer, J. F., Smalley, R. E. (2004). Macroscopic, neat, single walled carbon nanotube fibers, *Science*, *305*, 1447–1450.
258. Kozlov, M. E., Capps, R. C., Sampson, W. M., Ebron, V. H., Ferraris, J. P., Baughman, R. H. (2005). Spinning solid and hollow polymer-free carbon nanotube fibers, *Adv. Mater.*, *17*, 614–617.
259. Zhang, S. J., Koziol, K. K. K., Kinloch, I. A., Windle, A. H. (2008). Macroscopic fibers of well-aligned carbon nanotubes by wet spinning, *Small*, *4*, 1217–1222.
260. Ma, W. J., Song, L., Yang, R., Zhang, T. H., Zhao, Y. C., Sun, L. F., Ren, Y., Liu, D. F., Liu, L. F., Shen, J., Zhang, Z. X., Xiang, Y. J., Zhou, W. Y., Xie, S. S. (2007). Directly synthesized strong, highly conducting, transparent single walled carbon nanotube films, *Nano Lett.*, *7*, 2307–2311.
261. Ma, W. J., Liu, L. Q., Yang, R., Zhang, T. H., Zhang, Z., Song, L., Ren, Y., Shen, J., Niu, Z. Q., Zhou, W. Y., Xie, S. S. (2009). Monitoring a micromechanical process in macroscale carbon nanotube films and fibers, *Adv. Mater.*, *21*, 603–608.
262. Bogdanovich, A., Bradford, P., Mungalov, D., Fang, S. L., Zhang, M., Baughman, R. H., Hudson, S. (2007). Fabrication and mechanical characterization of carbon nanotube yarns, 3-D braids, and their composites, *SAMPE J.*, *43*, 6–19.
263. Zhang, S. J., Zhu, L. B., Wong, C. P., Kumar, S. (2009). Polymer-infiltrated aligned carbon nanotube fibers by *in situ* polymerization, *Macromol. Rapid Commun.*, *30*, 1936–1939.
264. Mora, R. J., Vilatela, J. J., Windle, A. H. (2009). Properties of composites of carbon nanotube fibres, *Compos. Sci. Technol.*, *69*, 1558–1563.
265. Wu, Z. C., Chen, Z. H., Du, X., Logan, J. M., Sippel, J., Nikolou, M., Kamaras, K., Reynolds, J. R., Tanner, D. B., Hebard, A. F., Rinzler, A. G. (2004). Transparent, conductive carbon nanotube films, *Science*, *305*, 1273–1276.
266. Gou, J. H. (2006). Single-walled nanotube bucky paper and nanocomposite, *Polym. Int.*, *55*, 1283–1288.
267. Spitalsky, Z., Tsoukleri, G., Tasis, D., Krontiras, C., Georga, S. N., Galiotis, C. (2009). High volume fraction carbon nanotube–epoxy composites, *Nanotechnology*, *20*, 405702.
268. Frizzell, C. J., in het Panhuis, M., Coutinho, D. H., Balkus, K. J., Minett, A. I., Blau, W. J., Coleman, J. N. (2005). Reinforcement of macroscopic carbon nanotube structures by polymer intercalation: the role of polymer molecular weight and chain conformation, *Phys. Rev. B*, *72*, 245420.
269. Xu, G. H., Zhang, Q., Zhou, W. P., Huang, J. Q., Wei, F. (2008). The feasibility of producing MWCNT paper and strong MWCNT film from VACNT array, *Appl. Phys. A*, *92*, 531–539.
270. Pham, G. T., Park, Y. B., Wang, S. R., Liang, Z. Y., Wang, B., Zhang, C., Funchess, P., Kramer, L. (2008). Mechanical and electrical properties of polycarbonate nanotube buckypaper composite sheets, *Nanotechnology*, *19*, 325705.
271. Park, J. G., Louis, J., Cheng, Q. F., Bao, J. W., Smithyman, J., Liang, R., Wang, B., Zhang, C., Brooks, J. S., Kramer, L., Fanchasis, P., Dorough, D. (2009). Electromagnetic interference shielding properties of carbon nanotube buckypaper composites, *Nanotechnology*, *20*, 415702.

272. Meng, C. Z., Liu, C. H., Fan, S. S. (2009). Flexible carbon nanotube/polyaniline paper-like films and their enhanced electrochemical properties, *Electrochem. Commun.*, *11*, 186–189.
273. Kim, Y., Minami, N., Zhu, W., Kazaoui, S., Azumi, R., Matsumoto, M. (2003). Homogeneous and structurally controlled thin films of single-wall carbon nanotubes by the Langmuir–Blodgett technique, *Synth. Met.*, *135*, 747–748.
274. Kim, Y., Minami, N., Zhu, W. H., Kazaoui, S., Azumi, R., Matsumoto, M. (2003). Langmuir–Blodgett films of single-wall carbon nanotubes: layer-by-layer deposition and in-plane orientation of tubes, *Jpn. J. Appl. Phys. 1*, *42*, 7629–7634.
275. Krstic, V., Duesberg, G. S., Muster, J., Burghard, M., Roth, S. (1998). Langmuir–Blodgett films of matrix-diluted single-walled carbon nanotubes, *Chem. Mater.*, *10*, 2338–2440.
276. Li, X. L., Zhang, L., Wang, X. R., Shimoyama, I., Sun, X. M., Seo, W. S., Dai, H. J. (2007). Langmuir–Blodgett assembly of densely aligned single-walled carbon nanotubes from bulk materials, *J. Am. Chem. Soc.*, *129*, 4890–4891.
277. Wei, J. Q., Zhu, H. W., Li, Y. H., Chen, B., Jia, Y., Wang, K. L., Wang, Z. C., Liu, W. J., Luo, J. B., Zheng, M. X., Wu, D. H., Zhu, Y. Q., Wei, B. Q. (2006). Ultrathin single-layered membranes from double-walled carbon nanotubes, *Adv. Mater.*, *18*, 1695–1700.
278. Lee, S. W., Kim, B. S., Chen, S., Shao-Horn, Y., Hammond, P. T. (2009). Layer-by-layer assembly of all carbon nanotube ultrathin films for electrochemical applications, *J. Am. Chem. Soc.*, *131*, 671–679.
279. Duggal, R., Hussain, F., Pasquali, M. (2006). Self-assembly of single-walled carbon nanotubes into a sheet by drop drying, *Adv. Mater.*, *18*, 29–34.
280. Du, F. M., Guthy, C., Kashiwagi, T., Fischer, J. E., Winey, K. I. (2006). An infiltration method for preparing single-wall nanotube/epoxy composites with improved thermal conductivity, *J. Polym. Sci. Polym. Phys.*, *44*, 1513–1519.
281. Zhu, H. W., Wei, B. Q. (2007). Direct fabrication of single-walled carbon nanotube macro-films on flexible substrates, *Chem. Commun.* 3042–3044.
282. Song, L., Ci, L., Lv, L., Zhou, Z. P., Yan, X. Q., Liu, D. F., Yuan, H. J., Gao, Y., Wang, J. X., Liu, L. F., Zhao, X. W., Zhang, Z. X., Dou, X. Y., Zhou, W. Y., Wang, G., Wang, C. Y., Xie, S. S. (2004). Direct synthesis of a macroscale single-walled carbon nanotube non-woven material, *Adv. Mater.*, *16*, 1529–1534.
283. Wang, D., Song, P. C., Liu, C. H., Wu, W., Fan, S. S. (2008). Highly oriented carbon nanotube papers made of aligned carbon nanotubes, *Nanotechnology*, *19*, 075609.
284. Chen, L. Z., Liu, C. H., Wang, J. P., Zhang, W., Hu, C. H., Fan, S. S. (2009). Auxetic materials with large negative Poisson's ratios based on highly oriented carbon nanotube structures, *Appl. Phys. Lett.*, *94*, 3.
285. Garcia, E. J., Hart, A. J., Wardle, B. L., Slocum, A. H. (2007). Fabrication of composite microstructures by capillarity-driven wetting of aligned carbon nanotubes with polymers, *Nanotechnology*, *18*, 165602.
286. Garcia, E. J., Hart, A. J., Wardle, B. L., Slocum, A. H. (2007). Fabrication and nanocompression testing of aligned carbon-nanotube–polymer nanocomposites, *Adv. Mater.*, *19*, 2151–2156.
287. Wardle, B. L., Saito, D. S., Garcia, E. J., Hart, A. J., de Villoria, R. G., Verploegen, E. A. (2008). Fabrication and characterization of ultrahigh-volume-fraction aligned carbon nanotube–polymer composites, *Adv. Mater.*, *20*, 2707–2714.
288. Cebeci, H., de Villoria, R. G., Hart, A. J., Wardle, B. L. (2009). Multifunctional properties of high volume fraction aligned carbon nanotube polymer composites with controlled morphology, *Compos. Sci. Technol.*, *69*, 2649–2656.
289. Vaddiraju, S., Cebeci, H., Gleason, K. K., Wardle, B. L. (2009). Hierarchical multifunctional composites by conformally coating aligned carbon nanotube arrays with conducting polymer, *ACS Appl. Mater. Interfaces*, *1*, 2565–2572.
290. Zhu, J., Imam, A., Crane, R., Lozano, K., Khabashesku, V. N., Barrera, E. V. (2007). Processing a glass fiber reinforced vinyl ester composite with nanotube enhancement of interlaminar shear strength, *Compos. Sci. Technol.*, *67*, 1509–1517.
291. Bekyarova, E., Thostenson, E. T., Yu, A., Kim, H., Gao, J., Tang, J., Hahn, H. T., Chou, T. W., Itkis, M. E., Haddon, R. C. (2007). Multiscale carbon nanotube–carbon fiber reinforcement for advanced epoxy composites, *Langmuir*, *23*, 3970–3974.

292. Fan, Z. H., Advani, S. G. (2008). Capillary effect of multi-walled carbon nanotubes suspension in composite processing, *J. Nanosci. Nanotechnol.*, *8*, 1669–1678.
293. Chen, W., Shen, H. B., Auad, M. L., Huang, C. Z., Nutt, S. (2009). Basalt fiber–epoxy laminates with functionalized multi-walled carbon nanotubes, *Compos. Part A*, *40*, 1082–1089.
294. Sager, R. J., Klein, P. J., Lagoudas, D. C., Zhang, Q., Liu, J., Dai, L., Baur, J. W. (2009). Effect of carbon nanotubes on the interfacial shear strength of T650 carbon fiber in an epoxy matrix, *Compos. Sci. Technol.*, *69*, 898–904.
295. Qian, H., Bismarck, A., Greenhalgh, E. S., Kalinka, G., Shaffer, M. S. P. (2008). Hierarchical composites reinforced with carbon nanotube grafted fibers: the potential assessed at the single fiber level, *Chem. Mater.*, *20*, 1862–1869.
296. Kepple, K. L., Sanborn, G. P., Lacasse, P. A., Gruenberg, K. M., Ready, W. J. (2008). Improved fracture toughness of carbon fiber composite functionalized with multi walled carbon nanotubes, *Carbon*, *46*, 2026–2033.
297. Mathur, R. B., Chatterjee, S., Singh, B. P. (2008). Growth of carbon nanotubes on carbon fibre substrates to produce hybrid/phenolic composites with improved mechanical properties, *Compos. Sci. Technol.*, *68*, 1608–1615.
298. Garcia, E. J., Hart, A. J., Wardle, B. L. (2008). Long carbon nanotubes grown on the surface of fibers for hybrid composites, *AIAA J.*, *46*, 1405–1412.
299. Pegel, S., Pötschke, P., Petzold, G., Alig, I., Dudkin, S. M., Lellinger, D. (2008). Dispersion, agglomeration, and network formation of multiwalled carbon nanotubes in polycarbonate melts, *Polymer*, *49*, 974–984.
300. Krause, B., Petzold, G., Pegel, S., Pötschke, P. (2009). Correlation of carbon nanotube dispersability in aqueous surfactant solutions and polymers, *Carbon*, *47*, 602–612.
301. Krause, B., Mende, M., Pötschke, P., Petzold, G. (2010). Dispersability and particle size distribution of CNTs in an aqueous surfactant dispersion as a function of ultrasonic treatment time, *Carbon*, *48*, 2746–2754.

CHAPTER 4

EFFECTS OF CARBON NANOTUBES ON POLYMER PHYSICS

4.1 OVERVIEW

The effect of a solid surface on the static and dynamic characteristics of polymer molecules has long been an active area of research because of the large number of applications where such interactions are important. With the introduction of solids having very small dimensions into polymers to form nanocomposites, this area has recently gained enhanced importance since geometric considerations force the conclusion that a large fraction of the polymer is "close" to a surface, and hence the contribution of the "interfacial fraction" to the global properties of the polymer should be expected to become significant. This interfacial fraction is clearly applicable to amorphous polymers, that is, a liquid or a solid that contains no crystalline component. The interfacial fraction concept is also important for semicrystalline polymers, that is, polymers that have both crystalline and amorphous regions. However, if the surface is able to initialize crystallization, that is, nucleate crystallization, then a second complication is added. The effect of nanotubes on amorphous polymers will be discussed in Section 4.2, while that for semicrystalline polymers will be discussed in Section 4.3. Both static and dynamic effects will be considered.

On a molecular level, there are a number of reasons for the configuration (conformation) of a polymer to change when a solid surface is introduced into an amorphous polymer. Impenetrability of the surface causes alterations in polymer chain configuration; changes in free energy and hence changes in configuration due to impenetrability can be well described mathematically. Gauging the static and dynamic properties of the molecules near the interface that occur as a result of configuration-change energies and repeat unit-surface interaction energies is non-trivial, but molecular dynamic simulations in particular are helping to answer some of these questions, at least for amorphous polymers. The application of well-known polymer thermodynamic concepts relevant to molecular weight should be able to explain perturbations in surface interactions caused by the fact that polymers are not

Carbon Nanotube–Polymer Composites: Manufacture, Properties, and Applications, First Edition.
Brian P. Grady.

monodisperse. The chemical identity of end groups could play a role if these groups have different interaction energies with the surface than does the repeat unit; however, as with other properties, only at very low molecular weights would such effects be expected to be important. One characteristic that is exceedingly difficult to account for in any theory is induced compositional or structural changes of the solid surface induced by the polymer. Of course, nanotubes will not change surface composition due to introduction of a polymer unless of course a covalent bond forms. Overall, a strongly interacting surface does change the conformation of a chain near its surface; such effects are more clearly seen in solution. In bulk, these effects are more difficult to discern. This chapter will discuss how nanotubes change the conformation of a polymer in bulk and in solution.

This molecular view of how a solid surface, in this case nanotubes, affects amorphous polymer behavior can be complemented by a thermodynamic, that is, macroscopic, view. For small-molecule liquids, characteristics of the solid–liquid interaction depend only on the interfacial energies of the liquid and the solid, which in turn are a function of the atomic level interactions between the surface and the liquid. Favorable interfacial energies, that is, the interaction of the polymer with the surface is stronger than the interaction of the polymer with itself, lead to adhesion of a bulk polymer and adsorption if the polymer is dissolved in a liquid. In general, high interfacial energies result from high surface energies, and high surface energies result from polar and/or polarizable functional groups being present on a surface or polymer. The addition of functional groups that increase the surface energy of nanotubes is an important strategy to strengthen the attraction between a polymer and a surface. Even more important is the compatibilization of nanotubes with groups that can react with the polymer or functionalization with the same type of polymer to provide possible entanglements with the bulk polymer.

The dynamics of polymer chains are affected as well as chain configuration. The glass transition is the simplest measure of polymer chain dynamics. The effects of carbon nanotubes on the glass transition temperature (T_g) are by no means universal for all types of polymers; depending on polymer identity both increases (indicating a reduction in chain mobility) and decreases (indicating an increase in chain mobility) have been measured. The amount of material participating in the glass transition can also decrease with the introduction of nanotubes, suggesting that at least some interfacial material is phase separated in a dynamic sense. The polymer chain diffusion coefficient is a molecular-level measure of chain mobility. A first-order approach to diffusion coefficient changes in the presence of nanotubes would be to assume a change in the average diffusion constant of the polymer, while a second-order approach consistent with dynamic phase separation would be to model the system as having two spatial regions with one region where the diffusion constant is unaffected by the solid and the other where the diffusion constant is significantly affected. More complicated approaches are of course possible. Experimentally, as will be described later in Section 4.2.2, in some situations the diffusion constant shows a decrease at low nanotube content followed by an increase at higher concentrations, which can be explained by an anisotropic diffusion coefficient.

The fundamental building block of the crystalline region, the unit cell, is unaffected by the introduction of nanotubes. However, the rate of formation of crystals, which in turn affects properties including crystallite size, shape, and so on, can be affected by the introduction of nanotubes. Crystallization is conveniently divided into two parts, nucleation and growth. Nucleation phenomena are not well understood even in the case of monodisperse homopolymers in the absence of any other added fillers. Nucleation in the presence of a solid substrate is even less understood. For example, enumerating surface characteristics that will induce nucleation cannot be done reliably, or, alternatively, whether nucleation of a certain polymer by a certain surface will occur cannot be predicted either. Clearly, the molecular level understanding of nucleation in the presence of nanotubes is poor. However, as will become clear, nanotubes do nucleate crystallization in most systems having a profound effect on the resultant crystalline morphology. The effect of a solid surface on crystal growth is better understood than nucleation. A solid surface affects growth both dynamically and spatially. Dynamically, the ability of polymer chains to diffuse to the growing crystal face will be affected. Spatially, a growing crystal encountering a solid has three choices: stop growing, change its growth direction, or force a change in the position of the solid. The effect of nanotubes on the growth rate of crystallization will also be explored.

Some of this chapter will concern evidence gleaned from surfaces having a chemical or geometric structure different from nanotubes. Common chemical surface identities studied include silicon dioxide, aluminum oxide, clays, and graphite; results from the latter, in particular, should likely be very applicable to nanotubes. The two most important geometries are flat 2D surfaces, which have been heavily studied because of the wide variety of characterization techniques available, and porous solids, where confinement is almost in all three directions and for which bulk measurements can be made. As might be expected, studies of polymers confined in porous solids are rare because of the difficulty of confining a polymer inside a pore. Although a nanotube surface in a composite is different from either of these geometries, it is certainly expected that the characteristic behavior of polymers near a surface gleaned from these studies could be applicable to nanotube–polymer composites. In fact, polymer–nanotube interactions do not seem to be affected, in most cases, by the radius or length of the nanotube. This statement arises from the fact that the fundamental interaction is at the repeat unit/hexagonal carbon length scale, and the curvature of the tubes is usually of a length scale much larger than the interaction distance. However, there are some instances where nanotube geometry plays a critical role; two of the most important are wrapping and crystallization in oriented systems. Wrapping describes how certain polymers wrap around a carbon nanotube adopting a helical-type configuration; only a small diameter cylinder can force this type of configuration. For crystallization, nucleation by a solid causes growth perpendicular to the surface (in this case, perpendicular to the nanotube long axis); hence, a small diameter cylinder will cause some very unique morphologies.

Each of the next two sections will begin with a short discussion that will introduce relevant concepts in polymer physics. With that discussion serving as a basis, the remainder of the section will explore the important effects of nanotubes on the amorphous and crystalline regions of a polymer.

4.2 AMORPHOUS POLYMERS

This section will discuss polymers having no crystalline regions, and includes polymer melts, polymers in solution, and solid amorphous (glassy) polymers. This section does not address what effect nanotubes have on the amorphous regions of semicrystalline polymers. If nanotubes do not nucleate crystallinity, then nanotubes must be excluded from the crystalline regions. A number of questions can be posed about the relationship between the amorphous regions of a semicrystalline polymer and nanotubes; for example, is there some preference for the interfacial region of the crystalline/amorphous boundary to preferentially interact with carbon nanotubes versus the noninterfacial amorphous region? However, detailed analyses of the interactions of nanotubes with the amorphous region in semicrystalline polymers have not yet been performed in a significant manner. These analyses have not been performed not because of the lack of expected importance; the characteristics of the amorphous region of a semicrystalline polymer, including material in the crystalline/amorphous boundary, are important for mechanical properties in particular. However, such analyses are difficult and hard to perform. The few studies that have given information on the amorphous regions in semcrystalline polymers will be described in Section 4.3 on crystalline polymers.

4.2.1 Statics: Adsorption and Chain Configuration

A solid surface will change the conformation of a polymer chain under most conditions. Figure 4.1 shows the conformation of a polymer chain not at a surface calculated from a self-avoiding random walk. The conformation of a polymer chain is normally quantified in terms of its radius of gyration (R_g) both because R_g can be measured experimentally and because R_g is a convenient way to quantify conformation. The radius of gyration is defined as the root mean square distance of a polymer mass segment from the overall polymer center of mass. For a homopolymer, the radius of gyration can be calculated from

$$\langle R_g^2 \rangle = \frac{1}{N}\sum_{k=1}^{N}(\vec{r}_k - \vec{r}_{\text{mean}})^2 = \frac{1}{2N^2}\sum_{i,j}(\vec{r}_i - \vec{r}_j)^2 \tag{4.1}$$

where $\vec{r}_i$, $\vec{r}_j$, and $\vec{r}_k$ are vectors from an arbitrary point to a point that contains the center of mass of the i, j, or k repeat units, respectively, $\vec{r}_{\text{mean}}$ is the vector from the same arbitrary point to the center of mass of the polymer chain, and N is the total number of repeat units. The angle brackets represent the "ensemble average," which is a quantum mechanical way of saying the average for a large number of structurally identical chains. For a polymer chain in an amorphous, unoriented pure sample, $\langle R_g^2 \rangle^{1/2}$ (the square root of the ensemble average radius of gyration squared) is typically between 10 and 200 nm and defines a sphere for which about 60% of the mass of that polymer chain is contained. However, the mass concentration of repeat units from a single chain in its R_g sphere is quite small, between 1% and 10% for most polymers (the amount increases as the molecular weight decreases). In other words, within the sphere defined by R_g, most of the volume is occupied by repeat units from other chains (in bulk) or solvent (in solution).

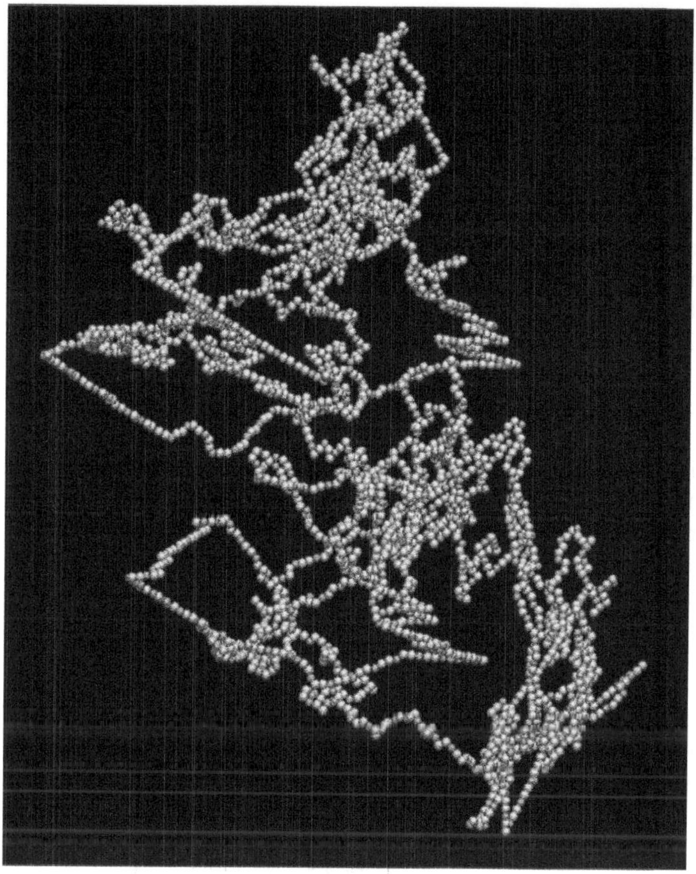

Figure 4.1 Conformation of an amorphous polymer chain. One thousand repeat unit poly(ethylene oxide) is shown and a self-avoiding random walk was used to create this configuration. Courtesy of Liu Shu and Alberto Striolo.

Under "theta" conditions, $\langle R_g^2 \rangle$ is proportional to N, the number of repeat units. "Theta" conditions exist in neat amorphous polymers, either melts or glasses, and also occur for polymers in solvents relatively close to their precipitation temperature. The radius of gyration is perhaps not as simple to understand as the end-to-end distance (i.e., the distance from one end of a chain to the other); under "theta" conditions, the ensemble average end-to-end distance squared $= 6\langle R_g^2 \rangle$. Not only do the radius of gyration and the end-to-end distance scale with $N^{1/2}$ under theta conditions, but a number of other thermodynamic quantities also exhibit simple scaling behavior or have specific values under "theta conditions." The thermodynamic consequences of "theta" conditions are beyond the scope of this book; an introductory polymer physics textbook will describe these consequences in more detail.

Adsorption by polymer chains onto nanotubes where both are found in a liquid will be considered first. In the case of an attractive interaction between polymer and nanotube, polymer molecules can adsorb on the surface of a nanotube; that is, some or all of the repeat units will be in close proximity to the nanotube surface. Other than covalent bonding (which is not adsorption in the context of this book since covalent bonding to a surface is termed functionalization), the strongest forces involved in adsorption are electrostatic this situation occurs in a high dielectric constant liquid where the charge of the adsorbing species has a charge opposite that of the surface. By far, the most important case of electrostatic adsorption occurs in water, and in fact electrostatic adsorption is a very important mechanism of polymer adsorption to functionalized nanotubes. Other types of forces relevant to adsorption on a nanotube surface include dispersion forces and various other forces, the most important of which for nanotubes is $\pi-\pi$ stacking. The solvent, and its interaction with both the polymer and the surface, is also a key to the behavior of the adsorbed polymer at the interface since a given polymer chain can stay surrounded by solvent or adsorb on the surface.

The general configuration of an adsorbed chain from solution onto a flat surface falls into one of three patterns: pancake, mushroom, or brush. The pancake pattern of a single polymer chain at a flat surface has three different types of segmental conformations: trains, tails, and loops. These different characteristics are schematically shown in Figure 4.2. Although developed for flat surfaces, these qualitative descriptions should be valid for polymers adsorbed to nanotubes. The following is generally found for homopolymer adsorption on a relatively flat surface:

- Polymer adsorption tends to be temperature insensitive.
- Polymer adsorption increases with molecular weight in a poor solvent, but is rather insensitive to molecular weight in a good solvent.

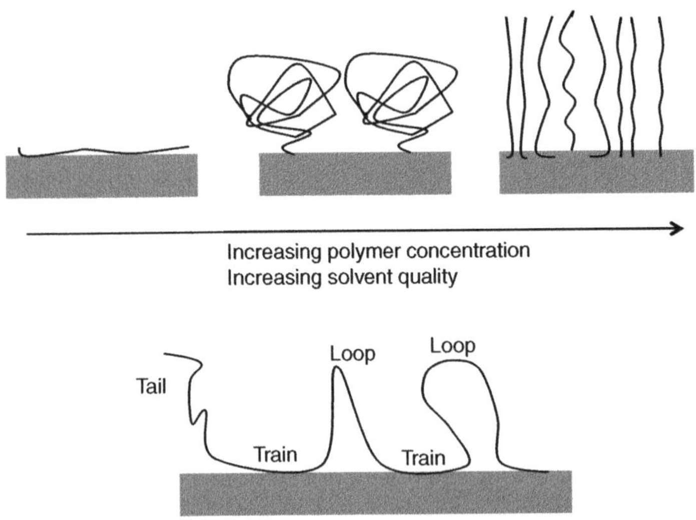

Figure 4.2 Classification of the way in which polymer chains can interact with a surface. Upper diagram represents the three main classifications of morphology for adsorbed polymers (from left to right pancake, mushroom or brush) while the lower represents, for a single chain, possible features of adsorption.

- Polymer adsorption increases as solvent quality decreases.
- Although polymer adsorption occurs on the timescale of minutes, to reach the final equilibrium state can take many hours, or even days, due to a great deal of rearrangement that can occur at a surface as well as replacement of longer chains by shorter chains.
- Lower molecular weight species tend to adsorb at a surface because there is less loss of configurational entropy for the smaller chains.
- Polymer adsorption can usually be thought of as "effectively irreversible" in many situations. Specifically, adding more solvent, or even a small molecule that will adsorb on the surface, does not cause desorption of the polymer even in cases where the polymer will stay in solution if not already adsorbed. The reason is that deadsorption requires all adsorbed polymer segments to release from intimate contact with the surface at the same time, and the probability of this happening is small even when energetics favor deadsorption. A notable exception is when the adsorption is due to oppositely charged polymer/surface; changing the charge can cause deadsorption. Another notable exception is when a more strongly adsorbing polymer is added, because the new polymer literally burrows under the adsorbed polymer.
- A reasonable value for the adsorption of polymer molecules is 1 mg/m^2, indicating that the thickness of the adsorbed layer is on the order of 1 nm assuming a bulk density. However, an adsorbed polymer chain can extend tens to hundreds of nanometers into solution depending on molecular weight and solvent quality indicating that a large amount of solvent is involved.

In general, homopolymer adsorption from solution tends to be of the pancake type, unless of course the end group of the polymer has a very high affinity for the surface while the polymer does not. To achieve mushroom and brush configurations usually (but not always!) requires a polymer with at least two different types of moieties, which is most often done either with a reacting group at a chain end or with a block copolymer where one block is adsorptive and the other is not.

The theory of pancake-type adsorption was considered in the seminal work of DeGennes,[1,2] who considered a weakly attracting wall, with the wall–repeat unit interaction length scale equal to A, which is also the repeat unit length. The concentration was semidilute, that is, the concentration was slightly greater than the overlap concentration. The volume fraction of repeat units was separated into three regions based on the proximity to the interface: a region $z < A$ (z is the distance) termed the proximal region where volume fraction of repeat units depends strongly on the interaction energy, the central region that is self-similar where the number of entanglements and hence the concentration scales as $z^{-4/3}$, and finally the distal region where the concentration is controlled by a few large loops and tails and where the concentration scales as e^{-z}.

Clearly in the pancake type of adsorption from solution, the chain would be expected to spread in the two directions parallel to a flat surface, while contracting in the direction perpendicular to the surface. Whether this change occurs and the magnitude of this change depend on the strength of the interaction of the polymer with the surface. This change is quantified by the radius of gyration in the parallel direction relative to that in the perpendicular direction. Simulations[3,4] indicate that the radius of gyration in the perpendicular directions can be one to two orders of magnitude lower than that in the parallel direction; however, the three-dimensional radius of gyration, which is easily measurable, will show little change because the sum of the square of the two radii of gyration is approximately constant.

Nanotubes in a bundled state will present a surface that has a roughness varying periodically in a manner that is a function of the nanotube diameter. The question that naturally arises is whether the nature of polymer adsorption from solution is altered by the periodic roughness presented by a nanotube bundle. The author is unaware of any study that examines this question, although a computational study on surfactant adsorption showed that such effects increase adsorbed surfactant amount.[5] For both individual nanotubes and small bundles, the diameter of the adsorbing surface will be less than the radius of gyration of a polymer chain in the direction perpendicular to the bundle long axis, in essence forcing the adsorbed chain to align in the direction of the bundle axis. Again, to the author's knowledge no studies have been done that might, for example, suggest that adsorption might be more facile on larger bundles since the orientation requirement is less severe. However, the bulleted points listed earlier and the theory of DeGennes are expected to hold in the case of adsorption to a nanotube bundle.

As stated previously, thermodynamics favor the adsorption of lower molecular weight species because of entropic considerations. However, two authors found that rheological measures such as melt strength and storage modulus decrease with the introduction of nanotubes at very low nanotube concentrations consistent with the selective adsorption of high molecular weight species.[6,7] Such decreases are not unique to nanocomposites made with carbon nanotubes; nanocomposites with polystyrene spheres[8] also show this effect. In fact, using a more narrow molecular weight distribution polymer eliminated the decrease in storage modulus.[7] The explanation for this phenomenon is likely kinetic. Although adsorption of shorter molecular weight chains is thermodynamically favored, displacing adsorbed longer polymer chains is very difficult due to the necessity of removing all contacts with the surface.

As the nanotube diameter gets small, a unique configuration is possible for the polymer chain; that is, the theory of DeGennes will not apply and some of the bullet points listed earlier will be violated. Specifically, the polymer chain can wrap around the nanotube and form a helical-type structure as shown in Figure 4.3. Most of the work has been performed with biological polymers, for example, DNA and proteins, because of the possible importance of nanotubes in biological applications. However, synthetic polymers have been shown to wrap around nanotubes as well. To directly image this wrapping requires scanning tunneling microscopy, which is capable of providing the molecular level resolution necessary to distinguish wrapping from some other chain configuration. Scanning tunneling microscopy uses an electric current between a tip and a surface to determine topological variation, and has significantly more resolution that atomic force microscopy that uses mechanical force between a tip and a surface. DNA[9] and poly(3-hexylthiophene)[10] have been shown to wrap around single-walled carbon nanotubes. STM allows the determination of the chiral angle and the polymer spacing along the nanotube. For DNA and a

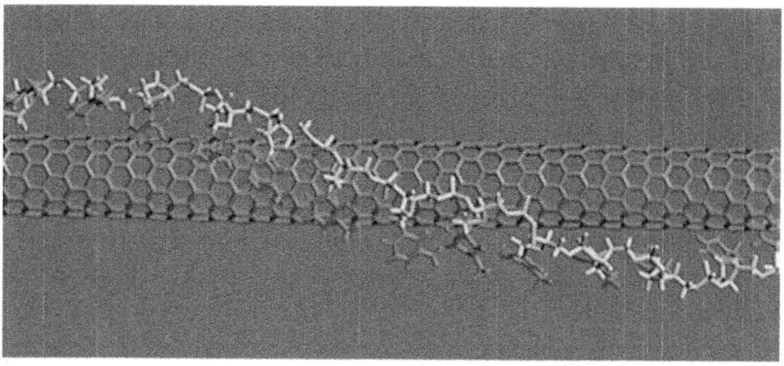

Figure 4.3 A schematic that represents how DNA might wrap around a (10,0) single-walled carbon nanotube. Copyright Nature Publishing Group. Reproduced with permission from Ref. 266.

(6,5) tube, the angle and distance were ~63° and ~3.3 nm, respectively, which agreed well with computer simulations of the structure. For poly(3-hexylthiophene), the angle and distance were ~45° and ~1.7 nm, respectively. However, in the DNA study,[9] there was a regular periodic height variation along the DNA chain (e.g., the distance from the DNA to the nanotube was not constant) that could not be explained and was not predicted by computer simulations. One interesting question is the maximum diameter where wrapping is possible. DNA has been shown to wrap MWCNTs of diameter ~10 nm; of course, the exact specifications depend on the polymer and nanotube.[11] The same study found another interesting adsorption conformation: polymer could be inserted inside a nanotube. In this study, DNA was found to insert using STM into specially synthesized MWCNTs with an inner diameter of ~0.35 nm. For insertion, tubes must have larger openings than the typical 0.1–0.2 nm diameter given the cross-sectional area of most polymer chains.

The reader should be cautioned because a great number of papers claim wrapping where the evidence for such a configuration is lacking. In some studies, the claim is given that wrapping is occurring even without any evidence that individually dispersed tubes or at least bundles with a very low number of tubes are present! AFM studies cannot be used to determine whether wrapping occurs because the probes do not have sufficient resolution to distinguish wrapping from some other type of arrangement such as adsorption as a linear chain rather than a helix. Computer simulations[12,13] have been of great utility in predicting the nature of polymer–nanotube adsorption; however, under certain conditions, both random[14,15] and inserted polymers[16,17] have been found to be the lowest energy adsorptive state. In the author's opinion, the number of experimental studies that have confirmed such predictions is not large enough to fully trust such computer simulations. However, the author believes that such studies will eventually prove that computer simulations are generally correct and it really is not necessary to prove experimentally that wrapping is occurring if computer simulations indicate that wrapping is the lowest adsorptive energy state.

Macroscopically, polymer adsorption from solution is described by an adsorption isotherm. An adsorption isotherm is a plot of adsorbed amount on the ordinate

(y-axis), ideally expressed in terms of mass of adsorbent/surface area of solid. Because the available surface area is typically unknown since the dispersion state is unknown for nanotubes, plotting the y-axis as weight of adsorbent/weight of nanotubes is acceptable. The abscissa (x-axis) should be plotted as concentration of polymer in solution after adsorption. If the amount of added polymer is plotted on the abscissa instead of the amount of polymer in solution after absorption, then the plot is only relevant for the solid amount, solution volume, and polymer concentration used in that experiment. Although such graphs have been published for surfactant adsorption to nanotubes,[18,19] studies of polymer adsorption have been quite rare. In one particularly imaginative study,[20] a layer of single-walled carbon nanotubes was sprayed onto a silica surface enabling a constant nanotube surface area. Optical reflectivity was used to determine the amount adsorbed; surprisingly, the roughness of the surface did not distort the laser light to levels that were unmanageable. Both adsorption rate and amount were monitored, although the nominal surface area, that is, not the nanotube surface area, was used for normalization. The authors showed that a protein monolayer on the order of 1 nm thick was formed under conditions near the isoelectric point of the protein; thinner layers formed when the protein was charged. Further, the characteristic shapes of adsorption isotherms were similar to those found with polymers on other surfaces as shown in Figure 4.4. In another study where the nanotubes were dispersed in solution,[21] the authors presented the ratio adsorbed polymer/total polymer (from which the adsorbed amount could be calculated); however, the x-axis was presented in terms

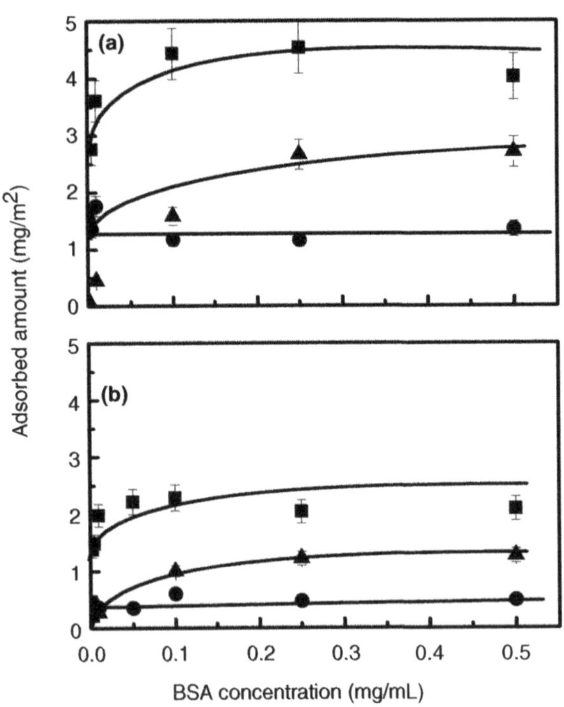

Figure 4.4 Adsorption of bovine serum albumin polymer on carbon nanotubes (a) and silica (b) at pH 3.0 (●), 4.8 (■), and 7.0 (▲). Copyright Elsevier Ltd. Reproduced with permission from Ref. 20.

of amount added instead of amount in solution after absorption, and the polymer/nanotube mass ratio was fixed at 1 : 1. The authors showed that there was a relative maximum in the ratio adsorbed polymer/total polymer at an added polymer or nanotube concentration of about 10^{-4} kg/m^3, while a relative minimum occurred at about 10^{-3} kg/m^3. The former would seem to indicate a maximum dispersion level under the conditions of this experiment, which is simple to understand since diffusion of the polymer plays an important role in debundling kinetics and higher polymer concentrations lead to lower diffusion coefficients due to higher viscosities and entanglements. The relative minimum is harder to understand, unless this is due to better sonication efficiency at higher viscosities. The authors analyzed the data using an adsorption/desorption model, which is likely irrelevant to the physics of the problem since desorption of a polymer is unlikely except in cases driven by molecular weight fractionation. This study does illustrate the difficulty of measuring adsorption isotherms of polymers on nanotubes; the changing surface area driven by different debundling levels will make the adsorption isotherms not follow the "true" isotherms because the adsorption surface area is not constant.

In the absence of a solvent, for example, in a polymer melt or glassy polymer, how does the introduction of nanotubes affect the radius of gyration of a polymer chain? This question is hard to answer because of the lack of experimental tools that are able to address this question, and the author is not aware of any study that measures the radius of gyration for polymers adsorbed on nanotubes. Some studies have been made on thin films adsorbed to flat surfaces. Unfortunately, such studies generally have two interfaces: a polymer–air interface and a polymer–solid interface. Further, the structures are, in most cases, inherently nonequilibrium structures since surface tension forces would tend to drive the film into droplets, and it is only due to the high viscosities that such droplets do not form. In such studies, the results have been mixed; some studies have found no change in the radius of gyration in the plane of the surface for poly(methyl methacrylate) on glass[22] and polystyrene on hydrogen-passivated silicon,[23] while others have found a significant lengthening for polystyrene on glass.[24,25] A very detailed and precise study on nanoscale silica with no agglomeration exhibited no changes in chain dimensions upon the addition of filler.[26] The latter paper critically examined the literature and concludes that with proper conditions, in particular monodisperse chains, no chain extension should occur. However, these authors do not consider nanocomposites that have two nanoscale dimensions such as nanotubes. The authors do acknowledge a "correlation hole" effect in filled systems, which occurs because of a decreased interpenetration/entanglement of chains near a solid interface.[27,28] Although a direct connection has not been established, it is very likely in the author's opinion that this "correlation hole" is related to some of the dynamic phenomena discussed in the next section.

4.2.2 Dynamics: Glass Transition and Diffusion Coefficient

The effect of filler on chain mobility is difficult to predict. A filler is an impediment to movement, since the filler is not permeable to the polymer. However, movement along the surface of the filler could be enhanced relative to movement in the bulk polymer. Enhancement can be easily rationalized in the case of an unfavorable energetic interaction of the polymer with the filler; the polymer is able to "slide by" the surface more easily. However, even in the case of a favorable interaction, enhancement could occur if a layer at the surface is pinned by the solid, allowing chains next to the pinned layer to move more easily since the entanglement density with the pinned polymer can reasonably be expected to be smaller than that in the bulk system, that is, the "correlation hole" effect discussed above.

The diffusion coefficient (D) is the measure of the speed of a molecule to move through a stationary medium and is defined in one dimension by Fick's law at steady state:

$$J = -D\frac{\mathrm{d}c}{\mathrm{d}x} \tag{4.2}$$

where c is the concentration of the diffusing species, x is the direction of diffusion, and J is the flux (molecules/area $\times$ time). Fick's law states that the rate of movement will be proportional to the gradient in concentration; that is, a difference in concentration causes molecules to move from one place to another and the greater the difference, the faster the movement. How do nanotubes affect the diffusion coefficient of a polymer? In two recently published studies by Winey and coworkers,[29,30] the authors found a minimum in diffusion constant with nanotube content in the case of single-walled and multiwalled nanotubes. Experiments were performed with a polydisperse high molecular weight polystyrene sample and the diffusion of deuterated tracer polystyrene molecules of various molecular weights was followed. To explain the minimum, the authors postulated that there was a decrease in diffusion coefficient at low concentrations due to the impeding effect of the nanotubes. An increase in diffusion coefficient occurred at higher nanotube concentrations because polymers diffuse more quickly along the nanotube surface, and, when a percolating network is formed, diffusion over long distances in directions defined by the nanotube surface is now possible. In fact, at high enough concentration, the diffusion coefficient is higher than that in pure polymer. Faster diffusion than in the pure polymer suggests that the interaction of the deuterated chains is not with the nanotube, but rather with polymer that is immobilized on the surface of the nanotube since the interaction between nanotubes and polystyrene is known to be adhesive. This observation of a minimum in diffusion coefficient held when R_g of the deuterated polymer was greater than the average radius of the nanotube (or nanotube bundle); when R_g was less than the average radius, the diffusion constant was independent of nanotube level. The independence of diffusion constant with nanotube addition is even more surprising given the increase in viscosity (although perhaps the viscosity does not increase in a local sense as pointed out by one group of authors)[31] as well as the increase in path tortuousity. The authors explain this result as a lack of entanglements for the lower molecular weight deuterated polymer with the nanotube. However, this explanation suggests that the lower molecular weight material does not interact well with the nanotubes, which is counter to the known effect that lower molecular weight polymers tend to adsorb better than higher molecular weight polymers; although deuteration might affect the relative strength of these interactions and/or nonequilibrium, higher molecular weight adsorption might be occurring. Two other studies also found no change in diffusion constant with the addition of nanotubes.[31,32]

The glass transition is another way to measure polymer dynamics. The glass transition is a change from an amorphous solid to an amorphous liquid. The most common technique used to characterize the glass transition is differential scanning calorimetry (DSC). DSC involves heating or cooling a sample at a constant rate, that is, °C/min, and precisely measuring a quantity that is proportional to the amount of energy being added or removed from a sample, which in turn allows for the calculation of the heat capacity. In the absence of a transition, the heat capacity of a polymer increases approximately linearly with temperature; in the glass transition region, there is a steep increase in the heat capacity. The change in the heat capacity over the temperature range that defines the glass transition is proportional to the amount of amorphous material that changes from an amorphous solid to an amorphous liquid. Other thermodynamic measurements such as volume, as well as dynamic measurements such as viscosity and diffusion coefficient, can also be used to characterize the glass transition.

Nanotube composites are one example of the more general area of the behavior of the glass transition for polymers under spatial confinement. In fact, nanotube composites are not the ideal system for studying such effects because of the difficulties in precisely characterizing dispersion. Even in easier systems to study, as pointed out in the review by Alcoutlabi and McKenna,[33] observations of the glass transition under confinement are not consistent and in fact sometimes contradictory to one another. This section will give some background about T_g of confined materials in order to be able to discuss how nanotubes affect the glass transition in polymers. Two geometries are of particular interest: molecules confined in pores and molecules in thin films. Both geometries have limitations relative to nanotube composites that will become apparent during the discussion; however, both of these geometries can be defined more precisely than nanotube composites.

Theoretical[34] and experimental studies[35,36] of the glass transition of small-molecule glass formers in pores indicate that there are likely two species, one which shows a higher T_g than the bulk T_g, and another that shows a T_g that is lower than or equivalent to the bulk T_g. The obvious explanation is that some material is strongly adsorbed to the surface raising the T_g of that material relative to the bulk. The unadsorbed material can have the same or lower T_g than the bulk, likely depending on the characteristics of the adsorbed layer. There is some controversy on the separated glass transition temperatures; some studies show a decrease in a single T_g upon confinement,[34,37] while others show a decrease followed by an increase with decreasing pore size.[38,39] To reconcile the studies with only one T_g with those that show two T_g values, it is only necessary to postulate that in some cases the surface adsorbed phase is either so strongly adsorbed or adsorbed in such small quantity so as to not give a measurable T_g. Unfortunately, studies of polymers inside of pores are difficult because of the difficulty of filling pores with a high-viscosity polymer.

The diversity of behavior, as well as the quantity of data, is much larger in studies of the glass transition in polymer thin films. The preparation of polymer thin films is almost always done in a nonequilibrium matter, so the nonequilibrium state of the film is convoluted with the nonequilibrium nature of the glass transition! This simple statement is perhaps part of the reason for the very inconsistent nature of the results that will be described. In particular, the choice of solvent used to create thin films can have an effect on the observed behavior; of course, annealing for a long time can remove such effects but recent work has shown that annealing to reach an "equilibrium" state can be quite long at temperatures far above the glass transition.[40] Further, calorimetry has been infrequently used as a measurement of the glass transition temperature in thin films; a full discussion of the different methods used to measure the glass transition temperature in thin films is beyond the scope of this text and the interested reader is encouraged to consult Ref. 33.

Free surfaces (i.e., the polymer–air interface) are an interesting case of surface confinement, and free surfaces can be reasonably assumed to be the most highly repulsive surfaces available. In fact, the term "repulsive surface" functionally means that poor wetting causes the presence of voids of some unspecified character near the solid, which confirms why free surfaces can be considered as being the most highly repulsive surfaces. Polystyrene has been the most studied polymer in thin film studies of the glass transition. T_g decreases at thicknesses roughly below 100 nm in freestanding films (i.e., two free surfaces) or for samples supported on one side by a nonstrongly adhesive surface. The decreases become larger as the film thickness becomes smaller reaching levels as much as 50–100 K.[41,42] A representative example of the drop in T_g is shown in Figure 4.5. Poly(methyl methacrylate) and poly(vinyl acetate) showed a much smaller drop[43] or almost no drop in T_g,[44] respectively, with film thickness in such situations.

An increase in glass transition temperature caused by a surface that interacts strongly with the polymer, either because of grafting of polymer to the surface[45] or because of strong adsorption of the polymer to surface,[46] has been found. One criticism of this field is the lack of measurement of interfacial energies for various polymers and substrates, which does not allow for quantitative relationships to be developed; however, it is not entirely clear the temperature and/or concentration (which is relevant depends on the method of making the samples) at which such measurements should be made. In general, the number of studies that found a

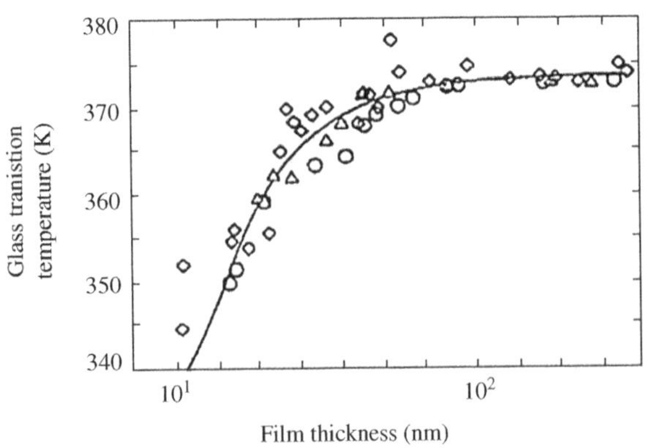

Figure 4.5 Changes of T_g with film thickness with three narrow molecular weight polystyrenes: (Δ) 120,000 g/mol, (○) 500,800 g/mol, and (◇) 2,900,000 g/mol. Copyright IOPscience. Reproduced with permission from Ref. 41.

decrease in T_g is significantly larger than the number of studies that have found an increase in T_g, although the presence of one free surface in these experiments skews the results. A group of studies that illustrate the subtleties inherent in these experiments were performed by Grohens et al.[47,48] for supported isotactic and syndiotactic PMMA films. Using spin-coated films on silicon and aluminum and infrared spectroscopy to quantify polymer–surface interactions, the authors found a decrease in T_g for the more lightly bound syndiotactic PMMA, and an increase for the more strongly interacting isotactic PMMA. Hence, changes in stereochemistry, which presumably would not significantly alter polymer–surface interactions, do in fact alter those interactions to such an extent so as to cause a qualitatively different response in the effect of film thickness on T_g.

One of the key questions that have been examined in thin films is whether the glass transition temperature changes as the polymer becomes closer to the substrate. Flat surfaces offer an interesting way to study that question, since depth profiling measurements of the T_g might be possible. The use of fluorescence probes has allowed for such an ability, and these probes have been used to show that in at least one case, there is a depression in T_g near the free surface and increase in T_g near the solid surface.[49] This type of spatial separation, that is, two T_g values, has not been noted in most studies of the T_g in thin polymer films. However, most probes of the glass transition in thin films do not allow for distinguishing two glass transitions, even if such separation occurs.

Nanocomposites offer an interesting geometry that is intermediate between that of thin polymer films and polymers in pores. With respect to thin polymer films, nanocomposites are a simpler system since the contribution from the free surface can be safely ignored. Nanofillers having only one dimension that is nanoscale, that is, platelet fillers such as graphene sheets or nanoclays, have a geometry that is in a practical sense identical to that offered by the surfaces used in the studies of thin polymer films, except perhaps for roughness. Polymers filled with spherical nanoscale objects have a geometry that is qualitatively the mirror image of porous systems. Nanotubes, of course, are the case where the nanofiller has two nanoscale dimensions. This discussion will, of course, focus on nanotubes; however, all nanofillers tend to affect the glass transition in a similar manner, although not necessarily identically, as will be pointed out when appropriate.

The simplest hypothesis is that the interaction of the filler with the surface will determine whether the glass transition temperature increases, decreases, or does not change. Unfortunately, in nanotubes, it is impossible to vary the surface energy of the filler and/or nanotube and know that the dispersions are identical, so the assignment of changes to a particular variable are difficult. Nanospheres are much easier to disperse without aggregation, and hence the interfacial energy can be varied while maintaining perfect dispersion. One study with nanospherical silica showed that the three types of changes in T_g could be obtained with three different polymers and hence three different surface energies: poly(2-vinyl pyridine) showed an increase in T_g, PMMA showed a decrease in T_g, and polystyrene showed no change in T_g.[50] With nanotubes, a similar study was performed where the surface was changed rather than the polymer. Using a dispersion–reaction scheme (see Figure 3.2 and the associated discussion), a polyimide was mixed with unfunctionalized, acid-functionalized, and amine-functionalized tubes.[51] In the former case, the T_g decreased, while in the latter two cases, the T_g increased. As an approximate measure of dispersion, the percolation threshold in conductivity was between 2% and 3% for all materials. Similar differences in the change of T_g with nanotube surface chemistry have been found elsewhere as well.[52,53]

Although the surface energy of the nanotubes can be altered in order to change the interaction between the polymer and the nanotube, another mechanism is also available as described in a unique study by Koning and coworkers.[54] For a low molecular weight polystyrene having a very low T_g of 90°C (about 10°C below that of a polystyrene sample with a more typical molecular weight), there is an increase in T_g of about 8°C/wt% nanotube with added nanotubes until a plateau is reached at a nanotube content of 2 wt%. For a material of more typical molecular weight, there is essentially no change in the glass transition temperature with added nanotubes after the effect of surfactant is accounted for (surfactant is used to disperse the tubes). Finally, adding a small amount of low molecular weight material to the typical molecular weight material causes the glass transition temperature to increase by about 1°C/wt% nanotube. The authors attribute the increase to the ability of the low molar mass material to displace surfactant on the surface of the nanotubes; however, whether surfactant is necessary to see this effect has not been determined.

Table 4.1 presents the results of papers that have examined the effect of adding nanotubes on the glass transition temperature as a function of nanotube content and is believed to be comprehensive. The column "T_g/wt% nanotube" is the amount of change in T_g for every weight percent nanotube; a linear relationship is assumed in this table. Within a good approximation, a linear relationship is correct for all studies. This linear relationship usually does not extend to higher concentrations; instead the glass transition temperature often plateaus and reaches a constant value. This value is also given in Table 4.1. The conductivity percolation threshold is listed to give the reader some idea of dispersion; for most if not all of the entries, the percolation threshold is a reasonable measure of dispersion for these materials. All entries suffer from some "nonequilibrium" effect; that is, none of the studies annealed samples for a long time to eliminate effects due to sample history. Eliminating such effects is expected to be more difficult in these materials than in polymer thin films, and, as stated earlier, eliminating such effects in thin films is very, very difficult.

The results shown in Table 4.1 have little, if any, discernible patterns with respect to surface or interfacial energies; the inconsistency of behavior represented by Table 4.1 is not unique to nanotubes as discussed in a recent review article on spherical silica nanofillers.[55] The most common shape of a curve of T_g versus wt% nanotube is an increase followed by a plateau at higher concentrations, an example of which is shown in Figure 4.6. This qualitative shape cannot be easily explained; the glass transition temperature should continue to increase (or decrease) with increasing nanotube content, or perhaps show a maximum or minimum as was found with the diffusion coefficient. A theory that fits experimental data rather well at low concentrations does not predict a plateau under any circumstances.[56] Perhaps other nanofillers might be able to provide some explanation for the plateau. To examine nanocomposites made with nanoclays is not appropriate, because the surfactant required to disperse the clays can cause plasticization of the polymer. Graphene

TABLE 4.1 Changes in the T_g of Amorphous Polymers with Nanotube Incorporation

Polymer	Nanotube type	Processing method	T_g change/wt% nanotube	Plateau (wt% nanotube)	Percolation threshold (wt% nanotube)[a]
Polystyrene[64]	Acid-functionalized SWCNTs	Dissolution–dispersion–precipitation	6°C/%	1%	~0.5%
Polystyrene[68]	Amine-functionalized SWCNTs	Dissolution–dispersion–precipitation	5°C/%	0.5%	NA
Polystyrene[69]	MWCNTs	Melt mixing	~1.5°C/%	NA–7%[b]	Not reported
Polystyrene[54] Low molar mass High molar mass	SWCNTs	Dispersion–dispersion–evaporation	 8°C/% No change	 2% NA	 0.2% 0.2%
PMMA[70]	SWCNTs	Dissolution–dispersion–precipitation	9°C/%	~2%	1–3%
PMMA[71]	Grafted from PMMA MWCNTs	Dispersion–reaction	3°C/%	3%	2–3%
Polycarbonate[72]	MWCNTs	Melt mixing	Increase to 1% followed by decrease due to masterbatch resin having lower molecular weight and possible resin degradation catalyzed by nanotubes	NA	1%

(continued)

TABLE 4.1 (*Continued*)

Polymer	Nanotube type	Processing method	T_g change/wt% nanotube	Plateau (wt% nanotube)	Percolation threshold (wt% nanotube)[a]
Polycarbonate[73]	MWCNTs	Dissolution–dispersion–precipitation	−50°C/%	0.5%	0.5%
Polycarbonate[74]	MWCNTs	Melt mixing	−0.75°C/%	NA–15%[b]	15%
Various acrylic terpolymers[75]	Lithium-reduced MWCNTs	Dissolution–dispersion–precipitation	−100°C/%	0.3%	0.27–0.4%
MMA–butyl acrylate (BA) copolymers[76c]	Lithium-reduced MWCNTs	Dissolution–dispersion–precipitation			
100/0 MMA/BA			−15°C/%	3%	0.6%
90/10 MMA/BA			−5°C/%	NA–10%	0.4%
80/20 MMA/BA			−4°C/%	5%	0.35%
70/30 MMA/BA			−3°C/%	NA–7%	0.35%
60/40 MMA/BA			No change	NA	0.35%
50/50 MMA/BA			0.5°C/%	NA–10%	0.4%
25/75 MMA/BA			−3°C/% to 3%, then 2°C/% to 10%	NA	0.45%

MMA–ethyl methacrylate copolymer[77]	Pristine MWCNTs	Dissolution–dispersion–precipitation	10°C/% to 1%, then 0.7–10%	None	Not reported
	Amine-functionalized MWCNTs		2°C/%	None	
Styrene–isoprene copolymer[78]	SWCNTs	Dispersion–dispersion–evaporation	No change	NA	0.2%
Polyetherimide–Vectra blend[79]	Acid-functionalized MWCNTs	Melt mixing	10°C/%	NA–1.8%	Not reported
Polyimide[51]	MWCNTs	Dispersion–reaction	−0.75°C/%	NA–7%[b]	Not reported
	Acid-functionalized		3°C/%	2.5%	
	Amine-functionalized		10°C/%	0.75%	
Polyurethane (soft segment)[80]	Grafted from polyurethane MWCNTs	Dispersion–reaction	−5°C/% to 1%; 2°C/% to 2%	NA–2%[b]	Not reported
Polyurethane (soft segment)[65]	Grafted from polyurethane SWCNTs	Dispersion–reaction	2°C/%	NA–0.7%[b]	Not reported
Polyurethane (soft segment)[81]	Grafted from polyurethane SWCNTs	Dispersion–reaction	−2°C/%	NA–2%[b]	Not reported

(continued)

TABLE 4.1 (*Continued*)

Polymer	Nanotube type	Processing method	T_g change/wt% nanotube	Plateau (wt% nanotube)	Percolation threshold (wt% nanotube)[a]
Poly(ethylene terephthalate) (PET) [82]	Grafted from PET MWCNTs	Dispersion–reaction	15°C/%	0.3%	Less than 0.3%
Epoxy[83]	MWCNTs	Dispersion–reaction	50°C/%	0.5%	0.6%
Epoxy[84]	MWCNTs	Dispersion–reaction	No change	NA	Not reported
	Grafted from epoxy MWCNTs		~200°C/%	0.1%	
Epoxy[85]	SWCNTs	Dispersion–reaction	No change	NA	Not reported
Epoxy[86] (two types)	MWCNTs	Dispersion–reaction	No change for three out of four tested	NA	Not reported
	SWCNTs		−30°C/% for one MWCNT	NA–0.5%[b]	
Epoxy[87]	MWCNTs	Dispersion–reaction	−200°C/% to 0.1%		Not reported
	SWCNTs		~40°C to 1%	NA	
Epoxy[88]	MWCNTs Grafted from epoxy MWCNTs	Dispersion–reaction	15°C/%[d]	NA– 0.8%[b]	Not reported

Epoxy[89]	MWCNTs	Dispersion–reaction	20°C/%	1%	Not reported
Epoxy[90]	Grafted from epoxy MWCNTs	Dispersion–reaction	−50°C/%	NA–0.25%[b]	Less than 0.1%
Epoxy[91]	MWCNTs with dispersing surfactant	Dispersion–reaction	0.4°C/%	NA–8%	3%
	MWCNTs		0.8°C/%	NA–12%	
Dicyclopentadiene[92]	Grafted from dicyclopentadiene MWCNTs	Dispersion–reaction	10°C/%	NA–0.4%[b]	Not reported

Only those studies that used multiple nanotube levels and used differential scanning calorimetry to characterize the transition are included.

[a]As calculated from electrical conductivity.

[b]These studies showed no plateau at high nanotube concentrations, so the maximum weight percent nanotubes added is given.

[c]Some of these samples showed an increase in T_g at 10% tube content.

[d]These data showed a great deal of scatter to the point where a linear assumption may not be accurate.

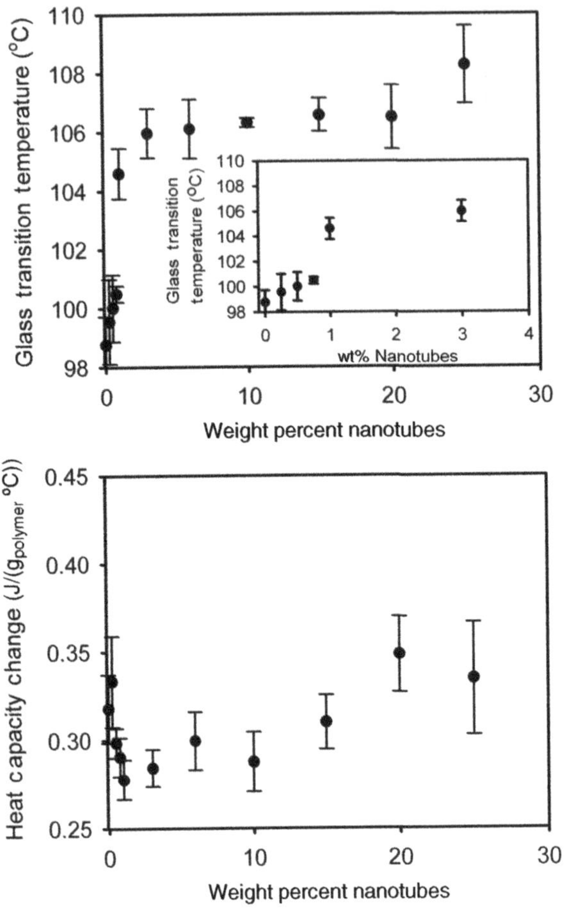

Figure 4.6 Effect of acid-functionalized SWCNTs on glass transition and change in heat capacity for polystyrene. Copyright 2009 American Chemical Society. Reprinted in part with permission from Ref. 64.

sheets offer an interesting parallel to nanotubes; one study using a dissolution–dispersion–precipitation mixing scheme did show an increase of 600°C/wt% graphene with a plateau at 0.05 wt%.[57] One obvious explanation for the plateau is that the amount of available surface area and hence the amount of polymer near a surface does not grow with an increase in filler content because of increased agglomeration; however, proving such a hypothesis is difficult. One study using gold nanoparticles in polystyrene, where the nanoparticles were imaged with transmission electron microscopy, did find both a plateau and definite evidence of agglomeration; however, evidence of agglomeration was found at a 1% nanoparticle level while the T_g plateau occurred at about 0.5%. Interestingly, the plateau and agglomeration weight percents did not shift upon changing the molecular weight of a grafted coating, although T_g shifted from a 10°C decrease to a 5°C increase.[58] Table 4.1 includes only changes that occur upon mixing a polymer with a nanotube and not covalently bonding a polymer to the nanotube; covalently attaching a polymer to the nanotube consistently increases the glass transition temperature of the attached polymer.[59,60]

Some of the large slopes found for some entries are due to an effect other than simply introducing a nanotube surface into a polymer including molecular weight reduction catalyzed by nanotubes that might be occurring for polycarbonate as noted in Table 4.1. Materials created via dispersion–reaction schemes could have effects on the T_g due to reaction completeness; that is, nanotubes may alter the chemical reaction that in turn causes changes in conversion that will affect the glass transition temperature. This effect was examined quantitatively by changing the cross-link density of an epoxy system by changing the ratio of mono- to difunctional amine and monitoring the change in glass transition temperature with added nanotubes.[61] At low cross-link densities, that is, when the mono/di ratio was greater than 1, the glass transition temperature increased at a rate of 5–10°C/wt% nanotubes. When the ratio was less than 1, the glass transition temperature decreased at approximately the same absolute rate. The authors explained this behavior as a complex interplay between surface immobilization, a change in the length (due to a change in cross-link density) over which a change in interfacial mobility can affect the system, and a change in reaction kinetics. As pointed out in a series of papers,[62,63] to properly determine the effect of adding carbon nanotubes to dispersion–reaction systems requires confirmation that the reaction itself is not affected by nanotubes. Possible interference is certainly more critical for thermosetting resins because of the strong effect of the crosslink density on the T_g, and more difficult to measure.

Nanotubes can have two other effects on the glass transition behavior of a polymer besides changing the glass transition temperature: nanotubes can change the broadness of the transition and/or decrease the amount of material participating in the glass transition. The latter will arise due to material that is strongly adsorbed to the interface, as was discussed previously. In other words, the dynamics have been altered for a fraction of the material to such an extent so as to cause a separation between regions of the polymer in a dynamic sense. This separation could cause a noticeable second glass transition at a higher temperature, or could cause no noticeable second glass transition if the second glass transition is above the degradation temperature of the polymer. Only a few papers have reported measuring the heat capacity change in amorphous polymer–nanotube composites, with a very detailed study recently published by the author's research group.[64] As Figure 4.6 shows, a decrease of about 20% in the heat capacity change was found at low nanotube levels (~1 wt%) in agreement with studies on polyurethanes[65] and polyisoprene,[66] but was much smaller than a factor of 3 decrease found in an epoxy composite.[52] Our work was unique in that we examined samples out to very high nanotube contents. At high nanotube contents, an increase in the heat capacity was found without any change in T_g; that is, the T_g was still in the plateau region. To our knowledge, this increase in heat capacity has not been reported in any other polymer nanocomposite. Further unpublished studies by the author's research group on other systems suggest that the increase at high nanotube contents can be due a poorer relative dispersion, but the possibility of a conformational rearrangement of isolated single-walled tubes as suggested in the paper cannot be ruled out.

Broadening (or narrowing) of the temperature range over which the steep increase in heat capacity occurs is a second effect that can be caused by the addition of nanotubes. None of the studies listed in Table 4.1 reported significant changes in the width of the glass transition. However, one study described the production of poly (vinyl alcohol) fibers with 20% nanotubes having such a broadened glass transition, as measured by the storage modulus; in fact the broadness became so large that the value of T_g was not clear.[67] Still, the preponderance of evidence suggests that the introduction of nanotubes does not cause a wide variety of environments for the polymer chains in a dynamic sense that in turn would lead to broadening of a single transition. Introducing nanotubes can, however, cause the formation of a second dynamic environment, which presumably is due to strongly adsorbed polymer in the immediate vicinity of the nanotubes.

4.3 SEMICRYSTALLINE POLYMERS

Figure 4.1 is a reasonable representation of a single polymer chain, and it is a simple matter to visualize how thousands upon thousands of these chains would form a bulk polymer. In the classroom, I often take 50 long pieces of string of various sizes, put them into a pile and announce to students "Here is what an amorphous polymer looks like." In other words, understanding, at least in a qualitative sense, the morphology of an amorphous polymer is not difficult. However, a complete qualitative description of semicrystalline polymer morphology is extremely difficult to generate. A semicrystalline material has regions that are both amorphous and crystalline, and a picture of a representative individual chain similar to Figure 4.1 does not exist since individual chains have many different configurations. To describe the morphology of a semicrystalline polymer, three length scales must be considered: the angstrom length scale, tens of nanometer length scale, and micron length scale.

Three relevant regions at the angstrom-level length scale must be considered. To a first approximation, the angstrom-level arrangement of molecules in the amorphous region is the same as that shown in Figure 4.1, and hence will not be discussed further. Of course, a single polymer chain can have part in a crystal and part not, but for that part of the chain in the amorphous region, Figure 4.1 is a reasonable representation. The angstrom-level arrangement in the crystalline regions is fundamentally no different from that for a small molecule, that is, atoms are found in discrete positions in one of seven types of "boxes" (e.g., lattice systems) and the same 230 space groups (i.e., symmetry operations related to the position of the atoms within one of the seven types of lattice systems) as for small molecule crystals. The description of unit cells found in any elementary materials science textbook is sufficient for understanding the angstrom-level length scale morphology of polymers at the level required for most scientists and engineers working with nanotube composites. The third region is the interfacial component, that is, the interface between the crystalline and amorphous regions. Chain folding, described in the next paragraph, is part of this interfacial region; relatively little is known about the interfacial region compared to other two regions.

Tens of nanometer is the next length scale to consider. Polymer crystals are usually lamellar, with the chains oriented and folded as shown in Figure 4.7. Thicknesses of lamellae are typically on the order of 10 nm, while the other two dimensions for the lamellae are on the order of 100–1000 nm. The frequency of chain folding determines the lamellar thickness with the preferred thermodynamic state

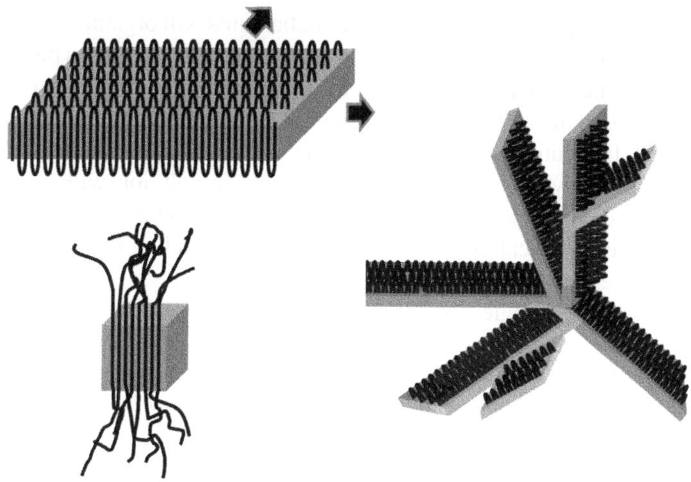

(a) (b)

Figure 4.7 Schematic showing morphology at two of the three length scales of crystallization: (a) tens of nanometer level (lamellar structure, top; fringed micelle, bottom) and (b) micron level (spherulite structure). The fringed micelle structure would extend in a direction orthogonal to the paper so that a granular-type morphology would result.

being no chain folding. Hence, treatments that allow longer time for crystallization and/or more molecular mobility will yield thicker crystals. How do chains, after folding, reenter the crystal? For simplicity, adjacent reentry is shown in Figure 4.7; that is, a chain reenters the crystal in the unit cell next to the one from which the polymer exited. The actual reentry point depends on the conditions under which the crystals are formed, but the best description seems to be one where the most energetically favored reentry is a couple of unit cells away from where the polymer exits the crystal. Not all polymer chains reenter the same crystal; the fraction of chains that do not is not a well-characterized parameter but does seem to depend very strongly on the processing method (and chains that span lamellae are believed to have a great deal of influence on large-strain mechanical properties!). Most of the discussion of this section will focus on the effect of nanotubes on the growth and shape of the lamellae. An alternate morphology to chain-folded lamallae that occurs in polymers in special situations is a fringed micelle.[93,94] As shown in Figure 4.7, fringed micelle crystals form without chain folding. During the first half of the twentieth century, most scientists believed that polymers crystallized into a fringed micelle rather than a lamellar morphology; today it is known that the fringed micelle morphology occurs only for polymers with a high level of substitution along the polymer backbone. However, the fringed micelle morphology has special significance with nanotubes, since nanotubes at sufficient concentrations in solution or in a polymer almost certainly have a fringed micelle structure.

The final length scale to consider is the micron length scale. The morphology on this length scale can be quite different depending on the conditions under which crystals are formed. With little or no external stress, lamellar units will organize into a spherulitic structure as depicted in Figure 4.7. In spherulites, lamellae begin at a point and radiate outward with the thickness direction of the lamellae being orthogonal (or close to orthogonal) to the radius of the sphere. Growth of lamellae in between radial arms is possible as well. In solution, spherulites will generally form in the absence of large stress fields. Under very special conditions, polymers in solution will form single crystals with a diamond-shaped morphology; these structures are very important in terms of the historical development of the understanding of polymer crystals but are of no relevance to this text.

Two other types of morphologies are possible with carbon nanotubes. One type occurs more frequently with nanotubes but only under special conditions with pure polymers, and the other type occurs frequently both with nanotubes and with other cylindrical fillers such as glass or carbon fibers. The first morphology is termed the shish-kebab morphology, where, in pure polymers, elongated chains serve as the shish and lamellae serve as the kebabs. This morphology occurs when polymer crystallization transpires under very high elongational stress fields and can occur either in solutions or in melts. With nanotubes, this shish-kebab morphology will form with or without a stress field; however, in this case, carbon nanotubes serve as the shish. Such structures are shown in Figure 4.8 and have been termed nanohybrid shish-kebabs (NHSKs). The second morphology is termed transcrystallinity, and has the crystal growth direction perpendicular to the nanotube axes just as in NHSKs, but in this case lamellae have a high enough density so as to overlap in the nanotube axis direction to the point where individual lamellae cannot be distinguished. Both shish-kebabs and transcrystallinity result from the same phenomena, nucleation of crystallinity by carbon nanotubes. The morphologies differ only in the nucleation density; in NHSKs nucleation sites along the nanotube axis are widely separated, while in transcrystallinity nucleation sites along the nanotube axis are close together. Further discussion of both morphologies is given in Section 4.3.1.

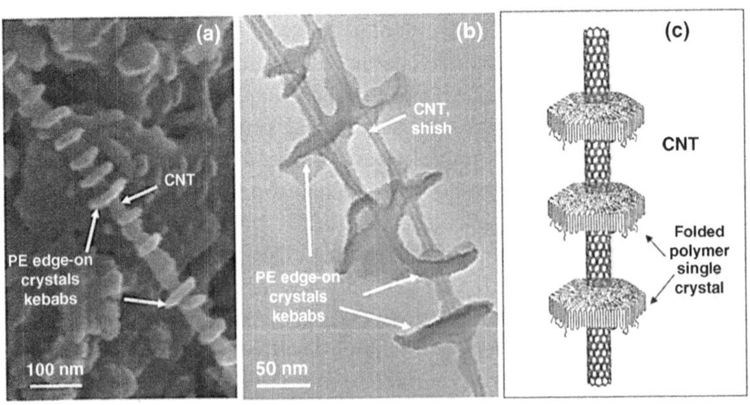

Figure 4.8 (a) SEM, (b) TEM, and (c) schematic representation of shish-kebab structure of polymer crystals growing on a surface of a nanotube. Parts (a) and (b) are micrographs of solution-crystallized polyethylene. Copyright Wiley-VCH Verlag GmbH & Co. KGaA. Reproduced with permission from Ref. 267.

Another important structural aspect of polymer crystallization is the equilibrium melting temperature. The equilibrium melting temperature is defined as the melting temperature for infinitely thick crystals, that is, crystals with no chain folding. Generally, the determination of the equilibrium melting temperature is done in one of the two ways. The first is to plot measured melting temperature (T_m) versus measured isothermal crystallization temperature (T_c). The intersection point of the extrapolation of this line with the $T_m = T_c$ line yields the equilibrium melting temperature. The second method is to plot isothermal crystallization temperature versus (1/lamellar thickness) and extrapolate a line back to $x = 0$ to determine the equilibrium melting temperature. The lamellar thickness is usually determined via small-angle X-ray diffraction.

The rate of the amorphous–crystal transition depends on the rate at which a new crystal begins to grow (nucleation) and the rate at which a crystal grows (growth). Because nanotubes nucleate crystallinity for a number of polymers, nanotubes can have a large effect on the nucleation rate. Nanotubes can also affect the growth rate of crystals, because nanotubes can alter chain mobility as well as provide spatial impediments to growth. The growth of crystals is usually divided into two regimes, a primary regime where crystallization is rather quick and due primarily to the growth of lamellae in the arms that radiate outward from the spherulitic centers and secondary crystallization due to growth of crystals between arms. In some cases, secondary growth leads to a bimodal distribution of crystal thicknesses. For some polymers, nanotubes have been shown to increase the overall amorphous–crystalline conversion, which is normally attributed to an increase in nucleation rate; in other cases, the growth rate has slowed, which is normally attributed to slower diffusion/impediments.

The rate at which crystals form in polymers is most often quantified in one of the two ways. DSC measures the fractional crystallinity versus time since the formation of crystals from an amorphous phase causes the release of energy. Hence, DSC measurements cannot be used to independently quantify changes in nucleation and growth rate. Optical microscopy follows growth only because spherulitic (or transcrystalline or kebab) dimensions with time are measured. Because of the fundamental difference with respect to what is measured, crystallization data from DSC and optical microscopy are analyzed differently. For the former, relative crystallinity (φ_c) is defined as the instantaneous fractional crystallinity (i.e., fractional crystallinity at some given time) normalized to the fractional crystallinity at infinite time. For a DSC crystallization experiment, φ_c is simply the fraction of heat released. The relative crystallinity plotted versus time for crystallization that occurs at constant temperature yields a curve with sigmoidal shape. For the temporal region corresponding exclusively to primary crystallization, which is usually a relative crystallinity between approximately 0.05 and 0.75, the Avrami equation is used to describe the crystallization rate. The Avrami equation assumes constant and random nucleation, spherulite growth that is isotropic and constant, and a constant melt composition. In practice, these assumptions are met for most polymers if temperature is constant. The Avrami equation has the form

$$\varphi_c = 1 - \exp(-k(t-\tau_0)^n) \tag{4.3}$$
$$\ln(-\ln(1-\varphi_c) = n(t-\tau_0) + \ln k$$

The linearized form shown in the second equation is the typical way in which data are plotted in order to more easily find the fitting parameters (n, k, and τ_0). The knowledgeable reader might note that the induction time, τ_0, is usually not shown in this equation; practically it is necessary to choose some time where crystallization begins and this form of the equation makes this necessary choice explicit. The value of n is indicative of the dimensionality of growth: theoretically $n = 4$ represents three-dimensional growth, $n = 3$ represents two-dimensional growth, and $n = 2$ represents one-dimensional growth. Practically, n should be viewed as a fitting parameter with a decrease in n representing a decrease in the dimensionality of growth. The rate constant, k, represents the rate at which crystals nucleate and grow. The crystallization half-time, defined as the time required for the relative crystallinity to reach a value of 0.5, is often used as a one-parameter measure to quantify the change in crystallization rate. Scattering, not used as frequently as calorimetry because of poorer time resolution, gives kinetic data almost equivalent to the kinetic data from calorimetry.

The Avrami treatment cannot distinguish between an increase in nucleation rate and an increase in growth rate, although constancy of n is usually taken as evidence that any rate changes are due to a change in nucleation rate. A clearer indication of an increase in nucleation rate via DSC experiments is given by nonisothermal experiments. In nonisothermal experiments, a polymer is cooled from the melt at a constant rate (°C/min) and the temperature corresponding to either when the crystallization exotherm begins or when it reaches its maximum value is recorded. A higher temperature with the addition of filler indicates nucleation by the filler. Nonisothermal data at constant cooling rate can be analyzed more quantitatively, for example, the Ozawa method, but such approaches are inherently more difficult and more susceptible to artifacts than isothermal experiments, so those methods are left for the interested reader to explore elsewhere.

Data from optical microscopy is analyzed in terms of the radial growth rate versus time. Until spherulites begin impinging on one another, that is, in the primary growth region, the radial growth rate tends to be linear at a given temperature. A plot of the logarithm of the radial growth rate versus crystallization temperature can be used to identify the way in which lamellae grow, as given by Hoffman's theory of crystallization. In general, a plot of radial growth rate versus temperature will show a maximum where the interplay of thermodynamic driving force (favored at lower temperatures) and fast chain diffusion (favored at higher temperatures) balance one another. A detailed description of Hoffman's theory is beyond this tome and is available in any standard polymer physics textbook, but Hoffman's theory is applicable to high crystallization temperatures, and a change in the slope of radial growth rate versus crystallization is used to distinguish the three regimes as defined by the theory. In short, at the highest temperatures, termed regime 1, growth of a lamellar length occurs after one nucleation event; at intermediate temperatures (regime 2), multiple nucleation events occur before growth reaches a lamellar length; and at lowest temperatures (regime 3), nucleation occurs as fast or faster than growth.

4.3.1 Statics: Unit Cells, Lamellae, Spherulites, and Shish-Kebabs

Polymorphism in polymers, that is, a polymer that will crystallize into more than one unit cell, is not uncommon. Nucleating agents are one very common way to induce the formation of one unit cell versus another, and since carbon nanotubes are nucleating agents for a number of polymers as evidenced by an increase in nucleating temperature in nonisothermal DSC experiments (see Table 4.2), it certainly is possible that nanotubes might promote one crystal form over another. For example, nanotubes favor the alpha crystalline form in polyamide 6 filled with MWCNTs,[95–98] which is opposite that of nanoclays that favor the gamma form.[99,100] Nanotubes favor the beta phase of poly(vinylidene fluoride) over the alpha phase for both pristine tubes[101,102] and fluorine-functionalized tubes.[103] In poly(1-butene), the transition from the kinetically favored tetragonal form to the thermodynamically favored hexagonal form was enhanced by MWCNTs.[104] In syndiotactic polypropylene, the addition of carbon nanotubes causes an increase in the more ordered helical form I of this polymer, as opposed to the disordered *trans*-planar form I.[105] An early paper by the present author suggested that nanotubes favored the beta phase in isotactic polypropylene based on multiple endotherms in DSC experiments,[106] while other papers suggested that the alpha form was favored.[107–110] Nanotubes can induce complicated melting–recrystallization phenomena, indicating that claims about polymorphism in isotactic polypropylene must be substantiated with X-ray diffraction measurements casting significant doubt on the conclusions reached by the present author in the early paper. Other than the specific polymers listed, the introduction of carbon nanotubes has not been reported to change other unit cells.

To discuss the role of carbon nanotubes on lamellae and spherulites characteristics, it is first important to discuss the nucleation mechanism. Carbon nanotubes cause the growth of crystals so that the chain axis direction is parallel to the nanotube axis; that is, the growth of crystals is in a direction perpendicular to the growth direction. Example micrographs of this type of crystallization are shown in Figures 4.8 and 4.9. In order to get large separation between lamellae as found in the NHSK structures shown Figure 4.8, the growth rate must be fast relative to the nucleation rate. In the transcrystalline morphology, lamellae growing in the normal direction to the fiber axis impinge upon one another in the direction along the fiber axis and hence the lamellar thickness is set by the nucleation density. Normally of course, interplay between thermodynamics and kinetics determines when chains fold, which in turn sets the lamellar thickness. Further, the characterization of transcrystalline morphologies in bulk is significantly more difficult since nanotubes are so small; with macroscopic fibers such as carbon or glass, such studies are typically done using optical microscopy with single fibers embedded in the polymer.

For shish-kebabs grown in solution, a very interesting observation is that the orientation of the chain is always perpendicular to the nanotube, irrespective of its chirality,[111] indicating that growth is likely induced by a geometrical constraint rather than an epitaxial (lattice matching) constraint. However, the lattice matching constraint is still possible if growth causes a "straightening" of the lamellae relative to the chain axis. As the diameter of the nanotube becomes larger, and again the

TABLE 4.2 Changes in the Crystallization Kinetics with Nanotube Incorporation for Both Nonisothermal and Isothermal Kinetic Studies

Polymer	Nanotube type	Processing method	Change in peak crystallization temperature (nonisothermal)	Ratio of crystallization half-times (isothermal)	Avrami exponent change (isothermal)	Nanotube levels investigated (wt%)
Isotactic polypropylene[106]	Grafted to octyldecylamine SWCNTs	Dissolution–dispersion–precipitation	2°C, 6°C	0.4, 0.4	Decrease	0.6%, 1.8%
Isotactic polypropylene[149]	MWCNTs	Dissolution–dispersion–precipitation	6°C	NA	NA	1%
Isotactic polypropylene[146,150,151]	SWCNTs	Melt mixing	13°C, 14°C, 15°C, 18°C	NA	NA	5%, 10%, 15%, 20%
Isotactic polypropylene[107]	SWCNTs	Melt mixing	4°C, 5°C, 6°C	NA	NA	0.25%, 0.5%, 1%
Isotactic polypropylene[152]	MWCNTs	Melt mixing with liquid-dispersed tubes	2°C, 4°C, 5°C, 6°C	NA	NA	0.1%, 0.5%, 1%, 3%
Isotactic polypropylene[153–156]	MWCNTs	Melt mixing	7°C	NA	NA	0.5%
Isotactic polypropylene[157]	SWCNTs	Melt mixing	9°C, 9°C, 9°C, 10°C, 11°C	NA	NA	0.25%, 0.5%, 0.75%, 1%, 2%
Isotactic polypropylene[158]	MWCNTs	Melt mixing	NA	0.1	Increase	0.1%

Isotactic polypropylene[159]	MWCNTs	Melt mixing	6°C, 10°C, 15°C, 13°C	NA	NA	1.5%, 2.7%, 4%, 8%
Isotactic polypropylene[159]	MWCNTs	Melt mixing	2°C, 5°C, 9°C, 8°C	NA	NA	2%, 2.7%, 4%, 8%
Isotactic polypropylene[160]	MWCNTs	Melt mixing	7°C, 7°C	NA	NA	0.1%, 0.3%
Isotactic polypropylene[144]	Acid-functionalized MWCNTs	Melt mixing	10°C, 10°C, 11°C, 11°C, 13°C	NA	NA	1%, 2%, 3%, 5%, 7%
Isotactic polypropylene[145]	C_{18} grafted chains	Dissolution–dispersion–precipitation	NA	Decrease to 0.2 at 1%, then slow decrease to 0.1 at 9%	Increase	Various to 9%
Isotactic polypropylene[109]	MWCNTs	Dissolution–dispersion–precipitation	9°C, 12°C	NA	NA	1%, 2%
Isotactic polypropylene[161]	MWCNTs	Melt mixing	0°C, 1°C, 5°C, 6°C	NA	NA	0.1%, 0.3%, 0.5%, 1%
Isotactic polypropylene[162]	Grafted from isotactic polypropylene MWCNTs	Dispersion–reaction	Increase to 8°C until plateau reached at 4%	Decrease to 0.2–0.4 at 4%	NA	Various to 25%

(continued)

TABLE 4.2 (*Continued*)

Polymer	Nanotube type	Processing method	Change in peak crystallization temperature (nonisothermal)	Ratio of crystallization half-times (isothermal)	Avrami exponent change (isothermal)	Nanotube levels investigated (wt%)
Isotactic polypropylene[163]	Acid-functionalized MWCNTs at various functionalization levels	Melt mixing	8–9°C	NA	NA	2.5%
Isotactic polypropylene[164]	Grafted from isotactic polypropylene MWCNTs	Dispersion–reaction	4°C, 8°C, 7°C, 7°C, 14°C, 12°C, 14°C	NA	NA	0.1%, 0.4%, 0.6%, 0.9%, 1.4%, 2.1%, 3.5%
Isotactic polypropylene[165]	MWCNTs	Dispersion–reaction	6–10°C	0.5–0.1[a]	Increase	Various to 0.1–5%
	Acid-functionalized MWCNTs		5–10°C	0.5–0.2	Increase	
	Grafted from isotactic polypropylene MWCNTs		5–9°C	0.5–0.3	Increase	
Isotactic polypropylene with grafted maleic anhydride[159]	MWCNTs	Melt mixing	11°C, 16°C, 18°C, 11°C	NA	NA	1.2%, 2.7%, 4%, 8%

Isotactic polypropylene[120]	MWCNTs with silica and silane coating	Melt mixing	NA	0.1, 0.075	Increase	0.5%, 1%
	MWCNTs			0.1, 0.075		
Isotactic polypropylene[147]	MWCNTs with various treatments, including surfactants	Dissolution–dispersion–precipitation	Increase between 3°C and 8°C	NA	NA	1%
Isotactic polypropylene[166]	MWCNTs	Melt mixing	8°C for all	NA	NA	1%, 2%, 3%
Isotactic polypropylene[127]	MWCNTs	Melt mixing	3°C, 6°C, 9°C	NA	NA	1%, 3%, 10%
	Acid-functionalized MWCNTs		7°C			10%
Isotactic polypropylene[167]	SWCNTs	Melt mixing	NA	0.1	Increase	0.8%
Isotactic polypropylene[168]	SWCNTs	Melt mixing	8°C, 8°C, 11°C, 12°C, 13°C	NA	NA	0.25%, 0.5%, 1%, 1.5%, 2.5%
Isotactic polypropylene[119]	MWCNTs Grafted to isotactic polypropylene	Melt mixing	NA	Not able to determine from the data given	Increase	1%, 1.5%, 2%
Isotactic polypropylene[169]	MWCNTs	Melt mixing	12°C, 13°C, 14°C, 17°C	NA	NA	1%, 2%, 4%, 8%
Isotactic polypropylene[170]	MWCNTs	Dissolution–dispersion–precipitation	NA	0.6	Decrease	0.5%

(*continued*)

TABLE 4.2 (*Continued*)

Polymer	Nanotube type	Processing method	Change in peak crystallization temperature (nonisothermal)	Ratio of crystallization half-times (isothermal)	Avrami exponent change (isothermal)	Nanotube levels investigated (wt%)
Isotactic polypropylene[134]	MWCNTs	Melt mixing	NA	0.1, 0.1, 0.05	NA	0.5%, 1%, 3.5%
		Shear pulverization		0.07, 0.03, 0.04		
Isotactic polypropylene[171]	MWCNTs	Melt mixing	11°C, 13°C, 15°C	NA	NA	1%, 2%, 4%
Isotactic polypropylene[110]	MWCNTs	Melt mixing	9°C, 10°C	NA	NA	3%, 5%
Isotactic polypropylene with 2% ethylene[172]	MWCNTs	Melt mixing	All 19 ± 1°C	NA	NA	1.1–5.3%
Isotactic polypropylene with grafted maleic anhydride[173]	MWCNTs with grafted maleic anhydride	Melt mixing	9°C	NA	NA	0.5°C
Isotactic polypropylene with grafted maleic anhydride[174]	Acid-functionalized MWCNTs	Dissolution–dispersion–precipitation	7°C	NA	NA	1%
	Acid-functionalized SWCNTs		5°C, 10°C			0.1%, 1%
Isotactic polypropylene with grafted maleic anhydride[108,114]	SWCNTs and MWCNTs	Dispersion–dispersion–evaporation	Increase to plateau of 10°C (MWCNTs) or 15°C (SWCNTs) at 0.75%	NA	NA	Various to 2%

Syndiotactic polypropylene[175]	Grafted from syndiotactic polypropylene MWCNTs	Dispersion–reaction	7°C, 9°C, 15°C	0.35, 0.16, too fast to measure	Increase	0.1%, 0.5%, 0.9%
Syndiotactic polypropylene[125]	MWCNTs	Melt blending	18°C, 19°C, 20°C	NA		1%, 3%, 10%
Syndiotactic polypropylene[164]	Grafted from syndiotactic polypropylene MWCNTs	Dispersion–reaction	18°C, 17°C, 18°C, 17°C, −1°C	NA	NA	0.1%, 0.4%, 0.7%, 1.5%, 2.5%
Low-density polyethylene[176]	MWCNTs	Melt mixing	All 4°C	NA	NA	0.5%, 1%, 2%, 5%
High-density polyethylene[116]	SWCNTs	Dissolution–dispersion–precipitation	All 5°C	Roughly 0.1	Decrease	1%, 2%, 5%
High-density polyethylene[113]	SWCNTs	Dissolution–dispersion–precipitation	No change	0.33–0.25, constant thereafter	NA	0.1–1%
High-density polyethylene[122,177]	Grafted to high-density polyethylene SWCNTs, MWCNTs, DWCNTs	Dispersion–reaction	2°C for almost all concentrations	Very large change (0.01), with SWCNTs < DWCNTs < MWCNTs	No change for 8% sample, large decrease for ~30% sample	Three concentrations from ~8% to 30% NT for each type of NT

(*continued*)

TABLE 4.2 (*Continued*)

Polymer	Nanotube type	Processing method	Change in peak crystallization temperature (nonisothermal)	Ratio of crystallization half-times (isothermal)	Avrami exponent change (isothermal)	Nanotube levels investigated (wt%)
High-density polyethylene[178]	SWCNTs	Dissolution–dispersion–precipitation	NA	0.05%, 0.15; 2% and 8%, 0.07	No change	0.02–8%
High-density polyethylene[179]	MWCNTs	Melt mixing	5°C	NA	NA	2.5%
High-density polyethylene[118]	MWCNTs	Melt mixing	4°C, 4°C, 5°C, 5°C	NA	NA	1%, 3%, 5%, 7%
High-density polyethylene[6]	MWCNTs grafted to high-density polyethylene	Melt mixing	1°C	0.6	NA	0.52%
Ultrahigh molecular weight polyethylene[180]	MWCNTs	Dissolution–dispersion–precipitation	No change within experimental error	NA	NA	0.5%, 1%, 2%
Ultrahigh molecular weight polyethylene[181]	SWCNTs with catalyst support	Powder mixing	No change within experimental error	NA	NA	0–0.7%
Ultrahigh molecular weight polyethylene[7]	SWCNTs	Variation on dissolution–dispersion–precipitation	NA	0.21, 0.14 (onset from rheology)	NA	0.1%, 0.6%

Ethylene–vinyl acetate copolymer[182]	SWCNTs	Dissolution–dispersion–precipitation	5°C	NA	NA	10%
Polyamide 6[183]	SWCNTs with encapsulated styrene–maleic anhydride (SMA) copolymer	Melt mixing	5°C (no SMA coating), 4°C (with SMA coating)	NA	NA	1.3%
Polyamide 6[184]	MWCNTs	Melt mixing	NA	0.35	Decrease	1%
	Amine-functionalized MWCNTs			0.5		
Polyamide 6[185]	Acid-functionalized MWCNTs	Dissolution–dispersion–precipitation	4°C, 5°C, 2°C	0.5, 0.25, 0.85	Decrease	0.25%, 1%, 3%
Polyamide 6[97]	Acid-functionalized MWCNTs	Melt mixing	10°C, 11°C, 11°C, 10°C	NA	NA	0.2%, 0.5%, 1%, 2%
Polyamide 6[186]	MWCNTs	Melt mixing	3°C	NA	NA	Both 2%
	Acid- functionalized MWCNTs		4°C			
Polyamide 6[98]	MWCNTs Grafted from polyamide 6	Dispersion–reaction	Low peak: 6°C; high peak: 19°C for both	NA	NA	1%, 2%, 3%, 5%, 7%

(*continued*)

TABLE 4.2 (*Continued*)

Polymer	Nanotube type	Processing method	Change in peak crystallization temperature (nonisothermal)	Ratio of crystallization half-times (isothermal)	Avrami exponent change (isothermal)	Nanotube levels investigated (wt%)
Polyamide 6[135]	MWCNTs	Melt mixing	Low peak: 3°C, 3°C, 5°C, 8°C; high peak: 12°C, 17°C, 18°C, 20°C	NA	NA	1%, 3%, 5%, 10%
Polyamide 6[96]	MWCNTs	Melt mixing	4°C for all	NA	NA	0.1%, 0.5%, 1%
	Amine-functionalized MWCNTs		3°C for all			
Polyamide 6[187]	Amine-functionalized MWCNTs	Melt mixing	5°C	NA	NA	1%
Polyamide 6[188]	MWCNTs	Melt mixing	5°C	NA	NA	All 1%
	Acid-functionalized MWCNTs		10°C			
	Amine-functionalized MWCNTs		14°C			
Polyamide 6[189]	MWCNTs	Melt mixing	3°C	NA	NA	1%

Polyamide 6[190]	SWCNTs	Melt mixing	All 4°C	NA	NA	0.1%, 0.25%, 0.5%, 1%
Polyamide 6[191]	MWCNTs	Melt mixing	5°C, 6°C, 8°C	NA	NA	1%, 3%, 10%
	Acid-functionalized MWCNTs		8°C			10%
Polyamide 6[192]	MWCNTs	Melt mixing	Low peak 4°C; high peak 17°C	NA	NA	2%
Polyamide 6,6[193]	Acid-functionalized MWCNTs (various functionalization levels)	Dissolution–dispersion–precipitation	NA	0.42, 0.29, 0.26	Decrease	0.5%, 1%, 3%
Polyamide 6,6[112]	MWCNTs	Dissolution–dispersion–precipitation	2°C, 1°C, −3°C	0.6, 1, 1.2	Decrease	0.1%, 0.5%, 2%
Polyamide 6,6[194]	MWCNTs	Dissolution–dispersion–precipitation	3°C, 3°C, 6°C	NA	NA	0.5%, 1%, 2%
	Amine-functionalized MWCNTs		5°C			0.5%
Polyamide 6,6[192]	MWCNTs	Melt mixing	10°C	NA	NA	2%
Polyamide 6,10[195]	Acid-functionalized MWCNTs	Melt mixing	All 3°C	NA	NA	0.1%, 0.5%, 1%, 2%

(continued)

TABLE 4.2 (*Continued*)

Polymer	Nanotube type	Processing method	Change in peak crystallization temperature (nonisothermal)	Ratio of crystallization half-times (isothermal)	Avrami exponent change (isothermal)	Nanotube levels investigated (wt%)
Polyamide 10,10[196]	SWCNTs	Melt mixing	11°C	0.18	Decrease	All 2°C
	MWCNTs		12°C	0.16		
	Acid-functionalized MWCNTs		13°C	0.21		
Polyamide 10,10[197]	MWCNTs	Melt mixing	All 6°C	0.19, 0.16, 0.09	Decrease	1%, 2%, 3%
Polyamide 12[123]	Grafted from styrene–maleic anhydride SWCNTs	Melt mixing	5°C, 5°C	NA	NA	3%, 6%
	SWCNTs		6°C, 6°C			
Polycaprolactone[198]	Hydroxy-functionalized MWCNTs	Melt mixing	3°C, 7°C, 9°C, 9°C, 10°C	NA	NA	0.5%, 1%, 2%, 3%, 5%
Polycaprolactone[199]	Acid-functionalized MWCNTs	Melt mixing	5°C, 6°C, 8°C	NA	NA	0.1%, 1%, 2%
Polycaprolactone[200]	Acid-functionalized MWCNTs	Dispersion–reaction	NA	1.1,1.2, 1.25	Decrease	0.1%, 1%, 5%

Polycaprolactone[148]	Acid-functionalized MWCNTs	Dissolution–dispersion–reaction	1% about 15°C	0.07, 0.06, 0.05	Decrease 0.25%, 0.5%, no change 1%	0.25%, 0.5%, 1%
Polycaprolactone[201]	Grafted from polycaprolactone MWCNTs	Dispersion–reaction	5°C	NA	NA	Not clear
Polycaprolactone[202]	MWCNTs	Melt mixing	10°C, 11°C, 11°C	NA	NA	0.5%, 1%, 2%
Polycaprolactone[203]	MWCNTs	Dispersion– reaction	3°C, 4°C, 5°C, 5°C, 4°C, 5°C	NA	NA	0.5%, 1%, 2%, 3%, 5%, 7%
	Grafted from polycaprolactone SWCNTs		3°C, 4°C, 5°C, 5°C, 6°C, 7°C			
Polycaprolactone[204]	Grafted from polycaprolactone SWCNTs	Dispersion–reaction	8°C, 9°C, 15°C	NA	NA	0.35%, 1.8%, 4.6%
Polycaprolactone[205]	MWCNTs	Dissolution–dispersion–precipitation	NA	0.2,0.15, 0.15	Decrease	0.25%, 0.5%, 1%
Poly(vinyl alcohol)[206]	DWCNTs	Dissolution–dispersion–precipitation	13°C	NA	NA	0.5%

(continued)

TABLE 4.2 (*Continued*)

Polymer	Nanotube type	Processing method	Change in peak crystallization temperature (nonisothermal)	Ratio of crystallization half-times (isothermal)	Avrami exponent change (isothermal)	Nanotube levels investigated (wt%)
Poly(vinyl alcohol)[141]	Various tubes	Dissolution–dispersion–precipitation	20–30°C with plateau below 0.1%; except one MWCNT type where curve was flat until 20°C at 0.3%	NA	NA	Various to 0.3%
Poly(vinyl alcohol)[207]	SWCNTs	Dissolution–dispersion–precipitation	4°C, 5°C	NA	NA	0.1%, 1%
Poly(ethylene oxide)[208]	MWCNTs	Dispersion–dispersion–evaporation	0°C, −11°C, −15°C, −20°C	NA	NA	0.4%, 1%, 2%, 4%
Poly(ethylene oxide)[209]	MWCNTs	Dissolution–dispersion–precipitation	1°C, 2°C, 1°C, 1°C	NA	NA	0.5%, 1%, 2%, 5%

Poly(ethylene oxide)[210]	MWCNTs	Dissolution–dispersion–precipitation	−9°C, −12°C, −28°C	1.1, 1.2, 1.8	Decrease	0.1%, 0.5%, 1%
	Acid-functionalized MWCNTs		−11°C, −20°C, −33°C	1, 1.1, 1.5		
	Hydroxy-functionalized MWCNTs		−10°C, −11°C, −28°C	1.1, 2, 2		
Poly(ethylene oxide)[147]	MWCNTs with various treatments, including surfactants	Dissolution–dispersion–precipitation	3–10°C	NA	NA	1%
Poly(ethylene oxide)[201]	Grafted from poly (ethylene oxide) MWCNTs	Dispersion– reaction	10°C	NA	NA	Not clear
Poly(ethylene oxide)[211]	SWCNTs with various surfactants	Dissolution–dispersion–precipitation	Decrease in T_c, but surfactant plays a role	NA	NA	Various to 0.2%
Poly(ethylene oxide)[212]	MWCNTs with salts of suberic acid	Dissolution–dispersion–precipitation	−11°C	NA	NA	2%
Poly(ethylene terephthalate)[213]	SWCNTs	Melt mixing	10°C, 12°C, 13°C, 14°C, 19°C	0.25, 0.2, 0.15, 0.13, 0.1	NA	0.03%, 0.1%, 0.3%, 1%, 3%

(*continued*)

TABLE 4.2 ***(Continued)***

Polymer	Nanotube type	Processing method	Change in peak crystallization temperature (nonisothermal)	Ratio of crystallization half-times (isothermal)	Avrami exponent change (isothermal)	Nanotube levels investigated (wt%)
Poly(ethylene terephthalate)[131]	Acid-functionalized MWCNTs	Melt mixing	13°C, 16°C, 19°C	NA	NA	0.5%, 2%, 4%
Poly(ethylene terephthalate)[214]	MWCNTs	Melt mixing	2°C, 6°C, 13°C, 17°C, 19°C	NA	NA	0.5%, 2%, 4%, 8%, 10%
Poly(ethylene terephthalate)[215]	MWCNTs	Melt mixing	20°C, 21°C, 23°C	NA	NA	0.5%, 1%, 2%
Poly(ethylene terephthalate)[216]	SWCNTs	Dissolution–dispersion–precipitation	23°C, 27°C, 31°C	NA	NA	0.3%, 1%, 3%
Poly(ethylene terephthalate)[217]	Acid-functionalized MWCNTs	Melt mixing	30°C, 31°C, 32°C, 34°C	NA	NA	0.1%, 0.5%, 1%, 2%
Poly(ethylene terephthalate)[218]	MWCNTs	Melt mixing	5°C, 6°C, 7°C, 8°C, 9°C	0.27, 0.15 (0.1% and 0.5% only)	Decrease	0.1%, 0.5%. 1%, 2.5%, 5%
		Dissolution–dispersion–precipitation	5°C, 6°C, 8°C, 11°C, 13°C			0.1%, 0.5%. 1%, 2.5%, 5%
Poly(ethylene terephthalate)[219]	MWCNTs	Melt mixing	See Figure 4.10			See Figure 4.10

Poly(ethylene terephthalate)[220]	MWCNTs	Dispersion–reaction	27°C, 21°C, 19°C, −11°C (decrease attributed to reaction)	NA	NA	0.25%, 0.5%, 1%, 2%
Poly(ethylene naphthalate)[221]	MWCNTs	Melt mixing	12°C	NA	NA	0.5%
	Acid-functionalized MWCNTs		14°C			
Poly(ethylene naphthalate)[222]	MWCNTs	Melt mixing	22°C, 24°C, 27°C, 27°C	0.2, 0.15, 0.11, 0.08	Decrease	0.1%, 0.5%, 1%, 2%
Poly(ethylene naphthalate)[223]	MWCNTs	Melt mixing	24°C, 24°C, 25°C, 26°C	NA	NA	0.1%, 0.5%, 1%, 2%
Poly(butylene terephthalate)[139]	SWCNTs	Dissolution–dispersion–precipitation	NA	0.2, 0.05, 0.05	NA	0.01%, 0.1%, 0.2%
Poly(butylene terephthalate)[137]	MWCNTs	Melt mixing	30°C, 31°C, 32°C	NA	NA	0.5%, 1%, 2%
Poly(butylene terephthalate)[224]	MWCNTs	Melt mixing	10°C, 12°C, 14°C, 15°C	NA	NA	1%, 3%, 5%, 7%
Poly(butylene terephthalate)[225]	MWCNTs	Melt mixing	14°C, 15°C, 16°C	NA	NA	0.5%, 1%, 2%

(continued)

TABLE 4.2 *(Continued)*

Polymer	Nanotube type	Processing method	Change in peak crystallization temperature (nonisothermal)	Ratio of crystallization half-times (isothermal)	Avrami exponent change (isothermal)	Nanotube levels investigated (wt%)
Poly(trimethylene terephthalate)[226]	MWCNTs	Dissolution–dispersion–precipitation	8°C, 9°C, 11°C, 12°C, 17°C, 15°C	NA	NA	0.05%, 0.1%, 0.5%, 1%, 5%, 7%
Polylactide[227]	MWCNTs	Dissolution–dispersion–precipitation	3°C, 6°C, 9°C, 18°C			1%, 2%, 5%, 10%
Polylactide[228]	MWCNTs	Dissolution–dispersion–precipitation	5°C	0.4	No change	1%
	Acid-functionalized MWCNTs		7°C	0.35	Increase	
Polylactide[229]	MWCNTs	Dissolution–dispersion–precipitation	−7°C, −1°C, 11°C, 11°C	0.25, 0.12 (varies a great deal with temperature)	Increase	0.002%, 0.008%, 0.02%, 0.08%
Polylactide[230]	MWCNTs	Dissolution–dispersion–precipitation	5°C, 10°C	0.45, 0.25	Increase	0.5%, 2%
Polyimide[231]	MWCNTs	Dispersion–reaction	NA	0.15	NA	1%

Polyurethane (soft segment)[232]	MWCNTs	Dissolution–dispersion–precipitation	6°C	NA	NA	10%
	Acid-functionalized MWCNTs		4°C, 4°C, 5°C, 7°C, 1°C			0.5%, 1%, 5%, 10%, 20%
Polyphenylene sulfide[233]	MWCNTs	Melt mixing	7°C 8°C, 9°C, 10°C	NA	NA	1%, 2%, 3%, 5%
Polyphenylene sulfide[234]	MWCNTs	Melt mixing	6°C, 6°C, 6°C, 10°C, 9°C	NA	NA	0.5%, 1%, 2%, 5%, 7%
Polyphenylene sulfide[235]	MWCNTs	Melt mixing	7°C, 10°C, 14°C, [illegible]4°C, 16°C, 16°C	NA	NA	0.2%, 0.5%, 1%, 2%, 5%, 7%
Syndiotactic polystyrene[236]	SWCNTs	Dissolution–dispersion–precipitation	4°C, 6°C, 7°C, 7°C, 9°C	NA	NA	0.1%, 0.2%, 0.3% 0.5%, 1%
	Grafted to polystyrene SWCNTs		1°C, 4°C, 5°C, 6°C, 8°C			0.2%, 0.3% 0.5%, 1%, 2%
Syndiotactic polystyrene[125]	MWCNTs	Melt mixing	19°C, 21°C, 22°C	NA	NA	1%, 3%, 10%
Poly ether ether ketone[237]	Laser-grown SWCNTs	Melt mixing	2°C, 4°C, 6°C	NA	NA	0.1%, 0.5%, 1% except for one that is 0.5% and 1% only
	Arc-grown SWCNTs		1°C, 3°C, 5°C			
	Laser-grown SWCNTs with polyimide		5°C, 9°C			
	Arc-grown SWCNTs with polyimide		3°C, 6°C, 10°C			

(continued)

TABLE 4.2 (*Continued*)

Polymer	Nanotube type	Processing method	Change in peak crystallization temperature (nonisothermal)	Ratio of crystallization half-times (isothermal)	Avrami exponent change (isothermal)	Nanotube levels investigated (wt%)
Poly(vinylidene fluoride)[238]	MWCNTs	Dissolution–dispersion–precipitation	10°C: ethanol used for precipitation; 3°C: water used for precipitation	NA	NA	5%
Poly(vinylidene fluoride)[239]	MWCNTs	Melt mixing	7°C	NA	NA	1.6%
	Grafted to PMMA MWCNTs		7°C, 4°C, 0°C			0.42%, 1.66%, 1.93%
Poly(vinylidene fluoride)[101]	Acid-functionalized MWCNTs	Dissolution–dispersion–precipitation	0°C, 0°C, 0°C, 0°C, 1°C, 2°C	NA	NA	0.25%, 0.5% 0.75%, 1%, 1.5%, 2%
Poly(vinylidene fluoride)[240]	MWCNTs	Melt mixing	0°C, 1°C, 3°C, 3°C, 4°C, 5°C	NA	NA	0.01%, 0.1%, 0.5%, 1%, 3%, 5%
Poly(vinylidene fluoride)[241]	MWCNTs	Melt mixing	8°C, 8°C, 11°C, 12°C, 13°C, 13°C	NA	NA	0.01%, 0.1%, 0.5%, 1%, 3%, 5%
Poly(vinylidene fluoride)[242]	MWCNTs	Melt mixing	0°C, 0°C, 3°C, 4°C, 5°C, 5°C, 6°C, 6°C	NA	NA	0.01%, 0.1%, 0.5%, 1%, 3%, 5%, 7%, 10%

Poly(ether–ester)[243]	MWCNTs	Melt mixing	14°C, 15°C, 15°C, 17°C, 13°C	NA	NA	0.1%, 0.3%, 0.5%, 1%, 2%
Polyoxymethylene[244]	MWCNTs	Melt mixing	7°C, 2°C	NA	NA	1%, 5%
Polyhydroxybutarate[245,246]	Acid-functionalized MWCNTs	Dissolution–dispersion–precipitation	4°C, 7°C, 10°C	0.75, 0.5 (no 2%)	Decrease	0.5%, 1%, 2%
Poly(1-butene)[136]	Amine-functionalized MWCNTs	Dissolution–dispersion–precipitation	NA	0.75, 0.55, 0.32 (onset from rheology)	NA	0.1%, 1%, 5%
Poly(4-methyl-1-pentene)[247]	MWCNTs	Melt mixing	All 1°C	NA	NA	1–10%
Poly(butylene succinate)[248]	Acid-functionalized MWCNTs	Melt mixing	7°C	0.25	Increase	2%
Poly(butylene succinate)[249]	MWCNTs	Melt mixing	10°C, 12°C	NA	NA	0.1%, 0.2%
Poly(butylene succinate) copolymer[250]	Acid-functionalized MWCNTs	Dissolution–dispersion–precipitation	5°C, 17°C, 22°C	0.35, 0.6, 0.65	NA	0.5%, 1%, 2%

Many studies used multiple cooling rates (nonisothermal) and multiple temperatures (isothermal); a rough average is given in these situations. Blends of two or more polymers are not included in this list. The numbers in the last four columns are listed in identical order from lowest to highest nanotube content.

[a]In this study, the pure polymer could not be reliably measured. Hence, the values are based on an extrapolation to zero nanotube content, which has a great deal of error associated with it.

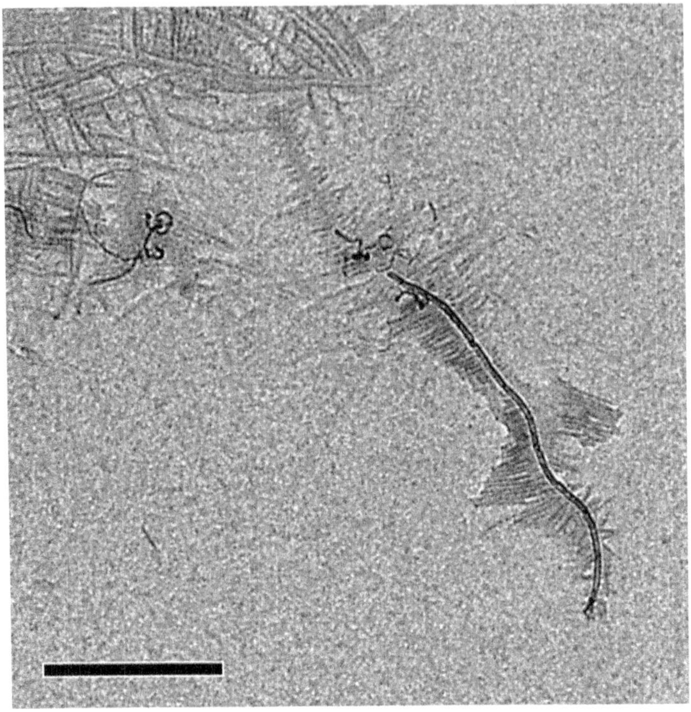

Figure 4.9 Bright-field TEM micrograph of isotactic polypropylene nonisothermally crystallized from the melt. Scale bar is 500 nm. Copyright 2008 American Chemical Society. Reprinted in part with permission from Ref. 108.

nucleation density is controlled, it is possible to get off-axis orientations of the lamellae relative to that of the fiber axis.[111] NHSKs grown in solution can then be used as fillers themselves for polymers crystallized in solution, which can lead to very unique morphologies.[112,113] Just as with polymeric shish-kebabs, the formation of NHSKs is not trivial and a great many variables (temperature, solvent quality, molecular weight) will control the exact type of morphology that results.

Whether shish-kebabs or transcrystallinity is the morphology in a melt-crystallized nanotube composite is unanswered in the general case; in other words, are the nucleation and growth rates such that the nucleation density along the fiber axis is low or high? In the most general case, the exact identity of the morphology for a given situation is not known. The author is not aware of microscopy showing a shish-kebab type morphology in a bulk material under quiescent cooling conditions; in isotactic polypropylene, a transcrystalline morphology was found.[108,114,115] Whether a transcrystalline or shish-kebab-type morphology results in melt-spun nanotube composites has been addressed for high-density polyethylene (HDPE). In unfilled linear polymers such as HDPE, shish-kebab structures from the melt form easily with materials with bimodal molecular weight distributions: a small amount of easy to orient high molecular weight material forms the shish and a larger amount of low molecular weight material forms the kebabs. The addition of a highly orientable nanotube should be expected to increase the ability to form shish, although the question of whether kebabs will form or transcrystallinity occurs depends on nucleation density as described earlier. Two studies[116,117] have examined HDPE–nanotube fiber crystallization behavior in detail, and both found behavior consistent with shish-kebabs rather than with transcrystallinity. In particular, scattering reflections in both small- and wide-angle X-ray scattering were consistent with an increase in the formation rate of kebabs. These experiments are not able to distinguish between polymer and nanotube shish. One key element in determining that nanotubes serve as nucleation sites in oriented systems is whether the shish-kebab morphology remains after melting and recrystallization; only in the case where nanotubes are nucleating will orientation remain. A quantitative examination of this issue indicated that the orientation did decrease after melting and recrystallization in high-density polyethylene but some orientation did remain.[116]

In theory, the nucleation rate on the surface of the carbon nanotubes could overwhelm other nucleation events and hence no spherulites would form in a quiescently crystallized polymer. A study that used an oscillatory shear field during cooling of injection molded bars and used acid to etch away the amorphous HDPE showed micrographs consistent with no spherulites.[118] In the case of spherulites, do carbon nanotubes have an effect on spherulites? Certainly, micron-sized fillers such as carbon or glass fibers arrest spherulitic growth. However, nanotubes, because of their small size, may not arrest growth; instead, the polymer could possibly simply grow around the nanotubes. At present, this question must be regarded as unanswered.

The introduction of carbon nanotubes should not affect the equilibrium melting temperature, yet three studies using a Hoffman–Weeks approach have found that the equilibrium melting temperature in the presence of nanotubes is lower than that for pure polypropylene,[108,119] and decreases with additional nanotubes.[120] This difference is not surprising since nanotubes likely act as a local heat sink lowering the temperature immediately next to the growing crystal and yielding thinner lamellae than in the case where such particles are not present. Such a reduction has also been found in nanoclay composites.[121] Under more normal industrial processing conditions, which is more closely represented by nonisothermal experiments, thicker crystals might be expected since crystal growth should occur at higher temperature due to the nucleation ability of carbon nanotubes. In one study on high-density polyethylene using specialized DSC experiments, an increase in crystal thickness was found as expected,[122] and peaks in small-angle scattering experiments generally shift toward lower angles indicative of thicker crystals with the introduction of nanotubes.[119,123] However, the number of studies that have looked at changes in crystal thickness with nanotube addition is small, and this issue should still be viewed somewhat as an open question.

4.3.2 Rate Effects: Glass Transition, Crystal Nucleation, and Growth

Nanotube addition could affect the glass transition region in a semicrystalline polymer. Of course, the cause of changes in T_g with the addition of carbon nanotubes is complicated by possible changes in either the size or fraction of crystals in the system; both of these can have significant effects on the glass transition of the amorphous phase independent of any effect nanotubes might have. Changes, if any, in the glass transition temperature with nanotube addition in semicrystalline polymers tend to be small. For example, in PET the T_g decreases very slightly (−1°C/wt% nanotube) and the transition breadth does not change.[124] A similar result has been found for polyamide 6; the glass transition was 3°C higher at 20 wt% MWCNT compared to that at 10 wt% MWCNT content.[95] For syndiotactic polypropylene, no change in the glass transition temperature was found with MWCNT addition.[125] In a study of three isotactic polypropylenes with three different melt indices, one of the samples showed a decrease in T_g of about 1°C/wt% nanotube, the second showed an increase of 7°C at 1% tubes, and then the increase dropped to about 5°C for concentrations from 3 to 10 wt%, and the third dropped 2°C with the addition of 1% nanotubes and was constant thereafter.[126] However, in one case for isotactic polypropylene filled with acid-modified tubes, a 16°C change with 10% tubes was found as measured by mechanical methods, while unmodified tubes showed a more typical rise of 5°C at 10% nanotube content.[127]

As described in Section 4.2.2, the introduction of nanotubes to an amorphous polymer can cause the formation of a portion of the material that does not participate in the glass transition. Crystallization can also cause the formation of material that does not participate in either the glass or melting transitions, which has been termed the rigid amorphous fraction (RAF). As described previously for amorphous polymers, the RAF corresponds to the situation where the dynamics of a portion of the amorphous phase have been altered to such an extent so as to cause a separation between parts of the amorphous regions of a semicrystalline polymer.[128] Quantification of the RAF amount is done by quantifying the heat capacity increase at the glass transition and the enthalpy of melting; material that does not participate in either of these transitions is assumed to be part of the RAF. As of the end of 2009, only one study to the author's knowledge has quantitatively examined changes in the RAF with the addition of nanotubes. In a study performed on electrospun crystalline poly (ethylene terephthalate) fibers with acid-functionalized MWCNTs, the RAF amount increased from 23% to 64% as the nanotube content increased from 0 to 2 wt%.[124] As-spun fibers were noncrystalline, and crystallinity in these systems was produced by annealing at 130°C, which is above T_g but below T_m. An increase of *trans* conformers in the semicrystalline PET with the addition of nanotubes was found along with the increase in the RAF; in previous work, the RAF has been correlated with the amount of *trans* conformers in the amorphous region.[129,130] Hence, this result suggests that the nanotubes cause alignment of polymer chains along the nanotube axis since amorphous *trans* conformers would be expected to be aligned along the nanotube axis. Interestingly, in the amorphous as-spun fibers, there was no increase in the number of *trans* conformers with nanotube addition. Further, a different study with poly(ethylene terephthalate)–nanotube composites manufactured using melt extrusion and compression molding showed an increase in the crystalline fraction at the expense of the *trans* conformer in the amorphous phase.[131] The RAF was not quantified in this study, however. Polyamide 6 showed a decrease in the heat capacity change at the glass transition, which could not be explained in terms of an increase in crystallinity and hence must be attributed to a growth in the RAF.[95]

The result presented in the previous paragraph that the introduction of nanotubes caused an increase in the RAF raises the question of whether the RAF and the material that disappears from the glass transition in an amorphous polymer are the same. Most authors believe that the RAF is due to material near the crystalline–amorphous interface. If the RAF and an amorphous polymer's "lost" material are different, then in a semicrystalline polymer calorimetry cannot determine whether material that does not participate in the glass transition or the melting transition is the material at the crystalline–amorphous interface or at the amorphous–nanotube interface. Further, dynamic phase separation of amorphous material is not limited to nanocomposites or semicrystalline polymers; ionomers that form very small phase-separated ionic regions can also cause a loss of material participating in the glass transition.[132] From a dynamic perspective, all of these phenomena appear identical. However, from a chain configuration/entanglement perspective, the phenomena are likely different; in particular, it is hard to imagine how immobilization of a chain on a surface is similar to the way in which amorphous chains interact with polymer crystals. Clearly details of phases that do not participate in the glass transition temperature or melting temperature need to be more fully investigated.

A related question is whether the presence of immobilized material on the surface of nanotubes affects crystal growth. With the application of a sinusoidal varying temperature profile on a typical isothermal hold in isothermal crystallization experiments, a reversible (i.e., melting and recrystallization occur on the timescale of the oscillations) and irreversible part to crystallization can be quantified. In one study, the authors found a decrease in the reversible part for a strongly polar ethylene–vinyl acetate copolymer, which was attributed to a decrease in polymer mobility. Functionalized nanotubes with moieties expected to interact with the copolymer showed a larger effect. Further, the effect with nanoclay was much more significant, not surprising given the much more polar nature, that is, higher surface energy, of the clay.[133]

For samples where nucleation occurs, the general shape of the crystallization half-time versus nanotube content plot (isothermal), or the crystallization temperature versus nanotube content plot (nonisothermal), typically has a steep increase at low nanotube contents followed by a plateau at high nanotube contents as indicated by the data in Table 4.2. The magnitude of the increase with nanotube content, as was seen with the glass transition, is not consistent for the same polymer, which is most likely a function of the inconsistency of dispersion. However, it is clear that poly (ethylene terephthalate) shows larger increases in the nucleation rate with nanotubes than other polymers on the list. Other than an aromatic ring being present, no obvious reason exists for such a difference. Certainly, it is unreasonable to believe that dispersion is that much better in PET and percolation measurements do not support such a conclusion.

The qualitative shape of an increase followed by a plateau is similar to that found for the glass transition temperature versus nanotube content. An important difference, as representatively shown in Figure 4.10, is that the plateau in crystallization kinetic data is usually better described as a slowly increasing nearly linear section, unlike T_g where the plateau is more frequently constant. This difference

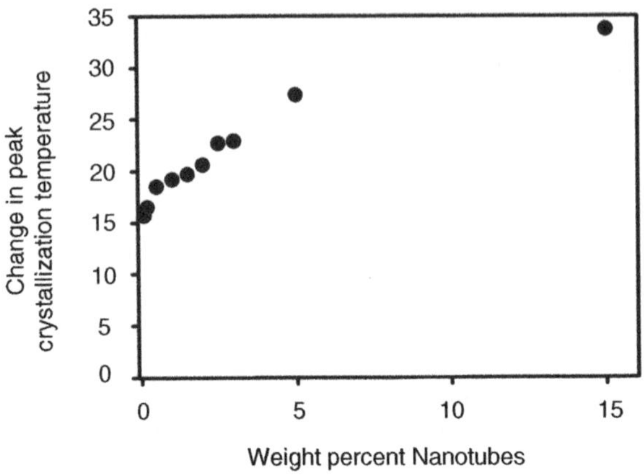

Figure 4.10 Graph that shows the representative qualitative change in peak crystallization temperature versus nanotube content. Data from Ref. 219.

might be due to the fundamental molecular difference in the glass and melting transitions. Alternatively, the explanation might be that the dispersion quality in semicrystalline polymers is generally lower because the less effective melt mixing dispersion methods are more common for semicrystalline polymers while the dissolution–dispersion–precipitation or dispersion–reaction methods are more common for amorphous polymers. Changes in both glass transition and crystallization behaviors are expected to scale with nanotube surface area, and perhaps less effective dispersion methods might see a relatively larger increase in nanotube surface area with an increase in nanotube content leading to a slow increase rather than a plateau. Improving dispersion in melt mixing by increasing either shear rate or mixing time has been shown to increase crystallization rate, and eventually a plateau with shear rate or mixing time was reached.[134]

Nucleation of crystallization by carbon nanotubes not only causes a shift to higher crystallization temperatures in nonisothermal crystallization experiments, but more often this shift is accompanied by a broadening of the exotherm as well. In a few studies with polyamide 6, the exotherm separates into low- and high-temperature peaks,[97,98] which in one case the authors convincingly showed was due to normal spherulitic crystallization and carbon nanotube nucleated crystallinity, respectively.[135] The latter study found that the ratio of the high- to low-temperature peak increased with an increase in dispersion quality. Generally separation, if present, is not enough to noticeably cause two peaks. A superposition of two characteristic nucleation peaks is understandable, but a single broader increase in the temperature range over which crystallization occurs is much more difficult to explain. Another observation from Table 4.2 is that, not surprisingly, in the majority of situations the Avrami exponent n decreases with the addition of nanotubes.

Another technique besides calorimetry, scattering, and microscopy used to measure crystallization kinetics is rheometry. The rise in the shear storage modulus is proportional to the fraction crystallized, although the relationship between amount crystallized and modulus is not generally known. Although some authors[7,136,137] have used the modulus to define the half-time of crystallization, this half-time *a priori* has no direct correspondence to the crystallization half-time determined via calorimetry or scattering. A quantitative relationship for crystallization rate determined from rheometry requires that the relationship between amount crystallized and a specific rheological parameter must be developed, which is essentially never done. Using the onset time for a rise in storage modulus under isothermal conditions to characterize crystallization rate should match the onset time from calorimetry. One advantage of rheometry is that crystallization after shear can be measured. For example, in isothermal polypropylene, the rheological half-time, as measured by when the storage modulus reached half its maximum value, reduced by 10–20% with preshear at higher isothermal crystallization temperatures in a nanotube-filled sample. Unfortunately, results for pure polypropylene were not given.[138] A similar study on poly(butylene terephthalate) found no change in crystallization rate with shear, although the addition of nanotubes caused a drop by a factor of 10–100 in the half-time as measured by small-angle X-ray scattering. X-ray scattering was used to show that shear did, however, increase the fraction of oriented material dramatically, although the orientation of that material did not change substantially with higher shear rates.[139]

The addition of nanotubes typically does not change the fractional crystallinity greatly for melt-mixed samples; the most common response is either no change or a slight decrease in fractional crystallinity. Although more commonly the addition of nanotubes does not change the fractional crystallinity for samples cast from solution, in a few cases samples cast from solution can show significant changes in fractional crystallinity with nanotube addition. These changes are more likely related to the interrelationship between drying conditions, viscosity, and nucleation by nanotubes. For example, poly(vinyl alcohol) cast from solution has shown substantial increases in fractional crystallinity with the addition of nanotubes;[140,141] in one case from 0.48 to 0.84.[142] Another exception was found for the semiconjugated polymer poly(*m*-phenylenevinylene-*co*-2,5-dioctyloxy-*p*-phenylenevinylene),[143] where an increase from a fractional crystallinity of 0.2 to 0.55 took place with the addition of 3% carbon nanotubes. Finally, polypropylene, for which numerous studies have shown only small changes in crystallinity when melt mixed,[107,144–146] showed an increase in fractional crystallinity from 0.35 to 0.50 with added nanotubes when the two were mixed in a dispersion–dissolution–precipitation manner.[147]

As stated previously, the spherulitic growth rate can be measured independently using optical microscopy. In a study of polymer spherulites, it was found that at high undercoolings (low temperatures) there was a drop of 25% in the radial growth rate with the addition of 0.2–1% MWCNTs to isotactic polypropylene; such a change corresponds to roughly a change of 2°C in the crystallization temperature. However, no change in spherulitic growth rate occurred at lower undercoolings (higher temperatures). This difference with undercooling was explained by a change in a longer range diffusive mechanism (at lower temperature) limiting crystal growth to a diffusive mechanism (at higher temperature) that is short range in character.[145] Another study at relatively low undercoolings on polycaprolactone showed a decrease of about 33% in spherulitic growth rate at 0.25 wt% versus the pure polymer, as opposed to an increase of to 25% and 100% at 0.5% and 1.0% nanotubes, respectively.[148] The observed nonmonotonic behavior suggests two competing effects: a likely hypothesis is that nanotubes reduce chain mobility as well as serve as heat sinks for the heat that is given off during the crystallization exotherm. A detailed study on single carbon nanotube fibers showed that the growth rate of the transcrystalline morphology for isotactic polypropylene approximately matched that measured for carbon or Kevlar fibers, indicating that the surface characteristics do not affect the growth rate.[115]

4.4 BLENDS AND BLOCK COPOLYMERS

This section focuses on changes in blend/block copolymer morphology and/or characteristics caused by the introduction of nanotubes. If nanotubes are concentrated in one of two components and the domain sizes are much larger than the size of the nanotubes, it is reasonable to assume that the changes in mobility, crystallization, and so on in that phase should be identical to those in the one-phase material. This section is more concerned with the location of nanotubes in a two-phase material, changes that nanotubes might have on the phase-separated morphology, and changes in nanotubes that might occur as the domain size of the preferred material begins to approach the nanotube length.

Most polymer blends are immiscible; that is, a blend of two polymers will spontaneously separate into two relatively pure phases whose sizes and shapes depend on composition, processing, and so on. In fact, there has only been one paper published on miscible polymer blends filled with nanotubes and no changes in miscibility were reported with the addition of nanotubes.[251] With immiscible blends, the first issue to consider is the location of nanotubes, which in turn will depend on the interfacial energies of the three surfaces (polymer 1–nanotube, polymer 2–nanotube, and polymer 1–polymer 2). The most direct way to examine this issue is through microscopy and this technique has been used to determine the location of nanotubes in blends.[252–254] Not surprisingly, carbon nanotubes with or without functionalization tend to be concentrated mostly or exclusively in the higher surface energy component.[255–258] This concentration is supported by theoretical work; even for very small differences in surface energies localization is expected to be much more significant for high aspect ratio fillers like nanotubes versus low aspect ratio fillers.[259]

If nanotubes thermodynamically are going to reside in one phase, how fast does the transfer occur from one phase to the other? If maleic anhydride-functionalized MWCNTs were initially mixed with one component of a two-phase polypropylene/ethylene vinyl acetate (PP/EVA) blend, the effect on crystallization kinetics of the PP was much larger when the nanotubes were dispersed in the PP than when dispersed in the EVA initially, indicating that the nanotubes did not migrate to their final equilibrium position in the EVA over the conditions of the twin-screw melt mixing used in this experiment.[260] However, this result contrasts with results in SAN/PC blends where the monomer fraction in the former was adjusted to give a very low interfacial energy between the two polymers. Complete transfer of nanotubes from the less favored component to the more favored component occurred in less than 5 min in a melt-mixed system.[261] The contact time in the extruder of the PP/EVA blend is not given; however, a few minutes is likely correct. Kinetics of transfer are likely to be heavily influenced by the aggregation state of the nanotubes, with more highly dispersed nanotubes being faster to transfer. Hence, the difference in these observations could simply be a difference in dispersion characteristics.

Transfer kinetics will be dependent on the diffusion constant of the nanotubes in the two polymers. A simple study examining diffusion constants used two films, a polypropylene film that contained nanotubes and a polycarbonate film that did not.[262] Since nanotubes prefer to reside in the polycarbonate, nanotubes diffused from the polypropylene film to the polycarbonate. According to the solution to Equation 4.2 with idealized boundary conditions, the thickness of the nanotube film in the polycarbonate should be proportional to time$^{1/2}$. This dependence was found, with an apparent diffusion constant of 6.5×10^{-9} m^2/s. The self-diffusion constant for a polymer above the entanglement molecular weight in a melt is typically significantly smaller than 6.5×10^{-9} m^2/s, while a nonentangled polymer can have a diffusion constant in this range or even higher. The molecular weight of polypropylene used in this experiment is almost certainly above the entanglement molecular weight, while the molecular weight of polycarbonate is likely not. For this experiment, it is not clear what the appropriate conditions are; that is, is the measured diffusion constant more of a reflection of the diffusion constant in polypropylene or polycarbonate? In simpler terms, this diffusion constant corresponds to a velocity of about 2 μm/min.

Nanotubes can actually reduce the tendency to phase separate.[263,264] In other words, nanotubes can reduce the size of the dispersed phase in a discrete blend system or reduce the size of both phases in a cocontinuous system. There are three mechanisms possible to reduce the tendency for phase separation. In the first, nanotubes can increase the viscosity of the low-viscosity phase (whose coalescence tends to regulate domain size). In the second, the nanotube could reduce the free energy of mixing. The third possibility is that the nanotube could serve as an interfacial stabilizer, e.g., a Pickering emulsion. In one study, it was found that nanotubes were selectively located at the interface after functionalization via TEM while without functionalization the nanotubes were only located in one of the two phases.[265] Another study did show some selective enhancement of nanotubes at the polymer–polymer interface.[253] In fact, polymer-functionalized nanotubes might be particularly effective as emulsion stabilizers if the functionalized polymer was soluble in the lower surface energy component.

Much of the same arguments can be made in block copolymers; however, in this case the size of the phases is much more constrained than in blends. Further, to the author's knowledge as of the end of 2009 there have been no experimental studies that have quantitatively looked at how nanotubes will partition in a block copolymer. A very interesting question is the following: if interfacial energies favor nanotubes being found in one phase and that phase is the dispersed phase and has a characteristic dimension much smaller than the nanotube, then what is the resulting morphology? In one case for a blend, small dispersed spheres of the phase that contained the nanotubes elongated and became irregular after the addition of nanotubes.[255] In block copolymers, domain sizes are primarily dictated by molecular architecture rather than processing conditions as for blends. So another question is whether it is possible for nanotubes to force miscibility in a block copolymer with a geometric constraint driving force? Such questions are very interesting and have not been explored to any substantial degree in the open literature.

4.5 CHALLENGES

The entire subject matter of the effect of confinement on the physics of amorphous polymers is not well understood. One area that certainly needs further study is the relationship between interfacial energies between the polymer and nanotube with respect to the observed behavior. Wrapping is certainly a phenomenon that arises because of the unique geometry of nanotubes. Related to wrapping, an unanswered question is how prevalent is wrapping, or entanglements in general, between polymers and nanotubes in bulk. A more general related question is how the conformation of the polymer chain is affected by an almost one-dimensional surface; does the general picture of loops, trains, and tails for two-dimensional surfaces still apply? Certainly, entanglements should be added to this description for nanotubes! Is there any preference for a single chain, adsorbed at one point to a nanotube, to preferentially be adsorbed at another point on the same nanotube?

Diffusion is also an interesting question, in terms of both how nanotubes affect the diffusion of the polymer and how fast nanotubes diffuse. The importance of diffusion to polymer–nanotube composites means that studies should become more plentiful in the future.

The concept of a rigid amorphous fraction for crystalline materials is well established. The concept of an immobilized fraction near an attractive surface is also well established. However, there is a great deal more unknown than known about these phenomena, especially the latter. How are the rigid amorphous fraction and immobilized fraction related? Is there a method to distinguish between chains immobilized at the nanotube interface and those immobilized at the crystalline interface? Does this immobilized fraction at the nanotube interface contribute in any meaningful way to mechanical properties? (As will be seen in Chapter 5, this phase does seem to affect some rheological properties.) How can an increase in diffusion constant be reconciled with this immobilized fraction? Why does the glass transition temperature reach a plateau in most cases?

Without question, nanotubes nucleate crystals in a large number of polymers. To state that nanotubes nucleate crystallinity in all polymers is a bit strong, but results that show little or no nucleation can for the most part be explained by poor dispersion. However, there is no question that in cases of excellent dispersion, the change in temperature with nanotube nucleation is much higher in poly(ethylene terephthalate) than in other polymers; what is the reason for this difference? Measurements of growth rate should be done more commonly; however, the few systems where such measurements have been done certainly do not paint a uniform picture of a drop in growth rate as expected. The question about whether nanotubes slow crystal growth rates significantly must still be considered unanswered.

Finally, work with understanding how nanotubes affect the morphology of blends and block copolymers is much further behind than in single-component systems. A host of questions related to these materials are either poorly understood or not studied at all; chief among these is how nanotubes might change morphology in systems where the nanotubes either are longer than maximum characteristic dimension of the higher surface energy phase or are functionalized with a polymer that is soluble in the lower surface energy phase. Block copolymers, where chain architecture is the primary determinant of domain size, are of particular interest.

REFERENCES

1. Degennes, P. G. (1981). Polymer solutions near an interface. 1. Adsorption and depletion layers, *Macromolecules, 14*, 1637–1644.
2. DeGennes, P. G. (1987). Polymers at an interface: a simplified view, *Adv. Colloid Interface Sci., 27*, 189–209.
3. Kallrot, N., Linse, P. (2007). Dynamic study of single-chain adsorption and desorption, *Macromolecules, 40*, 4669–4679.
4. Kallrot, N., Dahlqvist, M., Linse, P. (2009). Dynamics of polymer adsorption from bulk solution onto planar surfaces, *Macromolecules, 42*, 3641–3649.
5. Angelikopoulos, P., Bock, H. (2010). The differences in surfactant adsorption on carbon nanotubes and their bundles, *Langmuir, 26*, 899–907.
6. Vega, J. F., Martinez-Salazar, J., Trujillo, M., Arnal, M. L., Muller, A. J., Bredeau, S., Dubois, P. (2009). Rheology, processing, tensile properties, and crystallization of polyethylene/carbon nanotube nanocomposites, *Macromolecules, 42*, 4719–4727.
7. Zhang, Q. H., Lippits, D. R., Rastogi, S. (2006). Dispersion and rheological aspects of SWNTs in ultrahigh molecular weight polyethylene, *Macromolecules, 39*, 658–666.
8. Mackay, M. E., Dao, T. T., Tuteja, A., Ho, D. L., Van Horn, B., Kim, H. C., Hawker, C. J. (2003). Nanoscale effects leading to non-Einstein-like decrease in viscosity, *Nat. Mater., 2*, 762–766.
9. Yarotski, D. A., Kilina, S. V., Talin, A. A., Tretiak, S., Prezhdo, O. V., Balatsky, A. V., Taylor, A. J. (2009). Scanning tunneling microscopy of DNA-wrapped carbon nanotubes, *Nano Lett., 9*, 12–17.
10. Goh, R. G. S., Bell, J. M., Motta, N., Waclawik, E. R. (2006). Microscopic and spectroscopic study of self-ordering in poly(3-hexylthiophene)/carbon nanotubes nanocomposites, *J. Nanosci. Nanotechnol., 6*, 3929–3933.
11. Iijima, M., Watabe, T., Ishii, S., Koshio, A., Yamaguchi, T., Bandow, S., Iijima, S., Suzuki, K., Maruyama, Y. (2005). Fabrication and STM-characterization of novel hybrid materials of DNA/carbon nanotube, *Chem. Phys. Lett., 414*, 520–524.
12. Gao, H. J., Kong, Y. (2004). Simulation of DNA–nanotube interactions, *Annu. Rev. Mater. Res., 34*, 123–150.
13. Martin, W., Zhu, W. S., Krilov, G. (2008). Simulation study of noncovalent hybridization of carbon nanotubes by single-stranded DNA in water, *J. Phys. Chem. B, 112*, 16076–16089.
14. Zhao, X., Johnson, J. K. (2007). Simulation of adsorption of DNA on carbon nanotubes, *J. Am. Chem. Soc., 129*, 10438–10445.
15. Srebnik, S. (2008). Physical association of polymers with nanotubes, *J. Polym. Sci. Polym. Phys., 46*, 2711–2718.
16. Gao, H. J., Kong, Y., Cui, D. X., Ozkan, C. S. (2003). Spontaneous insertion of DNA oligonucleotides into carbon nanotubes, *Nano Lett., 3*, 471–473.
17. Kang, Y., Wang, Q., Liu, Y. C., Wu, T., Chen, Q., Guan, W. J. (2008). Dynamic mechanism of collagen-like peptide encapsulated into carbon nanotubes, *J. Phys. Chem. B, 112*, 4801–4807.
18. Matarredona, O., Rhoads, H., Li, Z. R., Harwell, J. H., Balzano, L., Resasco, D. E. (2003). Dispersion of single-walled carbon nanotubes in aqueous solutions of the anionic surfactant NaDDBS, *J. Phys. Chem. B, 107*, 13357–13367.
19. Xiao, Q., Wang, P. H., Ji, L. L., Tan, X. K., Ouyang, L. L. (2007). Dispersion of carbon nanotubes in aqueous solution with cationic surfactant CTAB, *J. Inorg. Mater., 22*, 1122–1126.

20. Valenti, L. E., Fiorito, P. A., Garcia, C. D., Giacomelli, C. E. (2007). The adsorption–desorption process of bovine serum albumin on carbon nanotubes, *J. Colloid Interface Sci.*, *307*, 349–356.
21. Coleman, J. N., Fleming, A., Maier, S., O'Flaherty, S., Minett, A. I., Ferreira, M. S., Hutzler, S., Blau, W. J. (2004). Binding kinetics and SWNT bundle dissociation in low concentration polymer–nanotube dispersions, *J. Phys. Chem. B*, *108*, 3446–3450.
22. Aoki, H., Morita, S., Sekine, R., Ito, S. (2008). Conformation of single poly(methyl methacrylate) chains in an ultra-thin film studied by scanning near-field optical microscopy, *Polym. J.*, *40*, 274–280.
23. Jones, R. L., Kumar, S. K., Ho, D. L., Briber, R. M., Russell, T. P. (2001). Chain conformation in ultrathin polymer films using small-angle neutron scattering, *Macromolecules*, *34*, 559–567.
24. Brulet, A., Boue, F., Menelle, A., Cotton, J. P. (2000). Conformation of polystyrene chain in ultrathin films obtained by spin coating, *Macromolecules*, *33*, 997–1001.
25. Kraus, J., Muller-Buschbaum, P., Kuhlmann, T., Schubert, D. W., Stamm, M. (2000). Confinement effects on the chain conformation in thin polymer films, *Europhys. Lett.*, *49*, 210–216.
26. Sen, S., Xie, Y. P., Kumar, S. K., Yang, H. C., Bansal, A., Ho, D. L., Hall, L., Hooper, J. B., Schweizer, K. S. (2007). Chain conformations and bound-layer correlations in polymer nanocomposites, *Phys. Rev. Lett.*, *98*, 128302.
27. Mukhopadhyay, M. K., Jiao, X., Lurio, L. B., Jiang, Z., Stark, J., Sprung, M., Narayanan, S., Sandy, A. R., Sinha, S. K. (2008). Thickness induced structural changes in polystyrene films, *Phys. Rev. Lett.*, *101*, 115501.
28. Li, Y. J., Wei, D. S., Han, C. C., Liao, Q. (2007). Dynamics of polymer melts confined by smooth walls: crossover from nonentangled region to entangled region, *J. Chem. Phys.*, *126*, 204907.
29. Mu, M. F., Composto, R. J., Clarke, N., Winey, K. I. (2009). Minimum in diffusion coefficient with increasing MWCNT concentration requires tracer molecules to be larger than nanotubes, *Macromolecules*, *42*, 8365–8369.
30. Mu, M. F., Clarke, N., Composto, R. J., Winey, K. I. (2009). Polymer diffusion exhibits a minimum with increasing single-walled carbon nanotube concentration, *Macromolecules*, *42*, 7091–7097.
31. Koo, J., Shin, K., Seo, Y. S., Koga, T., Park, S., Satija, S., Chen, X., Yoon, K., Hsiao, B. S., Sokolov, J. C., Rafailovich, M. H. (2007). Stabilizing, thin film polymer bilayers against dewetting using multiwalled carbon nanotubes, *Macromolecules*, *40*, 9510–9516.
32. Wu, D. F., Sun, Y. R., Zhang, M. (2009). Kinetics study on melt compounding of carbon nanotube/polypropylene nanocomposites, *J. Polym. Sci. Polym. Phys.*, *47*, 608–618.
33. Alcoutlabi, M., McKenna, G. B. (2005). Effects of confinement on material behaviour at the nanometre size scale, *J. Phys. Condens. Matter*, *17*, R461–R524.
34. Zhang, J., Liu, G., Jonas, J. (1992). Effects on confinement on the glass-transition temperature of molecular liquids, *J. Phys. Chem.*, *96*, 3478–3480.
35. Park, J. Y., McKenna, G. B. (2000). Size and confinement effects on the glass transition behavior of polystyrene/*o*-terphenyl polymer solutions, *Phys. Rev. B*, *61*, 6667–6676.
36. Melnichenko, Y. B., Schuller, J., Richert, R., Ewen, B., Loong, C. K. (1995). Dynamics of hydrogen-bonded liquids confined to mesopores: a dielectric and neutron spectroscopy study, *J. Chem. Phys.*, *103*, 2016–2024.
37. Jackson, C. L., McKenna, G. B. (1991). The glass-transition of organic liquids confined to small pores, *J. Non-Cryst. Solids*, *131*, 221–224.
38. Morineau, D., Xia, Y. D., Alba-Simionesco, C. (2002). Finite-size and surface effects on the glass transition of liquid toluene confined in cylindrical mesopores, *J. Chem. Phys.*, *117*, 8966–8972.
39. Alba-Simionesco, C., Dosseh, G., Dumont, E., Frick, B., Geil, B., Morineau, D., Teboul, V., Xia, Y. (2003). Confinement of molecular liquids: consequences on thermodynamic, static and dynamical properties of benzene and toluene, *Eur. Phys. J. E*, *12*, 19–28.
40. Simon, S. L., Bernazzani, P., McKenna, G. B. (2003). Effects of freeze-drying on the glass temperature of cyclic polystyrenes, *Polymer*, *44*, 8025–8032.
41. Keddie, J. L., Jones, R. A. L., Cory, R. A. (1994). Size-dependent depression of the glass transition temperature in polymer films, *Europhys. Lett.*, *27*, 59–64.
42. Dalnoki-Veress, K., Forrest, J. A., Murray, C., Gigault, C., Dutcher, J. R. (2001). Molecular weight dependence of reductions in the glass transition temperature of thin, freely standing polymer films, *Phys. Rev. E*, *63*, 031801.

43. Sharp, J. S., Forrest, J. A. (2003). Thickness dependence of the dynamics in thin films of isotactic poly (methylmethacrylate), *Eur. Phys. J. E*, *12*, S97–S101.
44. O'Connell, P. A., McKenna, G. B. (2005). Rheological measurements of the thermoviscoelastic response of ultrathin polymer films, *Science*, *307*, 1760–1763.
45. Tate, R. S., Fryer, D. S., Pasqualini, S., Montague, M. F., de Pablo, J. J., Nealey, P. F. (2001). Extraordinary elevation of the glass transition temperature of thin polymer films grafted to silicon oxide substrates, *J. Chem. Phys.*, *115*, 9982–9990.
46. Keddie, J. L., Jones, R. A. L., Cory, R. A. (1994). Interface and surface effects on the glass-transition temperature in thin polymer films, *Faraday Discuss.*, *98*, 219–230.
47. Grohens, Y., Brogly, M., Labbe, C., David, M. O., Schultz, J. (1998). Glass transition of stereoregular poly(methyl methacrylate) at interfaces, *Langmuir*, *14*, 2929–2932.
48. Grohens, Y., Hamon, L., Reiter, G., Soldera, A., Holl, Y. (2002). Some relevant parameters affecting the glass transition of supported ultra-thin polymer films, *Eur. Phys. J. E*, *8*, 217–224.
49. Ellison, C. J., Torkelson, J. M. (2003). The distribution of glass-transition temperatures in nanoscopically confined glass formers, *Nat. Mater.*, *2*, 695–700.
50. Rittigstein, P., Torkelson, J. M. (2006). Polymer–nanoparticle interfacial interactions in polymer nanocomposites: confinement effects on glass transition temperature and suppression of physical aging, *J. Polym. Sci. Polym. Phys.*, *44*, 2935–2943.
51. Yuen, S. M., Ma, C. C. M., Lin, Y. Y., Kuan, H. C. (2007). Preparation, morphology and properties of acid and amine modified multiwalled carbon nanotube/polyimide composite, *Compos. Sci. Technol.*, *67*, 2564–2573.
52. Prado, L., De La Vega, A., Sumfleth, J., Schulte, K. (2009). Noncovalent functionalization of multiwalled and double-walled carbon nanotubes: positive effect of the filler functionalization on high glass transition temperature epoxy resins, *J. Polym. Sci. Polym. Phys.*, *47*, 1860–1868.
53. Wang, J. G., Fang, Z. P., Gu, A. J., Xu, L. H., Liu, F. (2006). Effect of amino-functionalization of multi-walled carbon nanotubes on the dispersion with epoxy resin matrix, *J. Appl. Polym. Sci.*, *100*, 97–104.
54. Grossiord, N., Miltner, H. E., Loos, J., Meuldijk, J., Van Mele, B., Koning, C. E. (2007). On the crucial role of wetting in the preparation of conductive polystyrene–carbon nanotube composites, *Chem. Mater.*, *19*, 3787–3792.
55. Zou, H., Wu, S. S., Shen, J. (2008). Polymer/silica nanocomposites: preparation, characterization, properties, and applications, *Chem. Rev.*, *108*, 3893–3957.
56. Lee, K. J., Lee, D. K., Kim, Y. W., Choe, W. S., Kim, J. H. (2007). Theoretical consideration on the glass transition behavior of polymer nanocomposites, *J. Polym. Sci. Polym. Phys.*, *45*, 2232–2238.
57. Ramanathan, T., Abdala, A. A., Stankovich, S., Dikin, D. A., Herrera-Alonso, M., Piner, R. D., Adamson, D. H., Schniepp, H. C., Chen, X., Ruoff, R. S., Nguyen, S. T., Aksay, I. A., Prud'homme, R. K., Brinson, L. C. (2008). Functionalized graphene sheets for polymer nanocomposites, *Nat. Nanotechnol.*, *3*, 327–331.
58. Oh, H., Green, P. F. (2009). Polymer chain dynamics and glass transition in athermal polymer/nanoparticle mixtures, *Nat. Mater.*, *8*, 139–143.
59. Shanmugharaj, A. M., Bae, J. H., Nayak, R. R., Ryu, S. H. (2007). Preparation of poly(styrene-*co*-acrylonitrile)-grafted multiwalled carbon nanotubes via surface-initiated atom transfer radical polymerization, *J. Polym. Sci. A*, *45*, 460–470.
60. Yang, Y. K., Xie, X. L., Wu, J. G., Mai, Y. W. (2006). Synthesis and self-assembly of polystyrene-grafted multiwalled carbon nanotubes with a hairy-rod nanostructure, *J. Polym. Sci. A*, *44*, 3869–3881.
61. Putz, K. W., Palmeri, M. J., Cohn, R. B., Andrews, R., Brinson, L. C. (2008). Effect of cross-link density on interphase creation in polymer nanocomposites, *Macromolecules*, *41*, 6752–6756.
62. Miyagawa, H., Mohanty, A. K., Drzal, L. T., Misra, M. (2005). Nanocomposites from biobased epoxy and single-wall carbon nanotubes: synthesis, and mechanical and thermophysical properties evaluation, *Nanotechnology*, *16*, 118–124.
63. Miyagawa, H., Rich, M. J., Drzal, L. T. (2006). Thermo-physical properties of epoxy nanocomposites reinforced by carbon nanotubes and vapor grown carbon fibers, *Thermochim. Acta*, *442*, 67–73.
64. Grady, B. P., Paul, A., Peters, J. E., Ford, W. T. (2009). Glass transition behavior of single-walled carbon nanotube–polystyrene composites, *Macromolecules*, *42*, 6152–6158.

65. Xia, H. S., Song, M. (2006). Preparation and characterisation of polyurethane grafted single-walled carbon nanotubes and derived polyurethane nanocomposites, *J. Mater. Chem.*, *16*, 1843–1851.
66. Tonpheng, B., Yu, J. C., Andersson, O. (2009). Thermal conductivity, heat capacity, and cross-linking of polyisoprene/single-wall carbon nanotube composites under high pressure, *Macromolecules*, *42*, 9295–9301.
67. Miaudet, P., Derre, A., Maugey, M., Zakri, C., Piccione, P. M., Inoubli, R., Poulin, P. (2007). Shape and temperature memory of nanocomposites with broadened glass transition, *Science*, *318*, 1294–1296.
68. Pham, J. Q., Mitchell, C. A., Bahr, J. L., Tour, J. M., Krishanamoorti, R., Green, P. F. (2003). Glass transition of polymer/single-walled carbon nanotube composite films, *J. Polym. Sci. Polym. Phys.*, *41*, 3339–3345.
69. Choi, Y. J., Hwang, S. H., Hong, Y. S., Kim, J. Y., Ok, C. Y., Huh, W., Lee, S. W. (2005). Preparation and characterization of PS/multi-walled carbon nanotube nanocomposites, *Polym. Bull.*, *53*, 393–400.
70. Kumar, S., Rath, T., Khatua, B. B., Dhibar, A. K., Das, C. K. (2009). Preparation and characterization of poly(methyl methacrylate)/multi-walled carbon nanotube composites, *J. Nanosci. Nanotechnol.*, *9*, 4644–4655.
71. Cui, L., Tarte, N. H., Woo, S. I. (2009). Synthesis and characterization of PMMA/MWNT nanocomposites prepared by *in situ* polymerization with $Ni(acac)_2$ catalyst, *Macromolecules*, *42*, 8649–8654.
72. Pötschke, P., Bhattacharyya, A. R., Janke, A., Goering, H. (2003). Melt mixing of polycarbonate/multi-wall carbon nanotube composites, *Compos. Interfaces*, *10*, 389–404.
73. Jin, S. H., Choi, D. K., Lee, D. S. (2008). Electrical and rheological properties of polycarbonate/multiwalled carbon nanotube nanocomposites, *Colloids Surfaces A*, *313*, 242–245.
74. Schartel, B., Braun, U., Knoll, U., Bartholmai, M., Goering, H., Neubert, D., Pötschke, P. (2008). Mechanical, thermal, and fire behavior of bisphenol A polycarbonate/multiwall carbon nanotube nanocomposites, *Polym. Eng. Sci.*, *48*, 149–158.
75. Jin, S. H., Kang, I. H., Kim, Y. S., Park, C. Y., Lee, D. S. (2008). Thermal and electrical properties of nanocomposites based on acrylic copolymers and multiwalled carbon nanotube, *J. Nanosci. Nanotechnol.*, *8*, 5076–5079.
76. Jin, S. H., Lee, D. S. (2008). Preparation and properties of the nanocomposites based on poly(methyl methacrylate-*co*-butyl acrylate) and multiwalled carbon nanotube, *J. Nanosci. Nanotechnol.*, *8*, 4675–4678.
77. Yang, J. W., Hu, J. H., Wang, C. C., Qin, Y. J., Guo, Z. X. (2004). Fabrication and characterization of soluble multi-walled carbon nanotubes reinforced P(MMA-*co*-EMA) composites, *Macromol. Mater. Eng.*, *289*, 828–832.
78. Ha, M. L. P., Grady, B. P., Lolli, G., Resasco, D. E., Ford, W. T. (2007). Composites of single-walled carbon nanotubes and styrene–isoprene copolymer lattices, *Macromol. Chem. Phys.*, *208*, 446–456.
79. Roy, S., Sahoo, N. G., Mukherjee, M., Das, C. K., Chan, S. H., Li, L. (2009). Improvement of properties of polyetherimide/liquid crystalline polymer blends in the presence of functionalized carbon nanotubes, *J. Nanosci. Nanotechnol.*, *9*, 1928–1934.
80. Chen, X. H., Wang, J. F., Zou, J. G., Wu, X. L., Chen, X. J., Xue, F. (2009). Mechanical and thermal properties of functionalized multiwalled carbon nanotubes and multiwalled carbon nanotube–polyurethane composites, *J. Appl. Polym. Sci.*, *114*, 3407–3413.
81. Xia, H. S., Song, M. (2005). Preparation and characterization of polyurethane–carbon nanotube composites, *Soft Matter*, *1*, 386–394.
82. Mun, S. J., Jung, Y. M., Kim, J. C., Chang, J. H. (2008). Poly(ethylene terephthalate) nanocomposite fibers with functionalized multiwalled carbon nanotubes via *in-situ* polymerization, *J. Appl. Polym. Sci.*, *109*, 638–646.
83. Yuen, S. M., Ma, C. C. M., Wu, H. H., Kuan, H. C., Chen, W. J., Liao, S. H., Hsu, C. W., Wu, H. L. (2007). Preparation and thermal, electrical, and morphological properties of multiwalled carbon nanotube and epoxy composites, *J. Appl. Polym. Sci.*, *103*, 1272–1278.
84. Zou, W., Du, Z. J., Liu, Y. X., Yang, X., Li, H. Q., Zhang, C. (2008). Functionalization of MWNTs using polyacryloyl chloride and the properties of CNT–epoxy matrix nanocomposites, *Compos. Sci. Technol.*, *68*, 3259–3264.

85. Yan, Z., Yuexin, D., Lu, Y., Fengxia, G. (2009). The dispersion of SWCNTs treated by dispersing agents in glass fiber reinforced polymer composites, *Compos. Sci. Technol.*, *69*, 2115–2118.
86. Fidelus, J. D., Wiesel, E., Gojny, F. H., Schulte, K., Wagner, H. D. (2005). Thermo-mechanical properties of randomly oriented carbon/epoxy nanocomposites, *Compos. Part A*, *36*, 1555–1561.
87. Evseeva, L., Tanaeva, S. (2008). Influence of the concentration of carbon nanotubes (CNT) on the thermophysical properties of epoxy/CNT nanocomposites at low temperatures, *Mech. Compos. Mater.*, *44*, 487–494.
88. Gojny, F. H., Schulte, K. (2004). Functionalisation effect on the thermo-mechanical behaviour of multi-wall carbon nanotube/epoxy-composites, *Compos. Sci. Technol.*, *64*, 2303–2308.
89. Guadagno, L., Vertuccio, L., Sorrentino, A., Raimondo, M., Naddeo, C., Vittoria, V., Iannuzzo, G., Calvi, E., Russo, S. (2009). Mechanical and barrier properties of epoxy resin filled with multi-walled carbon nanotubes, *Carbon*, *47*, 2419–2430.
90. Prolongo, S. G., Gude, M. R., Urena, A. (2009). Synthesis and characterisation of epoxy resins reinforced with carbon nanotubes and nanofibers, *J. Nanosci. Nanotechnol.*, *9*, 6181–6187.
91. Cui, S., Canet, R., Derre, A., Couzi, M., Delhaes, P. (2003). Characterization of multiwall carbon nanotubes and influence of surfactant in the nanocomposite processing, *Carbon*, *41*, 797–809.
92. Jeong, W., Kessler, M. R. (2008). Toughness enhancement in ROMP functionalized carbon nanotube/polydicyclopentadiene composites, *Chem. Mater.*, *20*, 7060–7068.
93. Stephens, C. H., Yang, H., Islam, M., Chum, S. P., Rowan, S. J., Hiltner, A., Baer, E. (2003). Characterization of polyethylene with partially random chlorine substitution, *J. Polym. Sci. Polym. Phys.*, *41*, 2062–2070.
94. Wang, C., Chu, M. C., Lin, T. L., Lai, S. M., Shih, H. H., Yang, J. C. (2001). Microstructures of a highly short-chain branched polyethylene, *Polymer*, *42*, 1733–1741.
95. Logakis, E., Pandis, C., Peoglos, V., Pissis, P., Stergiou, C., Pionteck, J., Pötschke, P., Micusik, M., Omastova, M. (2009). Structure–property relationships in polyamide 6/multi-walled carbon nanotubes nanocomposites, *J. Polym. Sci. Polym. Phys.*, *47*, 764–774.
96. Li, J. A., Fang, Z. P., Tong, L. F., Gu, A. J., Liu, F. (2006). Polymorphism of nylon-6 in multiwalled carbon nanotubes/nylon-6 composites, *J. Polym. Sci. Polym. Phys.*, *44*, 1499–1512.
97. Phang, I. Y., Ma, J. H., Shen, L., Liu, T. X., Zhang, W. D. (2006). Crystallization and melting behavior of multi-walled carbon nanotube-reinforced nylon-6 composites, *Polym. Int.*, *55*, 71–79.
98. Saeed, K., Park, S. Y. (2007). Preparation of multiwalled carbon nanotube/nylon-6 nanocomposites by *in situ* polymerization, *J. Appl. Polym. Sci.*, *106*, 3729–3735.
99. Lincoln, D. M., Vaia, R. A., Wang, Z. G., Hsiao, B. S., Krishnamoorti, R. (2001). Temperature dependence of polymer crystalline morphology in nylon 6/montmorillonite nanocomposites, *Polymer*, *42*, 9975–9985.
100. Kim, G. M., Michler, G. H., Ania, F., Calleja, F. J. B. (2007). Temperature dependence of polymorphism in electrospun nanofibres of PA6 and PA6/clay nanocomposite, *Polymer*, *48*, 4814–4823.
101. Huang, X. Y., Jiang, P. K., Kim, C., Liu, F., Yin, Y. (2009). Influence of aspect ratio of carbon nanotubes on crystalline phases and dielectric properties of poly(vinylidene fluoride), *Eur. Polym. J.*, *45*, 377–386.
102. Levi, N., Czerw, R., Xing, S. Y., Iyer, P., Carroll, D. L. (2004). Properties of polyvinylidene difluoride–carbon nanotube blends, *Nano Lett.*, *4*, 1267–1271.
103. Manna, S., Nandi, A. K. (2007). Piezoelectric beta polymorph in poly(vinylidene fluoride)-functionalized multiwalled carbon nanotube nanocomposite films, *J. Phys. Chem. C*, *111*, 14670–14680.
104. Jog, J. P. (2006). Crystallisation in polymer nanocomposites, *Mater. Sci. Technol.*, *22*, 797–806.
105. Sarno, M., Gorrasi, G., Sannino, D., Sorrentino, A., Ciambelli, P., Vittoria, V. (2004). Polymorphism and thermal behaviour of syndiotactic poly(propylene)/carbon nanotube composites, *Macromol. Rapid Commun.*, *25*, 1963–1967.
106. Grady, B. P., Pompeo, F., Shambaugh, R. L., Resasco, D. E. (2002). Nucleation of polypropylene crystallization by single-walled carbon nanotubes, *J. Phys. Chem. B*, *106*, 5852–5858.
107. Leelapornpisit, W., Ton-That, M. T., Perrin-Sarazin, F., Cole, K. C., Denault, J., Simard, B. (2005). Effect of carbon nanotubes on the crystallization and properties of polypropylene, *J. Polym. Sci. Polym. Phys.*, *43*, 2445–2453.

108. Miltner, H. E., Grossiord, N., Lu, K. B., Loos, J., Koning, C. E., Van Mele, B. (2008). Isotactic polypropylene/carbon nanotube composites prepared by latex technology. Thermal analysis of carbon nanotube-induced nucleation, *Macromolecules*, *41*, 5753–5762.
109. Avila-Orta, C. A., Medellin-Rodriguez, F. J., Davila-Rodriguez, M. V., Aguirre-Figueroa, Y. A., Yoon, K., Hsiao, B. S. (2007). Morphological features and melting behavior of nanocomposites based on isotactic polypropylene and multiwalled carbon nanotubes, *J. Appl. Polym. Sci.*, *106*, 2640–2647.
110. Ganss, M., Satapathy, B. K., Thunga, M., Weidisch, R., Pötschke, P., Jehnichen, D. (2008). Structural interpretations of deformation and fracture behavior of polypropylene/multi-walled carbon nanotube composites, *Acta Mater.*, *56*, 2247–2261.
111. Li, L. Y., Li, B., Hood, M. A., Li, C. Y. (2009). Carbon nanotube induced polymer crystallization: the formation of nanohybrid shish-kebabs, *Polymer*, *50*, 953–965.
112. Li, L. Y., Li, C. Y., Ni, C. Y., Rong, L. X., Hsiao, B. (2007). Structure and crystallization behavior of Nylon 66/multi-walled carbon nanotube nanocomposites at low carbon nanotube contents, *Polymer*, *48*, 3452–3460.
113. Kodjie, S. L., Li, L. Y., Li, B., Cai, W. W., Li, C. Y., Keating, M. (2006). Morphology and crystallization behavior of HDPE/CNT nanocomposite, *J. Macromol. Sci. B*, *45*, 231–245.
114. Lu, K. B., Grossiord, N., Koning, C. E., Miltner, H. E., van Mele, B., Loos, J. (2008). Carbon nanotube/isotactic polypropylene composites prepared by latex technology: morphology analysis of CNT-induced nucleation, *Macromolecules*, *41*, 8081–8085.
115. Zhang, S., Minus, M. L., Zhu, L. B., Wong, C. P., Kumar, S. (2008). Polymer transcrystallinity induced by carbon nanotubes, *Polymer*, *49*, 1356–1364.
116. Haggenmueller, R., Fischer, J. E., Winey, K. I. (2006). Single wall carbon nanotube/polyethylene nanocomposites: nucleating and templating polyethylene crystallites, *Macromolecules*, *39*, 2964–2971.
117. Patil, N., Balzano, L., Portale, G., Rastogi, S. (2009). Influence of nanoparticles on the rheological behaviour and initial stages of crystal growth in linear polyethylene, *Macromol. Chem. Phys.*, *210*, 2174–2187.
118. Yang, J. H., Wang, C. Y., Wang, K., Zhang, Q., Chen, F., Du, R. N., Fu, Q. (2009). Direct formation of nanohybrid shish-kebab in the injection molded bar of polyethylene/multiwalled carbon nanotubes composite, *Macromolecules*, *42*, 7016–7023.
119. Causin, V., Yang, B. X., Marega, C., Goh, S. H., Marigo, A. (2009). Nucleation, structure and lamellar morphology of isotactic polypropylene filled with polypropylene-grafted multiwalled carbon nanotubes, *Eur. Polym. J.*, *45*, 2155–2163.
120. Zhou, Z., Wang, S. F., Lu, L., Zhang, Y., Zhang, Y. X. (2007). Isothermal crystallization kinetics of polypropylene with silane functionalized multi-walled carbon nanotubes, *J. Polym. Sci. Polym. Phys.*, *45*, 1616–1624.
121. Causin, V., Marega, C., Saini, R., Marigo, A., Ferrara, G. (2007). Crystallization behavior of isotactic polypropylene based nanocomposites, *J. Therm. Anal. Calorim.*, *90*, 849–857.
122. Trujillo, M., Arnal, M. L., Mueller, A. J., Bredeau, S., Bonduel, D., Dubois, P., Hamley, I. W., Castelletto, V. (2008). Thermal fractionation and isothermal crystallization of polyethylene nanocomposites prepared by *in situ* polymerization, *Macromolecules*, *41*, 2087–2095.
123. Bhattacharyya, A. R., Bose, S., Kulkarni, A. R., Pötschke, P., Haussler, L., Fischer, D., Jehnichen, D. (2007). Styrene maleic anhydride copolymer mediated dispersion of single wall carbon nanotubes in polyamide 12: crystallization behavior and morphology, *J. Appl. Polym. Sci.*, *106*, 345–353.
124. Chen, H., Liu, Z., Cebe, P. (2009). Chain confinement in electrospun nanofibers of PET with carbon nanotubes, *Polymer*, *50*, 872–880.
125. Gorrasi, G., Romeo, V., Sannino, D., Sarno, M., Ciambelli, P., Vittoria, V., De Vivo, B., Tucci, V. (2007). Carbon nanotube induced structural and physical property transitions of syndiotactic polypropylene, *Nanotechnology*, *18*, 275703.
126. Teng, C. C., Ma, C. C. M., Huang, Y. W., Yuen, S. M., Weng, C. C., Chen, G. H., Su, S. F. (2008). Effect of MWCNT content on rheological and dynamic mechanical properties of multiwalled carbon nanotube/polypropylene composites, *Compos. Part A*, *39*, 1869–1875.
127. Sahoo, N. G., Thet, N. T., Tan, Q. H., Li, L., Chan, S. H., Zha, J. H., Yu, S. Z. (2009). Effect of carbon nanotubes and processing methods on the properties of carbon nanotube/polypropylene composites, *J. Nanosci. Nanotechnol.*, *9*, 5910–5919.

128. Wunderlich, B. (2003). Reversible crystallization and the rigid-amorphous phase in semicrystalline macromolecules, *Prog. Polym. Sci.*, *28*, 383–450.
129. Huang, J. M., Chu, P. P., Chang, F. C. (2000). Conformational changes and molecular motion of poly (ethylene terephthalate) annealed above glass transition temperature, *Polymer*, *41*, 1741–1748.
130. Kawaguchi, T., Mamada, A., Hosokawa, Y., Horii, F. (1998). ^{2}H NMR analysis of the phenylene motion in different poly(ethylene terephthalate) samples, *Polymer*, *39*, 2725–2732.
131. Tzavalas, S., Drakonakis, V., Mouzakis, D. E., Fischer, D., Gregoriou, V. G. (2006). Effect of carboxy-functionalized multiwall nanotubes (MWNT-COOH) on the crystallization and chain conformations of poly(ethylene terephthalate) PET in PET–MWNT nanocomposites, *Macromolecules*, *39*, 9150–9156.
132. Grady, B. P. (2008). Review and critical analysis of the morphology of random ionomers across many length scales, *Polym. Eng. Sci.*, *48*, 1029–1051.
133. Peeterbroeck, S., Breugelmans, L., Alexandre, M., Nagy, J., Viville, P., Lazzaroni, R., Dubois, P. (2007). The influence of the matrix polarity on the morphology and properties of ethylene vinyl acetate copolymers–carbon nanotube nanocomposites, *Compos. Sci. Technol.*, *67*, 1659–1665.
134. Pujari, S., Ramanathan, T., Kasimatis, K., Masuda, J., Andrews, R., Torkelson, J. M., Brinson, L. C., Burghardt, W. R. (2009). Preparation and characterization of multiwalled carbon nanotube dispersions in polypropylene: melt mixing versus solid-state shear pulverization, *J. Polym. Sci. Polym. Phys.*, *47*, 1426–1436.
135. Brosse, A. C., Tence-Girault, S., Piccione, P. M., Leibler, L. (2008). Effect of multi-walled carbon nanotubes on the lamellae morphology of polyamide-6, *Polymer*, *49*, 4680–4686.
136. Iervolino, R., Somma, E., Nobile, M. R., Chen, X. M., Hsiao, B. S. (2009). The role of multi-walled carbon nanotubes in shear enhanced crystallization of isotactic poly(1-butene), *J. Therm. Anal. Calorim.*, *98*, 611–622.
137. Mago, G., Fisher, F. T., Kalyon, D. M. (2008). Effects of multiwalled carbon nanotubes on the shear-induced crystallization behavior of poly(butylene terephthalate), *Macromolecules*, *41*, 8103–8113.
138. Wang, K., Tang, C. Y., Zhao, P., Yang, H., Zhang, Q., Du, R. N., Fu, Q. (2007). Rheological investigations in understanding shear-enhanced crystallization of isotactic poly(propylene)/multi-walled carbon nanotube composites, *Macromol. Rapid Commun.*, *28*, 1257–1264.
139. Garcia-Gutierrez, M. C., Hernandez, J. J., Nogales, A., Pantine, P., Rueda, D. R., Ezquerra, T. A. (2008). Influence of shear on the templated crystallization of poly(butylene terephthalate)/single wall carbon nanotube nanocomposites, *Macromolecules*, *41*, 844–851.
140. Coleman, J. N., Cadek, M., Ryan, K. P., Fonseca, A., Nagy, J. B., Blau, W. J., Ferreira, M. S. (2006). Reinforcement of polymers with carbon nanotubes. The role of an ordered polymer interfacial region. Experiment and modeling, *Polymer*, *47*, 8556–8561.
141. Ryan, K. P., Cadek, M., Nicolosi, V., Blond, D., Ruether, M., Armstrong, G., Swan, H., Fonseca, A., Nagy, J. B., Maser, W. K., Blau, W. J., Coleman, J. N. (2007). Carbon nanotubes for reinforcement of plastics? A case study with poly(vinyl alcohol), *Compos. Sci. Technol.*, *67*, 1640–1649.
142. Minus, M. L., Chae, H. G., Kumar, S. (2006). Single wall carbon nanotube templated oriented crystallization of poly(vinyl alcohol), *Polymer*, *47*, 3705–3710.
143. Ryan, K. P., Lipson, S. M., Drury, A., Cadek, M., Ruether, M., O'Flaherty, S. M., Barron, V., McCarthy, B., Byrne, H. J., Blau, W. J., Coleman, J. N. (2004). Carbon-nanotube nucleated crystallinity in a conjugated polymer based composite, *Chem. Phys. Lett.*, *391*, 329–333.
144. Wu, D. F., Sun, Y. R., Wu, L., Zhang, M. (2008). Linear viscoelastic properties and crystallization behavior of multi-walled carbon nanotube/polypropylene composites, *J. Appl. Polym. Sci.*, *108*, 1506–1513.
145. Xu, D. H., Wang, Z. G. (2008). Role of multi-wall carbon nanotube network in composites to crystallization of isotactic polypropylene matrix, *Polymer*, *49*, 330–338.
146. Valentini, L., Biagiotti, J., Kenny, J. M., Santucci, S. (2003). Effects of single-walled carbon nanotubes on the crystallization behavior of polypropylene, *J. Appl. Polym. Sci.*, *87*, 708–713.
147. Vaisman, L., Marom, G., Wagner, H. D. (2006). Dispersions of surface-modified carbon nanotubes in water-soluble and water-insoluble polymers, *Adv. Funct. Mater.*, *16*, 357–363.
148. Wu, T. M., Chen, E. C. (2006). Isothermal and nonisothermal crystallization kinetics of poly(epsilon-caprolactone)/multi-walled carbon nanotube composites, *Polym. Eng. Sci.*, *46*, 1309–1317.

149. Assouline, E., Lustiger, A., Barber, A. H., Cooper, C. A., Klein, E., Wachtel, E., Wagner, H. D. (2003). Nucleation ability of multiwall carbon nanotubes in polypropylene composites, *J. Polym. Sci. Polym. Phys.*, *41*, 520–527.
150. Valentini, L., Biagiotti, J., Kenny, J. M., Santucci, S. (2003). Morphological characterization of single-walled carbon nanotubes–PP composites, *Compos. Sci. Technol.*, *63*, 1149–1153.
151. Valentini, L., Biagiotti, J., Lopez-Manchado, M. A., Santucci, S., Kenny, J. M. (2004). Effects of carbon nanotubes on the crystallization behavior of polypropylene, *Polym. Eng. Sci.*, *44*, 303–311.
152. Yin, C. L., Liu, Z. Y., Yang, W., Yang, M. B., Feng, J. M. (2009). Crystallization and morphology of iPP/MWCNT prepared by compounding iPP melt with MWCNT aqueous suspension, *Colloid Polym. Sci.*, *287*, 615–620.
153. Razavi-Nouri, M. (2009). Study of non-isothermal crystallization kinetics of single-walled carbon nanotubes filled polypropylene using Avrami and Mo models, *Iran. Polym. J.*, *18*, 167–178.
154. Fereidoon, A., Ahangari, M. G., Saedodin, S. (2009). Study of the nonisothermal crystallization kinetics and melting behaviors of polypropylene reinforced with single-walled carbon nanotubes nanocomposite, *J. Macromol. Sci. B*, *48*, 25–40.
155. Fereidoon, A., Ahangari, M. G., Saedodin, S. (2009). Thermal and structural behaviors of polypropylene nanocomposites reinforced with single-walled carbon nanotubes by melt processing method, *J. Macromol. Sci. B*, *48*, 196–211.
156. Fereidoon, A., Ahangari, M. G., Saedodin, S. (2009). A DSC study on the nonisothermal crystallization kinetics of polypropylene/single-walled carbon nanotube nanocomposite, *Polym. Plast. Technol. Eng.*, *48*, 579–586.
157. Razavi-Nouri, M., Ghorbanzadeh-Ahangari, M., Fereidoon, A., Jahanshahi, M. (2009). Effect of carbon nanotubes content on crystallization kinetics and morphology of polypropylene, *Polym. Test.*, *28*, 46–52.
158. Reyes-de Vaaben, S., Aguilar, A., Avalos, F., Ramos-de Valle, L. F. (2008). Carbon nanoparticles as effective nucleating agents for polypropylene, *J. Therm. Anal. Calorim.*, *93*, 947–952.
159. Peneva, Y., Valcheva, M., Minkova, L., Micusik, M., Omastova, M. (2008). Nonisothermal crystallization kinetics and microhardness of PP/CNT composites, *J. Macromol. Sci. B*, *47*, 1197–1210.
160. Hou, Z. C., Wang, K., Zhao, P., Zhang, Q., Yang, C. Y., Chen, D. Q., Du, R. N., Fu, Q. (2008). Structural orientation and tensile behavior in the extrusion-stretched sheets of polypropylene/multi-walled carbon nanotubes' composite, *Polymer*, *49*, 3582–3589.
161. Bao, S. P., Tjong, S. C. (2008). Mechanical behaviors of polypropylene/carbon nanotube nanocomposites: the effects of loading rate and temperature, *Mater. Sci. Eng. A*, *485*, 508–516.
162. Kaminsky, W., Funck, A., Klinke, C. (2008). *In-situ* polymerization of olefins on nanoparticles or fibers by metallocene catalysts, *Top. Catal.*, *48*, 84–90.
163. Bikiaris, D., Vassiliou, A., Chrissafis, K., Paraskevopoulos, K. M., Jannakoudakis, A., Docoslis, A. (2008). Effect of acid treated multi-walled carbon nanotubes on the mechanical, permeability, thermal properties and thermo-oxidative stability of isotactic polypropylene, *Polym. Degrad. Stabil.*, *93*, 952–967.
164. Kovalchuk, A. A., Shevchenko, V. G., Shchegolikhin, A. N., Nedorezova, P. M., Klyamkina, A. N., Aladyshev, A. M. (2008). Isotactic and syndiotactic polypropylene/multi-wall carbon nanotube composites: synthesis and properties, *J. Mater. Sci.*, *43*, 7132–7140.
165. Funck, A., Kaminsky, W. (2007). Polypropylene carbon nanotube composites by *in situ* polymerization, *Compos. Sci. Technol.*, *67*, 906–915.
166. Bao, H. D., Guo, Z. X., Yu, J. (2009). Electrical resistivity, crystallization and mechanical properties of polypropylene/multi-walled carbon nanotube/calcium carbonate composites prepared by melt mixing, *Chin. J. Polym. Sci.*, *27*, 393–398.
167. Bhattacharyya, A. R., Sreekumar, T. V., Liu, T., Kumar, S., Ericson, L. M., Hauge, R. H., Smalley, R. E. (2003). Crystallization and orientation studies in polypropylene/single wall carbon nanotube composite, *Polymer*, *44*, 2373–2377.
168. Castell, P., Medel, F. J., Martinez, M. T., Puertolas, J. A. (2009). Influence of gamma irradiation on carbon nanotube-reinforced polypropylene, *J. Nanosci. Nanotechnol.*, *9*, 6055–6063.

169. Kaganj, A. B., Rashidi, A. M., Arasteh, R., Taghipoor, S. (2009). Crystallisation behaviour and morphological characteristics of poly(propylene)/multi-walled carbon nanotube nanocomposites, *J. Exp. Nanosci.*, *4*, 21–34.
170. Sandler, J., Broza, G., Nolte, M., Schulte, K., Lam, Y. M., Shaffer, M. S. P. (2003). Crystallization of carbon nanotube and nanofiber polypropylene composites, *J. Macromol. Sci. B*, *42*, 479–488.
171. Rahmatpour, A., Aalaie, J. (2008). Steady shear rheological behavior, mechanical properties, and morphology of the polypropylene/carbon nanotube nanocomposites, *J. Macromol. Sci. B*, *47*, 929–941.
172. Deng, H., Skipa, T., Zhang, R., Lellinger, D., Bilotti, E., Alig, I., Peijs, T. (2009). Effect of melting and crystallization on the conductive network in conductive polymer composites, *Polymer*, *50*, 3747–3754.
173. Gao, Y., Wang, Y., Shi, J., Zhu, S. B., Bao, Y. L., Bai, H. W., Li, Y. L. (2009). Effect of functionalized SWCNTs on microstructure of PP-*g*-MA/OMMT/f-SWCNTs nanocomposite, *J. Appl. Polym. Sci.*, *112*, 2413–2424.
174. Lee, G. W., Jagannathan, S., Chae, H. G., Minus, M. L., Kumar, S. (2008). Carbon nanotube dispersion and exfoliation in polypropylene and structure and properties of the resulting composites, *Polymer*, *49*, 1831–1840.
175. Wiemann, K., Kaminsky, W., Gojny, F. H., Schulte, K. (2005). Synthesis and properties of syndiotactic poly(propylene)/carbon nanofiber and nanotube composites prepared by *in situ* polymerization with metallocene/MAO catalysts, *Macromol. Chem. Phys.*, *206*, 1472–1478.
176. Rizvi, R., Kim, J. K., Naguib, H. (2009). Synthesis and characterization of novel low density polyethylene–multiwall carbon nanotube porous composites, *Smart Mater. Struct.*, *18*, 104002.
177. Trujillo, M., Arnal, M. L., Muller, A. J., Laredo, E., Bredeau, S., Bonduel, D., Dubois, P. (2007). Thermal and morphological characterization of nanocomposites prepared by *in-situ* polymerization of high-density polyethylene on carbon nanotubes, *Macromolecules*, *40*, 6268–6276.
178. Jeon, K., Lumata, L., Tokumoto, T., Steven, E., Brooks, J., Alamo, R. G. (2007). Low electrical conductivity threshold and crystalline morphology of single-walled carbon nanotubes–high density polyethylene nanocomposites characterized by SEM, Raman spectroscopy and AFM, *Polymer*, *48*, 4751–4764.
179. Chrissafis, K., Paraskevopoulos, K. M., Tsiaoussis, I., Bikiaris, D. (2009). Comparative study of the effect of different nanoparticles on the mechanical properties, permeability, and thermal degradation mechanism of HDPE, *J. Appl. Polym. Sci.*, *114*, 1606–1618.
180. Xie, X. L., Aloys, K., Zhou, X. P., Zeng, F. D. (2003). Ultrahigh molecular mass polyethylene/carbon nanotube composites: crystallization and melting properties, *J. Therm. Anal. Calorim.*, *74*, 317–323.
181. Grady, B. P., Arthur, D. J., Ferguson, J. (2009). Single-walled carbon nanotube/ultrahigh-molecular-weight polyethylene composites with percolation at low nanotube contents, *Polym. Eng. Sci.*, *49*, 2440–2446.
182. Li, S. N., Li, Z. M., Yang, M. B., Hu, Z. Q., Xu, X. B., Huang, R. (2004). Carbon nanotubes induced nonisothermal crystallization of ethylene–vinyl acetate copolymer, *Mater. Lett.*, *58*, 3967–3970.
183. Bhattacharyya, A. R., Pötschke, P., Haussler, L., Fischer, D. (2005). Reactive compatibilization of melt mixed PA6/SWNT composites: mechanical properties and morphology, *Macromol. Chem. Phys.*, *206*, 2084–2095.
184. Li, J., Fang, Z. P., Zhu, Y., Tong, L. F., Gu, A. J., Uu, F. (2007). Isothermal crystallization kinetics and melting behavior of multiwalled carbon nanotubes/polyamide-6 composites, *J. Appl. Polym. Sci.*, *105*, 3531–3542.
185. Chen, E. C., Wu, T. M. (2008). Isothermal and nonisothermal crystallization kinetics of nylon 6/functionalized multi-walled carbon nanotube composites, *J. Polym. Sci. Polym. Phys.*, *46*, 158–169.
186. Giraldo, L. F., Lopez, B. L., Brostow, W. (2009). Effect of the type of carbon nanotubes on tribological properties of polyamide 6, *Polym. Eng. Sci.*, *49*, 896–902.
187. Meng, H., Sui, G. X., Xie, G. Y., Yang, R. (2009). Non-isothermal crystallization kinetics of polyamide 6/diamine-modified MWNTs nanocomposite, *J. Mater. Sci. Technol.*, *25*, 145–150.

188. Meng, H., Sui, G. X., Fang, P. F., Yang, R. (2008). Effects of acid- and diamine-modified MWNTs on the mechanical properties and crystallization behavior of polyamide 6, *Polymer*, *49*, 610–620.
189. Li, J., Fang, Z. P., Tong, L. F., Gu, A. J., Liu, F. (2006). Effect of multi-walled carbon nanotubes on non-isothermal crystallization kinetics of polyamide 6, *Eur. Polym. J.*, *42*, 3230–3235.
190. Naffakh, M., Marco, C., Gomez, M. A., Ellis, G., Maser, W. K., Benito, A., Martinez, M. T. (2009). Crystalline transformations in nylon-6/single-walled carbon nanotube nanocomposites, *J. Nanosci. Nanotechnol.*, *9*, 6120–6126.
191. Sahoo, N. G., Cheng, H. K. F., Cai, J. W., Li, L., Chan, S. H., Zhao, J. H., Yu, S. Z. (2009). Improvement of mechanical and thermal properties of carbon nanotube composites through nanotube functionalization and processing methods, *Mater. Chem. Phys.*, *117*, 313–320.
192. Krause, B., Pötschke, P., Haussler, L. (2009). Influence of small scale melt mixing conditions on electrical resistivity of carbon nanotube–polyamide composites, *Compos. Sci. Technol.*, *69*, 1505–1515.
193. Lin, S. Y., Chen, E. C., Liu, K. Y., Wu, T. M. (2009). Isothermal crystallization behavior of polyamide 6,6/multiwalled carbon nanotube nanocomposites, *Polym. Eng. Sci.*, *49*, 2447–2453.
194. Sengupta, R., Ganguly, A., Sabharwal, S., Chaki, T. K., Bhowmick, A. K. (2007). MWCNT reinforced polyamide-6,6 films: preparation, characterization and properties, *J. Mater. Sci.*, *42*, 923–934.
195. Li, Z. Y., Xu, S. Z., Liu, W. T., He, S. Q., Zhu, C. S. (2009). Preparation and characterization of nylon610/functionalized multiwalled carbon nanotubes composites, *J. Appl. Polym. Sci.*, *113*, 2805–2812.
196. Wang, B., Sun, G. P., Liu, J. J., He, X. F., Li, J. (2006). Crystallization behavior of carbon nanotubes-filled polyamide 1010, *J. Appl. Polym. Sci.*, *100*, 3794–3800.
197. Wang, B., Sun, G. P., He, X. F., Liu, J. J. (2007). The effect of multiwall carbon nanotube on the crystallization, morphology, and rheological properties of nylon1010 nanocomposites, *Polym. Eng. Sci.*, *47*, 1610–1620.
198. Wu, D., Wu, L., Sun, Y., Zhang, M. (2007). Rheological properties and crystallization behavior of multi-walled carbon nanotube/poly(epsilon-caprolactone) composites, *J. Polym. Sci. Polym. Phys.*, *45*, 3137–3147.
199. Jana, R. N., Im, C., Bhunia, H. (2009). Effect of multiwalled carbon nanotubes on crystallization behavior of poly(epsilon-caprolactone)diol, *J. Thermoplast. Compos. Mater.*, *22*, 531–546.
200. Jana, R. N., Im, C. (2009). Isothermal crystallization behavior of poly(epsilon-caprolactone) diol/ functionalized-multiwalled carbon nanotube composites, *Int. J. Polym. Anal. Charact.*, *14*, 418–436.
201. Priftis, D., Sakellariou, G., Hadjichristidis, N., Penott, E. K., Lorenzo, A. T., Muller, A. J. (2009). Surface modification of multiwalled carbon nanotubes with biocompatible polymers via ring opening and living anionic surface initiated polymerization. Kinetics and crystallization behavior, *J. Polym. Sci. A*, *47*, 4379–4390.
202. Xu, G. Y., Du, L. C., Wang, H., Xia, R., Meng, X. C., Zhu, Q. R. (2008). Nonisothermal crystallization kinetics and thermomechanical properties of multiwalled carbon nanotube-reinforced poly(epsilon-caprolactone) composites, *Polym. Int.*, *57*, 1052–1066.
203. Saeed, K., Park, S. Y. (2007). Preparation and properties of multiwalled carbon nanotube/poly-caprolactone nanocomposites, *J. Appl. Polym. Sci.*, *104*, 1957–1963.
204. Mitchell, C. A., Krishnamoorti, R. (2005). Non-isothermal crystallization of *in situ* polymerized poly (epsilon-caprolactone) functionalized-SWNT nanocomposites, *Polymer*, *46*, 8796–8804.
205. Chen, E. C., Wu, T. M. (2007). Isothermal crystallization kinetics and thermal behavior of poly(epsilon-caprolactone)/multi-walled carbon nanotube composites, *Polym. Degrad. Stabil.*, *92*, 1009–1015.
206. Ryan, K. P., Cadek, M., Nicolosi, V., Walker, S., Ruether, M., Fonseca, A., Nagy, J. B., Blau, W. J., Coleman, J. N. (2006). Multiwalled carbon nanotube nucleated crystallization and reinforcement in poly(vinyl alcohol) composites, *Synth. Met.*, *156*, 332–335.
207. Probst, O., Moore, E. M., Resasco, D. E., Grady, B. P. (2004). Nucleation of polyvinyl alcohol crystallization by single-walled carbon nanotubes, *Polymer*, *45*, 4437–4443.
208. Abraham, T. N., Ratna, D., Siengchin, S., Karger-Kocsis, J. (2008). Rheological and thermal properties of poly(ethylene oxide)/multiwall carbon nanotube composites, *J. Appl. Polym. Sci.*, *110*, 2094–2101.

209. Narh, K. A., Jallo, L., Rhee, K. Y. (2008). The effect of carbon nanotube agglomeration on the thermal and mechanical properties of polyethylene oxide, *Polym. Compos.*, *29*, 809–817.
210. Jin, J., Song, M., Pan, F. (2007). A DSC study of effect of carbon nanotubes on crystallisation behaviour of poly(ethylene oxide), *Thermochim. Acta*, *456*, 25–31.
211. Chatterjee, T., Yurekli, K., Hadjiev, V. G., Krishnamoorti, R. (2005). Single-walled carbon nanotube dispersions in poly(ethylene oxide), *Adv. Funct. Mater.*, *15*, 1832–1838.
212. Ratna, D., Abraham, T. N., Siengchin, S., Karger-Kocsis, J. (2009). Novel method for dispersion of multiwall carbon nanotubes in poly(ethylene oxide) matrix using dicarboxylic acid salts, *J. Polym. Sci. Polym. Phys.*, *47*, 1156–1165.
213. Anand, K. A., Agarwal, U. S., Joseph, R. (2006). Carbon nanotubes induced crystallization of poly (ethylene terephthalate), *Polymer*, *47*, 3976–3980.
214. Li, Z. F., Luo, G. H., Wei, F., Huang, Y. (2006). Microstructure of carbon nanotubes/PET conductive composites fibers and their properties, *Compos. Sci. Technol.*, *66*, 1022–1029.
215. Kim, J. Y., Park, H. S., Kim, S. H. (2007). Multiwall-carbon-nanotube-reinforced poly(ethylene terephthalate) nanocomposites by melt compounding, *J. Appl. Polym. Sci.*, *103*, 1450–1457.
216. Anand, K. A., Agarwal, U. S., Nisal, A., Joseph, R. (2007). PET–SWNT nanocomposites through ultrasound assisted dissolution-evaporation, *Eur. Polym. J.*, *43*, 2279–2285.
217. Gao, Y., Wang, Y., Shi, J., Bai, H. W., Song, B. (2008). Functionalized multi-walled carbon nanotubes improve nonisothermal crystallization of poly(ethylene terephthalate), *Polym. Test.*, *27*, 179–188.
218. Hu, G. J., Feng, X. Y., Zhang, S. M., Yang, M. S. (2008). Crystallization behavior of poly(ethylene terephthalate)/multiwalled carbon nanotubes composites, *J. Appl. Polym. Sci.*, *108*, 4080–4089.
219. Tzavalas, S., Mouzakis, D. E., Drakonakis, V., Gregoriou, V. G. (2008). Polyethylene terephthalate–multiwall nanotubes nanocomposites: effect of nanotubes on the conformations, crystallinity and crystallization behavior of PET, *J. Polym. Sci. Polym. Phys.*, *46*, 668–676.
220. Antoniadis, G., Paraskevopoulos, K. M., Bikiaris, D., Chrissafis, K. (2009). Melt-crystallization mechanism of poly(ethylene terephthalate)/multi-walled carbon nanotubes prepared by *in situ* polymerization, *J. Polym. Sci. Polym. Phys.*, *47*, 1452–1466.
221. Kim, J. Y., Han, S. I., Kim, D. K., Kim, S. H. (2009). Mechanical reinforcement and crystallization behavior of poly(ethylene 2,6-naphthalate) nanocomposites induced by modified carbon nanotube, *Compos. Part A*, *40*, 45–53.
222. Kim, J. Y., Han, S. I., Kim, S. H. (2007). Crystallization behaviors and mechanical properties of poly (ethylene 2,6-naphthalate)/multiwall carbon nanotube nanocomposites, *Polym. Eng. Sci.*, *47*, 1715–1723.
223. Kim, J. Y., Park, H. S., Kim, S. H. (2006). Unique nucleation of multi-walled carbon nanotube and poly(ethylene 2,6-naphthalate) nanocomposites during non-isothermal crystallization, *Polymer*, *47*, 1379–1389.
224. Wu, D. F., Wu, L., Yu, G. C., Xu, B., Zhang, M. (2008). Crystallization and thermal behavior of multiwalled carbon nanotube/poly(butylenes terephthalate) composites, *Polym. Eng. Sci.*, *48*, 1057–1067.
225. Kim, J. Y. (2009). The effect of carbon nanotube on the physical properties of poly(butylene terephthalate) nanocomposite by simple melt blending, *J. Appl. Polym. Sci.*, *112*, 2589–2600.
226. Xu, Y., Jia, H. B., Piao, J. N., Ye, S. R., Huang, J. (2008). Crystallization behavior of poly (trimethylene terephthalate)/multi-walled carbon nanotube composites, *J. Mater. Sci.*, *43*, 417–421.
227. Shieh, Y. T., Liu, G. L. (2007). Effects of carbon nanotubes on crystallization and melting behavior of poly(L-lactide) via DSC and TMDSC studies, *J. Polym. Sci. Polym. Phys.*, *45*, 1870–1881.
228. Zhao, Y. Y., Qiu, Z. B., Yang, W. T. (2008). Effect of functionalization of multiwalled nanotubes on the crystallization and hydrolytic degradation of biodegradable poly(L-lactide), *J. Phys. Chem. B*, *112*, 16461–16468.
229. Xu, H. S., Dai, X. J., Lamb, P. R., Li, Z. M. (2009). Poly(L-lactide) crystallization induced by multiwall carbon nanotubes at very low loading, *J. Polym. Sci. Polym. Phys.*, *47*, 2341–2352.
230. Zhao, Y. Y., Qiu, Z. B., Yang, W. T. (2009). Effect of multi-walled carbon nanotubes on the crystallization and hydrolytic degradation of biodegradable poly(L-lactide), *Compos. Sci. Technol.*, *69*, 627–632.

231. Yudin, V. E., Feldman, A. Y., Svetlichnyi, V. M., Shumakov, A. N., Marom, G. (2007). Crystallization of R-BAPB type polyimide modified by carbon nano-particles, *Compos. Sci. Technol.*, *67*, 789–794.
232. Sahoo, N. G., Jung, Y. C., Yoo, H. J., Cho, J. W. (2006). Effect of functionalized carbon nanotubes on molecular interaction and properties of polyurethane composites, *Macromol. Chem. Phys.*, *207*, 1773–1780.
233. Wu, D. F., Wu, L. F., Zhou, W. D., Yang, T., Zhang, M. (2009). Study on physical properties of multiwalled carbon nanotube/poly(phenylene sulfide) composites, *Polym. Eng. Sci.*, *49*, 1727–1735.
234. Yu, S. Z., Wong, W. M., Hu, X., Juay, Y. K. (2009). The characteristics of carbon nanotube-reinforced poly(phenylene sulfide) nanocomposites, *J. Appl. Polym. Sci.*, *113*, 3477–3483.
235. Yang, J. H., Xu, T., Lu, A., Zhang, Q., Tan, H., Fu, Q. (2009). Preparation and properties of poly(*p*-phenylene sulfide)/multiwall carbon nanotube composites obtained by melt compounding, *Compos. Sci. Technol.*, *69*, 147–153.
236. Xiong, H., Gao, Y., Li, H. M. (2007). Non-isothermal crystallization kinetics of syndiotactic polystyrene–polystyrene functionalized SWNTs nanocomposites, *Express Polym. Lett.*, *1*, 416–426.
237. Diez-Pascual, A. M., Naffakh, M., Gomez, M. A., Marco, C., Ellis, G., Gonzalez-Dominguez, J. M., Anson, A., Martinez, M. T., Martinez-Rubi, Y., Simard, B., Ashrafi, B. (2009). The influence of a compatibilizer on the thermal and dynamic mechanical properties of PEEK/carbon nanotube composites, *Nanotechnology*, *20*, 315707.
238. Mago, G., Kalyon, D. M., Fisher, F. T. (2008). Membranes of polyvinylidene fluoride and PVDF nanocomposites with carbon nanotubes via immersion precipitation, *J. Nanomater.*, *8*, 759825.
239. Wang, M., Shi, J. H., Pramoda, K. P., Goh, S. H. (2007). Microstructure, crystallization and dynamic mechanical behaviour of poly(vinylidene fluoride) composites containing poly(methyl methacrylate)-grafted multiwalled carbon nanotubes, *Nanotechnology*, *18*, 235701–235701.
240. Kim, G. H., Hong, S. M. (2007). Structures and physical properties of carbon nanotube reinforced PVDF composites, *Mol. Cryst. Liq. Cryst.*, *472*, 551–559.
241. Mago, G., Fisher, F. T., Kalyon, D. M. (2009). Deformation-induced crystallization and associated morphology development of carbon nanotube–PVDF nanocomposites, *J. Nanosci. Nanotechnol.*, *9*, 3330–3340.
242. Hong, S. M., Nam, Y. W., Hwang, S. S., Chae, D. W. (2007). Physical properties of multi-walled carbon nanotube-filled PVDF composites prepared by melt compounding, *Mol. Cryst. Liq. Cryst.*, *464*, 777–785.
243. Broza, G., Schulte, K. (2008). Melt processing and filler/matrix interphase in carbon nanotube reinforced poly(ether–ester) thermoplastic elastomer, *Polym. Eng. Sci.*, *48*, 2033–2038.
244. Wang, F., Wu, J. K., Xia, H. S., Wang, Q. (2007). Polyoxymethylene/carbon nanotubes composites prepared by solid state mechanochemical approach, *Plast. Rubber Compos.*, *36*, 297–303.
245. Xu, C. L., Qiu, Z. B. (2009). Nonisothermal melt crystallization and subsequent melting behavior of biodegradable poly(hydroxybutyrate)/multiwalled carbon nanotubes nanocomposites, *J. Polym. Sci. Polym. Phys.*, *47*, 2238–2246.
246. Xu, C. L., Qiu, Z. B. (2009). Isothermal melt crystallization kinetics study of biodegradable poly(3-hydroxybutyrate)/multiwalled carbon nanotubes nanocomposites, *Polym. J.*, *41*, 888–892.
247. Patil, P. A., Wanjale, S. D., Jog, J. P. (2008). Poly(4-methyl-1-pentene)/MWNT nanocomposites, *E-Polymers*, 079.
248. Song, L., Qiu, Z. B. (2009). Crystallization behavior and thermal property of biodegradable poly (butylene succinate)/functional multi-walled carbon nanotubes nanocomposite, *Polym. Degrad. Stabil.*, *94*, 632–637.
249. Pramoda, K. P., Linh, N. T. T., Zhang, C., Liu, T. X. (2009). Multiwalled carbon nanotube nucleated crystallization behavior of biodegradable poly(butylene succinate) nanocomposites, *J. Appl. Polym. Sci.*, *111*, 2938–2945.
250. Qiu, Z. B., Zhu, S. Y., Yang, W. T. (2009). Crystallization kinetics and morphology studies of biodegradable poly(butylene succinate-*co*-butylene adipate)/multi-walled carbon nanotubes nanocomposites, *J. Nanosci. Nanotechnol.*, *9*, 4961–4969.
251. Antolin-Ceron, V. H., Gomez-Salazar, S., Soto, V., Avalos-Borja, M., Nuno-Donlucas, S. M. (2008). Polymer nanocomposites containing carbon nanotubes and miscible polymer blends based on poly [ethylene-*co*-(acrylic acid)], *J. Appl. Polym. Sci.*, *108*, 1462–1472.

252. Meincke, O., Kaempfer, D., Weickmann, H., Friedrich, C., Vathauer, M., Warth, H. (2004). Mechanical properties and electrical conductivity of carbon-nanotube filled polyamide-6 and its blends with acrylonitrile/butadiene/styrene, *Polymer*, *45*, 739–748.
253. Zhang, L. Y., Wan, C. Y., Zhang, Y. (2009). Investigation on the multiwalled carbon nanotubes reinforced polyamide 6/polypropylene composites, *Polym. Eng. Sci.*, *49*, 1909–1917.
254. Hom, S., Bhattacharyya, A. R., Khare, R. A., Kulkarni, A. R., Saroop, M., Biswas, A. (2009). Blends of polypropylene and ethylene octene comonomer with conducting fillers: influence of state of dispersion of conducting fillers on electrical conductivity, *Polym. Eng. Sci.*, *49*, 1502–1510.
255. Zhang, L. Y., Wan, C. Y., Zhang, Y. (2009). Morphology and electrical properties of polyamide 6/polypropylene/multi-walled carbon nanotubes composites, *Compos. Sci. Technol.*, *69*, 2212–2217.
256. Bose, S., Bhattacharyya, A. R., Bondre, A. P., Kulkarni, A. R., Pötschke, P. (2008). Rheology, electrical conductivity, and the phase behavior of cocontinuous PA6/ABS blends with MWNT: correlating the aspect ratio of MWNT with the percolation threshold, *J. Polym. Sci. Polym. Phys.*, *46*, 1619–1631.
257. Wu, M., Shaw, L. (2006). Electrical and mechanical behaviors of carbon nanotube-filled polymer blends, *J. Appl. Polym. Sci.*, *99*, 477–488.
258. Li, Y., Shimizu, H. (2008). Conductive PVDF/PA6/CNTs nanocomposites fabricated by dual formation of cocontinuous and nanodispersion structures, *Macromolecules*, *41*, 5339–5344.
259. Krasovitski, B., Marmur, A. (2005). Particle adhesion to drops, *J. Adhes.*, *81*, 869–880.
260. Liu, L., Wang, Y., Xiang, F. M., Li, Y. L., Han, L., Zhou, Z. W. (2009). Effects of functionalized multiwalled carbon nanotubes on the morphologies and mechanical properties of PP/EVA blend, *J. Polym. Sci. Polym. Phys.*, *47*, 1481–1491.
261. Goldel, A., Kasaliwal, G., Pötschke, P. (2009). Selective localization and migration of multiwalled carbon nanotubes in blends of polycarbonate and poly(styrene–acrylonitrile), *Macromol. Rapid Commun.*, *30*, 423–429.
262. Yoon, H., Okamoto, K., Yamaguchi, M. (2009). Carbon nanotube imprinting on a polymer surface, *Carbon*, *47*, 2840–2846.
263. Mukherjee, M., Das, T., Rajasekar, R., Bose, S., Kumar, S., Das, C. K. (2009). Improvement of the properties of PC/LCP blends in the presence of carbon nanotubes, *Compos. Part A*, *40*, 1291–1298.
264. Li, Z. M., Li, S. N., Xu, X. B., Lu, A. (2007). Carbon nanotubes can enhance phase dispersion in polymer blends, *Polym. Plast. Technol. Eng.*, *46*, 129–134.
265. Wu, D. F., Zhang, Y. S., Zhang, M., Yu, W. (2009). Selective localization of multiwalled carbon nanotubes in poly(epsilon-caprolactone)/polylactide blend, *Biomacromolecules*, *10*, 417–424.
266. Zheng, M., Jagota, A., Semke, E. D., Diner, B. A., McLean, R. S., Lustig, S. R., Richardson, R. E., Tassi, N. G. (2003). DNA-assisted dispersion and separation of carbon nanotubes, *Nat. Mater.*, *2*, 338–342.
267. Li, C. Y., Li, L. Y., Cai, W. W., Kodjic, S. L., Tenneti, K. K. (2005). Nanohybrid shish-kebabs: periodically functionalized carbon nanotubes, *Adv. Mater.*, *17*, 1198–1202.

CHAPTER 5

MECHANICAL AND RHEOLOGICAL PROPERTIES

5.1 OVERVIEW

A famous (some would argue infamous!) possibility for the use of polymer–nanotube composites is in the construction of a space elevator, that is, a cable that would travel from the earth to the moon. Although the premise sounds like science fiction, an annual ongoing contest (as of 2010) is in place to design systems that are precursors to achieving this goal, including a contest to design high-strength fibers (www.spaceward.org). Carbon nanotubes are universally recognized as being the only material currently available that could be considered for such an application. When nanotubes were first invented, mechanical property enhancement in composites was one of the first uses envisioned. With perfect dispersion, orientation, and load transfer, adding ~15% nanotubes to polypropylene or some other common low-performance polymer fiber would transform this material into having strength and stiffness similar to the highest strength polymer fibers known, and adding 10% nanotubes to a high-strength fiber would increase the strength and stiffness by a factor of 2 or more, with little or no increase in weight. As will be seen, such improvements have not been found. For high-viscosity polymers, results have been within a factor of 2 of expected performance at low volume fractions for stiffness. Deviations become worse at higher contents, especially above 1% nanotube content. In most cases, theoretical predictions far outstrip actual performance at much lower loading levels for strength; in fact, in many cases, strength decreases with nanotube addition. In thermoset systems using premade pure nanotube fibers or sheets, the technology is not as advanced but the results are so far much more promising.

The words "mechanical properties" and "rheological properties" encompass a great deal. Although certainly not universally defined in this manner, most would agree that the two differ in that mechanical properties are measured on solids, while rheological properties are measured on liquids. This convention will be used throughout the text.

Rheological or mechanical tests can be classified as one of the two types: oscillatory and nonoscillatory. Another classification system for rheological or mechanical tests exists: steady-state versus non-steady-state; that is, are the values

Carbon Nanotube–Polymer Composites: Manufacture, Properties, and Applications, First Edition.
Brian P. Grady.

measured time independent? Tensile tests are an example of a non-steady-state measurement; the sample is stretched at a constant rate and the force is measured; clearly, the force depends on strain and hence time. Non-steady-state mechanical tests are performed more frequently than steady-state tests. Conversely, more steady-state rheological tests are performed than non-steady-state tests. A common example of a steady-state rheological test is where a liquid is placed between two parallel disks and one of the disks is spun at a constant speed (e.g., revolutions per minute) and the torsional force is measured. If the measurement is made at long enough time after starting the rotation, the torque will be constant. Although not as frequently performed, non-steady-state rheological tests and steady-state mechanical tests will be discussed where appropriate.

A single mechanical or rheological test can thus be considered to be of one of the four types: oscillatory, steady-state; oscillatory, non-steady-state; nonoscillatory, steady-state; and nonoscillatory, non-steady-state. Geometry must also be specified (tension, torsion, bending, etc.). Mechanical measurements will be considered first.

There are common steady-state and non-steady-state oscillatory mechanical measurements. For the steady-state type, termed dynamical mechanical analysis (DMA), measurements are typically done at fixed frequency with a change in temperature using small strains so that no permanent sample deformation takes place. In this measurement, done in tension or bending, the modulus that is in phase with the applied strain is termed the storage modulus (E') and is a measure of energy stored, while the out-of-phase modulus is termed the loss modulus (E'') and is a measure of viscous loss in the sample. The ratio E''/E' is termed $\tan\delta$. The storage modulus is often used as a measure of stiffness instead of the Young's modulus; the two are numerically different, but usually trend the same (but not always, especially in filled systems). If the oscillatory measurement is done at high deformations, then the test is termed a fatigue test and is a non-steady-state test. One important parameter for a fatigue test is the number of cycles to failure that will depend on the amplitude and frequency of the oscillating stress (usually stress rather than strain is controlled in these tests). The other commonly measured parameter from fatigue tests is the crack growth rate, which is the rate (distance/cycle) at which a crack grows and obviously is a function of the stress amplitude. Normally the growth rate is assumed to be constant with time, which strictly speaking makes this a steady-state parameter. Fatigue tests are also done in geometries other than tension, although tests in tension are most commonly performed.

Nonoscillatory mechanical tests are almost always non-steady-state. Mechanical properties are most often measured in tension. Tensile properties for polymers are measured by stretching a dog-bone-shaped sample at a constant rate, for example, mm/min, at room temperature, and stress (force/beginning cross-sectional area) and strain (change in length/beginning length) are recorded. The latter is more precisely termed the engineering strain; the actual strain has a different definition, but the engineering strain is always reported unless noted otherwise. Important parameters from this test are the tensile (Young's) modulus (see Equation 2.4), the stress and strain at break (σ_{break} and $\varepsilon_{\text{break}}$), and the stress at yield (YS, if applicable). A yield stress occurs in a polymer if there is a relative maximum in a plot of stress versus

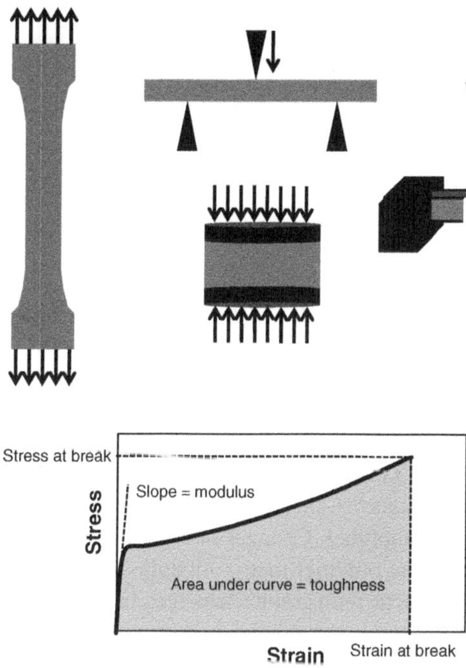

Figure 5.1 *Top:* Various geometries for mechanical tests. *Left to right*: Tensile test, three-point bend test (top), compression test (bottom), and impact test. *Bottom*: Possible representative result of tensile test of a polymer with various parameters indicated on the graph. This particular polymer does not have a clear yield stress; if one existed, it would appear as a relative maximum close to strain = 0.

strain. In most cases, the maximum stress, termed the tensile strength (TS), is the same as the stress at break, but in a few cases the yield stress is the maximum, that is, the TS. The area under the stress–strain curve, which has units of energy and is termed the toughness, is also an important parameter from tensile tests. Figure 5.1 graphically illustrates the various parameters from a tensile test.

Predictions of the tensile modulus, in particular, are quite simple and accurate for macrofillers in an amorphous polymer when all fillers are isolated, and these predictions should apply to nanotubes as well if the nanotubes are isolated. In the case of perfectly oriented fibrous fillers with high aspect ratio (>100), the modulus in the direction of fiber alignment can be described by this rather simple mixing law:

$$E = E_f V_f + E_p V_p \tag{5.1}$$

where E_f and E_p are moduli of the fiber (nanotube in this case) and polymer, respectively, and V_f and V_p are the respective volume fractions. Modifications to this expression for the case of fibers aligned randomly in a plane and for three-dimensionally randomly oriented fibers add a multiplicative term in front of the $E_f V_f$ where this term is 3/8 and 1/5, respectively. A more rigorous and accurate approach for anisotropic filler particles, with diameter D and length L, is the Halpin–Tsai theory that has the following form for perfect orientation in the stretch direction:[1]

$$\frac{E}{E_p} = \left[\frac{1+2(L/D)\eta_A V_f}{1-\eta_A V_f}\right], \quad \eta_A = \frac{(E_f/E_p)-1}{(E_f/E_p)+2(L/D)} \tag{5.2}$$

and the following for randomly oriented fibers:

$$\frac{E}{E_p}=\frac{3}{8}\left[\frac{1+2(L/D)\eta_L V_f}{1-\eta_L V_f}\right]+\frac{5}{8}\left[\frac{1+2\eta_T V_f}{1-\eta_T V_f}\right], \quad \eta_L=\frac{(E_f/E_p)-1}{(E_f/E_p)+2(L/D)},$$

$$\eta_T=\frac{(E_f/E_p)-1}{(E_f/E_p)+2} \tag{5.3}$$

A less commonly used model is the Mori–Tanaka model. The discussion of this model is beyond the scope of this book; however, the interested reader is referred to Ref. 2.

Equation 5.3 is often used to fit data from nanotube composites and a comparison between this theory and experimental results can be an excellent measure of dispersion quality. However, this comparison must be done somewhat carefully. First, manufacturing samples with no nanotube alignment is very difficult because of the propensity of nanotubes to align. Also, Equations 5.1–5.3 implicitly assume no change in the matrix modulus with the addition of nanotubes. Changes in crystallinity and/or changes in polymer alignment can change the polymer matrix modulus. If the glass transition temperature is near the measurement temperature, changes in T_g relative to the measurement temperature can also change the matrix modulus substantially. Some changes in modulus in the literature have been incorrectly attributed to good reinforcement/dispersion instead of simply changes in these other factors.

Equations 5.2 and 5.3 require a value for the aspect ratio of the nanotubes, which is not possible to determine after addition to a polymer except with very laborious electron microscopy measurements. AFM measurements on nanotube length prior to mixing are often not appropriate because, as pointed out in Chapter 3, processing methods can result in nanotube breakage. Hence, this value is usually assumed, or this value is allowed to vary and the aspect ratio from a fit to the data is compared with that for perfectly dispersed, unbroken tubes. A simpler approach, as described by Coleman et al. in his review article,[1] is to calculate the slope (dE/dV_F) at low volume fractions and use this value to compare processing methods, tube types, and so on. With the same nanotubes and polymer and assuming no changes in nanotube orientation or polymer modulus, changes in this value are indicative of changes in the filler aspect ratio due to either tube breakage or changes in dispersion. Note that the effect of two different starting polymer moduli can be ignored (assuming that moduli do not change with the addition of nanotubes) as derivations of Equations 5.1–5.3 show. In other words, the use of dE/dV_f allows for comparison between different polymers. Equation 5.1 with the given constants for the various orientations along with an assumed nanotube modulus of 1 TPa can be used to calculate a value between 200 GPa (unoriented) and 1 TPa (perfectly oriented) for dE/dV_F; these are also the limiting values from the Halpin–Tsai equations for high aspect ratio fillers. Comparisons using this formalism will be described in detail later in this chapter. However, there will be numerical differences between values presented in this text and in Coleman's paper because of differences in specific gravities assumed when weight percentages are reported; throughout this chapter, a specific gravity of 1.35 for nanotubes is assumed for SWCNTs and DWCNTs and 2.0

for MWCNTs. Also, in some cases, numbers had to be determined from graphs, and this procedure could also lead to numerical differences.

The value of dE/dV_F as a measure of modulus enhancement has a significant drawback in that the value does not give any information over the range that the linear approximation is valid. In order to exhibit significant application interest as a filler to increase stiffness, the increase would have to follow Equations 5.1–5.3 to a minimum of a few volume percent. In very simple terms, dispersed nanotubes have been very unsuccessful as mechanical property enhancers because improvements are significant only at low volume fractions (below 1%, sometimes far below 1%). The reader is meant to compare dE/dV_f values, but the concentration range over which these were calculated varied since not all papers tested extremely low volume fraction materials and hence the comparison is not always valid.

Schaefer and coworkers[3] raise an interesting countervailing view to the approach presented in the previous paragraphs. The fundamental assumption in Equations 5.1–5.3 and the associated discussion is that nanotubes are stretched during stretching of the composite. However, if nanotubes are curved, they can be straightened rather than stretched. In fact, straightening is exactly what happens with polymers; the modulus of a single perfectly elongated polymer molecule is far greater than the modulus of any polymer sample. If nanotubes are straightened rather than stretched, the bending modulus should be used in a formulation of the composite stiffness and the mathematical approaches given in Equations 5.1–5.3 are not appropriate. Schaefer and coworkers conclude that nanotubes are flexible through scattering experiments; certainly micrographs of nanotubes from other papers, see, for example, Figures 3.4 and 3.5, support the significant flexibility of carbon nanotubes. Certainly, the fact that Equations generally do a poor job in describing the observed behavior lends credence to this argument.

Predictions of tensile properties, other than the modulus, are very complicated and usually require a value for the interfacial adhesion (or the assumption of perfect adhesion, which is almost certainly not true in nanotube composites). Two methods exist to measure the interfacial adhesion experimentally. The first is to partially embed a single fiber in the polymer of interest and then measure the force required to separate the fiber from the matrix. The second, termed single-fiber fragmentation testing, is to totally embed a single fiber in a matrix, pull the matrix, and measure the length distribution of the fiber pieces that result when fiber breakage is complete. A somewhat quantitative measure can be made via Raman spectroscopy; the position of the Raman G′ band is sensitive to nanotube stress in SWCNTs.[4] This peak shifts to lower frequencies when a nanotube is stretched and the stress on the nanotube has been reported by one study to be 0.2 GPa/cm,[5] although this value likely varies greatly depending on the surrounding matrix. In other words, higher shifts should correspond to more efficient stress transfer; when the stress is lost, it could be taken as a measure of interfacial adhesion. A similar method uses the position of the G^+ band; nanotubes with higher interfacial bonding in an epoxy were found to have a higher slope for the change in peak position with heating, which in turn were much higher than the change in slope for pristine nanotubes.[6]

For filled stiff thermosets, bending geometry, such as three-point bend, is more often used than tension for a nonoscillatory non-steady-state mechanical test. In three-point bending tests, a stiff bar is set over two knife edge or cylindrical supports, and the same is pushed in the middle with a third pushing knife edge or nose. The sample experiences tensile, compressive, and shear stress components, which makes this test extremely complicated to model theoretically. The bending modulus and bending strength are calculated from this measurement. Another alternative to tension is compressive testing; a cylindrical sample is manufactured and the force required to compress the sample is measured. The analogous parameters for tensile testing are defined for compressive tests, for example, compressive modulus, compressive strength, and compressive strain at break. The compressive modulus and the tensile modulus are identical, while the latter two properties are not, especially in filled systems where the interfacial adhesion is more important in tensile strength measurements than in compressive strength measurements. This test is not the same as a hydrostatic compressive test where there is equivalent stress in all directions toward the center of the sample; this measurement yields the bulk modulus, which unfortunately is also sometimes termed the compressive modulus.

Other types of mechanical testing include the following:

- **Impact Testing (Nonoscillatory, Non-Steady-State):** Various methods exist to measure the response of a material to impact. The most common are Charpy or Izod testing; the sample is a rectangular bar and has dimensions similar to x:x:Ax, where A is much larger than 1. A notch of specified dimensions is cut in the sample in the plane defined by the two smallest dimensions. A pendulum swings and breaks the sample and the energy required to break the sample is calculated. All impact measurements, including but not limited to Charpy or Izod measurements, have in common that the important measured variable is the impact energy, that is, the amount of energy a sample absorbs during fracture.
- **Fracture Toughness (Nonoscillatory, Non-Steady-State):** The same type of sample is used as for impact testing, except the sample usually has more of a film-like geometry (although the thickness does need to be greater than a critical size). A notch is introduced into this sample as well; the plane of the notch contains the line that defines the thickness direction (think of tearing a piece of paper). Different types of stresses can be put on the sample, but in the most common termed mode I, the sample is then pulled apart in a tensile machine with the direction of pulling perpendicular to the notch plane. The important dependent variable is typically the fracture toughness (K_c, or K_{Ic} for mode I fracture toughness) that has units of pressure $\times$ (distance)$^{1/2}$. Crack growth rate can also be measured, which has units of distance/time. A related measured variable is the work of fracture, K_c^2/E, which has units of energy/distance^2.
- **Wear Testing (Oscillatory, Steady-State):** As the method name suggests, this test measures the ability to resist loss of mass via abrasion or wear. The test is performed by rubbing the sample with another object having specified characteristics with a constant normal force or pressure. The important dependent variable is usually mass loss/time.

- **Creep (Nonoscillatory, Non-Steady-State):** A constant force is applied to a sample, and the dimensions of the sample with time are measured. The important dependent variable is usually change in length/time.
- **Hardness Testing (Nonoscillatory, Non-Steady-State):** A small spherical or diamond-shaped indenter is forced into a surface using a specified force, and a parameter related to the depth of penetration, the geometry of the indenter, and the normal force is calculated. A relative number is assigned to the hardness that has meaning only when compared with a standard scale of other materials. This test is often used to quantify the scratch resistance of a surface (although scratch resistance testing, which involves dragging a sharp object along a surface, does exist).

Rheological properties, that is, properties measured on liquids, are also measured in oscillatory and nonoscillatory modes, although oscillatory, non-steady-state measurements are not performed. The steady-state oscillatory rheological measurements are done in torsion, and the relevant parameters are storage modulus (G'), loss modulus (G''), and $\tan\delta$ (G''/G') with analogous meanings to E', E'', and $\tan\delta$, respectively. Torsional geometry with a liquid means shear measurements, with the liquid between two parallel plates or between a cone and a plate (the axis of the three-dimensional cone is perpendicular to the plate). One plate or the cone is moved rotationally back and forth at a given angular frequency (ω) and the in-phase and out-of-phase parts of the force with respect to the strain are measured. Usually measurements are done at one temperature and ω is varied; measurements can be made at multiple temperatures if desired.

The nonoscillatory, steady-state measurements often use the same equipment, but in this case the plate or cone is spun at a constant rate, for example, revolutions per minute, which is converted to shear rate $\dot{\gamma}$. The important parameter in steady shear measurements is viscosity η, which is a proportionality constant between shear stress (τ) and shear rate: $\tau = \eta\dot{\gamma}$. The shear rate is not constant across the radius of the plates if parallel plates are used (zero in the center and a maximum at the outside) and hence an average shear rate is reported. For a cone-and-plate rheometer, the shear rate is constant (at least to within a very good approximation) and no such averaging is necessary. Averaging can cause distortion of results in polymers, since the stress does not in general change linearly with the strain. Again, temperature is generally held constant and the shear rate $\dot{\gamma}$ is varied.

The nonoscillatory, non-steady-state measurements are not nearly as popular as the steady-state measurements. If the stress, rather than the strain, is controlled and the strain is measured, then at low enough stresses the material may not move at all. If such a quantity exists, the minimum stress required to move the fluid is termed the yield stress; note this has no relation with the aforementioned mechanical property yield stress except for the unfortunate use of the same terminology. The non-steady-state measurements such as the time-dependent stress after the cessation of shear flow or the time-dependent stress at the start-up of shear flow can also be measured.

For steady-state shear measurements, a cone-and-plate or parallel plate rheometer is confined to an upper shear rate limit on the order of $10\ \mathrm{s}^{-1}$ because at higher shear rates the material is ejected from the rheometer. For higher shear rates, capillary rheometry is used in which a polymer is pushed through a die using a plunger at a constant volumetric flow rate. Capillary rheometry is restricted to shear rates above about $1\ s^{-1}$ because at lower shear rates the force between the barrel and the plunger is larger than the force required to push the polymer through the die. The same law, $\tau = \eta\dot{\gamma}$, applies, although the conversion of the volumetric flow rate to viscosity is not nearly as straightforward. Using a procedure based on capillary rheometry, the melt flow rate or melt flow index is a common commercial measure to characterize a polymer. In this case, a constant weight is used to push polymer through a capillary and the measured parameter is the amount of polymer that exits the tube for a given amount of time, for example, g/10 min.

Other geometries besides shear are less commonly used for rheological measurements. Elongational (tension) measurements in a nonoscillatory mode can be made; such measurements are not trivial to make and describing the specialized equipment required is beyond the scope of this text. The measured value in steady elongational measurements is the elongational viscosity $\widehat{\eta}$, which is a proportionality constant between elongational stress τ_{zz} and elongation rate $\dot{\varepsilon}$, that is, $\tau_{zz} = \widehat{\eta}\,\dot{\varepsilon}$. Performing elongational measurements for a long enough time so that steady state is reached is not trivial. Even though the measurements are much less frequently made, elongational rheological properties are extremely important. In particular, elongational rheology is extremely important in fiber formation, which is an important processing method for nanotubes since fibers represent an obvious application for nanotubes.

A particular liquid is classified as Newtonian or non-Newtonian; alternatively, the terms rheologically simple and rheologically complex are used. The most cited difference between the two types of fluids is that in the former the viscosity is independent of shear rate, although other differences exist including the fact that a Newtonian fluid cannot have a yield stress. One important characteristic of Newtonian fluids is Trouton's rule, which states that $\widehat{\eta} = 3\eta$. Polymers are one example of non-Newtonian fluids. A plot of viscosity versus shear rate generally has three regions for a polymer: a Newtonian region (i.e., the viscosity is constant) at low shear rates; a shear thinning region where the log(viscosity) versus log(shear rate) is approximately linear with a slope between -1 and 0 at high shear rates; and a transition region between the two. The viscosity in the low shear rate region is termed the zero-shear viscosity and is highly molecular weight dependent; the zero-shear viscosity increases by roughly an order of magnitude for every doubling of the molecular weight. The shear rate at which the viscosity begins to depend on shear rate is approximately equal to the inverse of the longest relaxation time of the polymer chain.

In the high shear rate regime, the relationship between shear stress and shear rate is typically written as

$$\eta = m\dot{\gamma}^{n-1} \tag{5.4}$$

with n termed the power-law index. Since n is less than 1, polymer melts and solutions are termed shear thinning; smaller values of the power-law index mean that the material thins more with stress. Shear thinning arises in polymers because the molecules become oriented in the direction of the flow field and the density of chain–chain entanglements decreases. A more shear thinning fluid represents a fluid where orientation effects are more important. This law is empirical, although almost all polymers melt and solutions follow this behavior. A shear thinning fluid is *not* easier to push as the shear rate increases; instead, the dependence of the increase in force is less than a linear dependence on the shear rate. Although polymers do not generally show such behavior because chain breakage cannot be avoided, at very high shear rates the flow will become Newtonian again. Molecularly, orientation is at a maximum and entanglements are at a minimum in this very high shear rate region. As will become clear, nanotubes individually dispersed in Newtonian fluids can reach this second Newtonian region at high shear rates.

Steady-state oscillatory rheometry gives important clues on the structure and behavior of a nanotube–polymer network. For a one-phase polymer melt, at low frequencies, the storage modulus will decrease with a slope of 2 ($G'(\omega)\alpha\omega^2$) and the loss modulus with a slope of 1 ($G''(\omega)\alpha\omega^1$), both of which are a result of the long-chain nature of polymers. Similar to the zero-shear viscosity, the frequency at which this terminal region begins is the inverse of the longest relaxation time, and hence the higher the temperature, the higher the critical frequency. In order to extend the frequency range presented on one plot, measurements may be made at different temperatures and then shifted horizontally on a modulus versus frequency plot so that overlap of the moduli measured at different temperatures occurs (in some cases, slight vertical shifts are necessary; in theory such shifts should only be necessary for density variations, but in practice such shifts are usually considered to be whatever is necessary to obtain data overlap). In well-dispersed filled systems with no interconnectedness of filler particles, the terminal slopes of 2 and 1 should not change, although the modulus value at a given frequency should be higher and the critical frequency might shift. Changes in slopes from 2 and 1 at low frequencies are a result of the emergence of an interconnected second phase, and have played an important role in the characterization of nanotube composites as will be described in Section 5.2.2.

An alternative way to present oscillatory shear data is in terms of storage and loss viscosities, η' and η''. These are related to the storage and loss moduli by the following: $\eta' = G''/\omega$ and $\eta'' = G'/\omega$. An important semiempirical relationship is the Cox–Merz law that has the following form:

$$\eta^*(\omega) = \eta(\dot{\gamma}) \tag{5.5}$$

where $\eta^*(\omega)$ is the complex viscosity $[(\eta^*)^2 = (\eta')^2 + (\eta'')^2]$. In words, the Cox–Merz law states that the complex viscosity is the same as the steady shear viscosity when the oscillatory frequency is the same as the shear rate. A material that follows this law will also have the frequency where the terminal region is reached in oscillatory measurements be identical to the shear rate where the Newtonian region is reached in steady shear measurements. Although not universally obeyed, this law generally holds if steady shear does not change the structure of what is measured.

Overall, the number of papers that describe mechanical and rheological properties of nanotube–polymer composites is in the thousands for the former and in the hundreds for the latter. Hence, to describe all, or even a majority of, papers for the former is a task that would lengthen this book immeasurably. This chapter gives examples all the different kinds of mechanical and rheological properties measured and in this task tries to be complete. As stated in Chapter 1, unless otherwise noted, the author has tried to make all tables complete; an exhaustive table for mechanical properties does not appear because of the size of the task. Hence, with respect to mechanical properties, the studies that show large improvements in mechanical properties will also be highlighted, with particular attention devoted to the reproducibility by others of such results as well as the practical implications.

5.2 RHEOLOGICAL PROPERTIES (MEASUREMENT OF MELT AND SOLUTION PROPERTIES)

Before describing the rheology of polymer–nanotube suspensions, this section will describe the rheology of nanotube suspensions in nonpolymeric, Newtonian fluids. Nanotubes suspended in Newtonian fluids have also complicated rheological behavior, which is important to understand prior to considering the more complicated case of nanotubes suspended in rheologically complex polymers.

From a rheological perspective, three concentration regions are usually considered: the dilute, the semidilute, and the concentrated. These regions are defined by entanglements between different tubes; in the dilute region, there are no tube–tube entanglements, and in the concentrated regime, each tube is involved in at least one entanglement. In terms of volume, in the dilute regime, the number of nanotubes in solution is significantly less than the solution volume/sphere volume ratio, where the sphere volume is defined by a diameter that is equal to the length of a nanotube. In the concentrated region, the number of nanotubes is significantly more than this ratio. The semidilute region is where the number of nanotubes in solution is approximately equal to the solution volume/sphere volume ratio. Two characteristics relevant to nanotubes complicate this distinction dramatically. First, concentration region characterization ignores nanotube aggregation; the definition implicitly assumes that all nanotubes are isolated. As was clearly stated in Chapter 3, if significant efforts are made to disperse nanotubes perfectly, then only under very special conditions (e.g., superacids)[7] will the nanotubes stay suspended indefinitely. Second, nanotubes are not precisely rigid rods; nanotubes, especially single-walled nanotubes, have the ability to curve[8] and a stress field can cause a change in nanotube curvature. Hence, the end-to-end length, not the contour length, should be used to define concentration regions and the end-to-end length will change depending on conditions.

In the dilute regime, the viscosity has been shown to linearly increase with concentration for SWCNTs suspended in water with surfactant, where great care was taken to ensure mostly if not totally isolated tubes were present, agreeing with the behavior predicted for rod-like fillers.[9] Extreme care must be taken to prevent aggregation, since with aggregation little or no viscosity enhancement will be seen.[10,11] Even in the dilute regime, nanotube solutions are shear thinning.[9] A reduction in viscosity at high shear rates occurs because nanotubes can become oriented if the shear rate is sufficient; the shear rate of the transition is a measure of how fast the tubes are able to rotate due to random Brownian forces. The behavior in the high shear region was well described by a relationship where the viscosity dropped linearly with the logarithm of the shear rate.[9] Note the difference with power-law behavior where the viscosity drops logarithmically with the logarithm of the shear rate. Since the power-law model is largely empirical, the difference in dependency is of note but not concern.

In the semidilute regime, nanotubes begin to overlap significantly, but not to the point where the orientation of one nanotube affects the orientation of a neighboring nanotube. The start of this region is similar to a percolation threshold (see Section 5.2.2) in the sense that properties change dramatically at concentrations bounded by this concentration, although the normal structural definition of a percolation threshold does not apply to the dilute to semidilute transition. In the semidilute region, one of the most interesting behaviors, of extreme importance as mentioned in Section 3.4, is aggregation caused by a change in rheological conditions. Changing from low or no shear to a higher shear state can induce aggregation in a rheologically simple liquid[12]; recall that in Section 3.4 the opposite was described, that is, aggregation induced by reducing the shear field. In this Newtonian case, aggregation is literally caused by moving nanotubes encountering one another and sticking together. A different study found the opposite effect, the aggregation decreased as the strain rate increased.[13] The problem with such studies, and making comparisons, is that the starting dispersion states are not generally identical; in a well-dispersed initial system, flow could induce aggregation, while in a poorly dispersed initial system, flow could increase dispersion. However, at high shear rates, dispersion should decrease or stay the same irrespective of the starting state because aggregates will be eliminated due to hydrodynamic forces. As a corollary to these statements, changing aggregation state can lead to τ versus $\dot{\gamma}$ behavior significantly more complicated than the three-region approach given earlier.[14] Finally, from an orientation perspective, cessation of flow was found to reduce orientation over a very long timescale, which was attributed to changes in the bending dynamics, rather than reorientation, of the carbon nanotubes.[13]

One extremely interesting study with a liquid that is a Newtonian epoxy monomer deserves special mention.[15] In this study, nanotubes of two different lengths were studied. Differences in the continuous networks of the two cases are shown in Figure 5.2. The slopes for the rheological measurements shown in Figure 5.2 indicate that with short nanotubes, the aggregates of nanotubes change size/shape in response to stress, while with long nanotubes, it is the nanotubes themselves that change their curvature in response to stress. A yield stress occurs with the formation of a continuous network, and non-steady-state shear stress versus shear rate curves using a step–strain methodology (the result is similar to a modulus) for different nanotube levels can be scaled using the yield stress to give a master curve. Similar scalings for storage and loss moduli are possible with the constant low-frequency storage modulus as the scaling parameter. An important point to note from Figure 5.2c is that at about 10% nanotubes, the moduli from short and long tubes would be the same if extrapolation holds, leading to the surprising conclusion that shorter nanotubes may yield higher modulus nanotube networks at sufficiently high nanotube contents. This conclusion is possible because the aggregated nanotube networks are the reinforcing moiety, not the individual nanotubes. Length effects were also found to be important in a steady shear experiment performed in a different

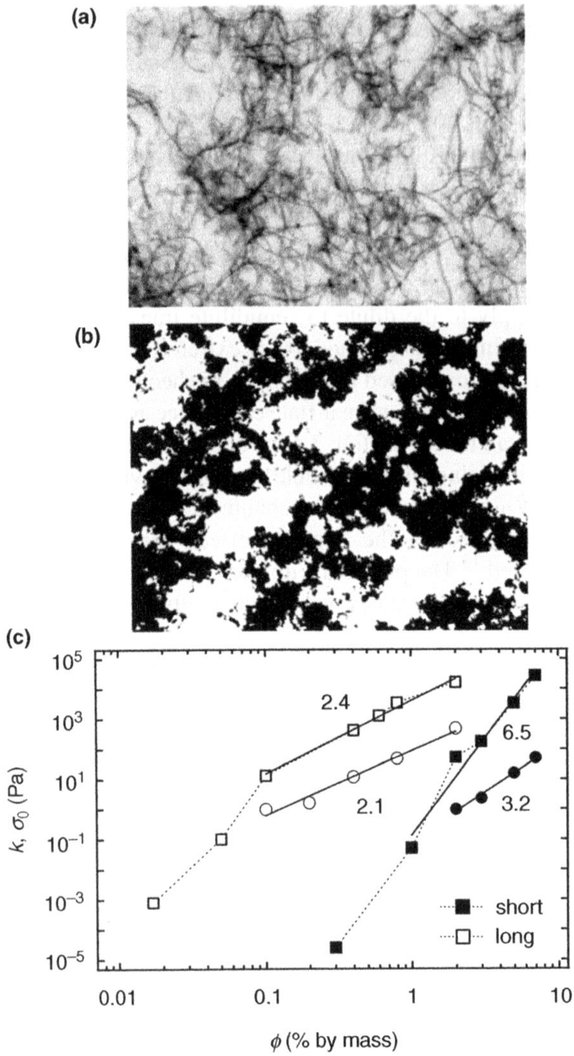

Figure 5.2 Optical microscopy images of nanotubes suspended in epoxy monomer: (a) 60 μm MWCNTs (volume fraction 0.05%, width of image 225 μm); (b) 4 μm MWCNTs (volume fraction 2.0%, width of image 350 μm). (c) Concentration dependence of the linear elastic shear modulus at low frequency (κ, squares) and the yield stress (σ_0, circles) with power-law fits, as indicated. Copyright Wiley–VCH Verlag GmbH & Co. KGaA. Reproduced with permission from Ref. 15.

Newtonian fluid: short nanotubes showed an almost Newtonian behavior, while long nanotubes showed power-law behavior.[16]

Oscillatory steady-state behavior is also quite interesting for nanotube networks in the semidilute regime. Measurements of the scaling exponent at high frequency for the storage modulus ($G'(\omega)\alpha\omega^a$) have been performed. Different values were found: 0.63,[17] something slightly greater than 0.75,[15] and 0.7[18] (the latter value was found for a fluid that is almost, but not quite, rheologically simple since G' is measurable at high frequencies). A value of 0.75 is the theoretical result for a material that is inextensible along its contour and redistributes in length via redistribution of its bending modes.[19–21] The increase in storage modulus at low frequencies has been used to determine a percolation threshold for nanotubes suspended in rheologically simple fluids;[17,22] however, this phenomenon was originally observed for nanotubes in polymers and hence this discussion will be delayed until Section 5.2.2. One study found that the Cox–Merz relationship holds for all shear rates at concentrations below the percolation threshold, and holds for concentrations above the percolation threshold only at high shear rates.[23] In another study, the temperature of measurement affected the oscillatory rheological behavior, which the authors attributed to a change in the nanotube network structure.[24] Temperature-induced changes in dispersion are common in polymers where kinetic effects can be extremely important; for such an effect to occur in a low-viscosity Newtonian fluid is a bit surprising.

At high concentrations, nanotubes would seem to be ideal for forming liquid crystalline phases, for example, nematic phases. Theoretical arguments predict that a monodisperse suspension of rigid rod particles with no attractive or repulsive interactions will spontaneously transition to a biphasic mixture (both isotropic and nematic phases) at a volume fraction of about $3.3D/L$ and a 100% nematic liquid crystal at a volume fraction of about $4.2D/L$.[25] A nematic phase has all nanotubes aligned in one direction; the fringed micelle model described in Section 4.3 for polymer crystallization is a nematic phase. Polydispersity or attractive interactions will tend to widen the biphasic phase especially toward higher concentrations. Of course, attractive interactions will also tend to form gel phases, for example, phases that show some local alignment, but overall the nanotubes are in an isotropic network. This tendency to form gel phases is further exacerbated by the inherent flexibility of nanotubes. In other words, a nematic phase may not form because nanotubes are trapped in a nonequilibrium gel structure. This problem of gelation is a substantial one, and the number of true observations of nematic phases has been relatively small.[7,26–28] Normally, such observations are made optically; however, rheologically such phases are of great interest because if formed this would reduce the viscosity dramatically and allow for much easier processing of nanotubes. Such phases could be of great interest in nanotube suspensions in low-viscosity solvents used for the production of nanotube fibers, but are unlikely to be of importance in polymers since gel-like phases tend to dominate.

In most situations, the state of aggregation is unknown. In other words, the concentration regime of interest (dilute, semidilute, or concentrated) cannot be determined precisely since a single structure of aggregated nanotubes would be considered to be an individual filler particle when determining the concentration regime of interest. In order to perform measurements that do not change in time due to a change in aggregation state, the measurements must be performed quickly enough so that the aggregation state does not change or the system must be allowed to reach an apparent equilibrium under the conditions of the experiment. Of course, in superacids such concerns are irrelevant since nanotubes are truly solubilized in this system and the phase diagram (and hence aggregate structure) can be predicted wholly through thermodynamic considerations.[7] In one experiment with a fluid having a viscosity of 10 Pa s, the characteristic timescale for clustering as measured rheologically was estimated at 45 min,[29] which seems quite long given the timescales found for clustering using percolation measurements of much higher viscosity polymers.[30]

As noted earlier, shear thinning behavior at high shear rates was found for a study where the nanotubes were mostly or completely isolated, and a power law did not accurately describe the data. Other studies have found that a power law is a good fit to data where the state of dispersion is unknown, and the power-law index decreases with an increase in nanotube content (i.e., the sample becomes more shear thinning).[16,31] In other cases, a power-law expression is not a good fit to the data; the slope on a log(viscosity)–log(shear rate) plot becomes more flat as the shear rate increases, as representatively shown in the top part of Figure 5.3. Further, at high shear rates ($\sim$100–1000 s^{-1}), a Newtonian region is eventually recovered that has a viscosity approximately equivalent to the pure fluid.[23,32] A flat region is evidence that the nanotubes are maximally aligned and do not rotate appreciably in this shear rate region. As mentioned earlier, a Newtonian region at high shear rates is theoretically possible in polymers, but only at shear rates orders of magnitude higher than the 100 s^{-1} found for nanotubes (such a region is never reached in a polymer because of molecular weight reduction). The time evolution of rheological properties can be used to monitor dispersion in Newtonian fluids.[33,34]

Elongational viscosity measurements on a nanotube suspension with two different nanotube lengths have been made.[16] As with shear measurements, the onset of non-Newtonian behavior occurs at much higher volume fraction for short tubes. More interestingly, fitting the equations to a Hershel–Buckley fluid (yield stress + power law) yields a very different yield stress for elongation and shear measurements, which the authors interpreted as a failure of the yield stress to be a fundamental rheological parameter of nanotube suspensions. In a different study in an epoxy Newtonian fluid, only functionalized nanotubes could be measured reliably in elongation because of agglomeration.[32]

5.2.1 Nonoscillatory Measurements

The number of papers that have examined steady shear behavior of a nanotube-filled polymer is smaller than the number of papers that have examined nanotubes suspended in Newtonian fluids. The zero-shear viscosity increases with nanotube concentration in most cases as was found for nanotubes in a Newtonian fluid. A harder to predict response is whether the critical shear rate, that is, the shear rate where a polymer fluid transitions from Newtonian to non-Newtonian behavior, changes with the addition of nanotubes. Studies have shown a clear shift toward lower critical shear rates with the addition of nanotubes.[35–38] In fact, as shown by the representative viscosity versus shear rate plot in the bottom part of Figure 5.3, two power-law regions were found in the plot: at shear rates in what was previously part of the Newtonian region for the pure polymer and at high shear rates where the power law applied to the pure polymer. As expected, shear thinning at lower shear rates was lower than shear thinning at high shear rates (higher n in the former).

Another key question is whether power-law behavior is maintained in the high shear region, and, if so, what change is there in the power-law index with the addition of nanotubes? A comprehensive oscillatory and steady shear study was made on carbon nanotubes suspended in poly(ethylene oxide) with the addition of surfactant to help maintain dispersion.[39] The shear rates investigated were in the power-law region only, and the slope of the power-law region did not depend on the nanotube

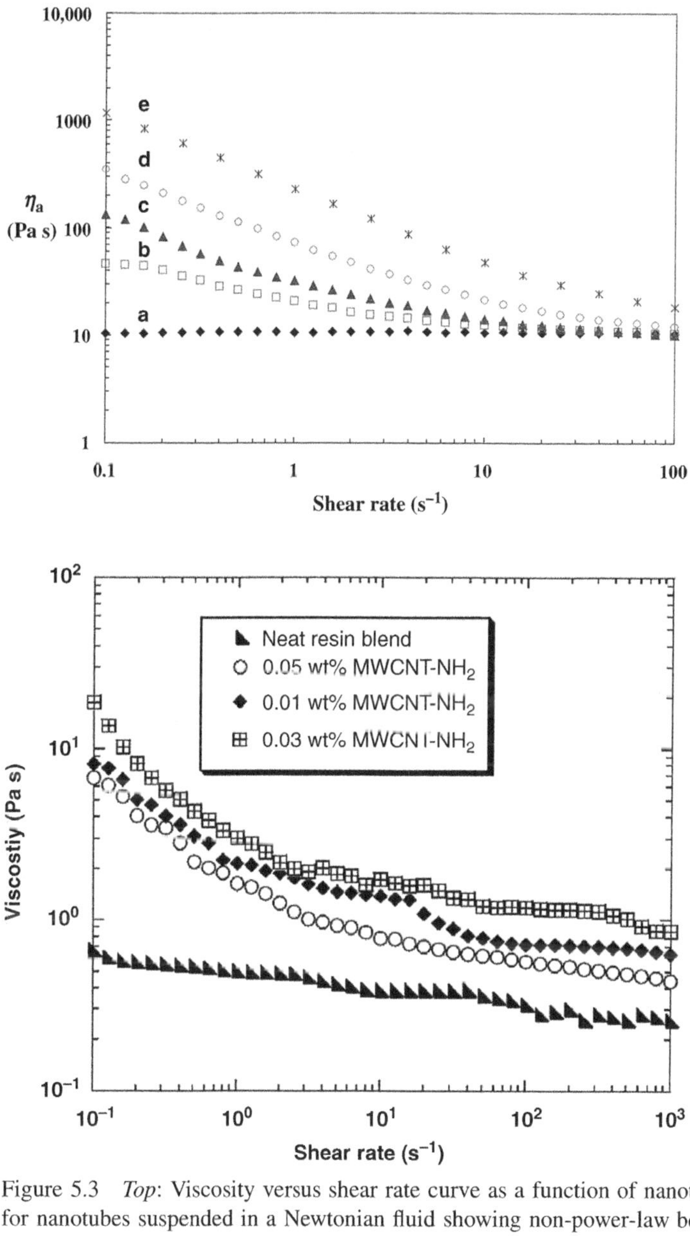

Figure 5.3 *Top*: Viscosity versus shear rate curve as a function of nanotube content for nanotubes suspended in a Newtonian fluid showing non-power-law behavior, which is the characteristic qualitative shape for some fluids. Different letters represent different nanotube contents (not given whether weight or volume percents): (a) pure fluid, (b) 0.05%, (c) 0.1%, (d) 0.25%, and (e) 0.5%. Copyright Springer. Reproduced with permission from Ref. 32. *Bottom*: Viscosity versus shear rate curve as a function of nanotube content for nanotubes suspended in a non-Newtonian fluid with nanotube levels given on graph. Copyright Elsevier Ltd. Reproduced with permission from Ref. 40.

concentration at concentrations three to eight times that of the percolation threshold for this material. Note that this behavior does not agree with experiments in Newtonian fluids where the shear thinning behavior was a strong function of the nanotube concentration. The Cox–Merz law did not hold for this material, and shear thinning was less for the nonoscillatory steady-state experiments as opposed to the oscillatory steady-state experiments, which the authors interpreted as a reduction in the number of stress-bearing junctions with the application of nonoscillatory shear. Transient, that is, non-steady-state, data also suggested that changes in the network with the application or removal of shear tended to dominate the response. The behavior of the pure resin was not given in these experiments, so the comparison with the pure resin could not be made. Another experiment on a very weakly shear thinning fluid[40] ($n \sim 0.9$) found that the addition of nanotubes eliminated the linear power-law relationship on a log–log plot of viscosity versus shear rate and instead yielded a curve qualitatively similar to that for a Newtonian fluid filled with nanotubes, as shown in Figure 5.3 (top). From the bottom of Figure 5.3, it is not clear whether the viscosity at high shear rates would reach that of the pure resin; however, the viscosity of the non-Newtonian fluid in this study is quite low and may affect the ability of the fluid to align and disperse the nanotubes. Two different studies with polypropylene, both having a power-law index of about 0.4, showed power-law behavior with the addition of nanotubes coupled with a decrease in n with an increase in nanotube concentration, that is, an increase in shear thinning behavior.[41,42] A similar decrease was found for linear low-density polyethylene, also from a starting value of about 0.4.[43] Another study, with n closer to 0.5, found essentially no change in n with nanotube addition, although the viscosities between the filled and neat systems differed by only a very small amount.[44]

Higher shear thinning in nanotube-filled systems versus the unfilled system suggests that eventually the viscosities of the filled and unfilled polymer should overlap at high shear rates (remember that the starting points are higher for the nanotube-containing systems because the zero-shear viscosities are higher). In some cases,[35,37] the viscosities overlap in the shear rate region below $10\,s^{-1}$. Overlap of the viscosities for the nanotube-filled and unfilled systems at high shear rates appears to be generally true, although this statement relies a great deal on extrapolation that may not be accurate. However, the shear rate where such overlap will occur varies substantially depending on the system investigated.

Although the general observation is that the viscosity increases with the addition of nanotubes, in a few cases the steady shear viscosity decreased.[45,46] This effect was attributed to selective adsorption of the high molecular weight fraction of the polymer. Lowering occurred only at low volume fractions of tubes; at higher volume fractions, the viscosity was higher than that of the pure polymer.[45] Further, this lowering was eliminated with the use of a polymer that had a more monodisperse molecular weight distribution.[46] Power-law behavior was still found with the addition of tubes even though the viscosity was lower.[45]

Non-steady-state elongational viscosities were measured using polypropylene and compared with functionalized and unfunctionalized tubes, where the former had better dispersion than the latter. The viscosity was certainly higher in the nanotube-filled samples, and the treatment that led to better dispersion led to very weak strain hardening, that is, a higher viscosity at higher strain times. The elongational and shear viscosities were not compared quantitatively.[37] In another study with poly(ether ether ketone) where the percolation threshold was found to be about 1 wt%, the elongational viscosity at 1 s^{-1} increased roughly linearly with nanotube content. At 5 wt% nanotube concentration, strain softening occurred with no evidence of this behavior at 2 wt% and lower concentrations. Trouton's rule was followed for the low nanotube content composites, while the elongational viscosity was much less than that predicted by Trouton's rule at 5 wt% nanotubes. The same data used melt strength data and a rather complicated model to estimate the elongational viscosity as a function of elongation rate; a maximum in viscosity was found at about 0.5 s^{-1} (e.g., strain hardening) followed by a power-law type decrease with more thinning for the sample containing more nanotubes.[47]

Another interesting rheological behavior of polymer fluids is normal stress. When a polymer is sheared in one direction, in the two orthogonal directions a stress occurs that molecularly is due to contraction of molecules; the stress is in the opposite direction of the contraction (i.e., the stress is due to the molecules wanting to return to the uncontracted dimension). The proper term is normal stress difference, and only one of the two normal stress differences is simple to measure. The most common positive normal stress difference pushes against the plates in a torsional rheometer and is also the cause of die swell, that is, the increase in dimensions of a polymer melt that exits a die. In both carbon nanotube dispersions in low-viscosity Newtonian fluids[29,48] and in polymers,[49] *negative* normal stress differences were found with the value in the polymer orders of magnitude higher than that in the Newtonian fluids. Negative normal stress differences are quite rare. A simple mechanical model, which involved the rotation around nanotube crossover points, was used to explain this behavior.[49] This simple model was confirmed in that short nanotubes (aspect ratio ~60), which do not have the capability of significant rotational distortion, showed a positive normal stress difference when dispersed in a polymer rather than a negative normal stress difference.[35]

As stated in the section on Newtonian fluids, the addition or removal of shear can cause a change from an electrically insulating to an electrically conductive state. For polymers, the more common result is that the application of shear causes a loss of conduction; that is, the material is conducting under quiescent conditions and the application of a shear field causes loss of conduction.[30,50–52] An example is shown in Figure 5.4 that shows that the ability to disrupt the network depends on the shear rate applied. Upon cessation of the stress field, the conductivity recovers. The qualitative difference between this and Newtonian fluids, that is, the application of a shear field causes a loss rather than a gain in conductivity, is likely not due to the non-Newtonian nature of polymers. Rather the difference is because the forces involved in high-viscosity polymers are much higher and the higher forces are able to disrupt the nanotube network. At slow steady shear rates, a conductive network did form upon application of shear with a time frame much faster than that occurring with no flow. Clearly in this case, the viscous forces were small enough not to break all agglomerates as they formed due to "sticking." Even in this study, cessation of the shear force, after raising the conductivity from its starting value, caused the conductivity of the network to rise even higher.[53]

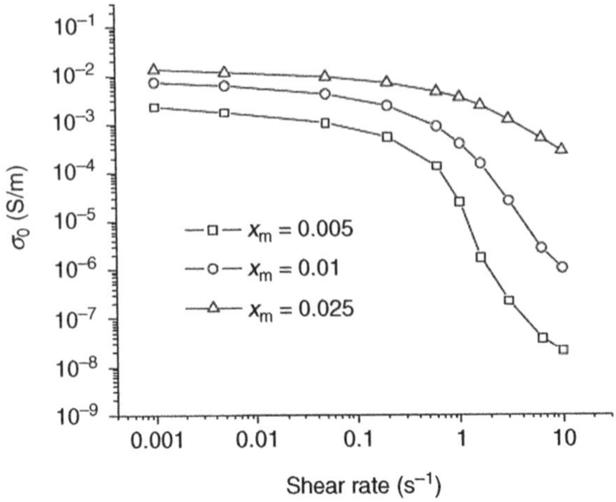

Figure 5.4 Effect of shear rate on the conductivity of a polypropylene filled with MWCNTs. The experiments were done in a specially modified torsional rheometer that allowed for the simultaneous measurement of rheological and electrical properties. Copyright 2007 American Physical Society. Reprinted with permission from Ref. 50.

The combination of these two effects, plus the agglomeration effect that occurs with nanotubes, has been captured in a simple model proposed by Alig et al.:

$$\frac{dC_a}{dt} = k_0(C_{a\infty} - C_a)^n + k_1(\dot{\gamma})(C_{a\infty} - C_a) - k_2(\dot{\gamma})C_a \tag{5.6}$$

C_a is the volume concentration of agglomerates and $C_{a\infty}$ is the volume concentration at infinite time. On the right-hand side, the rate constants and mathematical terms represent from left to right quiescent agglomeration, agglomeration due to shear, and deagglomeration due to shear. The utility of the model comes from a comparison of the first two rate constants assuming that $n = 1$; in one paper it was shown that k_1 was two orders of magnitude larger than k_0 demonstrating how much faster shear-induced agglomeration can be.[53] The issue of agglomeration that occurs with annealing time will be considered in more detail in Section 6.2.3, which describes the percolation threshold as measured by electrical conductivity. Section 5.2.2 discusses the percolation threshold as measured by rheology.

5.2.2 Oscillatory Measurements and the Percolation Threshold

Unlike the case with steady shear experiments, the number of papers that describe oscillatory measurements of polymers with nanotubes is quite high, primarily because such measurements can be used to determine the rheological percolation threshold. The concept of percolation is a very important concept and arises in both rheological and electrical measurements of carbon nanotube composites. A brief discussion of the general concept of percolation is presented here; further discussion appears in Section 6.2.

Percolation arises in polymer science in two general areas. One of these is composites; a material is said to be percolated if there is at least one continuous network pathway of filler particles, no matter how tortuous, in a given sample. The other area is gelation, that is, reacting polymer systems where at least some of the monomers are able to react more than one time, which eventually leads to a cross-linked network. Common theories of gelation, for example, the critical conversion where a system will form a gel, are more generally percolation theories. Percolation theory requires the definition of a critical fraction p_c (e.g., volume fraction of filler or fraction of bonds reacted) at which prior to this value being reached the property in question increases slowly or not at all. Above this critical fraction, the following formula applies:

$$\text{property} \sim (p - p_c)^{\beta} \tag{5.7}$$

The value of β is theoretically indicative of the dimensionality of the network, with a value of 1.1–1.3 for a 2D network and a value of 1.6–2.0 for a 3D network.[54] These values are true only for the case where the filler is infinitely conductive and the matrix is infinitely insulative. In practice, this parameter is viewed simply as a fitting parameter since higher values (more steeply increasing functions) are frequently found. A more steeply sloping function is indicative of a smaller number of nanotube–nanotube contacts per nanotube. For randomly packed spherical objects having the same diameter, the volume fraction of filler for the percolation threshold is 0.16. For nonspherical particles, the volume fraction of filler required to achieve percolation in the case of random orientation is based on excluded volume concepts, that is, the volume around an object where another object's center is not allowed to penetrate. Using this argument, an approximate scaling relationship for the critical volume fraction is given by $p_c = 0.5D/L$, where L/D is the aspect ratio of the filler.[55]

In rheology experiments, the property typically used is the storage modulus (or equivalently a variant of the complex viscosity) and the fraction is the volume fraction of tubes,[56] although other rheological properties such as the zero-shear viscosity[35] can be used as well. In other words, the storage modulus shows a very large change with concentration above the percolation threshold, as shown in Figure 5.5. The question of what frequency is used to evaluate the storage modulus is a legitimate one; typically low frequencies are used because the solid-like response will be evident in this region at lower concentrations. The mathematical procedure used to determine the percolation threshold is actually not trivial, since three fitting parameters are required: p_c, β, and a scaling prefactor. Since only the scaling prefactor changes, it is perfectly reasonable to use mass fractions rather than volume fractions. It is not *a priori* clear how many data points should be included in the fit since the choice of p_c will influence the number of data points used for a given data set. Further, there is an upper limit in concentration where this equation would no longer be expected to apply because of the emergence of a cocontinuous morphology, although practically in almost all cases this limit is far beyond the concentration region tested. The author believes that linearizing the equation and determining the best fit parameters based on finding the maximum in the correlation coefficient with different p_c's is the best procedure, and private conversations with Petra Pötschke indicated that her group uses a very similar procedure.

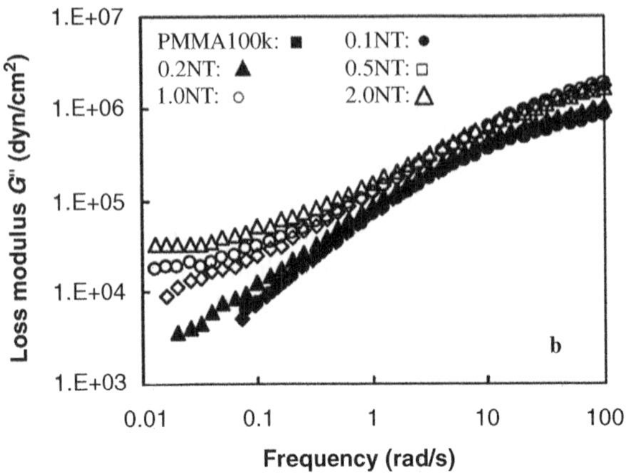

Figure 5.5 Graph showing storage and loss modulus as a function of frequency for SWCNTs in poly(methyl methacrylate). The percolation threshold of this material from the data presented in the top graph was 0.11% or 0.12% depending on the frequency chosen. Copyright 2004 American Chemical Society. Reprinted in part with permission from Ref. 56.

The first paper that noted that oscillatory rheology was sensitive to nanotube network formulation was by Pötschke et al.[57] followed shortly by a paper from Krishnamoorti and coworkers.[58] Although both authors noted the extremely large change in modulus, which had been seen previously in silicate nanocomposites,[59] neither applied the quantitative treatment represented by Equation 5.7. The first to perform such a quantitative analysis were Winey and coworkers[56] and many others have subsequently used this procedure. At about the same time, Pötschke et al. showed that the volume fraction where the large change occurred was found to be strongly dependent on the temperature chosen for analysis.[60] This observation led to the conclusion that the rheological network included both polymers and nanotubes, although the possibility of a change in aggregation state upon heating was not explicitly examined by testing whether the results were the same on heating and cooling. As an alternative quantitative approach to Equation 5.7, the identification of gelation through the Winter–Chambon method[61] was used[62] prior to the Winey and coworkers' paper. The Winter–Chambon method defines the critical concentration as being that where $\tan \delta$ is independent of frequency at low frequencies. The Winter–Chambon method was developed for liquid–solid transitions (gelation, crystallization) and admittedly whether this theory should be applied to composites is arguable. Still, it is very interesting that Equation 5.7 has come to dominate the field in terms of quantitative analysis rather than the Winter–Chambon method, likely because of the analogous equation used for electrical conductivity. Table 5.1 lists papers that have data from oscillatory rheological measurements at the appropriate and at enough nanotube concentrations to determine the percolation threshold. About half of the papers listed have the data but did not fit Equation 5.7.

Generally, the electrical percolation threshold has been found to be higher than the rheological percolation threshold,[47,56,90,97,109,112,128] although the opposite behavior has been found.[71,96] A lower percolation threshold for the rheological

TABLE 5.1 Oscillatory Rheological Measurements Made at Various Volume Fractions That Were Used (or Could Have Been Used) to Determine the Percolation Threshold

Polymer	References	Polymer	References
Polystyrene	58,63–70	High-density polyethylene	71–75
Polycarbonate	60,76–82	Low-density polyethylene	73,83,84
Ultrahigh molecular weight polyethylene	46	Medium-density polyethylene	85
Poly(methyl methacrylate)	56,64,86–89	Polypropylene	35,37,38,90–93
Poly(ethylene oxide)	36,94–96	Poly(ethylene terephthalate)	97–100
Epoxy	101	Poly(ethylene naphthalate)	102
Polycaprolactone	103–105	Poly(butylene terephthalate)	106,107
Poly(phenylene sulfide)	72,108,109	Liquid crystal polyester polymer	110,111
Poly(ethylene glycol-*co*-cyclohexane-1,4-dimethanol terephthalate)	112	Polyoxymethylene	113
Polyamide 6	114,115	Poly(vinylidene fluoride)	116–118
Polyamide 11	119	Poly(4-methyl-1-pentene)	120
Polyetherimide	121	Poly(propylene fumarate)	122
Poly(ether ether ketone)	47	Blend: polyamide 6/acrylonitrile–butadiene–styrene terpolymer	123–125
Blend: polypropylene/ethylene–octene copolymer	126	Blend: polypropylene/acrylonitrile–butadiene–styrene terpolymer	127
Styrene–Isoprene copolymer	128		

measurements represents a difference in the distance required between nanotubes to be considered to be part of a network; that is, the distance required is shorter for electrical conductivity than for rheology. The tube–tube distance for electrical conductivity is the distance required for electron tunneling, which is on the order of 3 nm. If the distance for rheological interaction is approximately the radius of gyration of a polymer molecule, then a distance larger than 3 nm would be correct. However, an important complicating consideration is that the rheological percolation threshold is measured on a melt, while the electrical percolation threshold is measured after solidification, and hence the nanotube networks might not be the same due to a changed aggregation state. The rheological percolation threshold has been found to decrease with temperature during heating,[60,77] indicative of structural changes due to agglomeration while the material exists as a melt. Hence, comparing the electrical and rheological percolation thresholds is not appropriate unless the electrical percolation threshold is found to be the same before and after melting.

The same arguments against comparison of percolation thresholds can be made for comparisons of the exponent, but still the exponents are interesting to compare in light of their possible geometric meaning. Given the lower rheological percolation threshold attributed to a larger effective particle, the critical exponent should be larger for rheological percolation since a larger nanotube would be more three dimensional. In cases where quantitative comparisons have been made, the exponent has been found to be significantly lower for rheological percolation[56,96,97,128] supporting the idea that the critical exponent in Equation 5.7 should be viewed as a fitting parameter with little or no physical significance.

5.3 MECHANICAL PROPERTIES (MEASUREMENT OF SOLID PROPERTIES)

The mechanical properties of a single carbon nanotube were discussed at length in Chapter 2; to summarize, carbon nanotubes have a measured tensile modulus of about 1000 GPa, a strength of about 50 GPa, and a strain at break of about 15%. The first number is fairly accurate and reproducible, with the second varying widely, as much as an order of magnitude lower and a factor of 3 higher. The measured tensile strength and strain at break are significantly less than calculated values for no-defect tubes of around 150 GPa and 30%, respectively.

Measured values for nanotube fibers and mats are far less than values for single tubes. As described in Section 3.7.1, nanotube fibers can be made in one of the four ways: from a gas, from nanotube forests, from solution, and from twisting a very thin nanotube sheet into a fiber. The first method has yielded the highest strengths of all the methods reported: a strength of 9 GPa and a stiffness of 350 GPa were reported for the best fiber having a diameter of about 10 μm, with average values of about 4.5 and 175 GPa, respectively.[129] The former measurements are the highest ever reported for nanotube fibers as of the end of 2009, and are significantly larger than the typical values for a high-strength polymer fibers, as shown in Table 5.2.[130] These values are even higher than those for carbon fibers; high-modulus carbon fibers have moduli ~375 GPa, while high-strength carbon fibers have strengths ~6 GPa.[131] These

TABLE 5.2 Modulus and Tensile Strengths of Various Forms of Pure Nanotubes*

Description	Modulus/tensile strength (GPa)	Description	Modulus/tensile strength (GPa)
Individual nanotube	1000/50	Kevlar®	150/3.6
Nanotube fibers from gas	175/4.5	Ultrahigh molecular weight polyethylene	80/3
Nanotube fibers from forest	240/0.41	Poly(*p*-phenylene benzobisoxazole)	240/5.8
Nanotube fibers from superacids	120/0.12	Buckypaper via filtration	10/0.1
Nanotube fibers from film	18/0.85	Layer-by-layer process to form nanotube film	12.2/0.75
Commercial nanotube film	–/1.2	Commercial nanotube fiber	–/3
High-modulus carbon fiber	375/–	High-strength carbon fiber	–/6

See the text for references where appropriate and for more detail on the size of fibers.

measurements were made on nanotube fibers having very small 1 mm lengths; at 2 cm lengths, the average strengths were a much lower 1 GPa. This difference points out a very important issue in the production of nanotube fibers; mechanical properties must be reported as a function of fiber length since manufacturing control is generally quite poor and longer fiber lengths increase the probability of defects. Another gas process used by a different group yielded strength and modulus values of about 0.8 and 60 GPa, respectively, on fibers 5–10 μm in diameter and 1 cm in length.[132]

The highest strengths and moduli reported via drawing from nanotube forests are 1.9 and 330 GPa, respectively, for a fiber with a diameter of 3 μm and a test length of 1 cm.[133] The highest mechanical properties were achieved only after adding a twist to the drawing process; the twist was thought to increase tube–tube interactions. Larger diameter fibers (10 μm) had values of 0.41 and 240 GPa, respectively. Drawing from solution generally does not yield high-modulus or high-strength fibers because the nanotubes do not stay well suspended during the processing, indicating that it is necessary to prevent agglomeration until absolutely necessary and have agglomeration occur on a very rapid timescale. Using superacids where the nanotubes are thermodynamically solubilized yields fibers having an average strength of 0.12 GPa and a modulus of 120 GPa. The fibers in this process had a specific gravity of 1.11, which is the highest ever reported, a diameter on the order of 100 μm, and the test length was not reported.[134] Finally, using a twisting process from a film, strengths as high as 0.85 GPa and modulus as high as 18 GPa were reported for fibers having a length of a few centimeters and a diameter of ~40 μm.[135] Although solution and sheet processes have not been as successful at producing high-strength materials as gas and drawing from nanotube forests, the quality of nanotubes likely has as much or more effect on the modulus and stiffness achievable versus the processing method used. In other words, it is entirely possible that higher strength and moduli fibers could be achieved in the latter two methods if higher quality tubes were used. Long, defect-free tubes are very important for such fibers, and thin-walled tubes are better because of higher contact areas between adjacent fibers due to wall distortion. Processing primarily affects orientation and tube–tube morphology/contact adhesion.

Mechanical properties of a nanotube film, for example, buckypaper, which is normally produced by filtering a solution of nanotubes, produce films with poor mechanical properties. In fact, if the samples are thin enough (one paper reports approximately 1 μm as the critical value[136]), then the films are not self-supporting and cannot be lifted off intact from the substrate. A recent paper shows that the tensile strength of buckypaper made via filtration (0.1 GPa) does not significantly depend on whether SWCNTs or MWCNTs are used, although the density normalized value does.[137] A list of mechanical properties for buckypapers from different publications appears in Ref. 138 with tensile strengths between 0.01 and 0.1 GPa and moduli between 1 and 10 GPa. Special efforts must be made in order to produce pure nanotube films with good mechanical properties. An LBL process, described more completely in Section 3.7.2, yielded a strength and modulus of 0.75 and 12.2 GPa, respectively.[139] A commercial sheet produced by Nanocomp Technologies using very long (1–2 mm) mostly single-walled nanotubes gave a strength and modulus of 0.67 and 25 GPa, respectively, in the stretch direction after stretching for a centimeter-scale sample.[140] On the Nanocomp Technologies corporation web site, www.nanocomptech.com/html/nanocomp-what-we-do.html, strength values as high as 1.2 GPa for films are claimed. Furthermore, the company can manufacture fibers with strengths of 3 GPa according to the site (accessed June 26, 2010). In another study, the strength of a nanotube film using a wide variety of nanotubes was investigated and the authors found that the strength and toughness scale linearly with the number of interbundle junctions per unit volume as calculated from the porosity and bundle size. They were able to use this observation and a simple model to show that the average energy to break an interbundle junction was approximately the same as the nanotube surface energy.[141]

The remainder of this chapter describes the mechanical properties of polymer–nanotube composites. Both impregnation/infusion processes, that is, process where low-viscosity thermoset resins are added to already formed fibers and sheets and are cured to make a part, and processes where the polymer and nanotubes are mixed will be described. Because of the large number of papers that describe mechanical measurements on composites, a table that tries to be exhaustive as has been done for other measurements will not be prepared, since such a table would incorporate many hundred, if not a thousand, papers as of the end of 2009. Instead, the author has chosen to probe cause and effect relations, and provide some key references to illustrate these relations. Because tensile tests are the most commonly performed tests by a large margin, exploring these relationships will be primarily done by using tensile test examples. The focus of discussions of other measurements will be to give examples where such measurements have been made. More importantly, this chapter focuses on those aspects of other measurements that have different sensitivities to particular characteristics versus tensile tests. For example, bending tests of laminate composites are much more sensitive to interfacial adhesion than tensile tests. Before discussing the mechanical properties of the composites, fundamental measurements of polymer-nanotube interfacial strength will be described.

5.3.1 Interfacial Shear Strength

The surface energy of MWCNTs has been measured directly by measuring the contact angle of organic[142,143] and polymeric liquids[144] having known surface tensions. Since the surface energy as calculated via contact angle is expected to decrease with an increase in curvature,[145] the fact that the surface energies of the carbon nanotubes were measured to be approximately identical to that of a carbon fiber, 40–45 mJ/m^2, with polymeric liquids was quite surprising.[144] The authors postulated that the nanotube surface had less defects than the fiber that balanced the curvature effect. A lower value was found for organic liquids, 27.8 mJ/m^2.[142] No explanation has been offered for the difference between the two types of liquids; however, the difference is significant since the almost nonwetting surface poly (tetrafluoroethylene) has a surface energy of about 20 mJ/m^2, while 40–45 mJ/m^2 would be considered a relatively easy surface to wet.

Many mechanical property measurements depend on the interfacial energy; for example, the tensile strength depends largely on this parameter for composites unless specific requirements are met. These requirements are given by the classical formula

$$l_c = \frac{TS_{NT}d}{2\tau_{IC}} \tag{5.8}$$

where TS_{NT} is the tensile strength of the nanotube, d is the diameter, and τ_{IC} is the interfacial shear strength. The numerator in SI units has a magnitude on the order of 50 for SWCNT and 500 for MWCNT; values of interfacial shear strength are on the order of 10^8 in the same system of units; hence, the critical length (l_c) is roughly equivalent (1 μm) to the actual length of the nanotube. Hence, the interfacial shear strength could have a very large impact on large strain mechanical properties of polymer composites. The interfacial shear strength is related to the interfacial energy; however, the former contains contributions from the fiber diameter and embedded length that have no relevance for the interfacial energy. A more fundamental parameter than the interfacial shear strength is the interfacial fracture energy, which requires a model to calculate its value but is a better representation of the processes that occur during debonding. The classical description represented by Equation 5.8 is a continuum approach, and it may have limited applicability to nanoscale fillers such as nanotubes.

Measurements of the interfacial shear strength using a pullout technique have been made. In one study, an atomic force microscopy tip was used to manipulate nanotubes protruding from an epoxy resin. An interfacial shear strength of 300–400 MPa was estimated for short tubes, with values half an order of magnitude smaller for long tubes suggesting that end effects are important.[146] A more classic experiment by the same group that is often used for macroscopic fibers, where a MWCNT was pushed into a hot polymer film then removed after cooling, gave values similar to the results for longer tubes, 20–90 MPa. Smaller diameter tubes were measured at the higher value, while larger diameter tubes at the smaller value. The authors attributed the difference to possible covalent bonding to nanotube defects that are expected to be more prevalent in the former. The critical fracture energy could not be calculated precisely because the effective length of stress transference could not be determined; however, the value was of the same order of magnitude as for glass fibers in matrices that are considered to be strongly adhering such as polyamide 6.[147] Finally, a later pullout experiment using both carboxylic acid-functionalized nanotubes and unfunctionalized tubes in an epoxy matrix and found rough equivalents at higher embedded lengths, but significantly improved performance for low embedded lengths, as shown in Figure 5.6.[148] A model that properly accounts for the geometry of the test, that is, the fact that end effects are important for short embedded lengths, quantitatively predicts such a decrease; the authors, however, did not go back to their earlier paper[146] to confirm that using the appropriate mechanics could explain the earlier difference. The maximum interfacial shear stress calculated was 30 MPa for the unmodified tubes and 150 MPa for the modified tubes.

Measurement of the interfacial shear strength using the distribution of fiber lengths after fracture have been made for an epoxy.[149] Rather than the classical experiment as for macroscopic fibers where a single fiber is embedded in the matrix, in this case the distribution of sizes after three-roll calendaring was measured after

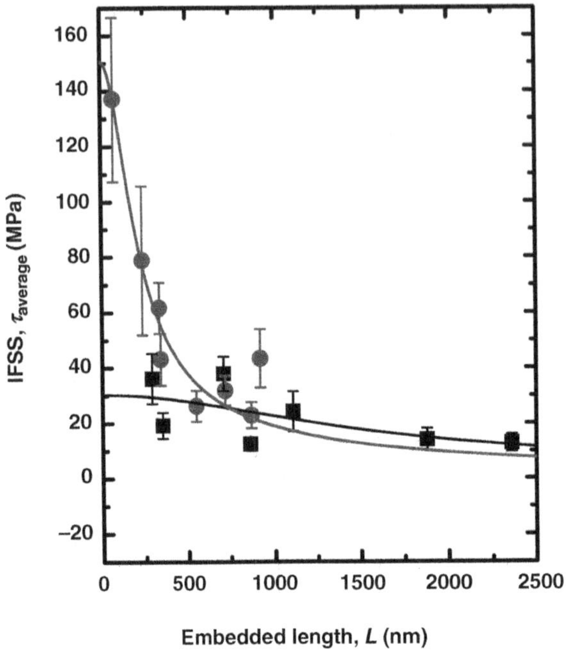

Figure 5.6 Results of single-fiber pullout tests from an epoxy matrix for functionalized (red) and pristine (black) MWCNTs. The solid lines are a fit to a shear-lag model that takes into account the geometry of the test; see the original text for details. The *y*-axis intercept is the maximum interfacial shear strength. Copyright Wiley–VCH Verlag GmbH & Co. KGaA. Reproduced with permission from Ref. 148.

being sure that the distribution was insensitive to further processing. Using the standard Weibull distribution to analyze the length, a value of 44 MPa was determined for the interfacial shear strength. This analysis also allowed the determination of the tensile strength of a single nanotube, which was determined as an extremely low 4 GPa. The authors did show that bending/flexural properties of nanotubes are essentially irrelevant when calculating ultimate properties; however, this analysis does not preclude such properties as having a large effect on small strain properties such as the modulus.

A recently published peel test technique[150,151] can measure the interfacial shear strength, although the values measured using this approach would not be expected to match those measured using pullout or single-fiber fragmentation testing. Although difficulties in quantifying contact area have prevented calculation of the interfacial shear strength, this test has been used as a relative measure to compare the strengths of interaction between a nanotube and a polyimide or a nanotube and an epoxy. A significant difference was found, but the authors were not able to conclude that such a difference was not due to a difference in contact area caused by a difference in stiffness.

5.3.2 Tensile, Compressive, and Bending Properties

Dispersion has a significant effect on reinforcement efficiency; in general, the better the dispersion, the better the reinforcement efficiency. Because dispersion is not usually well characterized, any sorts of comparisons between results are usually influenced by the question, "Is what I am seeing simply a difference in dispersion?" Consistency between papers is more the exception than the rule, especially with the rather large number of papers that report mechanical data.

Two other complications can also arise. First, as highlighted by Coleman et al.,[152] the introduction of nanotubes in some polymers also changes the fractional crystallinity, which in turn can be expected to substantially increase the modulus, especially if the crystallinity is low. Unfortunately, this is the case for poly(vinyl alcohol), and hence any study that has used poly(vinyl alcohol) should be considered extremely suspect unless careful crystallinity measurements are made. The rather large values of (dE/dV_F) reported for poly(vinyl alcohol) in the review paper by Coleman et al.[1] are almost certainly due to crystallization rather than nanotube reinforcement. Second, the modulus changes very steeply in the glass transition region (as much as 10 orders of magnitude in SI units for an amorphous polymer), so any study performed on a polymer at room temperature with a glass transition between 0 and 40°C should be examined carefully. As stated previously, the author has chosen to probe cause and effect relations and provides some key references to illustrate these relations.

5.3.2.1 Tensile Properties Generally, the reinforcement efficiency of nanotubes, for example, (dE/dV_F), is higher for weaker starting polymers. An argument is given by Schaefer and Justice[153] for this observation. If the effective modulus of the filler is less than the modulus of the polymer, then the filler will be totally ineffective in increasing the modulus of the polymer. As described previously, in an earlier paper, Schaefer pointed out that the bending modulus is more likely the critical modulus to use as opposed to the tensile modulus.[3] Another discussion point raised in this later paper[153] is that an aggregated nanotube cluster likely has a much smaller modulus than an unaggregated single nanotube. The effective modulus of the filler, which could be the modulus of the cluster or the bending modulus, is more likely to be smaller than the polymer modulus for the case of high-modulus polymers and that certainly would explain why nanotubes are generally more effective in increasing the moduli of flexible polymers.

Generally, improvements in the modulus are more significant in a relative sense than improvements in the tensile strength, with the latter often showing a decrease even at low added fractions of tubes. The strain at break often decreases, although perhaps 10% of the reports in the literature show no effect or even an increase in strain at break with nanotube addition. In this section, focus is given to the modulus and tensile strength. This decision is easy to criticize, since the toughness rather than these two parameters is often the critical design parameter. Further, a paper has recently been published suggesting that toughness increases in nanocomposites should be significantly larger than increases in composites made from conventional micron-sized fillers. The crux of the argument is twofold: first, breaking a 1 μm size fiber is much easier than breaking nanosize fibers with equivalent cross-sectional area; second, the close spacing of nanofillers can cause toughening mechanisms that are not relevant for micron-size fibers.[154] The most significant reason for not including the toughness is that most papers do not report values for toughness, while tensile strengths and moduli are almost always reported. In general, for nanotube composites, higher tensile strengths mean higher toughnesses, although certainly there are exceptions. Finally, fracture toughness and impact energy results are discussed in some detail, which reduces the need for reporting toughness measurements from tensile tests. Readers interested in toughness will have to consult the original references and, in many cases, hope that stress–strain curves are given from which a toughness could be calculated.

One clear piece of evidence that dispersion affects tensile properties is the fact that the modulus and tensile strength often increase at low tube volume fractions followed by a decrease at higher volume fractions. In an amorphous unoriented polymer without chemical reaction, there is no other possible explanation for such behavior other than a change in the relative dispersion at different volume fractions. This behavior also occurs in semicrystalline polymers, but in this case a reduction could in theory cause a reduction in tensile strength and modulus. Although certainly not a uniform value, 1% loading is a typical value where the maximum occurs. This qualitative behavior of a maximum in improvement is common, although certainly not universal; a small sampling where such behavior occurs in the modulus or tensile strength (or both) includes SWCNTs/phenolics,[155] SWCNTs/styrene–butadiene rubber,[128] MWCNTs/polyamide 6,[156] MWCNTs/polystyrene,[157] MWCNTs/polyamide 6,10,[158] MWCNTs/polypropylene,[159–161] MWCNTs/polyurethane,[162] MWCNTs/poly(methyl methacrylate-*co*-*n*-butyl acrylate),[163] and MWCNTs/poly(methyl methacrylate).[164] A representative example taken from the author's own work is shown in Figure 5.7. The remainder of this section will detail the effect of various independent variables on the tensile properties.

The influence of tube type on tensile properties has been explored. In poly(vinyl alcohol) composites produced using a dispersion–dissolution–precipitation method, it was found that the modulus increase, that is, dE/dV_f, was inversely proportional to the nanotube surface area. In other words, surface areas, e.g. smaller diameters, led to more effective reinforcement for four different types of MWCNTs and one DWCNT, although this conclusion should be questioned because the nanotube diameters were very much clustered. However, a SWCNT sample did not follow this trend, which the authors attributed to poorer dispersion.[165] In another paper using epoxy and dispersion–reaction methodology, three different types of tubes (MWCNTs, DWCNTs, and SWCNTs) were studied in an epoxy matrix, and the improvements in modulus were in general better for the DWCNTs versus the MWCNTs and again the SWCNTs were worse than an inverse surface area relationship would predict. The tensile strengths changed very little with nanotube incorporation in all cases.[166] A study on polyurethane using a dispersion–reaction methodology found a significant improvement in tensile strength for SWCNTs compared to MWCNTs, providing a counterexample to the previous statements.[167] Finally, a study on a dissolution–dispersion mechanism using drawn fibers showed best performance of the SWCNTs ($dE/dV_f = 157$ at 5% NT) in terms of modulus and best performance of the MWCNTs in terms of tensile strength (MWCNT:DWCNT:SWCNT:polymer = 412, 316, 335, 244 MPa). The small amount of data included in this paragraph show that tube type or diameter cannot be correlated with mechanical property reinforcement in any consistent manner, which is a valid conclusion based on more data than are presented here.

The inconsistency described in the previous paragraph could be a function of differing nanotube lengths for the different types of tubes; however, in the author's opinion, dispersion issues are a larger cause of. Studies with the same tubes, only with different lengths, have been the focus of a few studies. In one study on PMMA, only

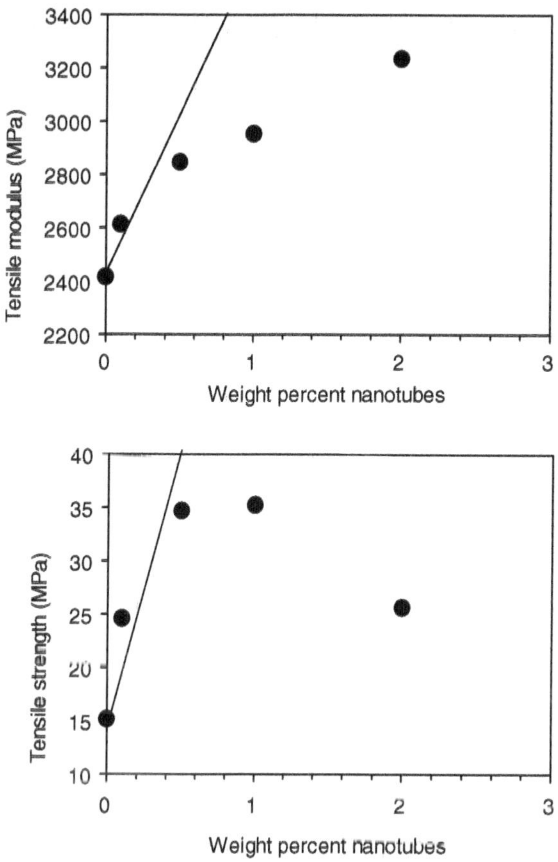

Figure 5.7 The most common qualitative behavior of the tensile strength and modulus for nanotube composites. If the nanotube fraction is increased to high enough values, the modulus will often increase with increasing nanotube content but quickly fall off from the theoretical maximum given by Equation 5.1 (line) or Equation 5.3, while the tensile strength will increase at low volume fractions and actually begin to decrease at high volume fractions. Data are taken from Ref. 128.

the longest MWCNTs (5–20 μm as opposed to less than 5 μm) were able to increase the toughness.[168] In poly(vinyl alcohol) filled with MWCNTs, image analysis was used to quantify aspect ratio in an electrospun fiber and it was found that below an aspect ratio of ~35, there was no effect of aspect ratio on modulus, while a very sharp increase occurred at an aspect ratio of ~42; the modulus increased from 4 to 8 GPa.[169] In an epoxy system filled with 0.5% MWCNTs, the effect of length was not clear since the intermediate length sample had the highest modulus and tensile strength, although the method used to reduce the length also reduced the largest agglomerates in the starting sample. (dE/dV_F) was quite high, 270 GPa, while the highest tensile strength change was at 0.5% loading and was from 28 to 41 MPa.[170]

There are four possible approaches to changing the surface chemistry of the nanotube in order to change interfacial energy and hence tensile properties. The first case is probably the most obvious: use covalent bonding between a modified nanotube and the polymer matrix. Table 5.3 lists a number of papers where functionalized and unfunctionalized tubes are compared (this table is not exhaustive). However, the improvement could be assigned either to an increase in interfacial adhesion or to an improvement in dispersion caused by covalent functionalization. The total number of papers that try to improve interfacial adhesion using functionalization (although many studies do not compare results with unfunctionalized tubes) is three to four times that shown in Table 5.3. The results presented in Table 5.3 indicate that, on the whole, functionalization has been shown to improve mechanical properties of composites.

The second approach to modify interfacial adhesion is to chemically modify the nanotube surface, but have no covalent bonding with the bulk polymer. One

TABLE 5.3 Tensile Properties of Composites with Covalent Bonds Between Polymer and Tube, Compared with Unfunctionalized Counterparts

Polymer	Nanotube type	dE/dV_f (GPa)[a]	Tensile strength at 1% NTs unless otherwise indicated (MPa)[b]
Epoxy[171]	Amine-functionalized SWCNTs	57, 82	74.1, 74.7
Epoxy[166]	Amine-functionalized DWCNTs	105, 128	63.8, 67.7, 69.1 (0.5%)
Epoxy[166]	Amine-functionalized MWCNTs	60, 83	63.8, 63.2, 63.6 (0.3%)
Epoxy[172]	Amine dendrimer-functionalized SWCNTs	46, 159	90, 104, 125 (0.5%)
Epoxy[173]	Polyacryloyl-functionalized MWCNTs	33, 75	46, 42, 49
Polypropylene[174]	Fluorinated SWCNTs	−0.5, 8	31, 29, 45 (2.5%)
MMA–ethyl methacrylate copolymer[175]	Amine-functionalized MWCNTs	68, 81	49, 53.5, 55
PMMA[176]	Acid-functionalized MWCNTs	86, 10	29, 41, 42
Polyamide 6[115]	Acid-functionalized MWCNTs	11, 17	18.9, 18.0, 21.6
Polyamide 6,6[177]	Amine-functionalized MWCNTs	40, 110	50, 60, 75
Polyamide 6,10[178]	Acid-functionalized MWCNTs	120, 68	36, 11, 54
Polyamide 12[179]	Anhydride-coated SWCNTs	1,3	37, 39, 41 (3%)
Poly(1-butene)[180]	Polypropylene-grafted MWCNTs	5, 37	31, 26, 45

This table is not exhaustive. When weight fractions were given in the paper, a density of 2 was assumed for MWCNTs and 1.35 for DWCNTs and SWCNTs.

[a]Order: pristine, functionalized.

[b]Order: no tubes, pristine, functionalized.

obvious choice is to bond the same polymer to the surface; although this approach seems obvious, only a relatively few number of papers have used this. For example, isotactic polypropylene-grafted nanotubes have been added to isotactic polypropylene, with a result (unmodified versus modified) of 35 versus 75 GPa for dE/dV_f, and 25, 31, and 53 MPa for tensile strength (unfilled, unmodified, and modified, respectively; 1% tubes).[181] Chlorinated polypropylene was grafted to MWCNTs via a grafting to approach and these nanotubes were added to chlorinated polypropylene via a dissolution–dispersion technique with the resulting $dE/dV_f = 50$ GPa and a tensile strength of 50 MPa at ~1 wt% tubes versus 10 MPa with no tubes. Unmodified tubes could not be dispersed in the solvent.[182] A grafting to process utilizing maleic anhydride-grafted polyethylene functionalized tubes mixed with linear low-density polyethylene led to a dE/dV_f of 20 versus 5 GPa for the unmodified tubes, while the tensile strength at 1% tube content showed a slight increase for the modified tubes and a significant decrease for the unmodified tubes.[183] A polymer not the same as the matrix polymer has also been used. Examples include coating tubes with high-density polyethylene via a grafting from process causing a dE/dV_f increase from 0.35 to 0.9 GPa (unmodified versus modified) in an ethylene–vinyl acetate copolymer, which the authors attributed to better dispersion.[184] Another example was maleic anhydride-grafted polypropylene functionalized tubes mixed with poly(1-butene), with a result (unmodified versus modified) of 5 versus 60 GPa for dE/dV_f, and 31, 26, and 45 MPa for tensile strength (unfilled, unmodified, and modified, respectively; 1% tubes).[180]

A third approach, which could affect interfacial adhesion but more likely improves dispersion, is to adsorb a molecule to the nanotube, which in turn covalently bonds to the polymer. One example was to coat styrene–maleic anhydride polymer onto unfunctionalized MWCNTs and then add these coated tubes to polyamide 12. The result was (uncoated versus coated) 1 versus 3 GPa for dE/dV_f, and 58, 36, and 44 MPa for tensile strength (unfilled, uncoated, and coated, respectively; 1% tubes).[179] The same approach was used for polyamide 6; in this case, the results were (uncoated versus coated) 7 versus 3 GPa for dE/dV_f, and 65, 50, and 48 MPa for tensile strength (unfilled, uncoated, and coated, respectively; 1% tubes).

The fourth and final approach is to use a molecule that does not covalently bond either to the polymer or to the nanotube; if the molecule is a polymer, such an approach can lead to improved mechanical properties.[185] However, surfactant-assisted processing is the most common example and with small molecules the effect is only to improve dispersion; further, if the small molecule is soluble in the polymer and does not stay on the surface of the nanotubes, then it will simply act as a plasticizer in the polymer.

Nanotube alignment will also affect mechanical properties. The effect of nanotube alignment on the mechanical properties of polyamide 12 fibers with reinforced nanotubes yielded a very interesting result. dE/dV_f was 10 and 16 GPa for entangled and aligned tubes, respectively, while the tensile strengths were basically unaffected. However, the strain at break for the aligned tubes was unchanged with the addition of nanotubes, while that for the entangled tubes decreased significantly from 400% to below 200% at 10% tubes. Although not given, the estimated toughness seems relatively similar for the two materials even though the strain at break decreased markedly. A study of draw ratio on

PMMA–MWCNT composites found an increase in tensile properties with drawing, but no significant differences between filled and unfilled samples.[168] Drawing a semidried epoxy film to large elongations of 50 times its initial length caused a dE/dV_f of 27 GPa and tensile strength change of 7 MPa (unfilled) to 13 MPa (1% tubes). In the perpendicular direction, the changes were about half of these values.[186] Improved nanotube alignment in a fiber by reducing the fiber diameter in melt-spun high-density polyethylene was shown via scattering; dE/dV_f was 20 GPa for the more highly aligned system, while for the less aligned system a value significantly less than half of 20 GPa was found. Unfortunately, any changes in crystallinity with orientation were not reported, so such results may simply be due to changes in crystallinity.[187] In another study on injection-molded high-density polyethylene filled with MWCNTs, dE/dV_f for the unoriented sample was about 7 GPa, while that for the oriented sample was approximately five times larger. The increase at 1 wt% tube content in the tensile strength was 25–30 MPa for the unoriented sample and 65–90 MPa for the oriented system; again however, crystallinity comparisons were not given.[188]

Alignment of nanotubes via a magnetic or electric field is a very good way to test the effect of nanotube alignment since nanotubes can be aligned without aligning the polymer. However, a magnetic field may also reduce reagglomeration perpendicular to the magnetic field gradient by reducing the van der Waals forces between nanotubes;[189] an opposite result, agglomeration in the perpendicular direction, was found for electric fields.[190–192] Although logical, results on nanotubes aligned in a very strong magnetic field (as high as 25 T, which is only available at national facilities) showed that in one epoxy resin the modulus went up with alignment, while in the other the modulus decreased. The viscosities of the two resins were roughly identical, and the authors were not able to provide a convincing explanation to the difference.[193] Results were much more encouraging for electric field alignment; in this case, the storage modulus increased roughly by 10% in the direction of the applied electric field.[190]

Fibers made with carbon nanotubes via melt, gel, solution, or electrospinning will generally have extremely high nanotube, as well as polymer, orientation. Melt spinning, unless the nanotubes are dispersed in the polymer first by a solution,[194] does not produce fibers with good mechanical properties almost certainly because of poor dispersion. Electrospinning cannot achieve high orientations and long fibers are difficult to produce. Modulus and tensile strength in PAN–CNT fibers, irrespective of whether MWCNTs or SWCNTs were used, seem to be better for gel-spun rather than solution-spun fibers according to one study.[195] Interestingly, small-angle X-ray scattering results on this material indicate that the superior mechanical properties of the gel-spun fibers are due to a smaller fraction of voids in the material.[196] Not unexpectedly, because of the high orientation, the dE/dV_f values can be quite high; for example, for gel-spun fibers from PAN the value was approximately 600 GPa.[197]

Experiments that can definitively answer the question of whether transcrystallinity improves mechanical strength are very difficult to perform because of the difficulty of suppressing transcrystallinity in a fiber that normally promotes this effect (or vice versa). Studies on single macroscopic fibers have led to contradictory results; that is, transcrystallinity has been shown to increase, decrease, or affect no change even for nominally identical fiber–polymer systems.[198] In one study on carbon nanotube fibers formed from a solution process, the transcrystalline layer and the composite showed a factor of ~2 and ~3 higher strength and 0.4 and 3 times higher modulus, respectively, than the isothermally crystallized bulk material.[199]

One advantage of infusion of already formed buckypaper mats or nanotube fibers is that high volume fractions can be achieved more easily than with other processing methods. In a paper involving epoxy and SWNT buckypaper formed via filtration, SWNT contents between 25 and 40 wt% were formed, with the variation caused by different numbers of stacked buckypapers. Only the storage modulus was reported, with an E' change at room temperature from 2.5 to 15 GPa.[200] A higher concentration, almost 50% by weight SWCNTs, was achieved with a polycarbonate solution, and the modulus was higher than the polycarbonate or buckypaper modulus, with the values of 1.7, 2.3, and 5.1 GPa for the polycarbonate, buckypaper, and composite, respectively.[201] Other studies looked at poly(ether ether ketone),[202] poly (vinyl alcohol), polystyrene, and poly(vinyl pyrrolidone)[203] as well as poly(ethylene oxide), poly(vinyl pyrrolidone), and poly(vinyl alcohol) solutions.[138] The latter showed that drawing the polymer through the buckypaper with a vacuum filtration process produced composites with significantly better mechanical properties than a simple soaking process. Various concentrations of epoxy–buckypaper composites were made using tubes treated with various solutions: nitric acid, potassium permanganate, and piranha solution (a mixture of concentration hydrogen peroxide and concentrated sulfuric acid). Although the dry sheets had different properties, the composite moduli and tensile strengths did not depend on the treatment, as shown in Figure 5.8.[204] Drawing of fibers from nanotube forests can give rise to a mat-like structure with good alignment in the draw direction; in one study, 1750 sheets were

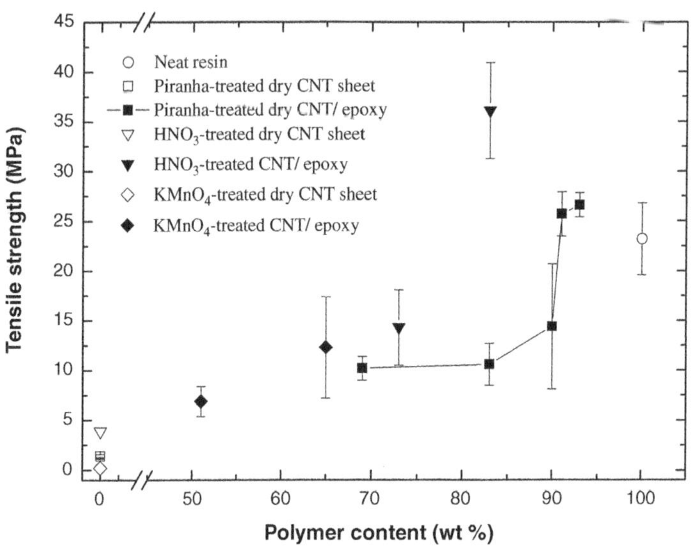

Figure 5.8 Graph showing the effect on tensile properties of buckypaper–epoxy composites prepared using nanotubes pretreated using various methods. Details are found in the text. Copyright IOPscience. Reproduced with permission from Ref. 204.

stacked with alignment directions the same; the modulus change in the stretch direction was linear until the maximum tested value of 8 wt% nanotubes with a dE/dV_f of 150 GPa. Although the dE/dV_f value is not different from what has been discussed earlier, the range of linearity was very high—much higher than any noninfusion process. Infusion processes generally have much higher linearity ranges in terms of modulus enhancement than noninfusion processes. The tensile strength was also linear until 8% tubes with a value that increased from 90 to 130 MPa. Smaller values were found for the case where the stacking was done at 90° to one another.[205]

Production of nanotube fibers followed by infusion of an epoxy has been studied. In one such study,[206] the fibers made with just SWCNTs had a modulus of 22 GPa and a tensile strength of 0.55 GPa, which is on the low end for the tensile properties of a pure nanotube fiber. Using a nanotube specific gravity of 1.35 rather than the measured specific gravity of the fibers (which was about 1), dE/dV_f was measured for a fiber containing more than 20% by weight nanotubes at 72 GPa, and the tensile strength was measured at 253 versus 43 MPa for the neat epoxy. Using a simple rule of mixtures yields a fiber modulus around 70 GPa instead of the measured 22 GPa; the authors attributed this to difficulties in precisely determining the density of the fiber. A similar study using methyl methacrylate monomer and a very weak undensified fiber ($E = 0.11$ GPa and TS = 1 MPa; the densified values were 25 GPa and 500 MPa, respectively[207]) led to a very high dE/dV_f of 440 GPa at 15% nanotube content with a tensile strength of 0.34 versus 0.02 GPa after polymerization.[208] A rule of mixtures applied to this composite yields a nanotube fiber modulus of 333 GPa instead of the measured 25 GPa. Note that in both studies, the effective modulus of the fibers increased with the addition of the fiber to a polymer; whether this is a real effect is not clear at this time. More importantly, infusion methods offer a way to get very high volume fractions of nanotubes while still maintaining reinforcing characteristics. The challenge is to create fibers that have high strength and stiffness initially, and more critically have a process where the variation along the length is small so that results do not depend on gauge length.

For fibers, the goal should be to produce nanotube-containing fibers that outperform commercially available high-strength polymeric fibers (strength ~5 GPa, modulus ~200 GPa). One strategy would be to add nanotubes to high-strength polymeric fibers to improve the strength even further. In one study using poly (*p*-phenylene benzobisoxazole), SWCNTs did significantly improve the modulus and tensile strength (values of 4.2 and 167 GPa were achieved with 10 wt% tubes); however, the pure polymer values were significantly lower than commercial materials indicating that the processing method was not optimized for the neat fibers.[209] In another study, Kevlar yarns were immersed in a nanotube-containing solution and the tensile strength increased from about 4 to 5 GPa with no effect on modulus.[210]

Because of their ability to be stretched and relaxed with little or no change in unrelaxed dimensions, cross-linked rubbers have unique mechanical behavior. One such behavior is the Mullins effect, characterized by a pronounced lowering of stress when filled elastomers are extended a second time after being stressed previously to high extensions. A study on filled styrene–butadiene rubber coupled with AFM experiments to image changes in the nanotube domains found a significant Mullins effect, and this stress softening was attributed to a loss of orientation, both polymer and nanotube, in the second stretch relative to the first caused by rotation of nanotube-rich domains.[211] Another investigation highlighted the high-strain behavior in an elastomer. The focus was on deriving a constitutive model for the enhancement of the large strain properties in the material. The authors found that the increase in strain hardening could be explained by stretching and rotating of the nanotubes.[212]

5.3.2.2 Compressive Properties The number of compression measurements on nanotube composites is orders of magnitude less than those made in tension, even though failure mechanisms are quite different (concrete being an incredibly important example!) because polymers are generally not used in applications where such stress fields are very important. One key difference between tension and compressive properties is that in the latter, nanotube buckling could possibly be an issue. In a very early study, it was found that simply cooling a thermoplastic polymer, which causes significant compressive strains on the nanotube due to the difference in thermal expansion coefficient, can cause nanotube buckling.[213] To the author's knowledge, however, only a few studies have appeared in the literature that conclusively show that nanotube buckling contributes to the response to compressive stress experimentally.[214,215] As pointed out in a paper that was able to directly image buckling under compression using TEM, the resistance to buckling by MWCNTs is going to be much larger than that of SWCNTs because of the large number of walls for the former involved in buckling of one nanotube.[216] Theoretical studies of buckling have been more common.[217–220] Another study found evidence of nanotube reorientation normal to the uniaxial compressive direction at low strains, which would eliminate the possibility of buckling at higher strains if such orientation were perfect. This same study interpreted changes in Raman signal at higher strains as due to debonding not buckling; unfortunately, no micrographs were taken to confirm this hypothesis.[221] The importance of buckling is likely a function of the interfacial energy; the higher the energy, the less contribution buckling will make to failure.

The studies under compression described above are part of only a handful of studies performed under compression. The effect of grafting a polymer onto the tubes and then dispersing the grafted tubes into the same polymer has been studied. Using polystyrene-grafted MWCNTs dispersed in polystyrene, a modest increase in modulus was found with the addition of nanotubes ($dE/dV_f = 40$) with essentially no difference between grafted and ungrafted tubes. However, high strain properties showed significant differences in that the ones made with grafted tubes showed a ~10% larger yield stress and significant strain hardening at higher strains. The author explains this observation, along with accompanying micrographs, as evidence of higher interfacial adhesion in the grafted material.[222] In a study with epoxy resin and MWCNTs, the bulk modulus was smaller for a composite where the nanotubes were not dispersed with the aid of a block copolymer, and was the same as the neat resin at 0.5 wt% NT content when dispersant was used. In both cases, the bulk failure strength increased, with an increase from 550 MPa for the neat polymer to 650 and 750 MPa at 1% NT content for the composites made without dispersant and composites made with dispersant, respectively.[223] In another study, a maximum in compression modulus was found at 0.05 wt% of tubes independent of whether the tubes were functionalized, with values of dE/dV_f of 380 and 620 GPa for unfunctionalized and functionalized tubes, respectively. No significant change in the compressive strength was found with nanotube addition.[122] Buckling stability is done under the same geometry as compressive testing, but in this case the sample is thin enough so that the sample buckles rather than fails. SWCNTs have been shown to increase the buckling stability of polycarbonate.[224]

Compression tests have been performed on samples prepared by infiltration. For nanotube fibers infused with epoxy resin, the modulus increase was approximately the same under tension and compression, with a dE/dV_f of about 65 GPa; however, the compressive strength increase was about half of the tensile strength increase.[206] A composite with very long nanotubes, where the nanotubes were as long as the sample, was made by infusing a vertically grown MWCNT forest with partially cured poly(dimethyl siloxane) followed by curing. The volume fraction of tubes was about 5%, and the increase in compressive modulus was rather small, with a dE/dV_f of about 0.4 GPa; however, the increase was about a factor of 3 larger than the partially filled polymer filled with nanotubes via melt mixing followed by curing. Hysteresis during repeated compression testing of the forest–polymer composite was also quite large, indicative of poor interfacial adhesion between the rubber and the nanotubes.[225]

5.3.2.3 Bending Properties Unless nanotubes are oriented in a plane and the direction of the deformation is perpendicular to the plane, the various relationships described in Section 5.3.2 for tensile tests should be equally valid. In other words, studying bending properties of composites with randomly oriented nanotubes offers no qualitative differences from studying tensile properties. Certainly though, the number of papers that have reported such measurements is extensive[122,226–231] (these references do not provide an exhaustive list!). This section will not describe the many measurements of bending properties except in the case of nanotube mats, simply because the strengths and moduli should differ from tensile strength and moduli in no significant way except numerically. In cases where nanotube mats are used and the stress of the composite is perpendicular to the plane of the mats, both the modulus and the tensile strength qualitative behavior can be different, in particular the latter. The reason is that the interfacial shear strength between the nanotubes and the polymer, and hence debonding, plays a much more important role in this type of stress field with this type of sample than in tension or compression.

In most applications, a single mat is not used to form a composite where the stress is applied perpendicular to the mat. Normally, composites from mats are made by placing a number of mats on top of one another and then impregnating these fibrous mats with low-viscosity resin. The resin is either forced through the mats using pressure or drawn through the mats using vacuum (or both!). In the case of a small number of mats, the bending forces applied on the composite laminate lead to tensile forces on the outer composite layers near the supports and compressive forces on the outer composite layer near the pushing nose. For thin laminates and large loading spans, the tensile and compressive stresses are much higher than the shear stress components. Under these conditions, failure strength due to bending is termed the flexural strength. In the case of a large number of mats, the bending of the laminate leads to much higher shear stresses through the thickness compared to tensile and compressive stresses. In particular, if the loading span is short enough, the laminate fails due to shear stresses and the failure strength is termed the interlaminar shear strength. The failure is almost always due to debonding of the resin from the fiber and shear failure of the resin that is found between mat layers near the laminate midplane. The latter is of interest in this discussion.

Nanotubes can be used to manufacture laminate composites where the nanotubes represent the only filler in the system. In one approach, a nanotube fiber can be drawn and then weaved in order to manufacture a mat. Although a mat has been formed in this manner,[232] to the author's knowledge, results from laminated composites made from a weaved nanotube mat have not been reported in the literature. With nanotube fibers, it is not clear whether an effective mat can be formed because the weaving process could degrade the mechanical properties of the individual fibers. The second approach is to use stacked individual buckysheets to form a laminate. Although buckysheets have been laminated to form a composite as described in Section 5.3.2, to the author's knowledge, bending tests have not been performed as of the end of 2009.

More commonly, nanotubes are used as an additive to already existing glass or carbon fiber mats or fibers. Carbon nanotubes are usually added to improve the interlaminar shear strength of the resulting continuous and aligned fiber-reinforced composite because the tubes are more isotropically oriented than the fibers. Since the amount of material required for many applications of laminated sheets is often determined by interlaminar shear strength, improvements allow for less material with the same design criteria. Another purpose is to improve thermal or electrical conductivity in the thickness direction. The reader should note that the filler volume fraction should be known in order to properly compare properties of laminated composites, and filler volume fraction with and without added tubes is usually not reported.

The simplest approach is to add nanotubes to the infusing resin. Since the tubes are normally infused in the thickness direction, any residual nanotube alignment is in the thickness direction, which is the exact direction desired for improvements of interlaminar shear strengths. There are two problems with infusing nanotube-filled resin, which are entirely separate from the difficulties of achieving good nanotube dispersion in the resin prior to infusion. First, as Section 5.2.1 describes, the addition of well-dispersed nanotubes generally causes a large rise in viscosity, especially at the low shear rates required by the infusion process, and it becomes more difficult to properly wet the carbon or glass fiber mats with the filled resin. Second, nanotubes can be filtered by the mats leading to a spatially nonuniform nanotube concentration distribution[233] that has the high likelihood of reducing the reinforcement efficiency of the nanotubes. All these factors limit this approach to fairly low volume fractions of tubes, and either pressure or vacuum is usually required to draw the resin into the mats. Nevertheless, improvements (10–30%) in the interlaminar shear strength have been found for epoxy/glass fiber mat[234–236] and epoxy/carbon fiber mat[237] with one paper attributing the improvement to alignment of the tubes in the thickness direction.[238] In a study with epoxy/glass systems, functionalized tubes were found to be much more effective at increasing interlaminar shear strength than unfunctionalized tubes.[239] A nanotube-filled epoxy designed for infusion, as well as the prepreg (the term refers to the partially cured resin mat product), is commercially available for this application as of the end of 2009 from Zyvex Performance Materials.

Another approach is to draw the fiber through a resin that contains dispersed nanotubes and adjust conditions so that nanotubes become part of the coated fiber.[240] More resin is then added to the coated fibers, which are then cured. Adding nanotubes to an epoxy resin and then coating a glass fiber with the nanotube–epoxy followed by the addition of more epoxy to make a composite has been shown to improve tensile properties of individual fibers more than coating with the epoxy resin alone.[241] With an epoxy–carbon fiber composite, only functionalized MWCNTs showed no drop in the interlaminar shear strength, while three other types of nanotubes caused the interlaminar shear strength to drop.[242] For an epoxy–carbon fiber material with 5% carbon nanotubes in the epoxy (the amount of nanotubes in the final product was not reported), there was no change in tensile properties while the tensile properties at 90° to the unidirectional oriented fibers increased from 51.2 to 57.9 MPa.[243]

Other approaches are to spray nanotubes onto the fiber with an evaporating liquid; a maximum increase of 45% in interlaminar shear strength was found with a very small amount (0.015 wt%) of MWCNTs in a vinyl ester/glass composite.[244] Electrophoretic deposition of nanotubes on a carbon fiber mat led to a 30% increase in interlaminar shear strength in an epoxy system.[245] Growing nanotube forests on a flat surface and then transferring them onto a carbon fabric (details of how this transfer process was done were not given) caused no increase in the interlaminar shear strength.[246] Finally, using a carbon nanotube yarn as a weave around already existing fibers to improve properties in the secondary direction has also been proposed.[247]

Another strategy is to grow nanotubes directly on the fiber surface. The disadvantage of growing nanotubes is that the high temperature used to grow CNTs typically leads to fiber damage on glass or carbon fibers and a decrease in fiber properties, although one recent study found no effect on the flexural modulus after nanotube growth.[248] Regardless, significant improvements in interfacial shear strength for carbon fiber–epoxy systems of 60% via single fiber pullout tests[249] and 71% for random alignment and 11% for aligned nanotubes were found from single-fiber fragmentation testing.[250] In macroscopic testing, carbon nanotubes were grown on different carbon fiber substrates, namely, unidirectional carbon fiber tows, bidirectional carbon fiber cloth, and three-dimensional carbon fiber felt. These substrates were used as reinforcement in phenolic resin matrix; the flexural strength improved by 20% for tows, 75% for cloth, and 66% for 3D felt compared to that prepared by neat reinforcements.[251] To reduce fiber property degradation, one approach is to try and heal the fibers via graphitization (heating at high temperatures with inert gas); at present, there are very little data on this issue or approach. The use of not-as-common alumina fibers or silicon carbide fibers eliminates degradation concerns because of higher fiber stabilities; interlaminar shear strength increases of 69%[252] or 240%, respectively,[253] have been found.

5.3.3 Fracture Toughness and Crack Propagation

As the name implies, fracture toughness is a toughness measurement; that is, with respect to tensile tests, fracture toughness is related more to toughness than to strength, modulus, strain at break, and so on. However, the use of notches means that crack initiation is not an issue in fracture toughness experiments, and especially in continuous and aligned fiber-reinforced composites, fracture toughness experiments often give information slightly different from that given by the tensile test toughness.

Fracture toughness measurements on randomly oriented composites are much less common than bending measurements; there are tens of papers with such measurements. Some examples include mode I fracture toughness in an epoxy-filled sample showing effectively no difference between nanotube-filled and carbon black-filled samples.[254] Interestingly, the fracture toughness was significantly more negatively affected than the bending strength for an epoxy filled with functionalized and unfunctionalized tubes.[255] In another measurement, a maximum in the resistance to crack propagation for PP–MWCNT composites prepared by melt mixing was observed at 0.5 wt% MWCNT demonstrating enhanced toughness compared to pure PP, followed by a sharp decline as the MWCNT content was increased to 1.5 wt% revealing a ductile-to-semiductile transition. The work of fracture was also quantified in this study. Micrographs indicated a significant increase in agglomeration as the fraction of nanotubes in the sample increased.[159] A similar study was performed on polycarbonate by the same group; in this case, the maximum crack propagation resistance was found at 2 wt% nanotube content.[256] Nanotubes have also been shown to improve fracture toughness in immiscible blends.[257] Finally, qualitative agreement between tensile tests and fracture toughness measurements was found when comparing different types of tubes.[166,258]

Like bending tests, the information gained from fracture toughness tests on composites made with mats is slightly different from that for toughness measurements from tensile tests because the interfacial shear strength generally has a larger influence on the results. No studies have been performed on the fracture toughness of infused buckypapers as of the end of 2009. Studies have examined the result of adding nanotubes to mode I and mode II fracture toughness via infusion of a resin containing nanotubes into a composite containing a carbon or glass fiber mat and significant increases, for example, 60% in modes I and II,[259] 50% and 30% in modes I and II, respectively,[243] 0% and 11% in modes I and II, respectively,[260] and 33% in mode I[261] have been found. A theoretical study of this type of reinforcement with respect to mode I fracture toughness has been published.[262] A unique process, meant to mimic a scarf joint, involved applying nanotubes to an already cured vinyl ester–carbon fiber surface by spraying with a nanotube-containing acetone solution and then curing layers on top with the same vinyl ester–carbon fiber laminate to test the ability of nanotubes to improve fracture toughness. The addition of nanotubes increased the fracture toughness by 10% and 30% for modes I and II, respectively.[263] Adding nanotubes to a resin and then drawing carbon fibers through this resin followed by curing into a composite caused a substantial increase in mode I fracture toughness (between 20% and 75%) where the interlaminar shear strength showed a reduction or no change with nanotube addition.[242] Almost no change was found at room temperature for the fracture toughness, with a 30% increase at −150°C with the addition of nanotubes in a filament winding method.[264] The growth of nanotubes on carbon fiber led to a 50% increase in fracture toughness in an epoxy–carbon fiber composite without a negative effect on flexural modulus.[248]

5.3.4 Impact Energy

Impact energy is normally assessed by making a bar of a sample with a notch and then allowing a pendulum to break the bar and measuring the difference in initial and final height to calculate the energy of fracture (see Figure 5.1). This measurement is essentially a very short timescale toughness test, and impact is critically important for a number of applications. Some papers have found higher relative improvements in impact energy than in tensile strength,[161,265–267] although this behavior is not universal.[91,268,269] An example that represents the largest difference was found with polycaprolactone-modified tubes in polyvinyl chloride; an increase in tensile strength of about 35% was found, while that for the impact energy was over 400% (the relative increase in toughness was about the same as the increase in impact energy).[270] Covalent bonding via nanotube functionalization can improve impact energy just as with other mechanical properties.[266,271–274] For example, the impact energy with functionalized tubes was improved by nearly 100% at 0.5% nanotubes, with essentially no change in impact energy with the use of nonfunctionalized tubes.[275] Longer tubes are also more effective at increasing impact energy.[276] Low-temperature impact energy is typically critically important since the impact energy decreases with temperature. Both at room temperature and at −196°C, the addition of 0.5 wt% MWCNTs caused a 12 kJ/m^2 increase in impact energy, starting from unfilled values of 38 and 23 kJ/m^2, respectively. While 0.5 wt% represented a maximum at both temperatures, at lower and higher loadings the increase was not as significant. Somewhat surprisingly, the tensile strength at room temperature was not improved with nanotube addition, while at cryogenic temperatures the tensile strength rose from 92 to 120 MPa.[277]

5.3.5 Oscillatory Measurements

Oscillatory experiments on solid materials are often done at low strains, that is, the linear viscoelastic region. In this case, the tests are steady-state. Instead of doing in shear as is done for liquids, most oscillatory measurements on solids are done in tension or bending. Many times, the storage modulus instead of the Young's modulus is used to characterize stiffness; although the storage modulus and Young's modulus are not numerically equivalent, trends found in one are usually reflected in the other (but not always for filled materials!). Measurements as a function of temperature at constant frequency are extremely common in nanotube composites, that is, dynamical mechanical analysis. DMA is used to identify the glass transition temperature; methods exist for determining T_g from the storage modulus, the loss modulus, and $\tan \delta$ (the values will not be equivalent). The exact value of T_g depends not only on which measurement is used but also on the oscillation frequency. Although not as large as the number of papers listed in Table 4.1 (DSC requires significantly less sample, and a DSC instrument is significantly cheaper than a DMA instrument), a large number of papers have used DMA to characterize the change in glass transition temperature with nanotube addition. Nearly all, if not all, papers that have compared DSC and DMA glass transition measurements have found relative agreement between the two. A more or less random sampling of papers that used oscillatory testing to characterize the glass transition or the stiffness of a filled system could be described at this point; however, the fundamental understanding provided by these measurements is no different from that described in Section 5.3.2 or 4.2.2, so such a listing will not be performed.

However, a few interesting capabilities of DMA should be pointed out with respect to T_g determination. DMA does have a number of advantages versus DSC in determining the glass transition temperature. DMA is inherently more sensitive to the glass transition than the DSC, somewhere between a factor of 10 and 100 times more sensitive. One interesting use of DMA was to measure the rheological percolation threshold; it was immediately clear that the percolation threshold was temperature dependent.[128] DMA measurements also offer the possibility of measuring a directional rheological percolation threshold. By changing the frequency it is possible to determine whether a sample follows time–temperature superposition, that is, does the introduction of nanotubes cause the appearance of different large-scale relaxation mechanisms; such studies have been performed with other nanofillers.[278] To study time–temperature superposition in melt samples is extremely problematic because of the tendency of nanotubes to reagglomerate in the melt as described earlier. One study that tried to apply time–temperature superposition found an extremely narrow temperature window where time-temperature position applied; the narrowness was attributed to reagglomeration at other conditions.[107]

Non-steady-state oscillatory tests, that is, large strain fatigue tests, have also been performed. Results represented by this type of test have no equivalent to any other previously described test; the primary parameter of interest is resistance to crack growth. In the first of two very detailed studies by Koratkar and coworkers,[279] no significant difference in fatigue crack growth rates between MWCNTs and SWCNTs was found in a filled epoxy, although there was an order of magnitude decrease at low stress amplitudes of the crack growth rate with the introduction of 0.5% tubes versus the neat epoxy. In the second study that only utilized MWCNTs, the fatigue crack growth rates at constant nanotube content were reduced by as much as an order of magnitude via (i) reducing the nanotube diameter, (ii) increasing the nanotube length, and (iii) improving the nanotube dispersion via the coating of the nanotubes with poly(methyl methacrylate).[280] In both papers, changes were found only at low stress amplitudes; there were essentially no change in fatigue crack growth rates at high stress amplitudes for all samples, including those without nanotubes. In a different study of an epoxy reinforced with MWCNTs, the fatigue life increased by a factor of 10 with nanotube addition, while the fracture toughness increased only by about 50%.[281]

One type of large amplitude cyclic test that is very commonly used on rubbers does not have the number of cycles to failure or crack growth rate as the primary parameters of interest. Rubbers used for automobile tires undergo this type of testing routinely to be able to predict heat buildup and hysteresis (permanent changes in the rubber), both of which are highly undesirable for this application. Although the addition of MWCNTs to replace some of the carbon black resulted in improvements in other properties ideal for automobile tires, heat buildup and hysteresis increased dramatically, which the authors attributed to poor adhesion of the rubber to the nanotubes.[282]

5.3.6 Other Mechanical Properties

The hardness of a polymer increases with introduction of nanotubes.[283,284] Nanoindentation experiments use a probe that has nanoscale dimensions. This measurement can be done in both steady-state[283] and oscillating[285] modes, and has shown that nanotubes influence polymer properties in the same way as the macroscopic measurements indicate.

Creep experiments have shown that the introduction of nanotubes reduces creep.[286,287] The instantaneous creep response for a PP/MWCNT composite was observed to be controlled by interphase effects at shorter timescales, while at longer timescales it is mainly determined by the coefficient of thermal expansion.[288] Factors that affect the thermal expansion coefficient for a semicrystalline material are actually quite complicated, and the high thermal conductivity of nanotubes plays a role in changes in this coefficient.

The wear resistance of a polymer is usually improved via the addition of a filler, and nanotube fillers are no different. For example, the wear resistance of a carbon fiber-filled poly(ether ether ketone) was improved with the addition of carbon nanotubes.[289] Other examples where nanotubes improved wear resistance include ultrahigh molecular weight polyethylene,[290–292] high-density polyethylene,[293] polyimide,[294,295] poly(methyl methacrylate),[296] poly(tetraflouroethylene),[297] and epoxy.[298–300] In both ultrahigh molecular weight polyethylene[301] and poly(tetraflouroethlylene),[302] wear resistance was improved via the addition of nanotubes; however, in the former case, no fundamental changes in morphology were found, while in the latter case the morphology of the worn surface changed with the addition of nanotubes. The mechanism of improvement of wear properties may also be a self-lubricating effect; that is, nanotubes dislodged during abrasion are not removed from the surface and hence prevent direct contact reducing wear rate and friction coefficient.[303] Some of the same effects studied elsewhere have also been examined with respect to wear behavior, for example, effect of functionalization,[304–306] and nanotube type.[307] In one study, a significant improvement in wear resistance was found with a dispersion–reaction scheme (with added solvent so that the mixture was precipitated) versus a dispersion–dissolution–precipitation scheme with poly(methyl methacrylate), which the authors attributed to better dispersion in the former, although a plot of electrical conductivity versus fraction of tubes showed no differences.[308]

5.4 CHALLENGES

The author believes that with respect to the main primary reinforcing agent of a polymer, infusion processes appear to be most promising. Even after 10 years of trying to develop methods to improve dispersion in order to increase mechanical properties at high volume fractions, in the author's opinion it is unlikely that effort will be successful. Most of the processes at high volume fractions involve solvent processes; for the much more commercially relevant melt mixing, processing becomes extremely difficult because of the high viscosities. Also, nanotube fibers hold the greatest promise in the author's opinion; the loss in strength that occurs with a loss in orientation in making buckypapers is unacceptable. If more than one-dimensional reinforcement is required, then weaving or laying of mats in different directions is almost certainly a better approach.

In certain selected cases, the mechanical properties of a nanotube-containing fiber, either with or without a polymer, approach that of other high-performance fibers. Superior performance is likely required for nanotube-containing fibers to gain significant market share, although such fibers would not necessarily be as sensitive to hydrolysis as the (*p*-phenylene benzobisoxazole) or Kevlar® fibers. To achieve superior performance requires better starting nanotube fibers. At this point, it is not clear whether the length-dependent properties of nanotube fibers are due to defects in the nanotubes, or non-nanotube impurities, or defects in the fibers caused by processing. However, in either case, longer, purer, defect-free nanotubes will help.

The author has been one of the great proponents of creating single-walled nanotubes with a length that is sufficiently long so that nanotube–nanotube entanglements become relevant, which is likely on the order of centimeters. As any student of polymer science recognizes, if polymers could not be synthesized to molecular weights above the entanglement molecular weight, the polymers would be extremely brittle or sticky liquids, both of which would have limited applications. Although processing could be exceedingly difficult, nanotubes entangled on an individual level would, in the author's opinion, offer a new type of material with as yet unknown properties.

REFERENCES

1. Coleman, J. N., Khan, U., Blau, W. J., Gun'ko, Y. K. (2006). Small but strong: a review of the mechanical properties of carbon nanotube–polymer composites, *Carbon*, *44*, 1624–1652.
2. Benveniste, Y. (1987). A new approach to the application of Mori–Tanaka theory in composite materials, *Mech. Mater.*, *6*, 147–157.
3. Zhao, C. G., Hu, G. J., Justice, R., Schaefer, D. W., Zhang, S. M., Yang, M. S., Han, C. C. (2005). Synthesis and characterization of multi-walled carbon nanotubes reinforced polyamide 6 via *in situ* polymerization, *Polymer*, *46*, 5125–5132.
4. Hadjiev, V. G., Iliev, M. N., Arepalli, S., Nikolaev, P., Files, B. S. (2001). Raman scattering test of single-wall carbon nanotube composites, *Appl. Phys. Lett.*, *78*, 3193–3195.
5. Cooper, C. A., Young, R. J., Halsall, M. (2001). Investigation into the deformation of carbon nanotubes and their composites through the use of Raman spectroscopy, *Compos. Part A*, *32*, 401–411.
6. Wang, S. R., Liang, R., Wang, B., Zhang, C. (2008). Load-transfer in functionalized carbon nanotubes/polymer composites, *Chem. Phys. Lett.*, *457*, 371–375.
7. Davis, V. A., Parra-Vasquez, A. N. G., Green, M. J., Rai, P. K., Behabtu, N., Prieto, V., Booker, R. D., Schmidt, J., Kesselman, E., Zhou, W., Fan, H., Adams, W. W., Hauge, R. H., Fischer, J. E., Cohen, Y., Talmon, Y., Smalley, R. E., Pasquali, M. (2009). True solutions of single-walled carbon nanotubes for assembly into macroscopic materials, *Nat. Nanotechnol.*, *4*, 830–834.
8. Fakhri, N., Tsyboulski, D. A., Cognet, L., Weisman, R. B., Pasquali, M. (2009). Diameter-dependent bending dynamics of single-walled carbon nanotubes in liquids, *Proc. Natl. Acad. Sci. USA*, *106*, 14219–14223.
9. Parra-Vasquez, A. N. G., Stepanek, I., Davis, V. A., Moore, V. C., Haroz, E. H., Shaver, J., Hauge, R. H., Smalley, R. E., Pasquali, M. (2007). Simple length determination of single-walled carbon nanotubes by viscosity measurements in dilute suspensions, *Macromolecules*, *40*, 4043–4047.

10. Rahatekar, S. S., Koziol, K. K. K., Butler, S. A., Elliott, J. A., Shaffer, M. S. P., Mackley, M. R., Windle, A. H. (2006). Optical microstructure and viscosity enhancement for an epoxy resin matrix containing multiwall carbon nanotubes, *J. Rheol.*, *50*, 599–610.
11. Yang, Y., Grulke, E. A., Zhang, Z. G., Wu, G. F. (2005). Rheological behavior of carbon nanotube and graphite nanoparticle dispersions, *J. Nanosci. Nanotechnol.*, *5*, 571–579.
12. Ma, A. W. K., Mackley, M. R., Rahatekar, S. S. (2007). Experimental observation on the flow-induced assembly of carbon nanotube suspensions to form helical bands, *Rheol. Acta*, *46*, 979–987.
13. Pujari, S., Rahatekar, S. S., Gilman, J. W., Koziol, K. K., Windle, A. H., Burghardt, W. R. (2009). Orientation dynamics in multiwalled carbon nanotube dispersions under shear flow, *J. Chem. Phys.*, *130*, 214903.
14. Yang, Y., Grulke, E. A., Zhang, Z. G., Wu, G. F. (2006). Thermal and rheological properties of carbon nanotube-in-oil dispersions, *J. Appl. Phys.*, *99*, 114307.
15. Rahatekar, S. S., Koziol, K. K., Kline, S. R., Hobbie, E. K., Gilman, J. W., Windle, A. H. (2009). Length-dependent mechanics of carbon–nanotube networks, *Adv. Mater.*, *21*, 874–878.
16. Tiwari, M. K., Bazilevsky, A. V., Yarin, A. L., Megaridis, C. M. (2009). Elongational and shear rheology of carbon nanotube suspensions, *Rheol. Acta*, *48*, 597–609.
17. Kayatin, M. J., Davis, V. A. (2009). Viscoelasticity and shear stability of single-walled carbon nanotube/unsaturated polyester resin dispersions, *Macromolecules*, *42*, 6624–6632.
18. Hobbie, E. K., Fry, D. J. (2007). Rheology of concentrated carbon nanotube suspensions, *J. Chem. Phys.*, *126*, 124907.
19. Morse, D. C. (1998). Viscoelasticity of concentrated isotropic solutions of semiflexible polymers. 2. Linear response, *Macromolecules*, *31*, 7044–7067.
20. Morse, D. C. (1998). Viscoelasticity of tightly entangled solutions of semiflexible polymers, *Phys. Rev. E*, *58*, R1237–R1240.
21. Gittes, F., MacKintosh, F. C. (1998). Dynamic shear modulus of a semiflexible polymer network, *Phys. Rev. E*, *58*, R1241–R1244.
22. Hough, L. A., Islam, M. F., Janmey, P. A., Yodh, A. G. (2004). Viscoelasticity of single wall carbon nanotube suspensions, *Phys. Rev. Lett.*, *93*, 168102.
23. Fan, Z. H., Advani, S. G. (2007). Rheology of multiwall carbon nanotube suspensions, *J. Rheol.*, *51*, 585–604.
24. Echeverria, I., Urbina, A. (2006). Viscoelastic properties of multiwalled carbon nanotube solutions, *Eur. Phys. J. B*, *50*, 491–496.
25. Kayser, R. F., Raveche, H. J. (1978). Bifurcation in Onsager model of isotropic–nematic transition, *Phys. Rev. A*, *17*, 2067–2072.
26. Song, W. H., Kinloch, I. A., Windle, A. H. (2003). Nematic liquid crystallinity of multiwall carbon nanotubes, *Science*, *302*, 1363–1363.
27. Song, W. H., Windle, A. H. (2005). Isotropic–nematic phase transition of dispersions of multiwall carbon nanotubes, *Macromolecules*, *38*, 6181–6188.
28. Badaire, S., Zakri, C., Maugey, M., Derre, A., Barisci, J. N., Wallace, G., Poulin, P. (2005). Liquid crystals of DNA-stabilized carbon nanotubes, *Adv. Mater.*, *17*, 1673–1676.
29. Lin-Gibson, S., Pathak, J. A., Grulke, E. A., Wang, H., Hobbie, E. K. (2004). Elastic flow instability in nanotube suspensions, *Phys. Rev. Lett.*, *92*, 048302.
30. Alig, I., Lellinger, D., Engel, M., Skipa, T., Pötschke, P. (2008). Destruction and formation of a conductive carbon nanotube network in polymer melts: in-line experiments, *Polymer*, *49*, 1902–1909.
31. Yang, Z. T., Bahr, J., D'Souza, N. A. (2008). Magnetorheology of single-walled carbon nanotube dispersions in mineral oil, *J. Intell. Mater. Syst. Struct.*, *19*, 1143–1152.
32. Ma, A. W. K., Chinesta, F., Tuladhar, T., Mackley, M. R. (2008). Filament stretching of carbon nanotube suspensions, *Rheol. Acta*, *47*, 447–457.
33. Cotiuga, I., Picchioni, F., Agarwal, U. S., Wouters, D., Loos, J., Lemstra, P. J. (2006). Block-copolymer-assisted solubilization of carbon nanotubes and exfoliation monitoring through viscosity, *Macromol. Rapid Commun.*, *27*, 1073–1078.
34. Battisti, A., Skordos, A. A., Partridge, I. K. (2009). Monitoring dispersion of carbon nanotubes in a thermosetting polyester resin, *Compos. Sci. Technol.*, *69*, 1516–1520.

35. Xu, D. H., Wang, Z. G., Douglas, J. F. (2008). Influence of carbon nanotube aspect ratio on normal stress differences in isotactic polypropylene nanocomposite melts, *Macromolecules, 41*, 815–825.
36. Song, Y. S. (2006). Rheological characterization of carbon nanotubes/poly(ethylene oxide) composites, *Rheol. Acta, 46*, 231–238.
37. Lee, S. H., Cho, E., Jeon, S. H., Youn, J. R. (2007). Rheological and electrical properties of polypropylene composites containing functionalized multi-walled carbon nanotubes and compatibilizers, *Carbon, 45*, 2810–2822.
38. Lee, S. H., Kim, M. W., Kim, S. H., Youn, J. R. (2008). Rheological and electrical properties of polypropylene/MWCNT composites prepared with MWCNT masterbatch chips, *Eur. Polym. J., 44*, 1620–1630.
39. Chatterjee, T., Krishnamoorti, R. (2008). Steady shear response of carbon nanotube networks dispersed in poly(ethylene oxide), *Macromolecules, 41*, 5333–5338.
40. Seyhan, A. T., Gojny, F. H., Tanoglu, M., Schulte, K. (2007). Rheological and dynamic–mechanical behavior of carbon nanotube/vinyl ester–polyester suspensions and their nanocomposites, *Eur. Polym. J., 43*, 2836–2847.
41. King, J. A., Via, M. D., Keith, J. M., Morrison, F. A. (2009). Effects of carbon fillers on rheology of polypropylene-based resins, *J. Compos. Mater., 43*, 3073–3089.
42. Teng, C. C., Ma, C. C. M., Huang, Y. W., Yuen, S. M., Weng, C. C., Chen, G. H., Su, S. F. (2008). Effect of MWCNT content on rheological and dynamic mechanical properties of multiwalled carbon nanotube/polypropylene composites, *Compos. Part A, 39*, 1869–1875.
43. Aalaie, J., Rahmatpour, A., Maghami, S. (2007). Preparation and characterization of linear low density polyethylene/carbon nanotube nanocomposites, *J. Macromol. Sci. B, 46*, 877–889.
44. Rahmatpour, A., Aalaie, J. (2008). Steady shear rheological behavior, mechanical properties, and morphology of the polypropylene/carbon nanotube nanocomposites, *J. Macromol. Sci. B, 47*, 929–941.
45. Vega, J. F., Martinez-Salazar, J., Trujillo, M., Arnal, M. L., Muller, A. J., Bredeau, S., Dubois, P. (2009). Rheology, processing, tensile properties, and crystallization of polyethylene/carbon nanotube nanocomposites, *Macromolecules, 42*, 4719–4727.
46. Zhang, Q. H., Lippits, D. R., Rastogi, S. (2006). Dispersion and rheological aspects of SWNTs in ultrahigh molecular weight polyethylene, *Macromolecules, 39*, 658–666.
47. Bangarusampath, D. S., Ruckdaschel, H., Altstadt, V., Sandler, J. K. W., Garray, D., Shaffer, M. S. P. (2009). Rheology and properties of melt-processed poly(ether ether ketone)/multi-wall carbon nanotube composites, *Polymer, 50*, 5803–5811.
48. Davis, V. A., Ericson, L. M., Parra-Vasquez, A. N. G., Fan, H., Wang, Y. H., Prieto, V., Longoria, J. A., Ramesh, S., Saini, R. K., Kittrell, C., Billups, W. E., Adams, W. W., Hauge, R. H., Smalley, R. E., Pasquali, M. (2004). Phase behavior and rheology of SWNTs in superacids, *Macromolecules, 37*, 154–160.
49. Kharchenko, S. B., Douglas, J. F., Obrzut, J., Grulke, E. A., Migler, K. B. (2004). Flow-induced properties of nanotube-filled polymer materials, *Nat. Mater., 3*, 564–568.
50. Obrzut, J., Douglas, J. F., Kharchenko, S. B., Migler, K. B. (2007). Shear-induced conductor–insulator transition in melt-mixed polypropylene–carbon nanotube dispersions, *Phys. Rev. B, 76*, 195420.
51. Alig, I., Skipa, T., Lellinger, D., Pötschke, P. (2008). Destruction and formation of a carbon nanotube network in polymer melts: rheology and conductivity spectroscopy, *Polymer, 49*, 3524–3532.
52. Alig, I., Lellinger, D., Dudkin, S. M., Pötschke, P. (2007). Conductivity spectroscopy on melt processed polypropylene-multiwalled carbon nanotube composites: recovery after shear and crystallization, *Polymer, 48*, 1020–1029.
53. Skipa, T., Lellinger, D., Saphiannikova, M., Alig, I. (2009). Shear-stimulated formation of multi-wall carbon nanotube networks in polymer melts, *Phys. Status Solidi B, 246*, 2453–2456.
54. Weber, M., Kamal, M. R. (1997). Estimation of the volume resistivity of electrically conductive composites, *Polym. Compos., 18*, 711–725.
55. Kyrylyuk, A.V, van der Schoot, P. (2008). Continuum percolation of carbon nanotubes in polymeric and colloidal media, *PNAS, 105*, 8221–8226.

56. Du, F. M., Scogna, R. C., Zhou, W., Brand, S., Fischer, J. E., Winey, K. I. (2004). Nanotube networks in polymer nanocomposites: rheology and electrical conductivity, *Macromolecules*, *37*, 9048–9055.
57. Pötschke, P., Fornes, T. D., Paul, D. R. (2002). Rheological behavior of multiwalled carbon nanotube/polycarbonate composites, *Polymer*, *43*, 3247–3255.
58. Mitchell, C. A., Bahr, J. L., Arepalli, S., Tour, J. M., Krishnamoorti, R. (2002). Dispersion of functionalized carbon nanotubes in polystyrene, *Macromolecules*, *35*, 8825–8830.
59. Ren, J. X., Silva, A. S., Krishnamoorti, R. (2000). Linear viscoelasticity of disordered polystyrene–polyisoprene block copolymer based layered-silicate nanocomposites, *Macromolecules*, *33*, 3739–3746.
60. Pötschke, P., Abdel-Goad, M., Alig, I., Dudkin, S., Lellinger, D. (2004). Rheological and dielectrical characterization of melt mixed polycarbonate–multiwalled carbon nanotube composites, *Polymer*, *45*, 8863–8870.
61. Winter, H. H., Mours, M. (1997). Neutron Spin Echo Spectroscopy Viscoelasticity Rheology, Vol. *134*, Springer, Berlin, pp. 165–234.
62. Liu, C. Y., Zhang, J., He, J. S., Hu, G. H. (2003). Gelation in carbon nanotube/polymer composites, *Polymer*, *44*, 7529–7532.
63. Woo, D. K., Kim, B. C., Lee, S. J. (2009). Preparation and rheological behavior of polystyrene/multi-walled carbon nanotube composites by latex technology, *Korea-Aust. Rheol. J.*, *21*, 185–191.
64. Kashiwagi, T., Mu, M. F., Winey, K., Cipriano, B., Raghavan, S. R., Pack, S., Rafailovich, M., Yang, Y., Grulke, E., Shields, J., Harris, R., Douglas, J. (2008). Relation between the viscoelastic and flammability properties of polymer nanocomposites, *Polymer*, *49*, 4358–4368.
65. Zhang, Q. H., Fang, F., Zhao, X., Li, Y. Z., Zhu, M. F., Chen, D. J. (2008). Use of dynamic rheological behavior to estimate the dispersion of carbon nanotubes in carbon nanotube/polymer composites, *J. Phys. Chem. B*, *112*, 12606–12611.
66. Park, S. D., Han, D. H., Teng, D., Kwon, Y. (2008). Rheological properties and dispersion of multi-walled carbon nanotube (MWCNT) in polystyrene matrix, *Curr. Appl. Phys.*, *8*, 482–485.
67. Kota, A. K., Cipriano, B. H., Powell, D., Raghavan, S. R., Bruck, H. A. (2007). Quantitative characterization of the formation of an interpenetrating phase composite in polystyrene from the percolation of multiwalled carbon nanotubes, *Nanotechnology*, *18*, 505705.
68. Kota, A. K., Cipriano, B. H., Duesterberg, M. K., Gershon, A. L., Powell, D., Raghavan, S. R., Bruck, H. A. (2007). Electrical and rheological percolation in polystyrene/MWCNT nanocomposites, *Macromolecules*, *40*, 7400–7406.
69. Park, S. D., Han, D. H., Teng, D., Kwon, Y., Choi, G. Y. (2006). Phase transition of multi-walled carbon nanotube/polystyrene composites, *J. Korean Phys. Soc.*, *48*, 476–479.
70. Choi, H. J., Zhang, K., Lim, J. Y. (2007). Multi-walled carbon nanotube/polystyrene composites prepared by *in-situ* bulk sonochemical polymerization, *J. Nanosci. Nanotechnol.*, *7*, 3400–3403.
71. Kim, J., Hong, S. M., Kwak, S., Seo, Y. (2009). Physical properties of nanocomposites prepared by *in situ* polymerization of high-density polyethylene on multiwalled carbon nanotubes, *Phys. Chem. Chem. Phys.*, *11*, 10851–10859.
72. Han, M. S., Lee, Y. K., Lee, H. S., Yun, C. H., Kim, W. N. (2009). Electrical, morphological and rheological properties of carbon nanotube composites with polyethylene and poly(phenylene sulfide) by melt mixing, *Chem. Eng. Sci.*, *64*, 4649–4656.
73. Valentino, O., Sarno, M., Rainone, N. G., Nobile, M. R., Ciambelli, P., Neitzert, H. C., Simon, G. P. (2008). Influence of the polymer structure and nanotube concentration on the conductivity and rheological properties of polyethylene/CNT composites, *Physica E*, *40*, 2440–2445.
74. Li, W. C., Shen, L., Sun, J., Zheng, Q., Zhang, M. Q. (2006). Electrical and dynamic rheological behaviors of multiwalled nanotubes-filled high-density polyethylene composites, *Acta Polym. Sin.* 269–273.
75. Zhang, Q. H., Rastogi, S., Chen, D. J., Lippits, D., Lemstra, P. J. (2006). Low percolation threshold in single-walled carbon nanotube/high density polyethylene composites prepared by melt processing technique, *Carbon*, *44*, 778–785.
76. Richter, S., Saphiannikova, M., Jehnichen, D., Bierdel, M., Heinrich, G. (2009). Experimental and theoretical studies of agglomeration effects in multi-walled carbon nanotube–polycarbonate melts, *Express Polym. Lett.*, *3*, 753–768.

77. Abbasi, S., Carreau, P. J., Derdouri, A., Moan, M. (2009). Rheological properties and percolation in suspensions of multiwalled carbon nanotubes in polycarbonate, *Rheol. Acta, 48*, 943–959.
78. Abdel-Goad, M., Pötschke, P. (2005). Rheological characterization of melt processed polycarbonate–multiwalled carbon nanotube composites, *J. Non-Newtonian Fluid Mech., 128*, 2–6.
79. Lee, S. H., Kim, J. H., Kim, K. W., Youn, J. R., Choi, S. H., Kim, S. Y. (2009). Effects of filler geometry on internal structure and physical properties of polycarbonate composites prepared with various carbon fillers, *Polym. Int., 58*, 354–361.
80. Schartel, B., Braun, U., Knoll, U., Bartholmai, M., Goering, H., Neubert, D., Pötschke, P. (2008). Mechanical, thermal, and fire behavior of bisphenol A polycarbonate/multiwall carbon nanotube nanocomposites, *Polym. Eng. Sci., 48*, 149–158.
81. Leer, C., Carneiro, O. S., Covas, J. A., Maia, J. M., van Hattum, F. W. J., Bernardo, C. A., Biro, L. P., Horvath, Z. E., Kiricsi, I. (2006). Dispersion of carbon nanotubes in polycarbonate and its effect on the composite properties, *Mater. Sci. Forum, 514–516* 1125–1130.
82. Pötschke, P., Bhattacharyya, A. R., Janke, A., Goering, H. (2003). Melt mixing of polycarbonate/multi-wall carbon nanotube composites, *Compos. Interfaces, 10*, 389–404.
83. Rizvi, R., Kim, J. K., Naguib, H. (2009). Synthesis and characterization of novel low density polyethylene–multiwall carbon nanotube porous composites, *Smart Mater. Struct., 18*, 104002.
84. Xiao, K. Q., Zhang, L. C., Zarudi, I. (2007). Mechanical and rheological properties of carbon nanotube-reinforced polyethylene composites, *Compos. Sci. Technol., 67*, 177–182.
85. McNally, T., Pötschke, P., Halley, P., Murphy, M., Martin, D., Bell, S. E. J., Brennan, G. P., Bein, D., Lemoine, P., Quinn, J. P. (2005). Polyethylene multiwalled carbon nanotube composites, *Polymer, 46*, 8222–8232.
86. Zhou, Z., Wang, S. F., Lu, L., Zhang, Y., Zhang, Y. X. (2007). Preparation and rheological characterization of poly(methyl methacrylate)/functionalized multi-walled carbon nanotubes composites, *Compos. Sci. Technol., 67*, 1861–1869.
87. Kashiwagi, T., Du, F. M., Winey, K. I., Groth, K. A., Shields, J. R., Bellayer, S. P., Kim, H., Douglas, J. F. (2005). Flammability properties of polymer nanocomposites with single-walled carbon nanotubes: effects of nanotube dispersion and concentration, *Polymer, 46*, 471–481.
88. Kashiwagi, T., Du, F. M., Douglas, J. F., Winey, K. I., Harris, R. H., Shields, J. R. (2005). Nanoparticle networks reduce the flammability of polymer nanocomposites, *Nat. Mater., 4*, 928–933.
89. Choi, H. J., Zhang, K., Park, S. Y., Lee, B. Y. (2009). Effect of preparation method of multi-walled carbon nanotube/poly(methyl methacrylate) nanocomposite on its characteristics, *J. Nanosci. Nanotechnol., 9*, 6089–6095.
90. Lee, J. I., Yang, S. B., Jung, H. T. (2009). Carbon nanotubes–polypropylene nanocomposites for electrostatic discharge applications, *Macromolecules, 42*, 8328–8334.
91. Prashantha, K., Soulestin, J., Lacrampe, M. F., Krawczak, P., Dupin, G., Claes, M. (2009). Masterbatch-based multi-walled carbon nanotube filled polypropylene nanocomposites: assessment of rheological and mechanical properties, *Compos. Sci. Technol., 69*, 1756–1763.
92. Pujari, S., Ramanathan, T., Kasimatis, K., Masuda, J., Andrews, R., Torkelson, J. M., Brinson, L. C., Burghardt, W. R. (2009). Preparation and characterization of multiwalled carbon nanotube dispersions in polypropylene: melt mixing versus solid-state shear pulverization, *J. Polym. Sci. Polym. Phys., 47*, 1426–1436.
93. Micusik, M., Omastova, M., Krupa, I., Prokes, J., Pissis, P., Logakis, E., Pandis, C., Pötschke, P., Pionteck, J. (2009). A comparative study on the electrical and mechanical behaviour of multi-walled carbon nanotube composites prepared by diluting a masterbatch with various types of polypropylenes, *J. Appl. Polym. Sci., 113*, 2536–2551.
94. Abraham, T. N., Ratna, D., Siengchin, S., Karger-Kocsis, J. (2008). Rheological and thermal properties of poly(ethylene oxide)/multiwall carbon nanotube composites, *J. Appl. Polym. Sci., 110*, 2094–2101.
95. Ratna, D., Abraham, T. N., Siengchin, S., Karger-Kocsis, J. (2009). Novel method for dispersion of multiwall carbon nanotubes in poly(ethylene oxide) matrix using dicarboxylic acid salts, *J. Polym. Sci. Polym. Phys., 47*, 1156–1165.
96. Chatterjee, T., Yurekli, K., Hadjiev, V. G., Krishnamoorti, R. (2005). Single-walled carbon nanotube dispersions in poly(ethylene oxide), *Adv. Funct. Mater., 15*, 1832–1838.

97. Hu, G. J., Zhao, C. G., Zhang, S. M., Yang, M. S., Wang, Z. G. (2006). Low percolation thresholds of electrical conductivity and rheology in poly(ethylene terephthalate) through the networks of multi-walled carbon nanotubes, *Polymer*, *47*, 480–488.
98. Anand, K. A., Agarwal, U. S., Joseph, R. (2007). Carbon nanotubes-reinforced PET nanocomposite by melt-compounding, *J. Appl. Polym. Sci.*, *104*, 3090–3095.
99. Kim, J. Y., Park, H. S., Kim, S. H. (2007). Multiwall-carbon-nanotube-reinforced poly(ethylene terephthalate) nanocomposites by melt compounding, *J. Appl. Polym. Sci.*, *103*, 1450–1457.
100. Shin, D. H., Yoon, K. H., Kwon, O. H., Min, B. G., Hwang, C. I. (2006). Surface resistivity and rheological behaviors of carboxylated multiwall carbon nanotube-filled PET composite film, *J. Appl. Polym. Sci.*, *99*, 900–904.
101. Auad, M. L., Mosiewicki, M. A., Uzunpinar, C., Williams, R. J. J. (2009). Single-wall carbon nanotubes/epoxy elastomers exhibiting high damping capacity in an extended temperature range, *Compos. Sci. Technol.*, *69*, 1088–1092.
102. Kim, J. Y., Kim, S. H. (2006). Influence of multiwall carbon nanotube on physical properties of poly (ethylene 2,6-naphthalate) nanocomposites, *J. Polym. Sci. Polym. Phys.*, *44*, 1062–1071.
103. Mitchell, C. A., Krishnamoorti, R. (2007). Dispersion of single-walled carbon nanotubes in poly (epsilon-caprolactone), *Macromolecules*, *40*, 1538–1545.
104. Wu, D., Wu, L., Sun, Y., Zhang, M. (2007). Rheological properties and crystallization behavior of multi-walled carbon nanotube/poly(epsilon-caprolactone) composites, *J. Polym. Sci. Polym. Phys.*, *45*, 3137–3147.
105. Saeed, K., Park, S. Y. (2007). Preparation and properties of multiwalled carbon nanotube/poly-caprolactone nanocomposites, *J. Appl. Polym. Sci.*, *104*, 1957–1963.
106. Kim, J. Y. (2009). The effect of carbon nanotube on the physical properties of poly(butylene terephthalate) nanocomposite by simple melt blending, *J. Appl. Polym. Sci.*, *112*, 2589–2600.
107. Wu, D. F., Wu, L., Zhang, M. (2007). Rheology of multi-walled carbon nanotube/poly(butylene terephthalate) composites, *J. Polym. Sci. Polym. Phys.*, *45*, 2239–2251.
108. Wu, D. F., Wu, L. F., Zhou, W. D., Yang, T., Zhang, M. (2009). Study on physical properties of multiwalled carbon nanotube/poly(phenylene sulfide) composites, *Polym. Eng. Sci.*, *49*, 1727–1735.
109. Yu, S. Z., Wong, W. M., Hu, X., Juay, Y. K. (2009). The characteristics of carbon nanotube-reinforced poly(phenylene sulfide) nanocomposites, *J. Appl. Polym. Sci.*, *113*, 3477–3483.
110. Park, S. K., Kim, S. H., Hwang, J. T. (2008). Carboxylated multiwall carbon nanotube-reinforced thermotropic liquid crystalline polymer nanocomposites, *J. Appl. Polym. Sci.*, *109*, 388–396.
111. Kim, J. Y., Kim, D. K., Kim, S. H. (2009). Effect of modified carbon nanotube on physical properties of thermotropic liquid crystal polyester nanocomposites, *Eur. Polym. J.*, *45*, 316–324.
112. Kalgaonkar, R. A., Jog, J. P. (2008). Copolyester nanocomposites based on carbon nanotubes: reinforcement effect of carbon nanotubes on viscoelastic and dielectric properties of nanocomposites, *Polym. Int.*, *57*, 114–123.
113. Sun, Y., Bao, H. D., Jia, M. Y., Guo, Z. X., Yu, J. (2009). Effect of multi-walled carbon nanotubes on the properties of polyoxymethylene, *Acta Polym. Sin.* 684–688.
114. Wang, M., Wang, W. Z., Liu, T. X., Zhang, W. D. (2008). Melt rheological properties of nylon 6/ multi-walled carbon nanotube composites, *Compos. Sci. Technol.*, *68*, 2498–2502.
115. Saeed, K., Park, S. Y. (2007). Preparation of multiwalled carbon nanotube/nylon-6 nanocomposites by *in situ* polymerization, *J. Appl. Polym. Sci.*, *106*, 3729–3735.
116. Chen, G. X., Li, Y. J., Shimizu, H. (2007). Ultrahigh-shear processing for the preparation of polymer/ carbon nanotube composites, *Carbon*, *45*, 2334–2340.
117. Hong, S. M., Nam, Y. W., Hwang, S. S., Chae, D. W. (2007). Physical properties of multi-walled carbon nanotube-filled PVDF composites prepared by melt compounding, *Mol. Cryst. Liq. Cryst.*, *464*, 777–785.
118. Kim, G. H., Lee, J. S., Koo, C. M., Hong, S. M. (2009). Preparation and characterization of thermoplastic composite based on poly(vinylidene fluoride) and multiwalled carbon nanotube, *Compos. Interface.*, *16*, 507–518.
119. Huang, S., Wang, M., Liu, T. X., Zhang, W. D., Tjiu, W. C., He, C. B., Lu, X. H. (2009). Morphology, thermal, and rheological behavior of nylon 11/multi-walled carbon nanotube nanocomposites prepared by melt compounding, *Polym. Eng. Sci.*, *49*, 1063–1068.

120. Patil, P. A., Wanjale, S. D., Jog, J. P. (2008). Poly(4-methyl-1-pentene)/MWNT nanocomposites, *E-Polymers*, 079.
121. Isayev, A. I., Kumar, R., Lewis, T. M. (2009). Ultrasound assisted twin screw extrusion of polymer-nanocomposites containing carbon nanotubes, *Polymer*, *50*, 250–260.
122. Shi, X. F., Hudson, J. L., Spicer, P. P., Tour, J. M., Krishnamoorti, R., Mikos, A. G. (2005). Rheological behaviour and mechanical characterization of injectable poly(propylene fumarate)/single-walled carbon nanotube composites for bone tissue engineering, *Nanotechnology*, *16*, S531–S538.
123. Bose, S., Bhattacharyya, A. R., Kulkarni, A. R., Pötschke, P. (2009). Electrical, rheological and morphological studies in co-continuous blends of polyamide 6 and acrylonitrile–butadiene–styrene with multiwall carbon nanotubes prepared by melt blending, *Compos. Sci. Technol.*, *69*, 365–372.
124. Bose, S., Bhattacharyya, A. R., Bondre, A. P., Kulkarni, A. R., Pötschke, P. (2008). Rheology, electrical conductivity, and the phase behavior of cocontinuous PA6/ABS blends with MWNT: correlating the aspect ratio of MWNT with the percolation threshold, *J. Polym. Sci. Polym. Phys.*, *46*, 1619–1631.
125. Bose, S., Bhattacharyya, A. R., Kodgire, P. V., Misra, A., Pötschke, P. (2007). Rheology, morphology, and crystallization behavior of melt-mixed blends of polyamide 6 and acrylonitrile–butadiene–styrene: influence of reactive compatibilizer premixed with multiwall carbon nanotubes, *J. Appl. Polym. Sci.*, *106*, 3394–3408.
126. Hom, S., Bhattacharyya, A. R., Khare, R. A., Kulkarni, A. R., Saroop, M., Biswas, A. (2009). Blends of polypropylene and ethylene octene comonomer with conducting fillers: influence of state of dispersion of conducting fillers on electrical conductivity, *Polym. Eng. Sci.*, *49*, 1502–1510.
127. Khare, R. A., Bhattacharyya, A. R., Kulkarni, A. R., Sarp, M., Biswas, A. (2008). Influence of multiwall carbon nanotubes on morphology and electrical conductivity of PP/ABS blends, *J. Polym. Sci. Polym. Phys.*, *46*, 2286–2295.
128. Ha, M. L. P., Grady, B. P., Lolli, G., Resasco, D. E., Ford, W. T. (2007). Composites of single-walled carbon nanotubes and styrene–isoprene copolymer lattices, *Macromol. Chem. Phys.*, *208*, 446–456.
129. Koziol, K., Vilatela, J., Moisala, A., Motta, M., Cunniff, P., Sennett, M., Windle, A. (2007). High-performance carbon nanotube fiber, *Science*, *318*, 1892–1895.
130. Chae, H. G., Kumar, S. (2006). Rigid-rod polymeric fibers, *J. Appl. Polym. Sci.*, *100*, 791–802.
131. Toray web site, accessed July 18, 2010.
132. Zhu, H. W., Xu, C. L., Wu, D. H., Wei, B. Q., Vajtai, R., Ajayan, P. M. (2002). Direct synthesis of long single-walled carbon nanotube strands, *Science*, *296*, 884–886.
133. Zhang, X. F., Li, Q. W., Tu, Y., Li, Y. A., Coulter, J. Y., Zheng, L. X., Zhao, Y. H., Jia, Q. X., Peterson, D. E., Zhu, Y. T. (2007). Strong carbon-nanotube fibers spun from long carbon-nanotube arrays, *Small*, *3*, 244–248.
134. Ericson, L. M., Fan, H., Peng, H. Q., Davis, V. A., Zhou, W., Sulpizio, J., Wang, Y. H., Booker, R., Vavro, J., Guthy, C., Parra-Vasquez, A. N. G., Kim, M. J., Ramesh, S., Saini, R. K., Kittrell, C., Lavin, G., Schmidt, H., Adams, W. W., Billups, W. E., Pasquali, M., Hwang, W. F., Hauge, R. H., Fischer, J. E., Smalley, R. E. (2004). Macroscopic, neat, single-walled carbon nanotube fibers, *Science*, *305*, 1447–1450.
135. Ma, W. J., Liu, L. Q., Yang, R., Zhang, T. H., Zhang, Z., Song, L., Ren, Y., Shen, J., Niu, Z. Q., Zhou, W. Y., Xie, S. S. (2009). Monitoring a micromechanical process in macroscale carbon nanotube films and fibers, *Adv. Mater.*, *21*, 603–608.
136. Nasibulin, A. G., Ollikainen, A., Anisimov, A. S., Brown, D. P., Pikhitsa, P. V., Holopainen, S., Penttila, J. S., Helisto, P., Ruokolainen, J., Choi, M., Kauppinen, E. I. (2008). Integration of single-walled carbon nanotubes into polymer films by thermo-compression, *Chem. Eng. J.*, *136*, 409–413.
137. Hall, L. J., Coluci, V. R., Galvao, D. S., Kozlov, M. E., Zhang, M., Dantas, S. O., Baughman, R. H. (2008). Sign change of Poisson's ratio for carbon nanotube sheets, *Science*, *320*, 504–507.
138. Xu, G. H., Zhang, Q., Zhou, W. P., Huang, J. Q., Wei, F. (2008). The feasibility of producing MWCNT paper and strong MWCNT film from VACNT array, *Appl. Phys. A*, *92*, 531–539.
139. Wei, J. Q., Zhu, H. W., Li, Y. H., Chen, B., Jia, Y., Wang, K. L., Wang, Z. C., Liu, W. J., Luo, J. B., Zheng, M. X., Wu, D. H., Zhu, Y. Q., Wei, B. Q. (2006). Ultrathin single-layered membranes from double-walled carbon nanotubes, *Adv. Mater.*, *18*, 1695–1700.

140. Cheng, Q. F., Bao, J. W., Park, J., Liang, Z. Y., Zhang, C., Wang, B. (2009). High mechanical performance composite conductor: multi-walled carbon nanotube sheet/bismaleimide nanocomposites, *Adv. Funct. Mater.*, *19*, 3219–3225.
141. Blighe, F. M., Lyons, P. E., De, S., Blau, W. J., Coleman, J. N. (2008). On the factors controlling the mechanical properties of nanotube films, *Carbon*, *46*, 41–47.
142. Barber, A. H., Cohen, S. R., Wagner, H. D. (2004). Static and dynamic wetting measurements of single carbon nanotubes, *Phys. Rev. Lett.*, *92*, 186103.
143. Barber, A. H., Cohen, S. R., Wagner, H. D. (2005). External and internal wetting of carbon nanotubes with organic liquids, *Phys. Rev. B*, *71*, 115443.
144. Nuriel, S., Liu, L., Barber, A. H., Wagner, H. D. (2005). Direct measurement of multiwall nanotube surface tension, *Chem. Phys. Lett.*, *404*, 263–266.
145. Neimark, A. V. (1999). Thermodynamic equilibrium and stability of liquid films and droplets on fibers, *J. Adhes. Sci. Technol.*, *13*, 1137–1154.
146. Cooper, C. A., Cohen, S. R., Barber, A. H., Wagner, H. D. (2002). Detachment of nanotubes from a polymer matrix, *Appl. Phys. Lett.*, *81*, 3873–3875.
147. Barber, A. H., Cohen, S. R., Kenig, S., Wagner, H. D. (2004). Interfacial fracture energy measurements for multi-walled carbon nanotubes pulled from a polymer matrix, *Compos. Sci. Technol.*, *64*, 2283–2289.
148. Barber, A. H., Cohen, S. R., Eitan, A., Schadler, L. S., Wagner, H. D. (2006). Fracture transitions at a carbon-nanotube/polymer interface, *Adv. Mater.*, *18*, 83–87.
149. Fu, S. Y., Chen, Z. K., Hong, S., Han, C. C. (2009). The reduction of carbon nanotube (CNT) length during the manufacture of CNT/polymer composites and a method to simultaneously determine the resulting CNT and interfacial strengths, *Carbon*, *47*, 3192–3200.
150. Strus, M. C., Zalamea, L., Raman, A., Pipes, R. B., Nguyen, C. V., Stach, E. A. (2008). Peeling force spectroscopy: exposing the adhesive nanomechanics of one-dimensional nanostructures, *Nano Lett.*, *8*, 544–550.
151. Strus, M. C., Cano, C. I., Pipes, R. B., Nguyen, C. V., Raman, A. (2009). Interfacial energy between carbon nanotubes and polymers measured from nanoscale peel tests in the atomic force microscope, *Compos. Sci. Technol.*, *69*, 1580–1586.
152. Coleman, J. N., Cadek, M., Ryan, K. P., Fonseca, A., Nagy, J. B., Blau, W. J., Ferreira, M. S. (2006). Reinforcement of polymers with carbon nanotubes. The role of an ordered polymer interfacial region. Experiment and modeling, *Polymer*, *47*, 8556–8561.
153. Schaefer, D. W., Justice, R. S. (2007). How nano are nanocomposites? *Macromolecules*, *40*, 8501–8517.
154. Wichmann, M. H. G., Schulte, K., Wagner, H. D. (2008). On nanocomposite toughness, *Compos. Sci. Technol.*, *68*, 329–331.
155. Tai, N. H., Yeh, M. K., Peng, T. H. (2008). Experimental study and theoretical analysis on the mechanical properties of SWNTs/phenolic composites, *Compos. Part B*, *39*, 926–932.
156. Chen, G. X., Kim, H. S., Park, B. H., Yoon, J. S. (2006). Multi-walled carbon nanotubes reinforced nylon 6 composites, *Polymer*, *47*, 4760–4767.
157. Wong, M., Paramsothy, M., Xu, X. J., Ren, Y., Li, S., Liao, K. (2003). Physical interactions at carbon nanotube–polymer interface, *Polymer*, *44*, 7757–7764.
158. Li, Z. Y., Xu, S. Z., Liu, W. T., He, S. Q., Zhu, C. S. (2009). Preparation and characterization of nylon610/functionalized multiwalled carbon nanotubes composites, *J. Appl. Polym. Sci.*, *113*, 2805–2812.
159. Ganss, M., Satapathy, B. K., Thunga, M., Weidisch, R., Pötschke, P., Jehnichen, D. (2008). Structural interpretations of deformation and fracture behavior of polypropylene/multi-walled carbon nanotube composites, *Acta Mater.*, *56*, 2247–2261.
160. Zhou, Z., Wang, S. F., Lu, L., Zhang, Y. X., Zhang, Y. (2008). Functionalization of multi-wall carbon nanotubes with silane and its reinforcement on polypropylene composites, *Compos. Sci. Technol.*, *68*, 1727–1733.
161. Hemmati, M., Rahimi, G. H., Kaganj, A. B., Sepehri, S., Rashidi, A. M. (2008). Rheological and mechanical characterization of multi-walled carbon nanotubes/polypropylene nanocomposites, *J. Macromol. Sci. B*, *47*, 1176–1187.

162. Zhao, C. X., Zhang, W. D., Sun, D. C. (2009). Preparation and mechanical properties of waterborne polyurethane/carbon nanotube composites, *Polym. Compos.*, *30*, 649–654.
163. Xia, H. S., Qiu, G. H., Wang, Q. (2006). Polymer/carbon nanotube composite emulsion prepared through ultrasonically assisted *in situ* emulsion polymerization, *J. Appl. Polym. Sci.*, *100*, 3123–3130.
164. Blond, D., Barron, V., Ruether, M., Ryan, K. P., Nicolosi, V., Blau, W. J., Coleman, J. N. (2006). Enhancement of modulus, strength, and toughness in poly(methyl methacrylate)-based composites by the incorporation of poly(methyl methacrylate)-functionalized nanotubes, *Adv. Funct. Mater.*, *16*, 1608–1614.
165. Cadek, M., Coleman, J. N., Ryan, K. P., Nicolosi, V., Bister, G., Fonseca, A., Nagy, J. B., Szostak, K., Beguin, F., Blau, W. J. (2004). Reinforcement of polymers with carbon nanotubes: the role of nanotube surface area, *Nano Lett.*, *4*, 353–356.
166. Gojny, F. H., Wichmann, M. H. G., Fiedler, B., Schulte, K. (2005). Influence of different carbon nanotubes on the mechanical properties of epoxy matrix composites: a comparative study, *Compos. Sci. Technol.*, *65*, 2300–2313.
167. Xia, H. S., Song, M. (2005). Preparation and characterization of polyurethane–carbon nanotube composites, *Soft Matter*, *1*, 386–394.
168. Gorga, R. E., Cohen, R. E. (2004). Toughness enhancements in poly(methyl methacrylate) by addition of oriented multiwall carbon nanotubes, *J. Polym. Sci. Polym. Phys.*, *42*, 2690–2702.
169. Wong, K. K. H., Zinke-Allmang, M., Hutter, J. L., Hrapovic, S., Luong, J. H. T., Wan, W. (2009). The effect of carbon nanotube aspect ratio and loading on the elastic modulus of electrospun poly (vinyl alcohol)–carbon nanotube hybrid fibers, *Carbon*, *47*, 2571–2578.
170. Bai, J. B., Allaoui, A. (2003). Effect of the length and the aggregate size of MWNTs on the improvement efficiency of the mechanical and electrical properties of nanocomposites: experimental investigation, *Compos. Part A*, *34*, 689–694.
171. Sun, L., Warren, G. L., O'Reilly, J. Y., Everett, W. N., Lee, S. M., Davis, D., Lagoudas, D., Sue, H. J. (2008). Mechanical properties of surface-functionalized SWCNT/epoxy composites, *Carbon*, *46*, 320–328.
172. Che, J. F., Yuan, W., Jiang, G. H., Dai, J., Lim, S. Y., Chan-Park, M. B. (2009). Epoxy composite fibers reinforced with aligned single-walled carbon nanotubes functionalized with generation 0-2 dendritic poly(amidoamine), *Chem. Mater.*, *21*, 1471–1479.
173. Zou, W., Du, Z. J., Liu, Y. X., Yang, X., Li, H. Q., Zhang, C. (2008). Functionalization of MWNTs using polyacryloyl chloride and the properties of CNT–epoxy matrix nanocomposites, *Compos. Sci. Technol.*, *68*, 3259–3264.
174. McIntosh, D., Khabashesku, V. N., Barrera, E. V. (2006). Nanocomposite fiber systems processed from fluorinated single-walled carbon nanotubes and a polypropylene matrix, *Chem. Mater.*, *18*, 4561–4569.
175. Yang, J. W., Hu, J. H., Wang, C. C., Qin, Y. J., Guo, Z. X. (2004). Fabrication and characterization of soluble multi-walled carbon nanotubes reinforced P(MMA-*co*-EMA) composites, *Macromol. Mater. Eng.*, *289*, 828–832.
176. Velasco-Santos, C., Martinez-Hernandez, A. L., Fisher, F. T., Ruoff, R., Castano, V. M. (2003). Improvement of thermal and mechanical properties of carbon nanotube composites through chemical functionalization, *Chem. Mater.*, *15*, 4470–4475.
177. Sengupta, R., Ganguly, A., Sabharwal, S., Chaki, T. K., Bhowmick, A. K. (2007). MWCNT reinforced polyamide-6,6 films: preparation, characterization and properties, *J. Mater. Sci.*, *42*, 923–934.
178. Jeong, J. Y., Lee, H. J., Kang, S. W., Tan, L. S., Baek, J. B. (2008). Nylon 610/functionalized multiwalled carbon nanotube composite prepared from *in-situ* interfacial polymerization, *J. Polym. Sci. A*, *46*, 6041–6050.
179. Bhattacharyya, A. R., Pötschke, P., Abdel-Goad, M., Fischer, D. (2004). Effect of encapsulated SWNT on the mechanical properties of melt mixed PA12/SWNT composites, *Chem. Phys. Lett.*, *392*, 28–33.
180. Yang, B. X., Shi, J. H., Li, X., Pramoda, K. P., Goh, S. H. (2009). Mechanical reinforcement of poly(1-butene) using polypropylene-grafted multiwalled carbon nanotubes, *J. Appl. Polym. Sci.*, *113*, 1165–1172.

181. Yang, B. X., Shi, J. H., Pramoda, K. P., Goh, S. H. (2008). Enhancement of the mechanical properties of polypropylene using polypropylene-grafted multiwalled carbon nanotubes, *Compos. Sci. Technol.*, *68*, 2490–2497.
182. Coleman, J. N., Cadek, M., Blake, R., Nicolosi, V., Ryan, K. P., Belton, C., Fonseca, A., Nagy, J. B., Gun'ko, Y. K., Blau, W. J. (2004). High-performance nanotube-reinforced plastics: understanding the mechanism of strength increase, *Adv. Funct. Mater.*, *14*, 791–798.
183. Causin, V., Yang, B. X., Marega, C., Goh, S. H., Marigo, A. (2008). Structure–property relationship in polyethylene reinforced by polyethylene-grafted multi-walled carbon nanotubes, *J. Nanosci. Nanotechnol.*, *8*, 1790–1796.
184. Peeterbroeck, S., Laoutid, F., Taulemesse, J. M., Monteverde, T., Lopez-Cuesta, J. M., Nagy, J. B., Alexandre, M., Dubois, P. (2007). Mechanical properties and flame-retardant behavior of ethylene vinyl acetate/high-density polyethylene coated carbon nanotube nanocomposites, *Adv. Funct. Mater.*, *17*, 2787–2791.
185. Wang, T., Dalton, A. B., Keddie, J. L. (2008). Importance of molecular friction in a soft polymer–nanotube nanocomposite, *Macromolecules*, *41*, 7656–7661.
186. Wang, Q., Dai, J. F., Li, W. X., Wei, Z. Q., Jiang, J. L. (2008). The effects of CNT alignment on electrical conductivity and mechanical properties of SWNT/epoxy nanocomposites, *Compos. Sci. Technol.*, *68*, 1644–1648.
187. Haggenmueller, R., Zhou, W., Fischer, J. E., Winey, K. I. (2003). Production and characterization of polymer nanocomposites with highly aligned single-walled carbon nanotubes, *J. Nanosci. Nanotechnol.*, *3*, 105–110.
188. Yang, J. H., Wang, C. Y., Wang, K., Zhang, Q., Chen, F., Du, R. N., Fu, Q. (2009). Direct formation of nanohybrid shish–kebab in the injection molded bar of polyethylene/multiwalled carbon nanotubes composite, *Macromolecules*, *42*, 7016–7023.
189. Steinert, B. W., Dean, D. R. (2009). Magnetic field alignment and electrical properties of solution cast PET–carbon nanotube composite films, *Polymer*, *50*, 898–904.
190. Zhu, Y. F., Ma, C., Zhang, W., Zhang, R. P., Koratkar, N., Liang, J. (2009). Alignment of multiwalled carbon nanotubes in bulk epoxy composites via electric field, *J. Appl. Phys.*, *105*, 054319.
191. Wang, M. W., Hsu, T. C., Weng, C. H. (2008). Alignment of MWCNTs in polymer composites by dielectrophoresis, *Eur. Phys. J.* *42*, 241–246.
192. Park, C., Wilkinson, J., Banda, S., Ounaies, Z., Wise, K. E., Sauti, G., Lillehei, P. T., Harrison, J. S. (2006). Aligned single-wall carbon nanotube polymer composites using an electric field, *J. Polym. Sci. Polym. Phys.*, *44*, 1751–1762.
193. Camponeschi, E., Vance, R., Al-Haik, M., Garmestani, H., Tannenbaum, R. (2007). Properties of carbon nanotube–polymer composites aligned in a magnetic field, *Carbon*, *45*, 2037–2046.
194. Lee, G. W., Jagannathan, S., Chae, H. G., Minus, M. L., Kumar, S. (2008). Carbon nanotube dispersion and exfoliation in polypropylene and structure and properties of the resulting composites, *Polymer*, *49*, 1831–1840.
195. Wang, W. J., Murthy, N. S., Chae, H. G., Kumar, S. (2008). Structural changes during deformation in carbon nanotube-reinforced polyacrylonitrile fibers, *Polymer*, *49*, 2133–2145.
196. Wang, W. J., Murthy, N. S., Chae, H. G., Kumar, S. (2009). Small-angle X-ray scattering investigation of carbon nanotube-reinforced polyacrylonitrile fibers during deformation, *J. Polym. Sci. Polym. Phys.*, *47*, 2394–2409.
197. Chae, H. G., Minus, M. L., Rasheed, A., Kumar, S. (2007). Stabilization and carbonization of gel spun polyacrylonitrile/single wall carbon nanotube composite fibers, *Polymer*, *48*, 3781–3789.
198. Quan, H., Li, Z. M., Yang, M. B., Huang, R. (2005). On transcrystallinity in semi-crystalline polymer composites, *Compos. Sci. Technol.*, *65*, 999–1021.
199. Zhang, S., Minus, M. L., Zhu, L. B., Wong, C. P., Kumar, S. (2008). Polymer transcrystallinity induced by carbon nanotubes, *Polymer*, *49*, 1356–1364.
200. Wang, Z., Liang, Z. Y., Wang, B., Zhang, C., Kramer, L. (2004). Processing and property investigation of single-walled carbon nanotube (SWNT) buckypaper/epoxy resin matrix nanocomposites, *Compos. Part A*, *35*, 1225–1232.
201. Pham, G. T., Park, Y. B., Wang, S. R., Liang, Z. Y., Wang, B., Zhang, C., Funchess, P., Kramer, L. (2008). Mechanical and electrical properties of polycarbonate nanotube buckypaper composite sheets, *Nanotechnology*, *19*, 325705.

202. Song, L., Zhang, H., Zhang, Z., Xie, S. S. (2007). Processing and performance improvements of SWNT paper reinforced PEEK nanocomposites, *Compos. Part A*, *38*, 388–392.
203. Coleman, J. N., Blau, W. J., Dalton, A. B., Munoz, E., Collins, S., Kim, B. G., Razal, J., Selvidge, M., Vieiro, G., Baughman, R. H. (2003). Improving the mechanical properties of single-walled carbon nanotube sheets by intercalation of polymeric adhesives, *Appl. Phys. Lett.*, *82*, 1682–1684.
204. Spitalsky, Z., Tsoukleri, G., Tasis, D., Krontiras, C., Georga, S. N., Galiotis, C. (2009). High volume fraction carbon nanotube–epoxy composites, *Nanotechnology*, *20*, 405702.
205. Cheng, Q. F., Wang, J. P., Jiang, K. L., Li, Q. Q., Fan, S. S. (2008). Fabrication and properties of aligned multiwalled carbon nanotube-reinforced epoxy composites, *J. Mater. Res.*, *23*, 2975–2983.
206. Mora, R. J., Vilatela, J. J., Windle, A. H. (2009). Properties of composites of carbon nanotube fibres, *Compos. Sci. Technol.*, *69*, 1558–1563.
207. Zhang, S., Zhu, L., Minus, M. L., Chae, H. G., Jagannathan, S., Wong, C. P., Kowalik, J., Roberson, L. B., Kumar, S. (2008). Solid-state spun fibers and yarns from 1-mm long carbon nanotube forests synthesized by water-assisted chemical vapor deposition, *J. Mater. Sci.*, *43*, 4356–4362.
208. Zhang, S. J., Zhu, L. B., Wong, C. P., Kumar, S. (2009). Polymer-infiltrated aligned carbon nanotube fibers by *in situ* polymerization, *Macromol. Rapid Commun.*, *30*, 1936–1939.
209. Kumar, S., Dang, T. D., Arnold, F. E., Bhattacharyya, A. R., Min, B. G., Zhang, X. F., Vaia, R. A., Park, C., Adams, W. W., Hauge, R. H., Smalley, R. E., Ramesh, S., Willis, P. A. (2002). Synthesis, structure, and properties of PBO/SWNT composites, *Macromolecules*, *35*, 9039–9043.
210. O'Connor, I., Hayden, H., Coleman, J. N., Gun'ko, Y. K. (2009). High-strength, high-toughness composite fibers by swelling Kevlar in nanotube suspensions, *Small*, *5*, 466–469.
211. Bokobza, L. (2007). Multiwall carbon nanotube elastomeric composites: a review, *Polymer*, *48*, 4907–4920.
212. Cantournet, S., Boyce, M. C., Tsou, A. H. (2007). Micromechanics and macromechanics of carbon nano tube-enhanced elastomers, *J. Mech. Phys. Solid.*, *55*, 1321–1339.
213. Bower, C., Rosen, R., Jin, L., Han, J., Zhou, O. (1999). Deformation of carbon nanotubes in nanotube–polymer composites, *Appl. Phys. Lett.*, *74*, 3317–3319.
214. Hadjiev, V. G., Lagoudas, D. C., Oh, E. S., Thakre, P., Davis, D., Files, B. S., Yowell, L., Arepalli, S., Bahr, J. L., Tour, J. M. (2006). Buckling instabilities of octadecylamine functionalized carbon nanotubes embedded in epoxy, *Compos. Sci. Technol.*, *66*, 128–136.
215. Lourie, O., Cox, D. M., Wagner, H. D. (1998). Buckling and collapse of embedded carbon nanotubes, *Phys. Rev. Lett.*, *81*, 1638–1641.
216. Thostenson, E. T., Chou, T. W. (2004). Nanotube buckling in aligned multi-wall carbon nanotube composites, *Carbon*, *42*, 3015–3018.
217. Yang, H. K., Wang, X. (2007). Torsional buckling of multi-wall carbon nanotubes embedded in an elastic medium, *Compos. Struct.*, *77*, 182–192.
218. Wang, X. Y., Wang, X., Xia, X. H. (2007). Eccentric compression stability of multi-walled carbon nanotubes embedded in an elastic matrix, *Compos. Sci. Technol.*, *67*, 1406–1414.
219. Li, C. Y., Chou, T. W. (2006). Multiscale modeling of compressive behavior of carbon nanotube/polymer composites, *Compos. Sci. Technol.*, *66*, 2409–2414.
220. Namilae, S., Chandra, N. (2006). Role of atomic scale interfaces in the compressive behavior of carbon nanotubes in composites, *Compos. Sci. Technol.*, *66*, 2030–2038.
221. Lance, M. J., Hsueh, C. H., Ivanov, I. N., Geohegan, D. B. (2005). Reorientation of carbon nanotubes in polymer matrix composites using compressive loading, *J. Mater. Res.*, *20*, 1026–1032.
222. Fragneaud, B., Masenelli-Varlot, K., Gonzalez-Montiel, A., Terrones, M., Cavaille, J. Y. (2008). Mechanical behavior of polystyrene grafted carbon nanotubes/polystyrene nanocomposites, *Compos. Sci. Technol.*, *68*, 3265–3271.
223. Cho, J., Daniel, I. M. (2008). Reinforcement of carbon/epoxy composites with multi-wall carbon nanotubes and dispersion enhancing block copolymers, *Scr. Mater.*, *58*, 533–536.
224. Zhang, W., Suhr, J., Koratkar, N. A. (2006). Observation of high buckling stability in carbon nanotube polymer composites, *Adv. Mater.*, *18*, 452–456.
225. Ci, L., Suhr, J., Pushparaj, V., Zhang, X., Ajayan, P. M. (2008). Continuous carbon nanotube reinforced composites, *Nano Lett.*, *8*, 2762–2766.

226. Lau, K. T., Shi, S. Q., Cheng, H. M. (2003). Micro-mechanical properties and morphological observation on fracture surfaces of carbon nanotube composites pre-treated at different temperatures, *Compos. Sci. Technol.*, *63*, 1161–1164.
227. Lau, K. T., Lu, M., Lam, C. K., Cheung, H. Y., Sheng, F. L., Li, H. L. (2005). Thermal and mechanical properties of single-walled carbon nanotube bundle-reinforced epoxy nanocomposites: the role of solvent for nanotube dispersion, *Compos. Sci. Technol.*, *65*, 719–725.
228. Moniruzzaman, M., Du, F. M., Romero, N., Winey, K. I. (2006). Increased flexural modulus and strength in SWNT/epoxy composites by a new fabrication method, *Polymer*, *47*, 293–298.
229. Lau, K. T., Shi, S. Q., Zhou, L. M., Cheng, H. M. (2003). Micro-hardness and flexural properties of randomly-oriented carbon nanotube composites, *J. Compos. Mater.*, *37*, 365–376.
230. Zhou, Y. X., Pervin, F., Lewis, L., Jeelani, S. (2007). Experimental study on the thermal and mechanical properties of multi-walled carbon nanotube-reinforced epoxy, *Mater. Sci. Eng. A*, *452*, 657–664.
231. Geng, Y., Liu, M. Y., Li, J., Shi, X. M., Kim, J. K. (2008). Effects of surfactant treatment on mechanical and electrical properties of CNT/epoxy nanocomposites, *Compos. Part A*, *39*, 1876–1883.
232. Bogdanovich, A., Bradford, P., Mungalov, D., Fang, S. L., Zhang, M., Baughman, R. H., Hudson, S. (2007). Fabrication and mechanical characterization of carbon nanotube yarns, 3-D braids, and their composites, *SAMPE J.*, *43*, 6–19.
233. Fan, Z. H., Advani, S. G. (2008). Capillary effect of multi-walled carbon nanotubes suspension in composite processing, *J. Nanosci. Nanotechnol.*, *8*, 1669–1678.
234. Kim, M. S., Lee, S. E., Lee, W. J., Kim, C. G. (2009). Mechanical properties of MWNT-loaded plain-weave glass/epoxy composites, *Adv. Compos. Mater.*, *18*, 209–219.
235. Wichmann, M. H. G., Sumfleth, J., Gojny, F. H., Quaresimin, M., Fiedler, B., Schulte, K. (2006). Glass-fibre-reinforced composites with enhanced mechanical and electrical properties: benefits and limitations of a nanoparticle modified matrix, *Eng. Fract. Mech.*, *73*, 2346–2359.
236. Gojny, F. H., Wichmann, M. H. G., Fiedler, B., Bauhofer, W., Schulte, K. (2005). Influence of nano-modification on the mechanical and electrical properties of conventional fibre-reinforced composites, *Compos. Part A*, *36*, 1525–1535.
237. Bekyarova, E., Thostenson, E. T., Yu, A., Itkis, M. E., Fakhrutdinov, D., Chou, T. W., Haddon, R. C. (2007). Functionalized single-walled carbon nanotubes for carbon fiber–epoxy composites, *J. Phys. Chem. C*, *111*, 17865–17871.
238. Fan, Z. H., Santare, M. H., Advani, S. G. (2008). Interlaminar shear strength of glass fiber reinforced epoxy composites enhanced with multi-walled carbon nanotubes, *Compos. Part A*, *39*, 540–554.
239. Qiu, J. J., Zhang, C., Wang, B., Liang, R. (2007). Carbon nanotube integrated multifunctional multiscale composites, *Nanotechnology*, *18*, 275708.
240. Chen, W., Shen, H. B., Auad, M. L., Huang, C. Z., Nutt, S. (2009). Basalt fiber–epoxy laminates with functionalized multi-walled carbon nanotubes, *Compos. Part A*, *40*, 1082–1089.
241. Siddiqui, N. A., Sham, M. L., Tang, B. Z., Munir, A., Kim, J. K. (2009). Tensile strength of glass fibres with carbon nanotube–epoxy nanocomposite coating, *Compos. Part A*, *40*, 1606–1614.
242. Godara, A., Mezzo, L., Luizi, F., Warrier, A., Lomov, S. V., van Vuure, A. W., Gorbatikh, L., Moldenaers, P., Verpoest, I. (2009). Influence of carbon nanotube reinforcement on the processing and the mechanical behaviour of carbon fiber/epoxy composites, *Carbon*, *47*, 2914–2923.
243. Yokozeki, T., Iwahori, Y., Ishiwata, S., Enomoto, K. (2007). Mechanical properties of CFRP laminates manufactured from unidirectional prepregs using CSCNT-dispersed epoxy, *Compos. Part A*, *38*, 2121–2130.
244. Zhu, J., Imam, A., Crane, R., Lozano, K., Khabashesku, V. N., Barrera, E. V. (2007). Processing a glass fiber reinforced vinyl ester composite with nanotube enhancement of interlaminar shear strength, *Compos. Sci. Technol.*, *67*, 1509–1517.
245. Bekyarova, E., Thostenson, E. T., Yu, A., Kim, H., Gao, J., Tang, J., Hahn, H. T., Chou, T. W., Itkis, M. E., Haddon, R. C. (2007). Multiscale carbon nanotube–carbon fiber reinforcement for advanced epoxy composites, *Langmuir*, *23*, 3970–3974.
246. Abot, J. L., Song, Y., Schulz, M. J., Shanov, V. N. (2008). Novel carbon nanotube array-reinforced laminated composite materials with higher interlaminar elastic properties, *Compos. Sci. Technol.*, *68*, 2755–2760.

247. Bradford, P. D., Bogdanovich, A. E. (2008). Electrical conductivity study of carbon nanotube yarns, 3-D hybrid braids and their composites, *J. Compos. Mater.*, *42*, 1533–1545.
248. Kepple, K. L., Sanborn, G. P., Lacasse, P. A., Gruenberg, K. M., Ready, W. J. (2008). Improved fracture toughness of carbon fiber composite functionalized with multi walled carbon nanotubes, *Carbon*, *46*, 2026–2033.
249. Qian, H., Bismarck, A., Greenhalgh, E. S., Kalinka, G., Shaffer, M. S. P. (2008). Hierarchical composites reinforced with carbon nanotube grafted fibers: the potential assessed at the single fiber level, *Chem. Mater.*, *20*, 1862–1869.
250. Sager, R. J., Klein, P. J., Lagoudas, D. C., Zhang, Q., Liu, J., Dai, L., Baur, J. W. (2009). Effect of carbon nanotubes on the interfacial shear strength of T650 carbon fiber in an epoxy matrix, *Compos. Sci. Technol.*, *69*, 898–904.
251. Mathur, R. B., Chatterjee, S., Singh, B. P. (2008). Growth of carbon nanotubes on carbon fibre substrates to produce hybrid/phenolic composites with improved mechanical properties, *Compos. Sci. Technol.*, *68*, 1608–1615.
252. Garcia, E. J., Wardle, B. L., Hart, A. J., Yamamoto, N. (2008). Fabrication and multifunctional properties of a hybrid laminate with aligned carbon nanotubes grown *in situ*, *Compos. Sci. Technol.*, *68*, 2034–2041.
253. Veedu, V. P., Cao, A. Y., Li, X. S., Ma, K. G., Soldano, C., Kar, S., Ajayan, P. M., Ghasemi-Nejhad, M. N. (2006). Multifunctional composites using reinforced laminae with carbon-nanotube forests, *Nat. Mater.*, *5*, 457–462.
254. Gojny, F. H., Wichmann, M. H. G., Kopke, U., Fiedler, B., Schulte, K. (2004). Carbon nanotube-reinforced epoxy-composites: enhanced stiffness and fracture toughness at low nanotube content, *Compos. Sci. Technol.*, *64*, 2363–2371.
255. Ma, P. C., Kim, J. K., Tang, B. Z. (2007). Effects of silane functionalization on the properties of carbon nanotube/epoxy nanocomposites, *Compos. Sci. Technol.*, *67*, 2965–2972.
256. Satapathy, B. K., Weidisch, R., Pötschke, P., Janke, A. (2007). Tough-to-brittle transition in multiwalled carbon nanotube (MWNT)/polycarbonate nanocomposites, *Compos. Sci. Technol.*, *67*, 867–879.
257. Liu, L., Wang, Y., Li, Y. L., Wu, J., Zhou, Z. W., Jiang, C. X. (2009). Improved fracture toughness of immiscible polypropylene/ethylene-*co*-vinyl acetate blends with multiwalled carbon nanotubes, *Polymer*, *50*, 3072–3078.
258. Seyhan, A. T., Tanoglu, M., Schulte, K. (2009). Tensile mechanical behavior and fracture toughness of MWCNT and DWCNT modified vinyl-ester/polyester hybrid nanocomposites produced by 3-roll milling, *Mater. Sci. Eng. A*, *523*, 85–92.
259. Karapappas, P., Vavouliotis, A., Tsotra, P., Kostopoulos, V., Paipetis, A. (2009). Enhanced fracture properties of carbon reinforced composites by the addition of multi-wall carbon nanotubes, *J. Compos. Mater.*, *43*, 977–985.
260. Seyhan, A. T., Tanoglu, M., Schulte, K. (2008). Mode I and mode II fracture toughness of E-glass non-crimp fabric/carbon nanotube (CNT) modified polymer based composites, *Eng. Fract. Mech.*, *75*, 5151–5162.
261. Romhany, G., Szebenyi, G. (2009). Interlaminar crack propagation in MWCNT/fiber reinforced hybrid composites, *Express Polym. Lett.*, *3*, 145–151.
262. Blanco, J., Garcia, E. J., De Villoria, R. G., Wardle, B. L. (2009). Limiting mechanisms of mode I interlaminar toughening of composites reinforced with aligned carbon nanotubes, *J. Compos. Mater.*, *43*, 825–841.
263. Faulkner, S. D., Kwon, Y. W., Bartlett, S., Rasmussen, E. A. (2009). Study of composite joint strength with carbon nanotube reinforcement, *J. Mater. Sci.*, *44*, 2858–2864.
264. Kim, M. G., Hong, J. S., Kang, S. G., Kim, C. G. (2008). Enhancement of the crack growth resistance of a carbon/epoxy composite by adding multi-walled carbon nanotubes at a cryogenic temperature, *Compos. Part A*, *39*, 647–654.
265. Zhao, D. Y., Wang, S. H., Wu, J., Bai, X. D., Lei, Q. Q. (2009). The mechanical and impact properties of MWNTs/LDPE nanocomposites, *Pigm. Resin Technol.*, *38*, 305–309.
266. Yuan, J. M., Fan, Z. F., Chen, X. H. C., X.H., Wu, Z. J., He, L. P. (2009). Preparation of polystyrene–multiwalled carbon nanotube composites with individual-dispersed nanotubes and strong interfacial adhesion, *Polymer*, *50*, 3285–3291.

267. Xiao, Y., Zhang, X. Q., Cao, W., Wang, K., Tan, H., Zhang, Q., Du, R. N., Fu, Q. (2007). Dispersion and mechanical properties of polypropylene/multiwall carbon nanotubes composites obtained via dynamic packing injection molding, *J. Appl. Polym. Sci.*, *104*, 1880–1886.
268. Li, S. Q., Wang, F., Wang, Y., Wang, J. W., Ma, J., Xiao, J. (2008). Effect of acid and TETA modification on mechanical properties of MWCNTs/epoxy composites, *J. Mater. Sci.*, *43*, 2653–2658.
269. Kelar, K., Jurkowski, B. (2007). Properties of anionic polymerized epsilon-caprolactam in the presence of carbon nanotubes, *J. Appl. Polym. Sci.*, *104*, 3010–3017.
270. Wang, G. J., Wang, L. J., Mei, Z., Chang, Z. M. (2009). Reinforcement and toughening of poly(vinyl chloride) with poly(caprolactone) grafted carbon nanotubes, *Compos. Part A*, *40*, 1476–1481.
271. Yang, K., Gu, M. Y., Guo, Y. P., Pan, X. F., Mu, G. H. (2009). Effects of carbon nanotube functionalization on the mechanical and thermal properties of epoxy composites, *Carbon*, *47*, 1723–1737.
272. Wang, J. H., Liang, G. Z., Yan, H. X., He, S. B. (2009). Mechanical and thermal properties of functionalized multiwalled carbon nanotubes/cyanate ester composite, *Polym. Eng. Sci.*, *49*, 680–684.
273. Gu, A. J., Liang, G. Z., Liang, D., Ni, M. A. (2007). Bismaleimide/carbon nanotube hybrids for potential aerospace application. I. Static and dynamic mechanical properties, *Polym. Adv. Technol.*, *18*, 835–840.
274. Zheng, Y. P., Zhang, A. B., Chen, Q. H., Zhang, J. X., Ning, R. C. (2006). Functionalized effect on carbon nanotube/epoxy nano-composites, *Mater. Sci. Eng. A*, *435*, 145–149.
275. Yang, K., Gu, M. Y. (2009). The effects of triethylenetetramine grafting of multi-walled carbon nanotubes on its dispersion, filler–matrix interfacial interaction and the thermal properties of epoxy nanocomposites, *Polym. Eng. Sci.*, *49*, 2158–2167.
276. Zhang, H., Zhang, Z. (2007). Impact behaviour of polypropylene filled with multi-walled carbon nanotubes, *Eur. Polym. J.*, *43*, 3197–3207.
277. Chen, Z. K., Yang, J. P., Ni, Q. Q., Fu, S. Y., Huang, Y. G. (2009). Reinforcement of epoxy resins with multi-walled carbon nanotubes for enhancing cryogenic mechanical properties, *Polymer*, *50*, 4753–4759.
278. Maiti, M., Bhowmick, A. K. (2007). Dynamic viscoelastic properties of fluoroelastomer/clay nanocomposites, *Polym. Eng. Sci.*, *47*, 1777–1787.
279. Zhang, W., Picu, R. C., Koratkar, N. (2007). Suppression of fatigue crack growth in carbon nanotube composites, *Appl. Phys. Lett.*, *91*, 193109.
280. Zhang, W., Picu, R. C., Koratkar, N. (2008). The effect of carbon nanotube dimensions and dispersion on the fatigue behavior of epoxy nanocomposites, *Nanotechnology*, *19*, 285709.
281. Yu, N., Zhang, Z. H., He, S. Y. (2008). Fracture toughness and fatigue life of MWCNT/epoxy composites, *Mater. Sci. Eng. A*, *494*, 380–384.
282. Cataldo, F., Ursini, O., Angelini, G. (2009). MWCNTs elastomer nanocomposite. Part 1. The addition of MWCNTs to a natural rubber-based carbon black-filled rubber compound, *Fullerenes Nanotubes Carbon Nanostruct.*, *17*, 38–54.
283. Penumadu, D., Dutta, A., Pharr, G. M., Files, B. (2003). Mechanical properties of blended single-wall carbon nanotube composites, *J. Mater. Res.*, *18*, 1849–1853.
284. Yang, Z., Dong, B., Huang, Y., Liu, L., Yan, F. Y., Li, H. L. (2005). A study on carbon nanotubes reinforced poly(methyl methacrylate) nanocomposites, *Mater. Lett.*, *59*, 2128–2132.
285. Li, X. D., Gao, H. S., Scrivens, W. A., Fei, D. L., Xu, X. Y., Sutton, M. A., Reynolds, A. P., Myrick, M. L. (2004). Nanomechanical characterization of single-walled carbon nanotube reinforced epoxy composites, *Nanotechnology*, *15*, 1416–1423.
286. Zhang, W., Joshi, A., Wang, Z., Kane, R. S., Koratkar, N. (2007). Creep mitigation in composites using carbon nanotube additives, *Nanotechnology*, *18*, 185703.
287. Dutta, A. K., Penumadu, D., Files, B. (2004). Nanoindentation testing for evaluating modulus and hardness of single-walled carbon nanotube-reinforced epoxy composites, *J. Mater. Res.*, *19*, 158–164.
288. Ganss, M., Satapathy, B. K., Thunga, M., Weidisch, R., Pötschke, P., Janke, A. (2007). Temperature dependence of creep behavior of PP–MWNT nanocomposites, *Macromol. Rapid Commun.*, *28*, 1624–1633.
289. Li, J., Zhang, L. Q. (2009). The effects of adding carbon nanotubes to the mechanical and tribological properties of a carbon fibre reinforced polyether ether ketone composite, *Proc. Inst. Mech. Eng. Part C*, *223*, 2501–2507.

290. Bakshi, S. R., Balani, K., Laha, T., Tercero, J., Agarwal, A. (2007). The nanomechanical and nanoscratch properties of MWNT-reinforced ultrahigh-molecular-weight polyethylene coatings, *JOM 59*, 50–53.
291. Xue, Y., Wu, W., Jacobs, O., Schadel, B. (2006). Tribological behaviour of UHMWPE/HDPE blends reinforced with multi-wall carbon nanotubes, *Polym. Test*, *25*, 221–229.
292. Wei, Z., Zhao, Y. P., Ruan, S. L., Gao, P., Yu, T. X. (2006). A study of the tribological behavior of carbon-nanotube-reinforced ultrahigh molecular weight polyethylene composites, *Surf. Interface Anal.*, *38*, 883–886.
293. Kanagaraj, S., Varanda, F. R., Zhil'tsova, T. V., Oliveira, M. S. A., Simoes, J. A. O. (2007). Mechanical properties of high density polyethylene/carbon nanotube composites, *Compos. Sci. Technol.*, *67*, 3071–3077.
294. Cai, H., Yan, F. Y., Xue, Q. J. (2004). Investigation of tribological properties of polyimide/carbon nanotube nanocomposites, *Mater. Sci. Eng. A*, *364*, 94–100.
295. Satyanarayana, N., Rajan, K. S. S., Sinha, S. K., Shen, L. (2007). Carbon nanotube reinforced polyimide thin-film for high wear durability, *Tribol. Lett.*, *27*, 181–188.
296. Dong, B., Yang, Z., Huang, Y., Li, H. L., Liu, L., Yan, F. Y. (2005). Preparation and tribological properties of poly(methyl methacrylate)/multi-walled carbon nanotubes composites, *J. Mater. Sci.*, *40*, 4379–4382.
297. Vail, J. R., Burris, D. L., Sawyer, W. G. (2009). Multifunctionality of single-walled carbon nanotube–polytetrafluoroethylene nanocomposites, *Wear*, *267*, 619–624.
298. Jacobs, O., Xu, W., Schadel, B., Wu, W. (2006). Wear behaviour of carbon nanotube reinforced epoxy resin composites, *Tribol. Lett.*, *23*, 65–75.
299. Zhang, L. C., Zarudi, I., Xiao, K. Q. (2006). Novel behaviour of friction and wear of epoxy composites reinforced by carbon nanotubes, *Wear*, *261*, 806–811.
300. Chen, H. Y., Jacobs, O., Wu, W., Rudiger, G., Schadel, B. (2007). Effect of dispersion method on tribological properties of carbon nanotube reinforced epoxy resin composites, *Polym. Test*, *26*, 351–360.
301. Zoo, Y. S., An, J. W., Lim, D. P., Lim, D. S. (2004). Effect of carbon nanotube addition on tribological behavior of UHMWPE, *Tribol. Lett.*, *16*, 305–309.
302. Chen, W. X., Li, F., Han, G., Xia, J. B., Wang, L. Y., Tu, J. P., Xu, Z. D. (2003). Tribological behavior of carbon-nanotube-filled PTFE composites, *Tribol. Lett.*, *15*, 275–278.
303. Meng, H., Sui, G. X., Xie, G. Y., Yang, R. (2009). Friction and wear behavior of carbon nanotubes reinforced polyamide 6 composites under dry sliding and water lubricated condition, *Compos. Sci. Technol.*, *69*, 606–611.
304. Song, H. J., Zhang, Z. Z., Men, X. H. (2007). Surface-modified carbon nanotubes and the effect of their addition on the tribological behavior of a polyurethane coating, *Eur. Polym. J.*, *43*, 4092–4102.
305. Zheng, Y. P., Zhang, J. X., Xiaodong, Y., Chen, W. W., Wang, R. M. (2009). Effects of functionalized MWNTs with GMA on the properties of PMMA nanocomposites, *J. Appl. Polym. Sci.*, *112*, 1755–1761.
306. Zhang, A. B., Liu, W., Li, M., Zheng, Y. P. (2009). Wear performance of tetrazine modified multiwalled carbon nanotube/epoxy composites, *J. Reinforced Plast. Compos.*, *28*, 2405–2412.
307. Giraldo, L. F., Lopez, B. L., Brostow, W. (2009). Effect of the type of carbon nanotubes on tribological properties of polyamide 6, *Polym. Eng. Sci.*, *49*, 896–902.
308. Huang, Y. L., Yuen, S. M., Ma, C. C. M., Chuang, C. Y., Yu, K. C., Teng, C. C., Tien, H. W., Chiu, Y. C., Wu, S. Y., Liao, S. H., Weng, F. B. (2009). Morphological, electrical, electromagnetic interference (EMI) shielding, and tribological properties of functionalized multi-walled carbon nanotube/poly methyl methacrylate (PMMA) composites, *Compos. Sci. Technol.*, *69*, 1991–1996.

CHAPTER 6

ELECTRICAL PROPERTIES

6.1 OVERVIEW

The term "cheap plastic" is a term of derision that I personally have used when a polymer piece breaks that is the key to proper function of something important (like the holder for my dishwasher basket!). Along with their low cost and their maddening habit of breaking at inopportune times, polymers are also well known for their electrically insulative properties. There are a few polymers, the discovery of which led to the awarding of the Nobel Prize for Chemistry in 2000, that are electrically conductive or semiconductive, but the vast majority of polymers are electrically insulative. However, some applications, such as current limiters, electrostatic dissipation devices, and electromagnetic interference shielding, require significantly higher electrical conductivities than an electrically insulative polymer can provide. Although a highly conductive metal might be suitable for some of these applications, cost and weight considerations often preclude their use. Electrically conductive polymers also tend to be very costly and are extremely difficult to process and form into complex shapes. Further, electrically conductive polymers tend to be brittle and not very resistant to environmental conditions such as high humidity. For these reasons, when electrical conductivity is required for a polymer, a common procedure is to mix a polymer with conductive filler. Usually, carbon black or metals are used as conductive fillers. Carbon nanotubes are also a suitable filler for applications that require electrical conductivity.

Electrical conductivity (σ) has units of current/(voltage $\times$ distance), which in SI (metric) units is ampere/(volt $\times$ meter). The unit ampere/volt is termed siemens, so the SI units of electrical conductivity are siemens/meter (S/m). Two types of conductivities are relevant for filled polymers: bulk conductivity, which measures the ability of charge to pass through a sample and has units of S/m, and surface conductivity. Surface conductivity measures the ability of charge to move along the surface of a material. Surface conductivity is more commonly expressed in terms of the resistance called the sheet resistance. Sheet resistance has units of resistance and can be calculated from the resistivity of a sample divided by its thickness. To distinguish this quantity from the bulk resistance, the units of sheet resistance are typically stated as resistance per square or ohm/sq in SI units. Bulk conductivity can also be expressed in terms of the resistance; bulk conductivity is the inverse of the bulk resistivity, so the latter has units of resistance $\times$ distance or ohm $\times$ meter in SI units.

Carbon Nanotube–Polymer Composites: Manufacture, Properties, and Applications, First Edition.
Brian P. Grady.

Measurements of electrical conductivity are not trivial. All the methods involve contacting a probe to the sample and measuring the relationship between voltage and current. Three geometries are the most popular: a ring geometry, a line geometry, and a point geometry. For the former, two rings are required: if the rings are on the same side of the sample, then the surface conductivity is measured; if they are on the opposite sides of the sample, then the bulk conductivity is measured. In theory, only two probes are needed for line and point geometries as well although typically the voltage and current generation/measurement are separated so that four probes are typically used. Specifics of the geometric factors to apply to convert the measured voltages and resistances to conductivity or sheet resistance for each geometry are beyond the scope of this chapter. However, good electrical contact is a key concern for all electrical conductivity measurements since the calculation assumes negligible resistance between the probe and the sample. A very sharp point, silver-filled epoxy, and high compressive forces are all used to ensure that the resistance between the probe and polymer is negligible compared to the polymer resistance. The practitioner needs to be aware of these issues and test for poor contact by, for example, testing the same sample at two different probe separation distances. One unique noncontact method to measure electrical conductivity of nanotube samples using a modified SQUID magnetometer has also been reported.[1]

The actual conductivity required depends on the application and the requirements are discussed in detail in Section 8.2. In most applications, higher conductivities are preferred, and achieving high conductivities is a challenge for most filled materials. The inherent conductivity of the tubes, the quality and number of the tube–tube contacts along a given pathway, and the number of conductive pathways determine the conductivity. The inherent conductivity of the tubes is determined by the *n,m* characteristics of the tubes and the number, type, and location of defects. Measurements on individual MWCNTs have shown both metallic and semiconducting behavior, with conductivities between 2×10^7 and 8×10^5 S/m and a maximum band gap of 0.3 eV.[2,3] Silver and copper have conductivities of 6.33×10^7 and 5.95×10^7 S/m, respectively. Values measured for bundles of SWCNTs were comparable to the lower value measured for MWCNTs.[4] Smaller numbers of tube–tube contacts are preferred for higher conductivities. The number of tube–tube contacts along a given pathway and the number of pathways both depend on the length of the nanotubes (longer nanotubes are better) and the spatial distribution of nanotubes. As will be described later, spatially inhomogeneous distributions of nanotubes can lead to conductivity at lower fractions but do not necessarily lead to higher conductivity at high volume fractions. The number of pathways also depends on the nanotube concentration. The quality of tube–tube contacts is definitely a more complicated issue and is responsible for the much lower conductivities of nanotube composites versus the inherent conductivity of individual nanotubes.[5] Tube–tube contact issues are discussed in more detail below.

Because of the very low conductivity of the insulating polymer, a model that has *any* conduction of charge by the polymer would have a conductivity not very different from the polymer conductivity; hence, nanotubes must carry all the charge for high conductivities to be achieved. If nanotubes are in contact, then transfer of charge from one nanotube to another will offer much higher resistance than charge traveling along nanotubes, even for the case of semiconducting tubes. High resistances occur because of the limited contact area between two tubes. This type of resistance, termed constriction resistance, has been shown to be limiting because increasing compression forces[6,7] or decreasing the hardness of the filler[8] leads to higher conductivities; that is, constriction resistance is the limiting resistance. Measurements on individual crossed SWCNTs at low voltages indicated that this constriction resistance is at a minimum hundreds of kiloohms.[9]

Constriction resistance is most important at high volume fractions of filler; at lower volume fractions, tunneling resistance is more important. Tunneling resistance is a quantum mechanical process that allows charge to pass from one conductor to another without any conduction by the polymer. The distance between adjacent particles must be on the order of 5 nm or less for tunneling to occur; a recent publication using Monte Carlo methods suggests that the maximum tunneling distance in a polymer is about 1.8 nm.[10] Further, the amount of tunneling depends on the voltage applied; tunneling is a nonlinear effect and the tunneling current has a stronger dependence on voltage than the linear dependence of Ohm's law ($V = IR$); that is, if the voltage doubles, then the current more than doubles. Tunneling also has a much stronger dependence than linear for the change in current with distance between conducting particles, which has led to the development of metal-filled composites that change many orders of magnitude in resistance with pressure. The tunneling resistance does depend on the polymer, although not because of the small variations in dielectric constant, but because of the quality of the interface between the polymer and the conducting filler that in turn affects the distance between two nanotubes. A recent study made the argument that if the maximum conductivity is below a certain level, then it is likely that tunneling dominates conductivity;[11] this argument will be examined more closely in Section 6.2.

A simple geometric scaling argument can be made if tunneling is the dominant mechanism. As stated previously, tunneling is not linear with distance; the dependence is logarithmic. Since, for spherical particles, the distance between individual particles scales with the volume (or weight) fraction to the 1/3 power, the relationship between conductivity and filler volume fraction (V_f) can be written as

$$\ln \sigma = V_f^{-1/3} \tag{6.1}$$

Although not used nearly as extensively as Equation 6.2, this equation has been used to fit data to explore the assumption that tunneling resistance controls the conductivity.[12–14] Given the fact that nanotubes are not spherical, in the author's opinion the use of this equation to assess tunneling resistance is likely flawed. Still, however, the preponderance of evidence suggests that constriction resistance and tunneling resistance usually dominate the barriers to conduction in nanotube-filled systems and a number of studies have reached the same conclusion.[10,14–16]

The remainder of this chapter is divided into three sections. The first and second sections are divided according to the method of mixing of the nanotubes with polymers—the latter discusses infusion/impregnation methods while the former discusses a variety of nanotube–polymer mixing methods. The latter methods, discussed in Section 6.3, use already formed nanotube fibers or mats, and nanotubes run continuously from one side of the sample to the other. Hence, the conductivity is most importantly a function of the quality of the mat or the fiber (and the associated nanotube quality), and only secondarily does the resin have an effect. Note that these systems still have constriction resistance and tunneling resistance issues since nanotubes are not, in general, long enough to span an entire sample dimension. Dispersion is the key issue with respect to mixing methods discussed in Section 6.2; hence the properties of the resin and the mixing method have a large effect. In particular, reagglomeration processes are critically important in mixing methods, and both the processing method and the viscosity of the resin will have a large effect. Section 6.4 discusses mixtures of nanotubes with electrically conducting polymers. In mixtures with electrically conducting polymers, charge is not exclusively carried by nanotubes and although a number of concepts are directly transferable from the first two sections, the treatment is different and hence a separate section is required.

6.2 MIXED COMPOSITES

The concept of percolation was introduced earlier in Section 5.2.2, so the focus in this section will be on the behavior of electrical conductivity with filler volume fraction. In an insulating matrix, a typical plot of conductivity versus volume or weight fraction is shown schematically in Figure 6.1. The first part of the graph appears constant on a logarithmic scale; in fact, there is a slight increase in conductivity, which is essentially linear in filler volume fraction. The filler volume fraction (V_f) at which the concentration begins to increase rapidly is termed the percolation threshold (V_{fc}), and the sharply rising region between the percolation threshold and the plateau

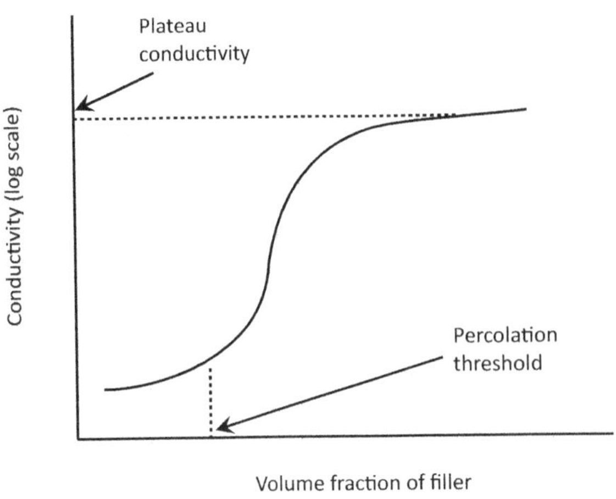

Figure 6.1 Electrical conductivity versus volume fraction for an insulating polymer filled with a conductive filler, such as carbon nanotubes. The increase in the percolation region is typically on the order of 10^{10}–10^{15} S/m.

region (the plateau region occurs), at high concentration where the curve in Figure 6.1 is approximately constant) is termed the percolation region. Within this region, Equation 6.2 describes the data

$$\sigma = A(V_{\mathrm{f}} - V_{\mathrm{fc}})^{\beta} \tag{6.2}$$

where A is a prefactor required for the fit. The value of β to a first approximation does not depend on whether volume or weight fractions are used at the low loading levels common for nanotube composites, so most often weight fractions are used to eliminate ambiguity with respect to the correct density. Also, many researchers have interpreted the value of A as the value corresponding to 100% nanotubes ($V - V_{\mathrm{fc}} \sim 1$); however, this extrapolation necessarily assumes that percolation behavior should extend far beyond the region over which such scaling behavior is appropriate to use.

The upper limit of where this equation should apply corresponds to the concentration where the morphology becomes cocontinuous, and is difficult to set with accuracy. Coleman and coworkers[17] showed that for very high volume fraction polystyrene–MWCNT composites prepared from a dissolution–dispersion precipitation regime a percolative-type scaling law, for example, Equation 6.2, was valid for volume fractions from 0.1 to 1 with a scaling exponent of 2.2. These authors interpreted this result as being due to a process that was more similar to an infusion process; that is, the nanotube network formed first, followed by deposition of the polymer. Infusion-like behavior is unique for a noninfusion process, and it likely occurred because of the extremely dilute solutions used coupled with the rather high stability of the nanotubes in the NMP solvent (e.g., polymer adsorption to nanotubes was likely negligible). As pointed out in Coleman's paper, the fact that the scaling behavior persisted well beyond the percolation region is in itself a bit surprising, although this was not the first study where such scaling behavior was seen to values well above 10 vol%.[18]

Equation 6.2 can only be rigorously applied to statistical percolation. Reagglomeration discussed in Chapter 3 means that percolation in nanotube composites can be characterized in two ways: statistical percolation that is governed by random placement of nanotubes and kinetic percolation that involves some reaggregation effect and hence concentration inhomogeneities of nanotubes over some reasonably long length scale. Although kinetic percolation does not have to follow a critical scaling law as exemplified by Equation 6.2, the extent of reagglomeration does not seem to affect the quality of fit of data to this equation. However, the geometric interpretation of the value for the β parameter for rigid rod fillers will not be accurate for kinetic percolation. According to one model for statistical percolation, the critical exponent based on geometric criteria is 1.1–1.3 for a 2D network and 1.6–2.0 for a 3D network,[19] and according to another, it is 2.0 for a 3D network.[20] The issue of kinetic percolation will be considered in more detail in Section 6.2.3.

Although many studies have measured conductivity of samples containing nanotubes, a much smaller number of studies have actually measured the percolation threshold with a sufficient number of data points to be able to fit Equation 6.2. Table 6.1 tabulates the cases where this equation has been applied to such data. Only data from mid-2007 to 2009 are included in this table because a similar table

TABLE 6.1 Measurements of the Percolation Threshold Fit with Equation 6.2

Polymer	Nanotube type	Processing method[b]	Percolation threshold (wt% unless otherwise noted)	Exponent	Plateau conductivity (S/m)
Polystyrene[21]	MWCNTs	Dissolution–dispersion–precipitation (electrospinning)	3%	0.795	0.01
	MWCNTs with adsorbed copolymer		3%	Not given, but larger	0.001
Polystyrene[22]	MWCNTs	Melt mixing	0.68%	1.2	1
Polystyrene[a23]	Acid-functionalized SWCNTs	Dissolution–dispersion–precipitation	0.17–0.3%	2, 2, 4.1	10^{-5}
	Polymer-coated SWCNTs		0.4–0.5%	2.9, 4.8, 4.5	10^{-8}–10^{-5}
Polystyrene[24]	SWCNTs	Dissolution–dispersion–precipitation	~0.1%	2.2	100
Polystyrene[25,26]	MWCNTs	Dissolution–dispersion–precipitation	1%	1.9	1
Polystyrene[27]	MWCNTs; polystyrene-grafted MWCNTs	Dissolution–dispersion–precipitation	0.4 vol%	2.4	1
Polystyrene[28]	Styrene block copolymer-grafted MWCNTs	Dissolution–dispersion–precipitation	0.3 vol%	1.9	100
Poly(styrene-*co*-butadiene)[27]	MWCNTs; polystyrene-grafted MWCNTs	Dissolution–dispersion–precipitation	0.9%	3.2	0.01
Polycarbonate[29]	MWCNTs	Dissolution–dispersion–precipitation	0.5%	1.6	10^4 (Ω/sq)

Poly(methyl methacrylate)[30]	SWCNTs	Dissolution–dispersion–precipitation	0.3 vol%	2.3	10
Poly(methyl methacrylate)[31]	SWCNTs	Dissolution–dispersion–precipitation	2.5%	1.8	1
Poly(methyl methacrylate)[32]	MWCNTs	Dispersion–reaction	1.8%	3.8	10
		Dissolution–dispersion–precipitation	7%	3.7	10
Poly(methyl acrylate)[28]	Methyl acrylate block copolymer-grafted MWCNTs	Dissolution–dispersion–precipitation	0.5 vol%	1.9	100
Ethylene–vinyl acetate copolymer[33]	SWCNTs	Melt mixing	1.8%	2.0	0.001
Various methyl methacrylate–butyl acrylate–acrylic acid copolymers[34]	Lithium-reduced MWCNTs	Dissolution–dispersion–precipitation	0.27–0.4 vol%	0.42–1.2	10^4 (Ω/sq)
High-density polyethylene[35]	MWCNTs	Melt mixing	1 vol%	4.5	10
High-density polyethylene[36]	MWCNTs	Melt mixing	2.8%	4.5	1
High-density polyethylene[37]	Grafted high-density polyethylene on MWCNTs	Dispersion–reaction	7.3%	2.0	1

(continued)

TABLE 6.1 (*Continued*)

Polymer	Nanotube type	Processing method[b]	Percolation threshold (wt% unless otherwise noted)	Exponent	Plateau conductivity (S/m)
High-density polyethylene[38]	SWCNTs	Dissolution–dispersion–precipitation	0.13%	3.5	100
Low-density polyethylene[39]	MWCNTs; vary film thickness 20–200 μm	Dissolution–dispersion–precipitation	1.1–4.4 vol%	1.1–1.8	All 10^{-3} to 10^{-2}
Ultrahigh molecular weight polyethylene[40]	MWCNTs	Dissolution–dispersion–precipitation	1.4 vol%	1.9	2
Ultrahigh molecular weight polyethylene[41]	SWCNTs	No fluid mixing	0.14%	2	10^{-4}
Ultrahigh molecular weight polyethylene[42]	MWCNTs	Dispersion–dispersion–evaporation	0.072 vol%	1.1	10
Ultrahigh molecular weight polyethylene[43]	SWCNTs (different types)	No fluid mixing, and dispersion–dispersion–evaporation	0.09–1.6%	2.2–2.3	0.001–1
	MWCNTs		0.05–0.3%	2.7	10
Polyethylene terephthalate[44]	SWCNTs	Melt mixing	0.024%	1.8	0.1 for both
		Dispersion–reaction	0.048%	1.9	
Polyvinyl acetate	MWCNTs (extremely low diameter)	Dispersion–dispersion–evaporation	0.01%	3.4	5×10^3
Polypropylene[45]	MWCNTs	Melt mixing	0.9 vol%	2.8	10
Polypropylene[46]	MWCNTs	Melt mixing	1.2%	3	10

Polyamide 6[45]	MWCNTs	Melt mixing	1.1 vol%	5	1
Polyurethane[12]	MWCNTs	Melt mixing	0.13%	4.6	1
Poly(vinyl chloride)[47]	MWCNTs	No fluid mixing	0.047%	3.3	10^{-2}
Poly(ether ether ketone)[48,49]	MWCNTs	Melt mixing	1.3% (0.9%)	1.2 (2.0)	1
Polyoxymethylene[45]	MWCNTs	Melt mixing	0.7 vol%	2.1	10
Polyurethane[50]	Amine-functionalized MWCNTs	Dissolution–dispersion–precipitation	0.35%	2.0	0.01
Poly(phenylene sulfide)[51]	MWCNTs	Melt mixing	1.6%	0.5	.1
Poly(phenylene sulfide)[52]	Acid-functionalized MWCNTs	Melt mixing	1.6%	3.3	1–10
	Imidazole-functionalized MWCNTs		1.3%	3.6	10–100
Poly(phenylene sulfide)[53]	MWCNTs	Dispersion–dispersion–evaporation	0.22 vol%	3.6	1000
Polyvinylidene fluoride[54]	DWCNTs	Dissolution–dispersion–precipitation	0.19%	1.9	10
Poly(vinylidene fluoride)[55]	MWCNTs	Dissolution–dispersion–precipitation	3.8%	3.2	Not reached; >0.1
Epoxy[56]	MWCNTs	Dispersion–reaction	0.02 vol%	1.8	.001

(*continued*)

TABLE 6.1 (*Continued*)

Polymer	Nanotube type	Processing method[b]	Percolation threshold (wt% unless otherwise noted)	Exponent	Plateau conductivity (S/m)
Epoxy[57]	MWCNTs	Dispersion–reaction	0.1 vol%	1.8	100
Epoxy[58]	MWCNTs	Dispersion–reaction	0.025%	3	Not reached; >0.1
Epoxy[59]	MWCNTs, varied types	Dispersion–reaction	0.08% to 0.002%	1.7–3.1	Not reached
Epoxy[60] Uniaxially drawn, parallel	SWCNTs	Dispersion–reaction	0.05%	2.9	10^{-3}
Uniaxially drawn, perpendicular			0.5%	2.0	10^{-7}
Epoxy[61]	Long SWCNTs	Dispersion–reaction	0.06%	2.7	10
	Short SWCNTs		0.32%	2.2	1
	Annealed SWCNTs		0.34%	1.7	10
Polyimide[62]	Mixture of SWCNTs, DWCNTs, and low-diameter MWCNTs	Dispersion–reaction	0.15%	1.6	10^{-2} to 10^{-3}
Polyaniline (undoped, i. e., insulating)[63]	MWCNTs	Dissolution–dispersion–precipitation	5.9%	2.4	1

Styrene–butadiene–styrene block copolymer[64]	MWCNTs	Dissolution–dispersion–precipitation	0.25%	1.6	100
50–50 Polyamide 6/acrylonitrile–butadiene–styrene blend [65]	Acid-functionalized MWCNTs	Melt mixing	0.15%	1.5	10^{-5}
45–55 Polypropylene/acrylonitrile–butadiene–styrene blend	MWCNTs	Melt mixing	0.4%	3.2	0.1
80–20 Polypropylene/acrylonitrile–butadiene–styrene blend[66]			2.8%	1.7	1

Only studies from mid-2007 to 2009 are included; this table complements one that appears in Ref. 11.

[a]Different manufacturer NTs were used in this study.

[b]Most dispersion–dissolution–precipitation and dispersion–dispersion–evaporation methods were followed by a compression molding step. The characteristics of this latter step, in particular the viscosity of the resin and the time of molding, are likely critically important since reagglomeration occurs in most, if not all, of the systems listed.

(for electrical conductivity only) that includes years prior to the mentioned period appears in an excellent review article of this subject by Bauhofer and Kovacs.[11] The plateau conductivity is also included in this table. The plateau conductivity does not actually reach a constant value as the term implies; in fact, the conductivity continues to increase slowly on the high concentration side of the S-shaped curve shown in Figure 6.1. Order of magnitude maximum values can be compared between different samples because the increase is small enough and the upper limits of nanotube content are not greatly different. In an order of magnitude sense, these conductivities though are critically important for many applications, and hence order of magnitude values are given in Table 6.1. Hence, in this chapter, the terms maximum conductivity and plateau conductivity are used interchangeably unless otherwise noted.

A great deal of inconsistency can be associated with the values in this table because of differences in dispersion, nanotube quality, and so on. An approach to understand effects of these differences is to develop computational models to describe the conductivity of nanotube-filled polymers as a function of nanotube content, nanotube characteristics, and so on. The simplest models treat nanotubes as rigid rods with fixed, monodisperse aspect ratio;[67–70] however, to quantitatively describe the conductivity requires expressions for tunneling resistance and contact resistance. Various additions such as flexibility or waviness[5,71–74] and penetrability (necessary to account for tunneling)[75] can be added. Models can be separated into two types: models that consider only statistical percolation and models that consider both statistical percolation and reagglomeration, with the latter often containing a time component as well. Models that are concerned only with the percolation threshold are of course much simpler than those that also attempt to quantify the conductivity. In this author's opinion, the problem with modeling efforts at the current time is more a function of the lack of critical experimental data, for example, the distribution of bundle sizes and an experimental examination of the relative influence of statistical percolation and reagglomeration. However, models have provided some important insights and will be discussed in detail in the following sections.

6.2.1 Maximum or Plateau Conductivity

Figure 6.2 shows the maximum conductivity as a function of the concentration, where the maximum conductivity was measured as compiled and graphed by Bauhofer and Kovacs.[11] With respect to the legend, nonentangled nanotubes are those that come from nanotube forests, that is, vertically grown nanotubes. Although composite conductivities from materials containing vertically grown nanotubes are higher, better dispersion cannot be assumed since longer, more defect-free tubes result from vertical growth. Longer tubes means the contribution of interfacial resistance to overall conductivity is lowered and defect-free means that these tubes likely have higher conductivities. A direct comparison between vertically grown tubes and commercial powder samples with everything else held constant yielded a plateau conductivity of 10^3 versus 10 S/m in polystyrene.[76] The lines on the plot represent a scaling relation derived from fitting a relationship $\sigma = \text{constant} \times V_f^a$; in the author's opinion, the data are too scattered to support one functional form over another. The

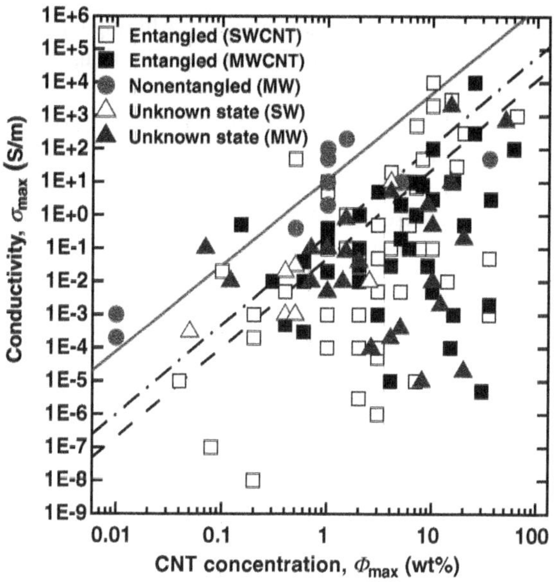

Figure 6.2 Compilation of data prior to approximately mid-2007 showing the maximum conductivity as a function of percolation threshold for mixed polymer–nanotube composites. Copyright Elsevier Ltd. Reproduced with permission from Ref. 11.

data might exhibit a slight positive slope on average; however, the conductivity in the plateau region increases slightly and might cause the slight increase. An important point that Bauhofer and Kovacs made from this graph is that tunneling is likely the dominant form of resistance below the dashed line; in the author's opinion, nanotube quality (aspect ratio, number of defects, *n*, *m* content) makes such a conclusion very questionable. There are also no clear trends in plateau or maximum conductivity with respect to polymer type. However, both Table 6.1 and the data in Bauhofer and Kovacs suggest that lower percolation thresholds generally lead to higher plateau conductivities, indicating that longer tubes and/or more optimally dispersed tubes for electrical conductivity have lower constrictive and/or tunneling resistances.

The properties of the tubes will affect plateau or maximum conductivity. In general, SWCNT composites have lower conductivities than MWCNT composites; this is due to the higher contact resistance in the former owing to a smaller diameter or perhaps the generally poorer nanoscale dispersion of SWCNTs. A recent modeling study showed that increasing the conductivity of the tubes would have a small effect on the conductivity of the composite.[16] However, experimentally doping MWCNTs with iodine was shown to increase the conductivity above the percolation threshold by about two orders of magnitude from 10^{-2} to 1 S/m in ultrahigh molecular weight polyethylene,[77] while a much smaller factor of 5 increase from ~100 to 500 S/m was found in a polycarbonate sample made with SWCNTs.[78] An increase in the interfacial conductivity can have a very large impact on the conductivity.[16] Similarly, reducing the number of interfacial contacts by increasing tube length will also have a large impact on conductivity. Assuming rigid rods with identical diameters and statistical dispersion, the plateau conductivity should go as $\sim L^{2.5}$ due to a reduction in the number of tube–tube contacts.[79] Experimentally, a significant change has been found with nanotube length; the conductivity increased by a factor of 68 with an increase in nanotube length of a factor of 10 in an epoxy.[80] Although this is substantial, the increase is significantly less than the rigid rod calculation, which is likely due to the fact that nanotubes are not rigid rods.

Functionalization of MWCNTs can have very little effect on the maximum conductivity[81,82] because only the outer shell of the MWCNTs is affected by functionalization; the ability of the inner shells to conduct electricity is unaffected. In fact, functionalization was found to increase the conductivity from 10^{-3} S/m to as high as 1 S/m in polypropylene melt-mixed composites with MWCNTs, which was attributed to significantly improved dispersion.[83] Using severe acid treatment on MWCNTs was found to reduce the conductivity only by roughly 1.5 orders of magnitude in a polyurethane composite, although dispersion was also affected.[84] Another study on blends of polycarbonate and acrylonitrile–styrene–butadiene with commercial nanotubes found approximately two orders of magnitude drop in conductivity with functionalization.[85] On the other hand, the conductivity of SWCNTs is greatly affected by functionalization, and hence the maximum conductivity of the composites can increase in the case of functionalization of semiconducting tubes with the appropriate atom (i.e., doping) or decrease in the case of organic functionalization.

Overall, there are no clear trends regarding the identity of the polymer and its effect on the plateau conductivity, although electrically conductive polymers have higher maximum conductivities as described in Section 6.4. Molecular weight can have a significant effect on the conductivity behavior. A study by Koning and coworkers showed that the presence of low molecular weight polystyrene caused two orders of magnitude increase in the plateau conductivity for a SWCNT–polystyrene manufactured via a dispersion–dissolution–evaporation method. Since little or no effect was found in the percolation threshold, the authors attributed the increase in plateau conductivity to lower constriction or tunneling resistance owing to preferential adsorption of low molecular weight material at the surface of the nanotubes.[86] This material also contained surfactant, the role of which is a bit unclear. A later study by the same authors[87] using the same procedures but different molecular weight distribution materials showed that having material with lower molecular weight caused a decrease in percolation threshold. The authors attributed this to a decrease in viscosity. However, the plateau conductivity was unchanged, suggesting that decreased viscosity caused agglomeration that in turn led to a lower percolation threshold without affecting the plateau conductivity.

The method of mixing will influence the maximum conductivity. For example, the mixing speed in a conical twin-screw extruder showed a maximum in electrical resistivity at an intermediate mixing speed, which the authors attributed to the competing effects of nanotube breakage (which tends to increase resistivity) as well as agglomerate formation (which the authors argued tends to decrease resistivity).[88] A unique method by which nanotubes were infused into a solvent-swollen polymer followed by evaporation of the solvent yielded a high conductivity of 66 S/m.[89] A dispersion–dispersion–evaporation method that left the nanotubes at the interface of latex particles yielded an even higher conductivity at 20 wt% tubes of almost 5000 S/m using a mixture of single-walled, double-walled and triple-walled tubes.[90] The effect of orientation on conductivity is difficult to predict; a study on drawn polypropylene showed little effect on the maximum conductivity,[46] while a study on an epoxy drawn to 100 times its initial length (drawing was done on a semidried film) showed a less than order of magnitude increase in conductivity, although the conductivity perpendicular to this direction dropped by two to three orders of magnitude.[60] A study on uniaxially drawn polyurethane films showed a logarithmic decrease in conductance up to about 100% strain.[50] Monte Carlo simulations showed that the plateau conductivity can go up or down depending on the degree of alignment and the flexibility of the nanotubes.[91] The influence of injection molding parameters on finished polycarbonate parts was studied in detail by Pötschke and coworkers.[92] High injection speeds lead to a highly oriented skin layer with well-separated MWCNTs, which acts as an electrically insulating layer due to a lack of network formation. In addition, the nanotube concentration tends to be lower near the core region than near the walls. Higher temperatures and lower injection speeds tend to improve nanotube dispersion because of the enhanced ability of nanotubes to agglomerate. *In situ* measurements during injection molding support the same general conclusions.[93]

The temperature dependence of the maximum conductivity has also been studied. For a well-dispersed system, the expectation is that the conductivity should decrease with an increase in temperature because of thermal expansion; such behavior was found in an ultrahigh molecular weight polyethylene–MWCNT composite in which the nanotubes were isolated between the fused polymer particles.[94] With better dispersed nanotubes, the opposite has been found for composites of MWCNT/high-density polyethylene,[36,38] MWCNT/polyurethane,[50,95] MWCNT/poly(ethylene oxide),[96] MWCNT/epoxy,[96] MWCNT/liquid crystal polymer,[97] and MWCNT/polyamide 6[98]. This seeming inconsistency is explained in terms of the tunneling resistance; the change in resistance with temperature was shown in the latter study to be well described by the thermal fluctuation-induced tunneling (TFIT) model as shown in Equation 6.3. T is temperature while T_0 is a reference temperature, and T_1 is a temperature that represents a barrier energy of $k_B T_1$ that electrons must overcome in order to tunnel.

$$\sigma = A e^{-T_1/(T + T_0)} \tag{6.3}$$

One study found that T_0 is extremely small for nanotubes (on the order of 5 K)[78], which is also the value for carbon black,[99] and hence can be ignored. In this case, Equation 6.3 reduces to a simple Arrhenius expression. In another study, the value was found to be much larger, that is, 173 K, which manifested itself in distinct curvature in an Arrhenius plot.[38] This latter study also fit a variable range hopping model to the data with no qualitative difference in fit from the TFIT model.

The glass transition caused no change in the conductivity as expected because no volume change occurs at the glass transition; however, the activation energy (i.e., T_1) was found to be much higher above the glass transition for the semicrystalline polymer polyamide 6.[98] The reason for this increase in T_1 is likely due to the increase in thermal expansion coefficient above the glass transition temperature. Melting has been shown to cause a substantial decrease in conductivity,[36,38] which can be attributed to either the macroscopic change in volume or a redistribution of nanotubes into regions that formally were crystalline. Similarly, a decrease in conductivity was seen during heating for the solid-to-nematic transition for nanotubes mixed with a liquid crystalline polymer. Above this transition, the electrical conductivity increased and was fit well with a thermal fluctuation-induced tunneling model even through the nematic–isotropic transition. Further confirming this model, the activation energy decreased when the MWCNT concentration was increased.[97]

6.2.2 Broadness of Percolation Region (Critical Exponent)

β (Equation 6.2) is a measure of how quickly (in a concentration sense) a sample develops electrical conductivity with nanotube addition after the first network is formed, and a larger value indicates slower development with concentration. The critical exponent β varies primarily between 1 and 4 with no obvious and consistent dependence on polymer type, preparation method, percolation threshold, nanotube properties (length, functionalization), or maximum conductivity. As stated previously, for statistical percolation, the value should range between 1 and 2, depending on whether a 2D or 3D network is formed; in the vast majority of cases in Table 6.1, a 3D network is formed. An argument has been made that a larger exponent indicates a broad distribution of tunneling resistance and hence a broad distribution in particle–particle distances.[100] In the author's opinion, both tube polydispersity and kinetic percolation argue against an exclusive geometric tunneling explanation. However, simulations have shown that tube flexibility will decrease, not increase, the exponent.[74]

β has been correlated to orientation. A study on solution-cast thin films of MWCNTs–low-density polyethylene showed that as film thickness became smaller and of the order of the length of a nanotube, the critical exponent dropped;[39] this results would be expected as a network shifts from three to two dimensions. Uniaxial orientation through drawing of a semidried epoxy film by a factor of 100 led to a critical exponent of 2.9 parallel to the draw direction and 2.0 perpendicular to it; unfortunately, the critical exponent for the undrawn film was not given.[60] Uniaxial drawing a film to relatively small strains caused a drop from 2.0 to 1.6 at 5% strain, but was essentially unaffected up to strains of 20%; the percolation threshold was essentially unchanged over this strain range.[50]

6.2.3 Percolation Threshold

The percolation threshold depends on the method of dispersion as shown in both Table 6.1 and the companion table in Bauhofer and Kovacs.[11] In cases with nominally homogeneous dispersion of nanotubes, dispersion–reaction and dispersion–dissolution–precipitation methods generally yield lower percolation thresholds than other methods such as melt mixing. The lower percolation threshold is likely due to the far superior nanoscale dispersion achievable through the use of sonication. The lowest percolation thresholds found have been in the 0.0020–0.0025 wt% range[59,101–103] and all have occurred in epoxy composites. A unique spray technique using dissolved perfluoroalkoxy polymer and dispersed MWCNTs (dispersion–dissolution–precipitation) was able to achieve a percolation threshold of 0.0017 vol%, which corresponds to a weight fraction only slightly higher than the range given above.[104] Dispersion–dispersion–evaporation methods can also lead to relatively low percolation thresholds with nominally homogeneous dispersion of nanotubes after compression molding, indicating that diffusive dispersion of nanoscale dispersed nanotubes is sufficient to achieve good mixing in a polymer.[105,106] Microscale inhomogeneous distribution of nanotubes, available through the use of no fluid mixing methods or dispersion–dispersion–evaporation, can also lead to percolation at very low nanotube fractions. The same sort of inhomogeneous distribution is possible if nanotubes are isolated at the interface of an immiscible blend.

The measured percolation threshold has an extremely strong dependence on the processing conditions, in particular the time allowed for agglomeration after dispersion, indicating that kinetic percolation is critical to understanding the behavior of electrical conductivity. With a nanotube diffusion constant measured that translates into a nanotube velocity on the order of μm/s,[107] the fact that agglomeration can occur is not surprising if the surface energies of the polymer and nanotubes are not sufficiently strong.

The electrical conductivity is usually used to quantify the time scale of reagglomeration; oscillatory rheology is used less frequently because of concerns that oscillations might change the agglomeration time. The time scale of the establishment of electrical conductivity was investigated in poly(vinylidene fluoride) filled with MWCNTs and carboxylated MWCNTs.[108] The authors chose various concentration values below the percolation threshold as measured in the solid at room temperature and monitored the development of conductivity with time at various temperatures above the melting temperature. A percolation time was defined as the time where the conductivity began to increase sharply; the times found were on the order of a few minutes. The authors found that activation energies as defined by an Arrhenius expression were independent of the nanotube concentration, although of course longer times were required for lower concentrations. The authors also found a ~20% larger activation energy for the functionalized nanotubes, which was attributed to a higher interfacial energy causing slower aggregation. A model was fit to the data that suggested that the equilibrium percolation concentration was higher for the unfunctionalized material, which is hard to justify since the functionalization procedure should not have changed the length of the nanotubes. A different model for polypropylene/MWCNT composites was fit to conductivity versus time data based on second-order kinetics of the agglomeration process with three different assumptions of the effect of agglomeration on conductivity. In general, the data fits were good, and none of the models could be said to outperform the others in all cases.[109] Three models were also fit to polycarbonate, again with no clear advantages of any model.[110]

The percolation threshold has been found to shift by a few weight percent with agglomeration (a rather significant change!) for melt-mixed polystyrene[22] and polycarbonate.[110] These observations are not limited to melt mixing, and a shift of a few percent was found for polystyrene–nanotube composites prepared via dispersion–dispersion–evaporation followed by compression molding.[105] Of course, this effect is not limited to high-viscosity thermoplastics; at low shear rate (50 rpm stirring) of an epoxy at elevated temperature (i.e., during the initial stages of curing), the percolation threshold was 0.01%, while at high stirring (2000 rpm stirring) the percolation threshold was an order of magnitude higher.[13] A comparison of no stir and low stir by the same group found that the percolation threshold could be reduced by an order of magnitude to values as low as 0.0024 wt%.[59]

The time scale of agglomeration is relatively well studied compared to the structure of the agglomerated objects. An interesting study using sonication time and temperature at constant nanotube content was able to vary the conductivity over many orders of magnitude and examined the resultant structure via optical microscopy and small-angle X-ray scattering. The authors proposed different microscale-type agglomerated structures: dendritic and tumbleweed-type structures, with the former being very effective for charge transport and the latter not nearly as effective. These structures are shown in Figure 6.3.[111]

Given the importance of kinetic percolation, the specifics of the processing method are also important with respect to percolation threshold. An interesting study with a thermoplastic polyurethane showed that an extruded strand had a very diffuse percolation region (large β) and still had not reached a plateau conductivity at 5% nanotube content, whereas a compression molded sample reached the plateau conductivity at about 1% nanotube content.[12] Uniaxial orientation would be expected to increase the percolation threshold in nanotube composites, which does occur, but the drop is not nearly as severe as for composites made with the more isotropic filler carbon black as was shown for composites made with polypropylene.[46] Certainly, the percolation threshold parallel to the draw direction would be

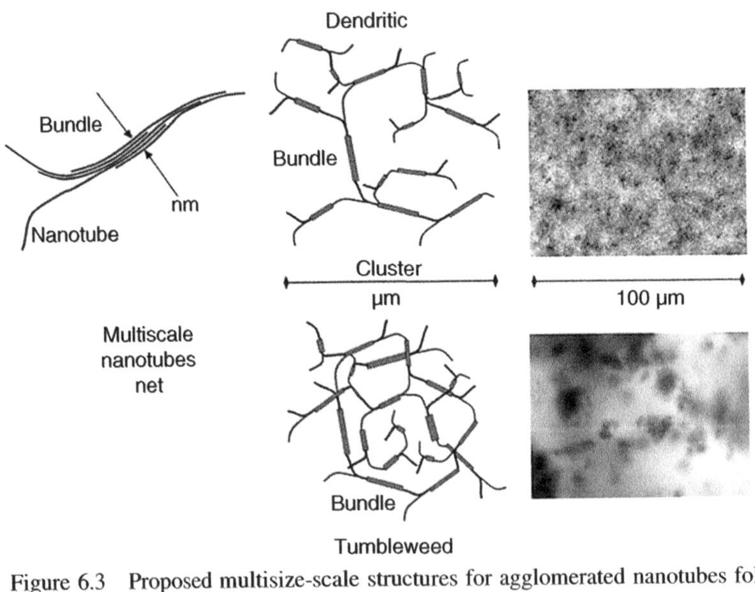

Figure 6.3 Proposed multisize-scale structures for agglomerated nanotubes following initial dispersion. Copyright American Institute of Physics. Reproduced with permission from Ref. 111.

expected to be smaller than that perpendicular to the draw direction as was found for a drawn epoxy.[60] Confining tubes to two rather than three dimensions by making composite films having thicknesses on the order of the length of a nanotube causes a reduction in percolation threshold as expected.[39] Surprisingly, the replacement of 50% of MWCNTs with carbon black caused a rather small increase in percolation threshold, much smaller than a simple mixing rule would suggest.[58]

Nanotube characteristics are also important. Functionalization has been shown to increase the percolation threshold,[81] reduce the percolation threshold,[84] and have no effect on the percolation threshold.[82] Functionalization generally improves nanoscale dispersion, which tends to reduce the percolation threshold; however, functionalization also generally reduces reagglomeration, which tends to increase the percolation threshold. Hence, it is not surprising that no consistent trends are found with respect to the influence of functionalization on the percolation threshold. Higher aspect ratio tubes are expected to reduce the percolation threshold based on statistical percolation arguments that assume no reagglomeration; however, longer tubes should have more difficulty in reagglomeration because of a presumably higher diffusion coefficient. For example, a maximum in percolation threshold was found at an intermediate aspect ratio for MWCNT/PVDF composites confirming the expected complex behavior,[112] while longer tubes showed a factor of 5 lower percolation threshold in an epoxy composite.[61]

As stated in Chapter 5, an electric field can cause alignment of nanotubes that in turn leads to increased aggregation in the perpendicular direction and alignment in the parallel direction. An AC field during curing of a polyimide resin led to a reduction in percolation threshold from 0.15% to 0.036% in the parallel direction with no change in plateau conductivity.[62] The time dependence of percolation network formation as a function of electric field was examined in a polycarbonate melt; times between 100 and 10,000 s were found. Alternatively, as shown in Figure 6.4, in the direction of the field the percolation threshold could be reduced by a factor of 5 with the application of an electric field. However, with increasing field a plateau was reached.[113] These experiments suggest that there is a limit to the lowest percolation threshold achievable via kinetic percolation for a given nanoscale dispersion of tubes since an aligned agglomeration process is expected to produce a lower percolation threshold than an isotropic agglomeration process.

The concept of "double percolation" is mentioned in Chapter 3 in relation to polymer blends. If nanotubes are isolated in the continuous phase of a two-component blend, then the percolation threshold can be reduced dramatically when the weight or volume fraction is normalized to the total amount of polymer. This effect has been seen for blends of polyamide/polypropylene,[114] polyamide/acrylonitrile–butadiene–styrene,[65,85,115,116] polypropylene/acrylonitrile–butadiene–styrene,[66] styrene–acrylonitrile/polycarbonate,[117] polyethylene/polycarbonate,[118] and poly(caprolactone)/polylactide.[119] A detailed examination of these studies suggests that the percolation threshold calculated on a volume fraction basis with the normalization being done to the volume of polymer that contains the nanotubes, rather than the entire amount of polymer present, does not change upon blending. In fact, a plot of electrical conductivity versus amount of nanotubes added should be similar, if not identical, if only the phase containing the nanotubes is considered. Such

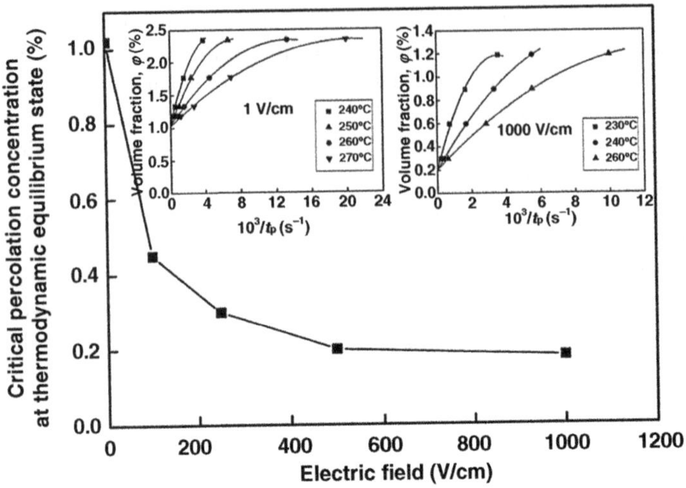

Figure 6.4 Dependence of percolation threshold on the applied electric field for MWCNTs in a polycarbonate melt. The two insets show the relationship between MWCNT volume fraction and the inverse of the time required to achieve maximum conductivity as a function of temperature. Copyright American Institute of Physics. Reproduced with permission from Ref. 113.

behavior was found when a second filler was added; for three polymers where composites were made with melt mixing and calcium carbonate filler was added, the percolation curves, when normalized to (polymer + nanotube) rather than (polymer + nanotube + $CaCO_3$), were identical.[45] Deviations from a volume exclusion argument might be expected when at least one dimension of the phase becomes very small, which could occur, for example, when the sample is not isotropic. Alternatively, adding clay to a nanotube-filled epoxy resulted in a much smaller percolation threshold than a simple volume exclusion effect; the authors showed that nanotubes tend to cluster around the clay that in turn helps to form a continuous network at a much lower value than a simple volume exclusion effect can explain.[120] Similarly, but in the opposite direction, the addition of cross-linked polymer particles to poly(methyl methacrylate) caused an increase in the percolation threshold, which was attributed to poorer dispersion induced by the polymer particles.[121]

6.2.4 Dielectric Constant

Alternating current measurement is used to calculate the dielectric constant, which is a measure of the charge stored rather than charge transported (which is what conductivity measures). The dielectric constant is the in-phase part of the measurement; the out-of-phase part of the measurement is the conductivity. At very low frequencies, the AC conductivity becomes equivalent to the direct current conductivity (most of the measurements in Table 6.1 were DC measurements although some were AC measurements at low frequency). The large difference in dielectric constant between the insulating polymer matrix and the conductive filler results in large dielectric constants at low frequencies. The behavior of the dielectric constant at low frequencies is qualitatively identical to that of electrical conductivity; that is, the dielectric constant increases abruptly at percolation and then reaches a plateau region. However, as more nanotubes are added, the sample becomes more conductive and the dielectric constant should decrease. Most studies on the dielectric constant never achieve this decreasing region, although there are exceptions.[82]

A critical scaling law is also an appropriate way to fit data to changes in the dielectric constant with nanotube concentration. Percolation thresholds calculated from the dielectric constant and conductivity are identical or nearly identical; however, the underlying physics of the dielectric constant are not the same as the conductivity, and hence the critical exponents will be different. In two studies on poly (vinylidene fluoride)/MWCNT composites, percolation thresholds were identical. However, for one study, the critical exponent for electrical conduction was 1.2 and that for dielectric constant 3.2,[122] while for another study, the critical exponent for electrical conduction was 3.2 and that for dielectric constant 1.1.[55]

A simple macroscopic view would argue against critical scaling behavior as exemplified by Equation 6.2 because, as stated above, the dielectric constant should decrease rather than increase with increasing conductivity. Further, the statistical approach that leads to a scaling law in the case of electrical conductivity is not really relevant for the dielectric constant. The anomalous behavior has been explained in terms of the existence of minicapacitors, e.g., parts of two nanotubes close together separated by a low dielectric constant medium.[123,124] One set of authors used simple composite theory and experimental data to show that the dielectric constant of pristine SWCNTs was about 2000,[125] agreeing with electrophoresis experiments[126] but disagreeing with theoretical studies.[127] Unlike electrical conductivity, chemical functionalization of tubes can increase the dielectric constant in a composite.[82,128] Although not yet proven experimentally, this procedure should be effective only for MWCNTs and perhaps DWCNTs since functionalization effectively makes the outer tube insulating, and hence the inner tube–outer tube/conductive–insulating arrangement can contribute to charge storage. With SWCNTs, such an effect is not possible. Because the dielectric constant is also a function of dispersion, it is not clear what functionalization strategies will be most effective. In fact, different functionalization chemistries have been shown to change the maximum dielectric constant achieved in MWCNT composites.[82] Finally, the dielectric constant of a composite made with a mixture of SWCNTs, DWCNTs, and MWCNTs has been tuned with an external voltage field.[129]

6.3 IMPREGNATED/INFUSED COMPOSITES

The fundamental picture of conduction in nonimpregnated composites is that the most important resistance is contact resistance that is composed of two parts: tunneling resistance and constriction resistance. Percolation theory is a key concept that must be understood in order to understand conductivity. In impregnated/infused composites, percolation is irrelevant since a continuous network has already been formed. However, contact resistance is still the most important source of resistance. Unless the monomer or polymer has penetrated and separated the nanotubes, tunneling resistance will also be irrelevant, and hence the only source of resistance will be constriction resistance. Given the small decreases in conductance that generally occur after infusion, as will be described later, and the generally much higher conductivities than those in nonimpregnated systems, tunneling resistance is likely a minor contributor in impregnated/infused composites.

As described in Chapter 3, the most common method to make mats or fibers of pure nanotubes is to initially disperse nanotubes in a liquid and then separate or evaporate the liquid in such a manner so as to leave a nanotube mat or fiber. The conductivity of such a mat or fiber is much larger than that of the composites described in Section 6.2 and shown in Table 6.1, with typical conductivities on the order of 10^2–10^5 S/m as shown in the table appearing in Ref. 130. Even if monomer or polymer is able to separate the nanotubes somewhat during the infusion process, the conductivity is still higher than that in the case where nanotubes are well dispersed in polymer or monomer.

A nanotube yarn made from drawing from a MWCNT nanotube forest and then braiding five or six of these into a 5–6-ply braid had an electrical conductivity of 9000 S/m. After infusion of epoxy resin, the conductivity dropped by about 10%.[131] Using a laminated-type structure of alternating polyethylene sheets and buckypapers with various conductivities (because different tubes were used) showed a constant factor of 2–4 decrease in conductivity after infusion.[132] Using a solution infiltration process with polycarbonate led to a factor of 1.5–3 decrease in conductivity at 40–60% nanotube content. This study also found that the nanotubes "filtered" the polycarbonate in that the polycarbonate concentration in the top part of the buckypaper was higher than that in the bottom part (the resin flowed top to bottom).[133] Another report compared the solution procedure and a vacuum method to better distribute the polymer and found that the latter decreased the conductivity by a much larger factor of 6 and higher although the nanotube content was not stated.[130] A quantitative relationship between nanotube content and conductivity for an infused epoxy resin into a MWCNT nanotube mat was developed by Galiotis and coworkers, and the data are represented graphically in Figure 6.5. The lowest nanotube content material still has an electrical conductivity of approximately three orders of magnitude higher than that of the composite made from the same materials but mixed in a dispersion–reaction scheme, even though the nanotube contents are the same.[134] These results, in aggregate, strongly suggest that in general tunneling resistance in composites made from impregnation/infusion processes is negligible or at least much lower than composites made from the methods described in the previous section. Consistent with this statement is that the temperature dependence of the conductivity of buckypaper is shown to be quite small unlike the behavior in mixed composites, where tunneling is known to be very important; the conductance of buckypaper was linear with temperature and increased with increasing temperature only by about 10% with a 200 K change in temperature centered around room temperature.[135] At significantly lower temperatures (approximately less than 100 K), the conductivity drops substantially,[136] while this slowly increasing function continues until about 450 K, at which point the conductivity rises substantially with temperature.[137]

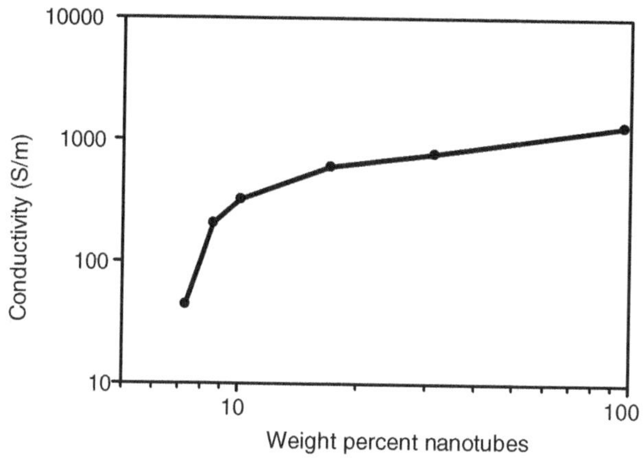

Figure 6.5 Dependence of electrical conductivity on MWCNT/epoxy composite as a function of nanotube content where epoxy resin is infused using acetone. The pure buckypaper was not 100% nanotubes likely due to incomplete removal of the catalyst. Copyright IOPscience. Data used with permission from Ref. 134.

To this author's knowledge, no studies have been published concerning the effect of temperature on conductivity in infused composites.

Nanotube properties are important in infused composites as well. An experimental study in an epoxy matrix found that the conductivity of a buckypaper/epoxy composite increased by a factor of 10 when the tube aspect ratio increased by a factor of 5.5.[138] Coating MWCNT buckypaper with copper with a very simple solution deposition process causes an increase in conductivity from a starting value about 3000 to 10,000 S/m at 20.5% copper content.[139]

6.4 COMPOSITES WITH ELECTRICALLY CONDUCTING POLYMERS

A small number of polymers are electrically conducting; that is, they have conductivities greater than 1 S/m, which is much greater than the less than 10^{-14} S/m for most polymers. The most important conducting polymers are polyacetylene, polyparaphenylenevinylene, polypyrrole, polythiophene, and polyaniline. Structures of these polymers are shown schematically in Figure 6.6. Unsaturation along the main chain (e.g., double or triple bonds) is common to all electrically conducting polymers, and the exact arrangement of the unsaturation is critical in leading to electrical conductivity. In their pure state, these polymers are semiconducting with very large band gaps, meaning that the room temperature conductivity is on the order of 10^{-8}–10^{-6} S/m. Doping is required to reach conductivities that allow this polymer to be classified as conductive, which usually means adding an electron-withdrawing small molecule or atom. Doping agents for electrically conducting polymers are generally oxidants and include sulfonic and other acids as well as group VII elements. Doping can be used to control the actual conductivity although precise control to

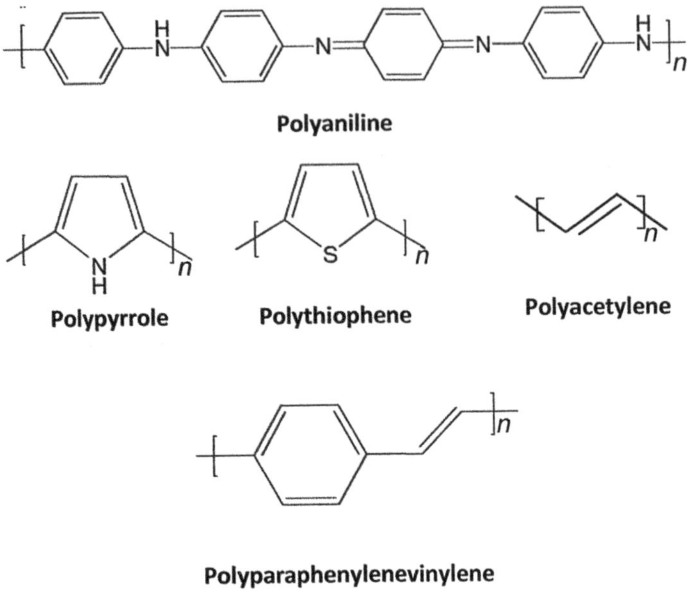

Figure 6.6 Structures of common electrically conducting polymers.

values less than ~1 S/m is difficult. The highest conductivity at the optimal doping level depends on the molecular organization; conduction is along the polymer chain, and hence the higher the chain orientation, the higher the conductivity in the orientation direction. Conductivity of unoriented films is typically between 1×10^4 and 5×10^4 S/m, with orientation increasing these values to as high as 2×10^5 S/m.[140] Their commercial importance is much less than the attention given to these materials in the literature; for example, the number of papers containing the keyword "polyaniline" is approximately 40% that of the number of papers containing the keyword "polypropylene,"[141] yet both the commercial production and the market value of the latter is many orders of magnitude larger than that of polyaniline.

The methods of production of nanotubes with electrically conducting polymers comprise all those shown in Figure 3.2 except melt mixing, since electrically conducting polymers will degrade prior to becoming a melt. As with thermosets, the most common method of preparation is dispersion–reaction although a solvent is always used with electrically conducting polymers. Section 3.4.1 describes a polymerization method not available with other monomers that takes advantage of conductivity, namely, electrochemical polymerization. Oxidative stability is an important issue for all electrically conducting polymers, since the unsaturation required for electrical conduction makes the polymer very susceptible to oxidation. Nanotubes have been shown to increase the decomposition temperature of electrically conducting polymers.[142] The mechanical properties of electrically conducting polymers tend to be quite poor in that these materials are extremely brittle; the addition of nanotubes has been found to both increase[143] and decrease[144] brittleness. Also, as would be expected owing to the conjugated nature of both nanotubes and electrically conducting polymers, there tend to be significant molecular level interactions between the two, leading to shifts in various spectra (Raman, UV–Vis, and others). A large number of studies describe the details of these changes, but these details are beyond the scope of this book. A large number of studies about nanotube/electrically conducting polymer composites concern potential applications, and these will be discussed in detail in Chapter 8.

Since electrical conduction is a key aspect of nanotubes, their composites with electrically conducting polymers are critically important. The vast majority of these studies report measurements of electrical properties, primarily in relation to a specific synthesis technique. This section discusses only a small fraction of the studies that have measured the electrical properties of nanotube/electrically conducting composites.

In general, the same sort of percolation law does not apply when nanotubes are mixed with a conducting polymer because a low-concentration region where conductivity increases slowly with nanotube amount does not exist (although there are exceptions to this generality[145]). Instead, conductivity increases immediately with the addition of nanotubes, followed by a region where the conductivity increases less quickly.[143,146–148] Typical increases in conductivity depend primarily on the starting conductivity of the conducting polymer, and the higher the starting conductivity, the lesser the increase. In the region where the conductivity increases more slowly, the conductivity would be termed a "plateau" region in composites with insulating polymers, and the fact is that the conductivity increase in this region for composites of nanotubes and conducting polymers is approximately the same as that found for composites of nanotubes and insulating polymers. However, the data do not appear as a plateau in plots, since the increase in conductivity with the addition of nanotubes at low nanotube concentrations is much less in a conducting polymer than in an insulating polymer. An example of such a plot is shown in Figure 6.7.

The temperature dependence of the conductivity for an electrically conducting polymer generally follows a $\sigma = \sigma_0 e^{(-T_0/T)}$ dependence. With the addition of nanotubes in the plateau concentration region, this formula has been found to still apply, but the dependence with temperature is significantly less. The drop in conductivity from room temperature to ~10 K is about one order of magnitude for an electrically conducting polymer containing nanotubes while that of the pure electrically conducting polymer the drop is 20 orders of magnitude. In the lower nanotube concentration region where the conductivity increases more strongly with nanotube addition, the dependence is more complicated. At high temperatures, the temperature behavior of the polyaniline dominates the temperature dependence of conductivity, while at low temperatures the temperature dependence of the nanotubes dominates the temperature dependence of the conductivity.[146]

A Schottky diode is a device that has a metal attached to a semiconductor, and it conducts current in only one direction. Especially at low doping levels, conducting polymers can actually be classified as semiconductors, and hence laboratory devices are built with the conducting polymer acting as the semiconductor. The key characteristics are the reverse voltage that leads to device breakdown, the forward voltage required to cause conduction, and the relationship between voltage and current at high forward voltages. As expected, the addition of nanotubes to an electrically conducting polymer that is acting as one part of a Schottky diode causes

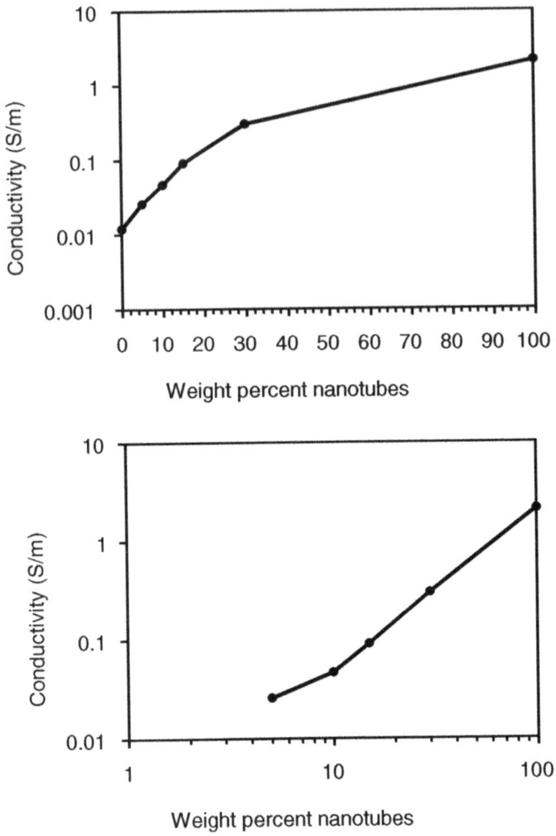

Figure 6.7 Representative (qualitatively, not necessarily quantitatively) plot of conductivity versus weight percent nanotubes for a composite made from nanotubes (in this case multiwalled) and an electrically conducting polymer (in this case doped polyaniline). Copyright Wiley-VCH Verlag GmbH & Co. KGaA. Data used with permission from Ref. 147.

an increase in conduction, although the nanotube level must be below a certain level so that the semiconductor does not become too conducting. At voltages greater than a value on the order of 1 V, the conductivity is no longer ohmic but rather increases more rapidly than a linear relationship would predict and is consistent with Child's law of space charge limited emission. Such behavior is the result of electrical defects in the system, and possibly better fabrication procedures could improve the performance.[149,150] Not surprisingly, the forward voltage required to cause conduction decreases with an increase in carbon nanotube content.[150,151] To the author's knowledge, no studies have examined what change, if any, nanotubes will cause in the reverse voltage necessary for device breakdown.

6.5 CHALLENGES

Although the electrical properties of polymers filled with nanotubes have already led to commercial applications, other applications will be limited unless higher conductivities can be achieved. The clear advantage of nanotube composites over those filled with carbon black is their low percolation threshold; a substantial disadvantage is that the number of filler–polymer junctions in a given conduction path is usually higher with nanotubes than with carbon black causing a larger effect for interfacial resistance. Reducing the number of filler–polymer junctions, both by lengthening the tubes and by improving orientation, is a key challenge. In fact, given the importance of orientation with respect to electrical conduction, the number of studies that have quantified orientation and electrical conductivity in various directions is surprisingly few. Schemes to reduce constriction or tunneling resistance would also be very attractive.

The influence of nanotubes on the electrical conductivity of composites with electrically conducting polymers is rather small, assuming the latter is doped well. As will be described in detail in Chapter 8, the number of applications that could result from such materials is quite high, which is represented by the fact that between 10% and 15% of all studies that describe nanotube–polymer composites also contain a keyword referring to an electrically conducting polymer. Challenges in this area from the point of view of this chapter are not many, although certainly higher conductivities are always desired. With electrically conducting polymers, tunneling resistance is a much lesser contributor to conductivity restriction, and hence tubes with higher conductivities could have more impact in terms of composite conductivity.

REFERENCES

1. Bonavolonta, C., Valentino, M., Meola, C., Carlomagno, G. M., Volponi, R., Rosca, I. D. (2009). Non-destructive testing of a carbon-nanotube-reinforced composite using HTS-SQUID and electromagnetic techniques, *Supercond. Sci. Technol.*, *22*, 095001.
2. Dai, H. J., Wong, E. W., Lieber, C. M. (1996). Probing electrical transport in nanomaterials: conductivity of individual carbon nanotubes, *Science*, *272*, 523–526.
3. Ebbesen, T. W., Lezec, H. J., Hiura, H., Bennett, J. W., Ghaemi, H. F., Thio, T. (1996). Electrical conductivity of individual carbon nanotubes, *Nature*, *382*, 54–56.
4. Thess, A., Lee, R., Nikolaev, P., Dai, H. J., Petit, P., Robert, J., Xu, C. H., Lee, Y. H., Kim, S. G., Rinzler, A. G., Colbert, D. T., Scuseria, G. E., Tomanek, D., Fischer, J. E., Smalley, R. E. (1996). Crystalline ropes of metallic carbon nanotubes, *Science*, *273*, 483–487.
5. Sun, X. X., Song, M. (2009). Highly conductive carbon nanotube/polymer nanocomposites achievable? *Macromol. Theory Simul.*, *18*, 155–161.
6. Carmona, F., Canet, R., Delhaes, P. (1987). Piezoresistivity of heterogeneous solids, *J. Appl. Phys.*, *61*, 2550–2557.
7. Yoshikawa, S., Ota, T., Newnham, R., Amin, A. (1990). Piezoresistivity in polymer–ceramic composites, *J. Am. Ceram. Soc.*, *73*, 263–267.
8. Ruschau, G. R., Yoshikawa, S., Newnham, R. E. (1992). Resistivities of conductive composites, *J. Appl. Phys.*, *72*, 953–959.
9. Fuhrer, M. S., Nygard, J., Shih, L., Forero, M., Yoon, Y. G., Mazzoni, M. S. C., Choi, H. J., Ihm, J., Louie, S. G., Zettl, A., McEuen, P. L. (2000). Crossed nanotube junctions, *Science*, *288*, 494–497.
10. Li, C. Y., Thostenson, E. T., Chou, T. W. (2007). Dominant role of tunneling resistance in the electrical conductivity of carbon nanotube-based composites, *Appl. Phys. Lett.*, *91*, 223114.
11. Bauhofer, W., Kovacs, J. Z. (2009). A review and analysis of electrical percolation in carbon nanotube polymer composites, *Compos. Sci. Technol.*, *69*, 1486–1498.
12. Zhang, R., Dowden, A., Deng, H., Baxendale, M., Peijs, T. (2009). Conductive network formation in the melt of carbon nanotube/thermoplastic polyurethane composite, *Compos. Sci. Technol.*, *69*, 1499–1504.
13. Kovacs, J. Z., Velagala, B. S., Schulte, K., Bauhofer, W. (2007). Two percolation thresholds in carbon nanotube epoxy composites, *Compos. Sci. Technol.*, *67*, 922–928.

14. Kilbride, B. E., Coleman, J. N., Fraysse, J., Fournet, P., Cadek, M., Drury, A., Hutzler, S., Roth, S., Blau, W. J. (2002). Experimental observation of scaling laws for alternating current and direct current conductivity in polymer–carbon nanotube composite thin films, *J. Appl. Phys.*, *92*, 4024–4030.
15. Foygel, M., Morris, R. D., Anez, D., French, S., Sobolev, V. L. (2005). Theoretical and computational studies of carbon nanotube composites and suspensions: electrical and thermal conductivity, *Phys. Rev. B*, *71*, 104201.
16. Yan, K. Y., Xue, Q. Z., Zheng, Q. B., Hao, L. Z. (2007). The interface effect of the effective electrical conductivity of carbon nanotube composites, *Nanotechnology*, *18*, 255705.
17. Blighe, F. M., Lyons, P. E., De, S., Blau, W. J., Coleman, J. N. (2008). On the factors controlling the mechanical properties of nanotube films, *Carbon*, *46*, 41–47.
18. Kim, H. M., Kim, K., Lee, C. Y., Joo, J., Cho, S. J., Yoon, H. S., Pejakovic, D. A., Yoo, J. W., Epstein, A. J. (2004). Electrical conductivity and electromagnetic interference shielding of multiwalled carbon nanotube composites containing Fe catalyst, *Appl. Phys. Lett.*, *84*, 589–591.
19. Weber, M., Kamal, M. R. (1997). Estimation of the volume resistivity of electrically conductive composites, *Polym. Compos.*, *18*, 711–725.
20. Gingold, D. B., Lobb, C. J. (1990). Percolative conduction in three dimensions, *Phys. Rev. B*, *42*, 8220–8224.
21. Mazinani, S., Ajji, A., Dubois, C. (2009). Morphology, structure and properties of conductive PS/CNT nanocomposite electrospun mat, *Polymer*, *50*, 3329–3342.
22. Cipriano, B. H., Kota, A. K., Gershon, A. L., Laskowski, C. J., Kashiwagi, T., Bruck, H. A., Raghavan, S. R. (2008). Conductivity enhancement of carbon nanotube and nanofiber-based polymer nanocomposites by melt annealing, *Polymer*, *49*, 4846–4851.
23. Tchoul, M. N., Ford, W. T., Ha, M. L. P., Chavez-Sumarriva, I., Grady, B. P., Lolli, G. L., Resasco, D. E., Arepalli, S. (2008). Composites of single-walled carbon nanotubes and polystyrene: preparation and electrical conductivity, *Chem. Mater.*, *20*, 3120–3126.
24. Blighe, F. M., Hernandez, Y. R., Blau, W. J., Coleman, J. N. (2007). Observation of percolation-like scaling—far from the percolation threshold—in high volume fraction, high conductivity polymer–-nanotube composite films, *Adv. Mater.*, *19*, 4443–4447.
25. Kota, A. K., Cipriano, B. H., Powell, D., Raghavan, S. R., Bruck, H. A. (2007). Quantitative characterization of the formation of an interpenetrating phase composite in polystyrene from the percolation of multiwalled carbon nanotubes, *Nanotechnology*, *18*, 505705.
26. Kota, A. K., Cipriano, B. H., Duesterberg, M. K., Gershon, A. L., Powell, D., Raghavan, S. R., Bruck, H. A. (2007). Electrical and rheological percolation in polystyrene/MWCNT nanocomposites, *Macromolecules*, *40*, 7400–7406.
27. Fragneaud, B., Masenelli-Varlot, K., Gonzalez-Montiel, A., Terrones, M., Cavaille, J. Y. (2007). Electrical behavior of polymer grafted nanotubes/polymer nanocomposites using N-doped carbon nanotubes, *Chem. Phys. Lett.*, *444*, 1–8.
28. Datsyuk, V., Billon, L., Guerret-Piecourt, C., Dagreou, S., Passade-Boupatt, N., Bourrigaud, S., Guerret, O., Couvreur, L. (2007). *In situ* nitroxide-mediated polymerized poly(acrylic acid) as a stabilizer/compatibilizer carbon nanotube/polymer composites, *J. Nanomater.* 74769.
29. Jin, S. H., Choi, D. K., Lee, D. S. (2008). Electrical and rheological properties of polycarbonate/multiwalled carbon nanotube nanocomposites, *Colloids Surf. A.*, *313*, 242–245.
30. Bonnet, P., Sireude, D., Garnier, B., Chauvet, O. (2007). Thermal properties and percolation in carbon nanotube–polymer composites, *Appl. Phys. Lett.*, *91*, 201910.
31. Das, N. C., Liu, Y. Y., Yang, K. K., Peng, W. Q., Maiti, S., Wang, H. (2009). Single-walled carbon nanotube/poly(methyl methacrylate) composites for electromagnetic interference shielding, *Polym. Eng. Sci.*, *49*, 1627–1634.
32. Shang, S. M., Li, L., Yang, X. M., Wei, Y. Y. (2009). Polymethylmethacrylate–carbon nanotubes composites prepared by microemulsion polymerization for gas sensor, *Compos. Sci. Technol.*, *69*, 1156–1159.
33. Das, N. C., Maiti, S. (2008). Electromagnetic interference shielding of carbon nanotube/ethylene vinyl acetate composites, *J. Mater. Sci.*, *43*, 1920–1925.
34. Jin, S. H., Kang, I. H., Kim, Y. S., Park, C. Y., Lee, D. S. (2008). Thermal and electrical properties of nanocomposites based on acrylic copolymers and multiwalled carbon nanotube, *J. Nanosci Nanotechnol.*, *8*, 5076–5079.

35. Linares, A., Canalda, J. C., Cagiao, M. E., Garcia-Gutierrez, M. C., Nogales, A., Martin-Gullom, I., Vera, J., Ezquerra, T. A. (2008). Broad-band electrical conductivity of high density polyethylene nanocomposites with carbon nanoadditives: multiwall carbon nanotubes and carbon nanofibers, *Macromolecules*, *41*, 7090–7097.
36. Liu, F., Zhang, X. B., Li, W. C., Cheng, J. P., Tao, X. Y., Li, Y., Sheng, L. (2009). Investigation of the electrical conductivity of HDPE composites filled with bundle-like MWNTs, *Compos. Part A*, *40*, 1717–1721.
37. Kim, J., Hong, S. M., Kwak, S., Seo, Y. (2009). Physical properties of nanocomposites prepared by *in situ* polymerization of high-density polyethylene on multiwalled carbon nanotubes, *Phys. Chem. Chem. Phys.*, *11*, 10851–10859.
38. Jeon, K., Lumata, L., Tokumoto, T., Steven, E., Brooks, J., Alamo, R. G. (2007). Low electrical conductivity threshold and crystalline morphology of single-walled carbon nanotubes: high density polyethylene nanocomposites characterized by SEM, Raman spectroscopy and AFM, *Polymer*, *48*, 4751–4764.
39. Fu, M., Yu, Y., Xie, J. J., Wang, L. P., Fan, M. Y., Jiang, S. L., Zeng, Y. K. (2009). Significant influence of film thickness on the percolation threshold of multiwall carbon nanotube/low density polyethylene composite films, *Appl. Phys. Lett.*, *94*, 012904.
40. Xi, Y., Yamanaka, A., Bin, Y. Z., Matsuo, M. (2007). Electrical properties of segregated ultrahigh molecular weight polyethylene/multiwalled carbon nanotube composites, *J. Appl. Polym. Sci.*, *105*, 2868–2876.
41. Grady, B. P., Arthur, D. J., Ferguson, J. (2009). Single-walled carbon nanotube/ultrahigh-molecular-weight polyethylene composites with percolation at low nanotube contents, *Polym. Eng. Sci.*, *49*, 2440–2446.
42. Gao, J. F., Li, Z. M., Meng, Q. J., Yang, Q. (2008). CNTs/UHMWPE composites with a two-dimensional conductive network, *Mater. Lett.*, *62*, 3530–3532.
43. Mierczynska, A., Mayne-L'Hermite, M., Boiteux, G., Jeszka, J. K. (2007). Electrical and mechanical properties of carbon nanotube/ultrahigh-molecular-weight polyethylene composites prepared by a filler prelocalization method, *J Appl. Polym. Sci.*, *105*, 158–168.
44. Hernandez, J. J., Garcia-Gutierrez, M. C., Nogales, A., Rueda, D. R., Kwiatkowska, M., Szymczyk, A., Roslaniec, Z., Concheso, A., Guinea, I., Ezquerra, T. A. (2009). Influence of preparation procedure on the conductivity and transparency of SWCNT–polymer nanocomposites, *Compos. Sci. Technol.*, *69*, 1867–1872.
45. Bao, H. D., Guo, Z. X., Yu, J. (2008). Effect of electrically inert particulate filler on electrical resistivity of polymer/multi-walled carbon nanotube composites, *Polymer*, *49*, 3826–3831.
46. Deng, H., Zhang, R., Bilotti, E., Loos, J., Peijs, T. (2009). Conductive polymer tape containing highly oriented carbon nanofillers, *J. Appl. Polym. Sci.*, *113*, 742–751.
47. Mamunya, Y., Boudenne, A., Lebovka, N., Ibos, L., Candau, Y., Lisunova, M. (2008). Electrical and thermophysical behaviour of PVC–MWCNT nanocomposites, *Compos. Sci. Technol.*, *68*, 1981–1988.
48. Bangarusampath, D. S., Ruckdaschel, H., Altstadt, V., Sandler, J. K. W., Garray, D., Shaffer, M. S. P. (2009). Rheology and properties of melt-processed poly(ether ether ketone)/multi-wall carbon nanotube composites, *Polymer*, *50*, 5803–5811.
49. Bangarusampath, D. S., Ruckdaschel, H., Altstadt, V., Sandler, J. K. W., Garray, D., Shaffer, M. S. P. (2009). Rheological and electrical percolation in melt-processed poly(ether ether ketone)/multi-wall carbon nanotube composites, *Chem. Phys. Lett.*, *482*, 105–109.
50. Zhang, R., Baxendale, M., Peijs, T. (2007). Universal resistivity–strain dependence of carbon nanotube/polymer composites, *Phys. Rev. B*, *76*, 195433.
51. Wu, D. F., Wu, L. F., Zhou, W. D., Yang, T., Zhang, M. (2009). Study on physical properties of multiwalled carbon nanotube/poly(phenylene sulfide) composites, *Polym. Eng. Sci.*, *49*, 1727–1735.
52. Jang, Y. K., Jang, P. G., Kim, J. K., Park, M., Yoon, H. G. (2009). Electrical properties of imidazole-modified MWNT/polyphenylenesulfide composites prepared by melt mixing, *J. Nanosci. Nanotechnol.*, *9*, 4180–4186.
53. Yang, J. H., Xu, T., Lu, A., Zhang, Q., Fu, Q. (2008). Electrical properties of poly(phenylene sulfide)/multiwalled carbon nanotube composites prepared by simple mixing and compression, *J. Appl. Polym. Sci.*, *109*, 720–726.

54. Almasri, A., Ounaies, Z., Kim, Y. S., Grunlan, J. (2008). Characterization of solution-processed double-walled carbon nanotube/poly(vinylidene fluoride) nanocomposites, *Macromol. Mater. Eng.*, *293*, 123–131.
55. Li, Q., Xue, Q. Z., Zheng, Q. B., Hao, L. Z., Gao, X. L. (2008). Large dielectric constant of the chemically purified carbon nanotube/polymer composites, *Mater. Lett.*, *62*, 4229–4231.
56. Chang, L., Friedrich, K., Ye, L., Toro, P. (2009). Evaluation and visualization of the percolating networks in multi-wall carbon nanotube/epoxy composites, *J. Mater. Sci.*, *44*, 4003–4012.
57. Hu, N., Masuda, Z., Yamamoto, G., Fukunaga, H., Hashida, T., Qiu, J. (2008). Effect of fabrication process on electrical properties of polymer/multi-wall carbon nanotube nanocomposites, *Compos. Part A*, *39*, 893–903.
58. Sumfleth, J., Adroher, X. C., Schulte, K. (2009). Synergistic effects in network formation and electrical properties of hybrid epoxy nanocomposites containing multi-wall carbon nanotubes and carbon black, *J. Mater. Sci.*, *44*, 3241–3247.
59. Kovacs, J. Z., Mandjarov, R. E., Blisnjuk, T., Prehn, K., Sussiek, M., Muller, J., Schulte, K., Bauhofer, W. (2009). On the influence of nanotube properties, processing conditions and shear forces on the electrical conductivity of carbon nanotube epoxy composites, *Nanotechnology*, *20*, 155703.
60. Wang, Q., Dai, J. F., Li, W. X., Wei, Z. Q., Jiang, J. L. (2008). The effects of CNT alignment on electrical conductivity and mechanical properties of SWNT/epoxy nanocomposites, *Compos. Sci. Technol.*, *68*, 1644–1648.
61. Huang, Y., Li, N., Ma, Y. F., Feng, D., Li, F. F., He, X. B., Lin, X., Gao, H. J., Chen, Y. S. (2007). The influence of single-walled carbon nanotube structure on the electromagnetic interference shielding efficiency of its epoxy composites, *Carbon*, *45*, 1614–1621.
62. Thuau, D., Koutsos, V., Cheung, R. (2009). Electrical and mechanical properties of carbon nanotube–polyimide composites, *J. Vac. Sci. Technol. B*, *27*, 3139–3144.
63. Shi, S. L., Zhang, L. Z., Li, J. S. (2009). Electrical and dielectric properties of multiwall carbon nanotube/polyaniline composites, *J. Polym. Res.*, *16*, 395–399.
64. Pedroni, L. G., Soto-Oviedo, M. A., Rosolen, J. M., Felisberti, M. I., Nogueira, A. F. (2009). Conductivity and mechanical properties of composites based on MWCNTs and styrene–butadiene–styrene block (TM) copolymers, *J. Appl. Polym. Sci.*, *112*, 3241–3248.
65. Bose, S., Bhattacharyya, A. R., Kulkarni, A. R., Pötschke, P. (2009). Electrical, rheological and morphological studies in co-continuous blends of polyamide 6 and acrylonitrile–butadiene–styrene with multiwall carbon nanotubes prepared by melt blending, *Compos. Sci. Technol.*, *69*, 365–372.
66. Khare, R. A., Bhattacharyya, A. R., Kulkarni, A. R., Sarp, M., Biswas, A. (2008). Influence of multiwall carbon nanotubes on morphology and electrical conductivity of PP/ABS blends, *J. Polym. Sci. Polym. Phys.*, *46*, 2286–2295.
67. Hu, N., Masuda, Z., Yan, C., Yamamoto, G., Fukunaga, H., Hashida, T. (2008). The electrical properties of polymer nanocomposites with carbon nanotube fillers, *Nanotechnology*, *19*, 215701.
68. Celzard, A., McRae, E., Deleuze, C., Dufort, M., Furdin, G., Mareche, J. F. (1996). Critical concentration in percolating systems containing a high-aspect-ratio filler, *Phys. Rev. B*, *53*, 6209–6214.
69. Natsuki, T., Endo, M., Takahashi, T. (2005). Percolation study of orientated short-fiber composites by a continuum model, *Physica A*, *352*, 498–508.
70. Wescott, J. T., Kung, P., Maiti, A. (2007). Conductivity of carbon nanotube polymer composites, *Appl. Phys. Lett.*, *90*, 033116.
71. Dalmas, F., Dendievel, R., Chazeau, L., Cavaille, J. Y., Gauthier, C. (2006). Carbon nanotube-filled polymer of electrical conductivity in composites. *Numerical simulation three-dimensional entangled fibrous networks, Acta Mater.*, *54*, 2923–2931.
72. Berhan, L., Sastry, A. M. (2007). Modeling percolation in high-aspect-ratio fiber systems. II. The effect of waviness on the percolation onset, *Phys. Rev. E*, *75*, 041121.
73. Li, C. Y., Chou, T. W. (2007). Continuum percolation of nanocomposites with fillers of arbitrary shapes, *Appl. Phys. Lett.*, *90*, 174108.
74. Li, C., Thostenson, E. T., Chou, T. W. (2008). Effect of nanotube waviness on the electrical conductivity of carbon nanotube-based composites, *Compos. Sci. Technol.*, *68*, 1445–1452.
75. Berhan, L., Sastry, A. M. (2007). Modeling percolation in high-aspect-ratio fiber systems. I. Soft-core versus hard-core models, *Phys. Rev. E*, *75*, 041120.

76. Grossiord, N., Loos, J., van Laake, L., Maugey, M., Zakri, C., Koning, C. E., Hart, A. J. (2008). High-conductivity polymer nanocomposites obtained by tailoring the characteristics of carbon nanotube fillers, *Adv. Funct. Mater.*, *18*, 3226–3234.
77. Bin, Y., Chen, Q., Tashiro, K., Matsuo, M. (2008). Electrical and mechanical properties of iodine-doped highly elongated ultrahigh molecular weight polyethylene films filled with multiwalled carbon nanotubes, *Phys. Rev. B*, *77*, 035419.
78. Sankapal, B. R., Setyowati, K., Chen, J., Liu, H. (2007). Electrical properties of air-stable, iodine-doped carbon nanotube–polymer composites, *Appl. Phys. Lett.*, *91*, 173103.
79. Balberg, I., Binenbaum, N. (1983). Computer study of the percolation threshold in a two-dimensional anisotropic system of conducting sticks, *Phys Rev B*, *28*, 3799–3812.
80. Li, J., Lumpp, J. K., Andrews, R., Jacques, D. (2008). Aspect ratio and loading effects of multiwall carbon nanotubes in epoxy for electrically conductive adhesives, *J. Adhes. Sci. Technol.*, *22*, 1659–1671.
81. Pötschke, P., Pegel, S., Claes, M., Bonduel, D. (2008). A novel strategy to incorporate carbon nanotubes into thermoplastic matrices, *Macromol. Rapid Commun.*, *29*, 244–251.
82. Li, Q., Xue, Q. Z., Hao, L. Z., Gao, X. L., Zheng, Q. B. (2008). Large dielectric constant of the chemically functionalized carbon nanotube/polymer composites, *Compos. Sci. Technol.*, *68*, 2290–2296.
83. Lee, S. H., Cho, E., Jeon, S. H., Youn, J. R. (2007). Rheological and electrical properties of polypropylene composites containing functionalized multi-walled carbon nanotubes and compatibilizers, *Carbon*, *45*, 2810–2822.
84. Jang, P. G., Suh, K. S., Park, M., Kim, J. K., Kim, W. N., Yoon, H. G. (2007). Electrical Behavior of polyurethane composites with acid treatment-induced damage to multiwalled carbon nanotubes, *J. Appl. Polym. Sci.*, *106*, 110–116.
85. Bose, S., Bhattacharyya, A. R., Bondre, A. P., Kulkarni, A. R., Pötschke, P. (2008). Rheology, electrical conductivity, and the phase behavior of cocontinuous PA6/ABS blends with MWNT: correlating the aspect ratio of MWNT with the percolation threshold, *J. Polym. Sci. Polym. Phys.*, *46*, 1619–1631.
86. Grossiord, N., Miltner, H. E., Loos, J., Meuldijk, J., Van Mele, B., Koning, C. E. (2007). On the crucial role of wetting in the preparation of conductive polystyrene–carbon nanotube composites, *Chem. Mater.*, *19*, 3787–3792.
87. Hermant, M. C., Smeets, N. M. B., van Hal, R. C. F., Meuldijk, J., Heuts, H. P. A., Klumperman, B., van Herk, A. M., Koning, C. E. (2009). Influence of the molecular weight distribution on the percolation threshold of carbon nanotube–polystyrene composites, *E-Polymers*, *22*, 1–13.
88. Krause, B., Pötschke, P., Haussler, L. (2009). Influence of small scale melt mixing conditions on electrical resistivity of carbon nanotube–polyamide composites, *Compos. Sci. Technol.*, *69*, 1505–1515.
89. O'Connor, I., De, S., Coleman, J. N., Gun'ko, Y. K. (2009). Development of transparent, conducting composites by surface infiltration of nanotubes into commercial polymer films, *Carbon*, *47*, 1983–1988.
90. Yu, C., Kim, Y. S., Kim, D., Grunlan, J. C. (2008). Thermoelectric behavior of segregated-network polymer nanocomposites, *Nano. Lett.*, *8*, 4428–4432.
91. Li, C. Y., Chou, T. W. (2009). Electrical conductivities of composites with aligned carbon nanotubes, *J. Nanosci. Nanotechnol.*, *9*, 2518–2524.
92. Villmow, T., Pegel, S., Pötschke, P., Wagenknecht, U. (2008). Influence of injection molding parameters on the electrical resistivity of polycarbonate filled with multi-walled carbon nanotubes, *Compos. Sci. Technol.*, *68*, 777–789.
93. Lellinger, D., Xu, D. H., Ohneiser, A., Skipa, T., Alig, I. (2008). Influence of the injection moulding conditions on the in-line measured electrical conductivity of polymer–carbon nanotube composites, *Phys. Status Solidi B*, *245*, 2268–2271.
94. Gao, J. F., Yan, D. X., Huang, H. D., Dai, K., Li, Z. M. (2009). Positive temperature coefficient and time-dependent resistivity of carbon nanotubes (CNTs)/ultrahigh molecular weight polyethylene (UHMWPE) composite, *J. Appl. Polym. Sci.*, *114*, 1002–1010.
95. Zhang, C. S., Ni, Q. Q., Fu, S. Y., Kurashiki, K. (2007). Electromagnetic interference shielding effect of nanocomposites with carbon nanotube and shape memory polymer, *Compos. Sci. Technol.*, *67*, 2973–2980.

96. Lazarenko, A., Vovchenko, L., Matsui, D., Prylutskyy, Y., Matzuy, L., Ritter, U., Scharff, P. (2008). Electrical and thermal conductivity of polymer–nanocarbon composites, *Mol. Cryst. Liq. Cryst.*, *497*, 397–407.
97. Lebovka, N., Dadakova, T., Lysetskiy, L., Melezhyk, O., Puchkovska, G., Gavrilko, T., Baran, J., Drozd, M. (2008). Phase transitions, intermolecular interactions and electrical conductivity behavior in carbon multiwalled nanotubes/nematic liquid crystal composites, *J. Mol. Struct.*, *887*, 135–143.
98. Logakis, E., Pandis, C., Peoglos, V., Pissis, P., Stergiou, C., Pionteck, J., Pötschke, P., Micusik, M., Omastova, M. (2009). Structure–property relationships in polyamide 6/multi-walled carbon nanotubes nanocomposites, *J. Polym. Sci. Polym. Phys.*, *47*, 764–774.
99. Sichel, E. K., Gittleman, J. I., Sheng, P. (1978). Transport properties of the composite material carbon–poly(vinyl chloride), *Phys. Rev. B*, *18*, 5712–5716.
100. Grimaldi, C., Balberg, I. (2006). Tunneling and nonuniversality in continuum percolation systems, *Phys. Rev. Lett.*, *96*, 066602.
101. Sandler, J. K. W., Kirk, J. E., Kinloch, I. A., Shaffer, M. S. P., Windle, A. H. (2003). Ultra-low electrical percolation threshold in carbon-nanotube–epoxy composites, *Polymer*, *44*, 5893–5899.
102. Martin, C. A., Sandler, J. K. W., Shaffer, M. S. P., Schwarz, M. K., Bauhofer, W., Schulte, K., Windle, A. H. (2004). Formation of percolating networks in multi-wall carbon-nanotube–epoxy composites, *Compos. Sci. Technol.*, *64*, 2309–2316.
103. Moisala, A., Li, Q., Kinloch, I. A., Windle, A. H. (2006). Thermal and electrical conductivity of single- and multi-walled carbon nanotube–epoxy composites, *Compos. Sci. Technol.*, *66*, 1285–1288.
104. Zhao, X., Koos, A. A., Chu, B. T. T., Johnston, C., Grobert, N., Grant, P. S. (2009). Spray deposited fluoropolymer/multi-walled carbon nanotube composite films with high dielectric permittivity at low percolation threshold, *Carbon*, *47*, 561–569.
105. Grossiord, N., Kivit, P. J. J., Loos, J., Meuldijk, J., Kyrylyuk, A. V., van der Schoot, P., Koning, C. E. (2008). On the influence of the processing conditions on the performance of electrically conductive carbon nanotube/polymer nanocomposites, *Polymer*, *49*, 2866–2872.
106. Ha, M. L. P., Grady, B. P., Lolli, G., Resasco, D. E., Ford, W. T. (2007). Composites of single-walled carbon nanotubes and styrene–isoprene copolymer lattices, *Macromol. Chem. Phys.*, *208*, 446–456.
107. Yoon, H., Okamoto, K., Yamaguchi, M. (2009). Carbon nanotube imprinting on a polymer surface, *Carbon*, *47*, 2840–2846.
108. Zhang, C., Zhu, J., Ouyang, M., Ma, C. A., Sumita, M. (2009). Conductive network formation and electrical properties of poly(vinylidene fluoride)/multiwalled carbon nanotube composites: percolation and dynamic percolation, *J. Appl. Polym. Sci.*, *114*, 1405–1411.
109. Deng, H., Skipa, T., Zhang, R., Lellinger, D., Bilotti, E., Alig, I., Peijs, T. (2009). Effect of melting and crystallization on the conductive network in conductive polymer composites, *Polymer*, *50*, 3747–3754.
110. Alig, I., Skipa, T., Lellinger, D., Bierdel, M., Meyer, H. (2008). Dynamic percolation of carbon nanotube agglomerates in a polymer matrix: comparison of different model approaches, *Phys. Status Solidi B*, *245*, 2264–2267.
111. Faiella, G., Piscitelli, F., Lavorgna, M., Antonucci, V., Giordano, M. (2009). Tuning the insulator to conductor transition in a multiwalled carbon nanotubes/epoxy composite at substatistical percolation threshold, *Appl. Phys. Lett.*, *95*, 153106.
112. Yao, S. H., Dang, Z. M., Jiang, M. J., Xu, H. P. (2007). Influence of aspect ratio of carbon nanotube on percolation threshold in ferroelectric polymer nanocomposite, *Appl. Phys. Lett.*, *91*, 212901.
113. Zhang, C., Zhu, J., Ouyang, M., Ma, C. A. (2009). Electric field controlled formation and dissociation of multiwalled carbon nanotube conductive pathways in a polymer melt, *Appl. Phys. Lett.*, *94*, 111915.
114. Zhang, L. Y., Wan, C. Y., Zhang, Y. (2009). Morphology and electrical properties of polyamide 6/polypropylene/multi-walled carbon nanotubes composites, *Compos. Sci. Technol.*, *69*, 2212–2217.
115. Meincke, O., Kaempfer, D., Weickmann, H., Friedrich, C., Vathauer, M., Warth, H. (2004). Mechanical properties and electrical conductivity of carbon-nanotube filled polyamide-6 and its blends with acrylonitrile/butadiene/styrene, *Polymer*, *45*, 739–748.
116. Bose, S., Bhattacharyya, A. R., Khare, R. A., Kulkarni, A. R., Patro, T. U., Sivaraman, P. (2008). Tuning the dispersion of multiwall carbon nanotubes in co-continuous polymer blends: a generic approach, *Nanotechnology*, *19*, 335704.

117. Goldel, A., Kasaliwal, G., Pötschke, P. (2009). Selective localization and migration of multiwalled carbon nanotubes in blends of polycarbonate and poly(styrene–acrylonitrile), *Macromol. Rapid Commun.*, *30*, 423–429.
118. Pötschke, P., Bhattacharyya, A. R., Janke, A. (2004). Carbon nanotube-filled polycarbonate composites produced by melt mixing and their use in blends with polyethylene, *Carbon*, *42*, 965–969.
119. Wu, D. F., Zhang, Y. S., Zhang, M., Yu, W. (2009). Selective localization of multiwalled carbon nanotubes in poly(epsilon-caprolactone)/polylactide blend, *Biomacromolecules*, *10*, 417–424.
120. Liu, L., Grunlan, J. C. (2007). Clay assisted dispersion of carbon nanotubes in conductive epoxy nanocomposites, *Adv. Funct. Mater.*, *17*, 2343–2348.
121. Jin, S. H., Lee, D. S. (2007). Electrical and rheological properties of double percolated poly(methyl methacrylate)/multiwalled carbon nanotube nanocomposites, *J. Nanosci. Nanotechnol.*, *7*, 3847–3851.
122. Wang, L., Dang, Z. M. (2005). Carbon nanotube composites with high dielectric constant at low percolation threshold, *Appl. Phys. Lett.*, *87*, 042903.
123. Pötschke, P., Dudkin, S. M., Alig, I. (2003). Dielectric spectroscopy on melt processed polycarbonate–multiwalled carbon nanotube composites, *Polymer*, *44*, 5023–5030.
124. Simoes, R., Silva, J., Vaia, R., Sencadas, V., Costa, P., Gomes, J., Lanceros-Mendez, S. (2009). Low percolation transitions in carbon nanotube networks dispersed in a polymer matrix: dielectric properties, simulations and experiments, *Nanotechnology*, *20*, 035703.
125. Perez, R. (2009). Prediction of the effective dielectric constant in SWNT polyimide nanocomposites using the Bruggemann model, *J. Appl. Polym. Sci.*, *113*, 2264–2270.
126. Li, J. Q., Zhang, Q., Peng, N., Zhu, Q. (2005). Manipulation of carbon nanotubes using AC dielectrophoresis, *Appl. Phys. Lett.*, *86*, 153116.
127. Kozinsky, B., Marzari, N. (2006). Static dielectric properties of carbon nanotubes from first principles, *Phys. Rev. Lett.*, *96*, 166801.
128. Kohlmeyer, R. R., Javadi, A., Pradhan, B., Pilla, S., Setyowati, K., Chen, J., Gong, S. Q. (2009). Electrical and dielectric properties of hydroxylated carbon nanotube–elastomer composites, *J. Phys. Chem. C*, *113*, 17626–17629.
129. Liu, L., Kong, L. B., Matitsine, S. (2008). Tunable effective permittivity of carbon nanotube composites, *Appl. Phys. Lett.*, *93*, 113106.
130. Xu, G. H., Zhang, Q., Zhou, W. P., Huang, J. Q., Wei, F. (2008). The feasibility of producing MWCNT paper and strong MWCNT film from VACNT array, *Appl. Phys. A*, *92*, 531–539.
131. Bradford, P. D., Bogdanovich, A. E. (2008). Electrical conductivity study of carbon nanotube yarns, 3-D hybrid braids and their composites, *J. Compos. Mater.*, *42*, 1533–1545.
132. Park, J. G., Louis, J., Cheng, Q. F., Bao, J. W., Smithyman, J., Liang, R., Wang, B., Zhang, C., Brooks, J. S., Kramer, L., Fanchasis, P., Dorough, D. (2009). Electromagnetic interference shielding properties of carbon nanotube buckypaper composites, *Nanotechnology*, *20*, 415702.
133. Pham, G. T., Park, Y. B., Wang, S. R., Liang, Z. Y., Wang, B., Zhang, C., Funchess, P., Kramer, L. (2008). Mechanical and electrical properties of polycarbonate nanotube buckypaper composite sheets, *Nanotechnology*, *19*, 325705.
134. Spitalsky, Z., Tsoukleri, G., Tasis, D., Krontiras, C., Georga, S. N., Galiotis, C. (2009). High volume fraction carbon nanotube–epoxy composites, *Nanotechnology*, *20*, 405702.
135. Wang, D., Song, P. C., Liu, C. H., Wu, W., Fan, S. S. (2008). Highly oriented carbon nanotube papers made of aligned carbon nanotubes, *Nanotechnology*, *19*, 075609.
136. Skakalova, V., Kaiser, A. B., Dettlaff-Weglikowska, U., Hrncarikova, K., Roth, S. (2005). Effect of chemical treatment on electrical conductivity, infrared absorption, and Raman spectra of single-walled carbon nanotubes, *J. Phys. Chem. B*, *109*, 7174–7181.
137. Kulesza, S., Szroeder, P., Patyk, J. K., Szatkowski, J., Kozanecki, M. (2006). High-temperature electrical transport properties of buckypapers composed of doped single-walled carbon nanotubes, *Carbon*, *44*, 2178–2183.
138. Rosca, I. D., Hoa, S. V. (2009). Highly conductive multiwall carbon nanotube and epoxy composites produced by three-roll milling, *Carbon*, *47*, 1958–1968.
139. Byrne, M. T., Hernandez, Y. R., Conaty, T., Blighe, F. M., Coleman, J. N., Gun'ko, Y. K. (2009). Preparation of buckypaper–copper composites and investigation of their conductivity and mechanical properties, *ChemPhysChem*, *10*, 774–777.

140. Panhuis, M. I. H. (2006). Carbon nanotubes: enhancing the polymer building blocks for intelligent materials, *J. Mater. Chem.*, *16*, 3598–3605.
141. Web of Science Search; papers up to 2009.
142. Yu, Y. J., Che, B., Si, Z. H., Li, L., Chen, W., Xue, G. (2005). Carbon nanotube/polyaniline core–shell nanowires prepared by *in situ* inverse microemulsion, *Synth. Met.*, *150*, 271–277.
143. Mottaghitalab, V., Spinks, G. M., Wallace, G. G. (2005). The influence of carbon nanotubes on mechanical and electrical properties of polyaniline fibers, *Synth. Met.*, *152*, 77–80.
144. Ramamurthy, P. C., Harrell, W. R., Gregory, R. V., Sadanadan, B., Rao, A. M. (2004). Mechanical and electrical properties of solution-processed polyaniline/multiwalled carbon nanotube composite films, *J. Electrochem. Soc.*, *151*, G502–G506.
145. Mottaghitalab, V., Spinks, G. M., Wallace, G. G. (2006). The development and characterisation of polyaniline–single walled carbon nanotube composite fibres using 2-acrylamido-2-methyl-1-propane sulfonic acid (AMPSA) through one step wet spinning process, *Polymer*, *47*, 4996–5002.
146. Long, Y. Z., Chen, Z. J., Zhang, X. T., Zhang, J., Liu, Z. F. (2004). Synthesis and electrical properties of carbon nanotube polyaniline composites, *Appl. Phys. Lett.*, *85*, 1796–1798.
147. Yilmaz, F., Kucukyavuz, Z. (2009). Conducting polymer composites of multiwalled carbon nanotube filled doped polyaniline, *J. Appl. Polym. Sci.*, *111*, 680–684.
148. Deng, J. G., Ding, X. B., Zhang, W. C., Peng, Y. X., Wang, J. H., Long, X. P., Li, P., Chan, A. S. C. (2002). Carbon nanotube–polyaniline hybrid materials, *Eur. Polym. J.*, *38*, 2497–2501.
149. Ramamurthy, P. C., Harrell, W. R., Gregory, R. V., Sadanadan, B., Rao, A. M. (2004). Polyaniline/carbon nanotube composite Schottky contacts, *Polym. Eng. Sci.*, *44*, 28–33.
150. Ramamurthy, P. C., Malshe, A. M., Harrell, W. R., Gregory, R. V., McGuire, K., Rao, A. M. (2004). Polyaniline/single-walled carbon nanotube composite electronic devices, *Solid State Electron.*, *48*, 2019–2024.
151. Nastase, C., Nastase, F., Vaseashta, A., Stan-Iatin, L. (2006). Nanocomposites based on functionalized nanotubes in polyaniline matrix by plasma polymerization, *Prog. Solid State Chem.*, *34*, 181–189.

CHAPTER 7

THERMAL CONDUCTIVITY

7.1 OVERVIEW

Around 2002 or so, when I was serving on an National Science Foundation review panel, a panelist who worked for a commercial firm made the statement that all of the work (and all of the proposals!) that were investigating electrical and mechanical properties of nanotube composites were somewhat misguided, since what was really needed was a nonmetal high thermal conductor, and nanotube composites seemed appropriate. In other words, he felt that the extremely high thermal conductivity of carbon nanotubes meant that the greatest commercial opportunity for carbon nanotubes was in the enhancement of thermal conductivity of composites. He was very likely correct, but unfortunately the thermal conductivity of a composite filled with nanotubes does not in general increase anywhere near to the point where such materials would be useful as thermal conductors as shown conclusively in Table 7.1, even though a mixing rule predicts such an improvement.

Measurements of thermal conductivity are more difficult than measurements of electrical conductivity. The fundamental difference is that thermally isolation is very difficult, while electrically isolation is relatively simple. For example, a probe can conduct heat that can be a large fraction of the total if a thermal insulator is being tested. A number of different methods have been used to measure thermal conductivity, and it is beyond the scope of this chapter to review each in detail. The most common methods include laser flash, hot wire, and hot disk. Most methods have in common that a known amount of energy is input, followed by the measurement of temperature as a function of time and position. Usually, the heat capacity is measured simultaneously or must be known, although for some techniques the shape of the time evolution of temperature can be used to eliminate this requirement.

The thermal conductivity of a carbon nanotube has been measured between 650 and 10,000 W/(m K).[1–5] Applying a simple rule of mixtures law to a typical polymer that has a thermal conductivity of approximately 0.3 W/(m K) yields a thermal conductivity for a composite in the tens of W/(m K) at 1% loading and in the hundreds at 10% loading. Given that the thermal conductivity of copper, which is a very high thermally conductive material (see http://www.all-clad.com/collections/copper-core/!!!), is about 400 W/(m K) at room temperature, the reason for the initial excitement was obvious. However, currently there are no applications in polymers designed to take advantage of the high thermal conductivity of nanotubes because

Carbon Nanotube–Polymer Composites: Manufacture, Properties, and Applications, First Edition.
Brian P. Grady.

TABLE 7.1 Measured Thermal Conductivities of Polymer–Nanotube Composites

Polymer	Nanotube type	Nanotube level (wt% unless otherwise noted)	Thermal conductivity (room temperature) (W/(m K))	Notes
Polystyrene[25]	SWCNTs	0%, 5%, 10%, 30%	0.15, 0.36, 0.55, 0.62	
Polystyrene[72]	MWCNTs	Various to 10%	0.14 (0%), 0.21 (1%), 0.43 (10%)	Linear increase between 1% and 10%
Polystyrene[73]	MWCNTs	0%, 1%, 4%, 6%, 11%, 20%	0.12, 0.17, 0.18, 0.19, 0.22, 0.32	Material was a foam
Poly(methyl methacrylate) (PMMA)[54]	SWCNTs	Various to 7.5 vol%	0.22 (0%), 0.34 (7.5%)	Approximate linear increase although authors fit curve
Poly(methyl methacrylate)[74]	Silane-functionalized MWCNTs	0%, 0.5%, 1%, 2%, 3%, 4%	1, 1.6, 1.87, 1.6, 1.6, 1.6	Normalized values on thermal conductivity (no units)
Poly(methyl methacrylate)[75]	SWCNTs SWCNTs	0%, 1%, 2%, 3%, 4%	0.21, 0.41, 0.58,0.7, 0.68 0.21, 0.6, 0.7, 0.8, 1.03	Unpurified tubes had significantly higher conductivities
Poly(methyl methacrylate)[15]	SWCNTs	0%, 1.75 vol%, 2.75 vol%, 4.5 vol%, 6.5 vol%, 9 vol%	0.18, 0.21, 0.24, 0.23, 0.42, 0.42	
Poly(methyl methacrylate)[76]	SWCNTs	Various to 7%	0.19 (0%), 0.45 (7%)	Linear increase
Low-density polyethylene[13]	SWCNTs	0%, 1 vol%, 20 vol%	0.26, 0.35, 1.7	
High-density polyethylene[68]	MWCNTs	Various to 40 vol%	0% (0.4), 40% (1.3)	Linear increase
High-density polyethylene[13]	SWCNTs	0%, 1 vol%, 3 vol%, 6.5 vol%, 20 vol%	0.5, 0.6, 0.7, 0.9, 3.4	
High-density polyethylene[77]	MWCNTs	Various to 12%	0.52 (0%), 0.83 (12%)	Linear increase
High-density polyethylene[78]	MWCNTs	Various to 11 vol%	0.27 (0%), 0.48 (11%)	Linear increase, although data were somewhat irregular and authors interpreted as percolation

Ethylene–vinyl acetate copolymer[79]	MWCNTs	4%	0.22–0.24 (0%), 0.9–1.18 (4%)	Varying vinyl acetate ratio; higher thermal conductivities are for higher vinyl acetate contents
Ethylene–vinyl acetate copolymer[62]	MWCNTs	0%, 20%, 30%	0.32, 0.56, 0.71	
	MWCNTs with dispersant	0%, 20%	0.32, 0.67	
Ethylene–vinyl acetate copolymer[62]	Aligned MWCNTs	0%, 20%, 30%	0.32, 2.3, 2.9 (parallel) 0.31, 0.36, 0.47 (perpendicular)	
Ethylene–vinyl acetate copolymer[62]	Aligned MWCNTs with dispersant	0%, 20%	0.32, 3.4 (parallel) 0.31, 0.57 (perpendicular)	
Polypropylene[14]	MWCNTs	Various to 15%	0.20 (0%), 0.45 (15%)	Linear increase
Polypropylene[57]	MWCNTs	0.1%, 0.5%, 2%	0.09, 0.23, 0.24	
	Acid-functionalized MWCNTs		0.10, 0.15, 0.40	
	Base-treated MWCNTs		0.09, 0.44, 0.48	
Polypropylene[69]	MWCNTs	0%, 2%, 10%, 15%	0.24, 0.26, 0.30, 0.31	
Polyamide 6[80]	MWCNTs	0%, 0.5%, 1%, 2%, 4%	0.10, 0.15, 0.19, 0.22, 0.28	
Polyamide 6[81]	MWCNTs	0.1%, 0.5%, 1%, 2%	0.22, 0.28, 0.28, 0.32, 0.39	
Polyamide 6 laminated with glass mat (30 vol% glass)[80]	MWCNTs	0%, 0.5%, 1%, 2%, 4%	0.041, 0.048, 0.049, 0.053, 0.058	
Poly(vinyl acetate)[82]	SWCNTs	Various to 3%	0.21–0.23	Linear increase, samples were made so that nanotubes were isolated to interface of latex particles
Polycarbonate[53]	MWCNTs	Various to 15%	Perhaps 0.2 (0%) to 0.5 (15%)	Percolation-type behavior with percolation about 1.5%

(continued)

TABLE 7.1 (*Continued*)

Polymer	Nanotube type	Nanotube level (wt% unless otherwise noted)	Thermal conductivity (room temperature) (W/(m K))	Notes
Polycarbonate[83]	MWCNTs	0%, 1.75%, 5%, 15%	0.14, 0.15, 0.16, 0.24	
Poly(vinyl acetate)[84]	MWCNTs, extremely low diameter	Various to 20%	0.2 (0%), 0.34 (20%)	Linear increase, nanotubes isolated to edge of latex particles
MMA–butyl acrylate (BA) copolymers[64]	MWCNTs	0%, 0.3%, 0.5%, 0.7%		Sample that had increase in thermal conductivity and had no T_g depression
50/50 MMA/BA			0.25, 0.6, 0.7, 0.65	
80/20 MMA/BA			0.15, 0.06, 0.07	
80/11/9 MMA/BA/acrylic acid			0.11, 0.12, 0.13, 0.13	
Poly(ether ether ketone)[85]	MWCNTs	Various to 17%	0.3 (0%), 0.7 (17%)	Linear increase
Poly(ether ether ketone)[56]	Arc-grown SWCNTs	0%, 0.1%, 0.5%, 1%	0.23, 0.3, 0.46, 0.5	Addition of PEI compatibilizer reduced thermal conductivity increase by about a factor of 2
	Laser-grown SWCNTs		0.23, 0.32, 0.48, 0.55	
Polylactide[47]	MWCNTs	0%, 1.2%	0.15, 0.42	
	Polylactide-grafted MWCNTs		0.15,0.4	
Polyurethane[86]	MWCNTs	0%, 1%, 3%	0.12, 0.18, 0.23	
Polyurethane[87]	MWCNTs	0%, 0.5%, 1%, 2%, 3%, 5%, 7%	0.15, 0.16, 0.18, 0.21, 0.21, 0.20, 0.27	
Polyurethane[88]	MWCNTs	Various to 7%	0.15 (0%), 0.42 (3%)	Linear increase to 3 wt%, followed by plateau
Poly(vinylidene fluoride)[16]	MWCNTs	Various to 49 vol%	0.23 (0), 0.54 (49)	Linear increase
Poly(vinylidene fluoride)[89]	MWCNTs	Various to 5%	0.28 (0%), 0.33 (5%)	Linear increase

Poly(vinyl chloride)[90]	MWCNTs	0%, 0.025%, 0.05%, 0.075%, 0.1%, 0.15%, 0.4%, 0.55%, 0.7%	0.16, 0.15, 0.15, 0.15, 0.15, 0.16, 0.16, 0.16	Notice minimum in thermal conductivity
Polyaniline with tetradecanol[91]	MWCNTs	Various to 5%	0.34 (0%), 0.43 (5%)	Linear increase
Epoxy[66]	SWCNTs	0%, 0.25%, 0.5%, 1%	0.2, 0.34, 0.32, 0.44	
Epoxy[92]	SWCNTs	Various to 0.5%	Decrease from 0.25 to 0.23 (ball milling) or 0.24 (ultrasonication) at 0.05% or 0.1% followed by plateau	Two dispersion methods used, ball milling and ultrasonication. The authors believe neither changed the chemical nature of the tubes, but the former shortened the tubes
Epoxy[63]	SWCNTs	0%, 3%	2.0 (0%), 5.2 and 6.1 (3%)	Larger value corresponds to alignment via magnetic field during curing
Epoxy[92]	MWCNTs	Various to 0.5%	0.25 (0%), 0.29 (0.5%)	Linear increase
Epoxy[46]	MWCNTs, DWCNTs, and SWCNTs and amine-modified versions of the first two	Various to 0.5%	0.24 to 0.24–0.25 (0.5%)	All show linear increase
Epoxy[42]	SWCNTs	0%, 0.5%	0.18, 0.21	
	Acid-functionalized SWCNTs		0.18, 0.19	
	Shortened SWCNTs		0.18, 0.25	
Epoxy[93]	MWCNTs	0%, 1%, 2%	0.25, 0.33, 0.45	
Epoxy[44]	MWCNTs	Various to 10 vol%	0.35 (0%), 0.8 (10)	Linear increase
	Amine-grafted MWCNTs		0.35 (0%), 1.2 (10)	

(continued)

TABLE 7.1 *(Continued)*

Polymer	Nanotube type	Nanotube level (wt% unless otherwise noted)	Thermal conductivity (room temperature) (W/(m K))	Notes
Epoxy[43]	SWCNTs	Various to 5 vol%	0.2 (0%), 0.4 (5%)	Linear increase
	Acid-functionalized SWCNTs	Various to 8 vol%	0.2 (0%), 0.9 (9%)	
Epoxy[45]	MWCNTs	Various to 30%	0.35 (0%), 1.2 (15%)	Linear to 15%, plateau after
	Amine-grafted MWCNTs		0.35 (0%), 1.4 (15%)	
Epoxy[58]	MWCNTs poorly dispersed	0, 0.5%, 1%, 1.5%	0.12, 0.13, 0.16, 0.18	Dispersion was altered by using solvent or without using solvent in dispersion–reaction scheme
	MWCNTs well dispersed		0.12, 0.21, 0.23, 0.26	
Epoxy[94]	MWCNTs	0%, 0.25%, 0.5%, 2%	0.252, 0.265, 0.290, 0.275	Also measured with glass fiber weave
Epoxy[67]	MWCNTs	0%, 0.05%, 0.1%, 0.3%, 0.5%, 0.7%, 1.0%, 3.0%	0.21, 0.24, 0.27, 0.27, 0.27, 0.27, 0.27, 0.24	
Epoxy[95]	MWCNTs	0%, 2%	0.25, 0.28	
Epoxy[11]	SWCNTs	Various to 10 vol%	0.20 (0%), 0.32 (10)	Linear increase, difference is in ways tubes were processed; see Ref. 11 for details
		Various to 0.5%	0.20 (0%), 0.25 (0.5)	
Epoxy[30]	MWCNTs	8 vol%	5.5	Nanotube forests, conductivity along nanotube long axis direction

Epoxy[96]	Peroxide-functionalized MWCNTs	Not given	0.2, 0.8, or 1.4	Nanotube forests, conductivity along nanotube long axis direction; difference in conductivity due to different curing procedures
Epoxy[26]	MWCNTs	Not given	250	Nanotube forests, conductivity along nanotube long axis direction; exposed nanotube tips and tips were coated with gold and then indium
Epoxy[76]	SWCNTs	2%	Factor of 3 increase over pure epoxy	Epoxy infused into nanotube aerogel
Epoxy filled with alumina[97]	MWCNTs	0%, 3 vol%	1, 1.3; 1.2, 1.8; 1.7, 2.6	Different pairs of data represent alumina volume fractions of 42%, 50%, and 60%, respectively
Polydimethylsiloxane[29]	MWCNTs	0%, 2%	0.2, 3.4, or 7.2	Nanotube forests, conductivity along nanotube long axis direction, varying synthesis conditions
Polydimethylsiloxane[28]	MWCNTs	Various to 3 vol%	0.56 (0%), 0.71 (0%)	Linear increase
PDMS[98]	MWCNTs	0%, 2%	0.11 (0%), 0.19, 0.19, 0.18, 0.19 (2%)	Various functionalization strategies
Polydimethylsiloxane[28]	MWCNTs	0%, 0.2 vol%, 0.4 vol%	0.56 (0), 1.21 (0.4)	Linear increase; nanotube forests, conductivity along nanotube long axis direction; exposed nanotube tips
Phenol resin[71]	MWCNTs	0%, 10%	0.29, 1.07	
Phenol resin[99]	MWCNTs	0%, 25%	0.29, 1.94	

(*continued*)

TABLE 7.1 (*Continued*)

Polymer	Nanotube type	Nanotube level (wt% unless otherwise noted)	Thermal conductivity (room temperature) (W/(m K))	Notes
Phenol resin[100]	MWCNTs	Various to 2.5%	0.5 (0%), 0.8 (2.5%)	Linear increase; compares surfactant-assisted dispersed with pristine and acid functionalized, very little difference but surfactant dispersed are best
Phenol resin: UV curable[101]	SWCNTs	Various to 5%	0.2 (0%), 0.5 (5%)	Linear increase
Styrene–butadiene rubber[102]	MWCNTs	1%, 2%, 3%, 4%, 5%	0.078, 0.11, 0.13, 0.10	
Polyisoprene (cross-linked and uncross-linked)[103,104]	SWCNTs	0%, 1%, 5%	0.15, 0.19, 0.33	Also measured pressure, temperature dependence

Units on percentages (wt% or vol%) are the same as those for the cell immediately to the left.

increases in thermal conductivity, instead of being measured in the tens or hundreds of W/(m K), are typically less than 1 W/(m K). The reason for this behavior is that the resistance to the transfer of heat at the nanotube–polymer boundary is extremely large, and hence the thermal conductivity increase is extremely small.

This boundary layer resistance is termed the Kapitza resistance, and was first measured in Soviet Russia in 1941 with experiments on metal–liquid helium interfaces. Pyotr Kapitza (also spelled Kapitsa) was interested in low-temperature physics and performed pioneering studies involving liquid helium. In fact, Kapitza was also the first to discover superfluidity and won a Nobel Prize in Physics in 1978 for his work with liquid helium and study of low-temperature physics. (The Physics Prize that year was also awarded to Arno Allan Penzias and Robert Woodrow Wilson, who won for work that used liquid helium approximately 25 years later to build very sensitive instruments to study radio noise from outer space. Those are pretty disparate subject matters to share the same Nobel Prize!). Kapitza first measured boundary layer resistance between bronze and liquid helium and later used platinum and copper as the solid in the experiments. This resistance is characterized by a temperature discontinuity across the interface; the magnitude of the discontinuity divided by the power per unit area flowing across the interface ($R_K = \Delta T/Q$) defines the interfacial resistance. In fact, the Kapitza resistance has an upper bound defined by an acoustic mismatch and a lower bound defined by a diffuse mismatch. The qualitative difference between these two processes is in the fact that the acoustic mismatch is precisely defined based on the properties of the two materials, while the diffuse mismatch is probabilistic in the sense all the phonons striking the interface are scattered to one of the adjoining substances with a probability proportional to the phonon density of states in each substance.[6,7] In all real cases, the resistance is never purely one or the other, but order of magnitude calculations show that resistance due to acoustic mismatch is the dominating resistance in nanotube composites.[8]

Detailed descriptions of the fundamental theory that describes the acoustic model or diffuse mismatch model are beyond the scope of this book. The former can be calculated based on classical wave propagation theory, and the end result is that the larger the resistance, the larger the difference between the rates at which sound is carried through the material, the larger the resistance. Since sound transmission rate is proportional to the modulus, the large modulus of nanotubes works against heat transfer in composites. The other issue is that the small dimensions of the nanotube lead to a large number of nanotube–polymer transfers within the composite.

Although only the upper and lower bounds of the resistance can be calculated precisely, it is possible to use continuum theories (also termed as effective medium theories) of thermal conductivity to calculate the Kapitza resistance from measured thermal conductivity values for a given filler geometry. The simplest theory of thermal conductivity was derived by Maxwell in the nineteenth century for spherical objects and has the form

$$\kappa = \kappa_p \left(\frac{1 + 2\beta v_f}{1 - \beta v_f} \right) \tag{7.1}$$

where κ is the thermal conductivity of the composite, $\beta = (\kappa_f - \kappa_p)/(\kappa_f + \alpha\kappa_p)$, κ_p and κ_f are the thermal conductivity of the polymer and filler, respectively,

$\alpha = 2\kappa_p R_K/d$ (d is the diameter), and v_f is the volume fraction of the filler. A simple extension to highly anisotropic, stiff rods with random orientation is given by[9–11]

$$\kappa = \kappa_p \left(\frac{3 + v_f(\beta_x + \beta_z)}{2 \text{ or } 3^* - \beta v_f} \right), \quad \beta_x = \frac{2(\kappa_{f,d} - \kappa_p)}{\kappa_{f,d} + \kappa_p}, \quad \beta_z = \frac{\kappa_{f,l}}{\kappa_p} - 1 \tag{7.2}$$

where $\kappa_{f,l}$ and $\kappa_{f,d}$ are the effective thermal conductivities parallel and perpendicular to the rod long axis (along the length or the diameter), respectively. These two values are different because of the contribution of the distance that the heat has to transfer through the rod relative to the interfacial resistance and can be written as

$$\kappa_{f,l} \text{ or } \kappa_{f,d} = \frac{\kappa_f}{1 + (2a_K/l \text{ or } d)(\kappa_f/\kappa_p)} \tag{7.3}$$

where a_K is the Kapitza radius and is equal to R_K/κ_p, where R_K is the Kapitza resistance.

The Kapitza resistance has been directly measured in water on a single nanotube to be 8.33×10^{-8} m^2 K/W.[12] Fitting of Equations 7.2 and 7.3 has led to values similar to those measured from bulk conductivity measurements in polymers: 1×10^{-8} m^2 K/W in high-density polyethylene,[13] and 2.4×10^{-9} and 2.6×10^{-8} m^2 K/W in an epoxy.[11] The difference in the latter measurements with the epoxy resin was the dispersion technique; however, neither technique should have affected the chemical characteristics of the tube. These R_K values correspond to distances in the tens of nanometers for thermal conductivities representative of an insulating polymer (e.g., 0.1–0.5 W/(m K)). This distance represents equivalence between the magnitude of the interfacial resistance and the thickness of an insulating layer of polymer coating around a nanotube. This approach is certainly not the only effective medium approach to model the thermal conductivity of composites; other models used to fit thermal conductivity data from nanotube composites include Nielsen's model,[14–16] Foygel's model,[13] and Bruggeman's model.[17] A different effective medium model has been developed for the case where the rods were perfectly aligned as well.[18] Surprisingly, this latter theory suggested that the interfacial resistance is unimportant, which certainly is not confirmed by experiment.

7.2 INTERFACIAL RESISTANCE AND THERMAL CONDUCTIVITY

Experimental evidence directly supports the supposition that interfacial resistance controls thermal conductivity. In one approach, the thermal conductivity above and below the glass transition temperature of the composite was measured. Below the glass transition temperature, the thermal conductivity of an amorphous polymer is monotonically increasing with temperature. The qualitative behavior of the thermal conductivity above the glass transition of an amorphous polymer is most often very slightly decreasing,[19–22] although measurements where it is flat or slightly increasing

*References 9 and 11 give this factor as 3, while Ref. 10 gives this factor as 2.

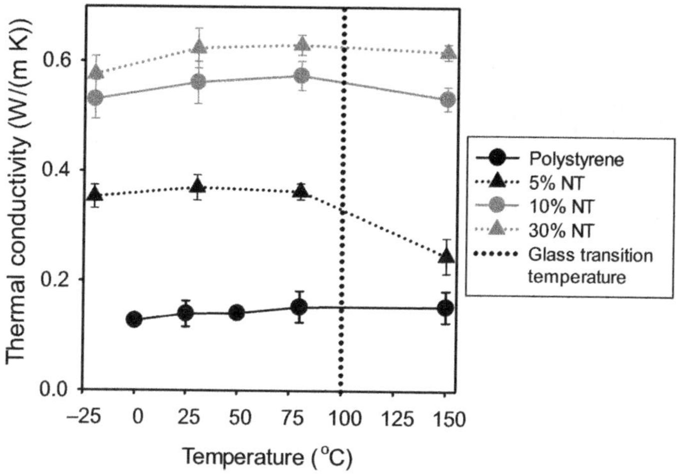

Figure 7.1 Effect of temperature on thermal conductivity of polystyrene–SWCNT composites showing the drop that occurs at the glass transition of the polymer.

have been made.[23,24] Regardless, there is very little difference in the thermal conductivities between the solid and the melt when the temperature is more than ~20°C removed from T_g. If the Kapitza resistance controls the thermal conductivity, then the thermal conductivity of the composite should fall dramatically as the polymer transitions from a glass to a liquid since the modulus of the polymer falls dramatically, increasing the modulus mismatch between the nanotubes and the polymer. Figure 7.1 shows that the thermal conductivity does indeed fall dramatically, which is clear evidence that the Kapitza resistance is critically important in nanotube composites.[25]

Another experimental study highlighted the critical role that interfacial resistance can have on heat transfer, and also represents the highest thermal conductivity ever recorded, to the author's knowledge, for a nanotube–polymer composite. A nanotube forest was grown, an epoxy resin was infused, and then the nanotube tips were exposed by plasma etching the polymer. The tips were coated with gold and then indium, and finally this material was placed between two plates where the indium was melted to form a good bond with the plates. A schematic of this configuration is shown in Figure 7.2. These processing steps led to an extremely high thermal conductivity, on the order of what a mixing law would suggest, that is, 250 W/(m K).[26,27] Two controls show that the layer between the tips and the plates is critical. The first is simply to etch away the polymer without coating the exposed tips. Without this layer to bridge the gap between the nanotube tips and the plates, the measured thermal conductivity was only slightly higher than 1 W/(m K).[28] The second is not to etch away the polymer and expose the tips, and in this case the measured thermal conductivity was between 1 and 10 W/(m K).[29,30]

A number of different types of theoretical approaches have been applied to study the problem of thermal conductance in nanotube composites from a nanoscopic or microscopic viewpoint rather than a continuum viewpoint. These approaches include molecular dynamic simulation,[8] boundary node method,[31,32] fast

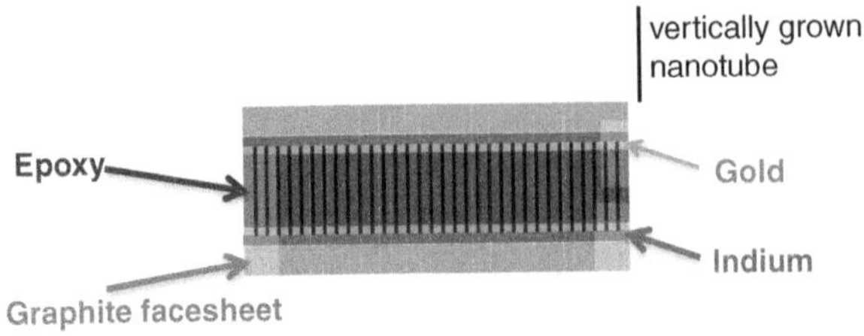

Figure 7.2 Schematic of the sample that yielded thermal conductivity of a nanotube–polymer composite of 250 W/(m K).

multipole boundary element method,[33] Monte Carlo simulation,[34,35] element-free Galerkin approach,[36,37] asymptotic expansion homogenization technique,[38,39] and micromechanics.[40] The most important information that could be gleaned from a simulation is to quantitatively estimate the interfacial resistance, and then ideally propose and validate possible strategies for reducing the interfacial resistance. In this regard, molecular dynamic simulations are extremely useful. The most obvious strategy to apply is to functionalize the tubes and change the interface. In one computational study, it was shown that covalently bonding the tubes to the matrix reduced the conductivity of the tubes to almost exactly the same fraction as the resistivity was reduced due to an improvement in interfacial heat transfer as a function of the number of functionalized atoms. At high functionalizations, however, there was some benefit to the formation of covalent bonds (overall, roughly a factor of 2 improvement).[41] A different molecular dynamic simulation with only grafting at the end of the tubes found a much larger improvement in thermal conductivity.[42]

Experimental comparison of the thermal conductivities for samples filled with functionalized and unfunctionalized tubes, where the functionalization group was chosen to be able to bond to the polymer, has been successful in achieving modest improvements (less than 50%) in thermal conductivity versus samples filled with unmodified MWCNTs.[43–45] In one study, functionalization led to a very slight decrease in the thermal conductivity for both DWCNTs and MWCNTs, although very low weight fractions were studied and the improvements with unmodified tubes were extremely slight.[46] Another study where the functionalized polymer matched the matrix polymer found a small drop in thermal conductivity with functionalization.[47] Since functionalization also generally changes dispersion, the moderate success in improving thermal conductivity in some cases could just as easily be due to improvements in dispersion, as opposed to reductions in Kapitza resistance.

For infused systems, nanotube–nanotube interfacial resistance could perhaps be much more important than nanotube–polymer interfacial resistance. The thermal conductivity of nanotubes that have been processed into a film and aligned in one direction was 225 W/(m K),[48] processed as a film but randomly aligned in a plane was 30 W/(m K)[48] and 100 W/(m K)[49], and processed as three-dimensional cubes was 0.13–0.2 W/(m K).[50] In fact, the Kapitza resistance calculated from molecular dynamic simulations at a nanotube–nanotube interface gives a Kapitza resistance on the order of 10^{-8} m^2 K/W, or approximately the same as that for the nanotube–polymer interface. The limited contact area is responsible for this high resistance, and hence the resistance decreases as the tube diameter increases, although the dependence is small (except for small-diameter SWCNTs) because the contact area does not change much with diameter.[51]

7.3 DISPERSION, PERCOLATION, AND THERMAL CONDUCTIVITY

As Table 7.1 indicates, most studies have found a linear increase in thermal conductivity with added nanotubes. This type of behavior was predicted by finite element analysis. The essential conclusion from this analysis is that even in the absence of interfacial resistance, the difference in thermal conductivities of the polymer and the nanotubes is not great enough to cause a percolation effect; that is, a rapid rise in conductivity with the formation of a continuous network, and hence a linear increase is expected. Kapitza resistance makes this possible discontinuity even smaller.[52]

Regardless, a few authors have claimed percolation behavior with respect to thermal conductivity. Percolation behavior was found for nanotubes in polycarbonate, with the percolation threshold between the higher electrical and lower rheological thresholds, and only slight shifts with temperature were found. Rheological measurements showed much larger shifts in percolation threshold over this same temperature range.[53] In another study, the percolation threshold determined via electrical conductivity was used to fit a power-law-type behavior to the thermal conductivity data, and for a few points at high nanotube contents, the slope was the same, within experimental error, as that from electrical conductivity.[54] Frankly though, except for possibly a few data points, a linear fit was adequate. Another study with high-density polyethylene found essentially linear behavior for the thermal conductivity at all measured volume fractions except the highest, but still fit a power-law expression to the data.[13] Overall, even with the small number of counter-examples, percolation behavior does not seem to describe well the conductivity behavior of nanotube-filled materials.

Although the relationship between dispersion and electrical conductivity is not well understood from a quantitative viewpoint (mostly related to difficulties in characterizing dispersion across the many length scales required), discussions in Chapter 6 indicate that the effects of dispersion on the electrical conductivity are well studied, and a significant level of understanding has been gleaned from those studies. The relationship has not been well studied for thermal conductivity, although the same factors important for electrical conductivity should be important for thermal conductivity. For example, for maximal thermal conductivity, there is also likely an optimal level of dispersion characterized by both significant inhomogeneity on the microscopic scale and significant nanoscale dispersion into individual tubes. This optimum is likely different for thermal conductivity versus electrical conductivity because there is no analogue in heat transfer to tunneling current. Experimental studies that have examined dispersion with respect to thermal conductivity are few. The most informative study, in this author's opinion, was done on oil rather than a polymer. An optimal shear rate for maximum dispersion as measured by the viscosity was found that also corresponded to a minimum in thermal conductivity. As sonication time increased, the thermal conductivity fell, which was thought to be due to a decrease in nanotube length with sonication. Microscopy images confirmed that samples with larger agglomerates had higher thermal conductivities.[55] With a polymer, improving dispersion via the addition of a polymeric compatibilizing agent significantly reduced the thermal conductivity of the composite. The authors postulated that a polymeric coating was forming on the tubes, but no direct evidence was presented to prove this hypothesis.[56] Functionalizing MWCNTs with acid or base, and then melt mixing with polypropylene, which would be expected to improve dispersion without changing the Kapitza resistance substantially (but could reduce tube thermal conductivity), led to a significant increase in thermal conductivity versus composites prepared with unfunctionalized tubes, suggesting that in this case increased dispersion benefits conductivity.[57] Similarly, using a solvent to improve dispersion in a dispersion–reaction scheme also showed a significant increase in thermal conductivity versus more poorly dispersed material.[58] However, in all cases, thermal conductivities were less than 1 W/(m K).

7.4 EFFECTS OF OTHER VARIABLES ON THERMAL CONDUCTIVITY

Kapitza resistance is not the only source of reduced thermal conductivity compared to a mixing rule in a composite. Recent molecular dynamic simulations indicate that the interaction of a single-walled nanotube with a flat substrate[59] or embedded in a medium[60,61] can lead to a reduction in the thermal conductivity of the nanotube, due to scattering processes at the walls of the nanotube. The magnitude of this effect relative to the magnitude of the interfacial resistance effect seems to be small, although experimental confirmation of this statement is lacking.

The reduction in conductivity as nanotubes change orientation from axis-aligned fibers or mats to planar-aligned mats, to unaligned powders detailed earlier highlights the very strong positive effect that alignment can have on thermal conductivity for pure nanotubes. Alignment has also been studied in composites. Using very high weight fractions of nanotubes (20% and 30%), the thermal conductivity in the alignment direction was as high as 3.4 W/(m K) in a thermoplastic ethylene–vinyl acetate polymer, while the thermal conductivity in the perpendicular direction was six to seven times smaller. The alignment was produced by simply extruding the material through a slit die at relatively high velocities. No details were given regarding take-up speeds, so it is not clear whether further alignment occurred due to drawing.[62] Since aligned polymers conduct heat in the direction of alignment better than unaligned polymers, some of the improvement could have been due to polymer, rather than nanotube, alignment. In fact, another study showed that thermal conductivity increased with alignment in high-density polyethylene; however, X-ray studies showed that the increase was due to the alignment of the crystalline polymer chains, rather than the alignment of the nanotubes.[13] Finally, in an epoxy where alignment was via magnetic field so that no polymer chain alignment would be expected, the thermal conductivity increased from 5.2 to 6.1 W/(m K) with alignment;[63] however, the values of the thermal conductivity were so high as to make the results of this study extremely suspect.

Table 7.1 indicates that the identity of the polymer has no obvious influence on the thermal conductivity of the nanotube-filled composite. One interesting study[64] showed large differences in thermal conductivities of acrylate copolymers depending on the monomers used. The percolation thresholds were different, but not remarkably so. The only other significant difference between the copolymers was that the copolymer that had the highest thermal conductivity also had no change in the glass transition with nanotubes, while all other samples showed a decrease. The authors interpreted this difference to mean that the sample that had no change in T_g with nanotube addition did not have an immobilized layer at the interface to retard thermal conductivity. However, the electrical conductivity above the percolation threshold was not affected, which argues against the explanation. At this point, these results are not fully understood.

Nanotube length would also be expected to have an important effect on the thermal conductivity. A molecular dynamics study highlighted the important role that nanotube length plays in the thermal conductivity as shown in Figure 7.3. The figure shows that in the typical range of nanotube lengths (1–10 μm) and interfacial resistances normally encountered (10^{-8} m^2 K/W), the thermal conductivity will increase by approximately a half order of magnitude for every order of magnitude increase in length.[10] Note that this effect is independent of the increase in inherent tube conductivity found with an increase in nanotube length,[59] so the real increase could be higher than that shown in Figure 7.3. This theoretical study ignored the effect of dispersion; in general, longer nanotubes are going to be more difficult to disperse and poorer dispersion will likely tend to lower the thermal conductivity. An experimental study in oil showed an increase of a factor of about 1.8 in thermal conductivity with an order of magnitude increase in aspect ratio.[55] A study in an epoxy showed a decrease, rather than an increase, in thermal conductivity with an increase in nanotube length, which was attributed to poorer dispersion of the longer tubes.[42]

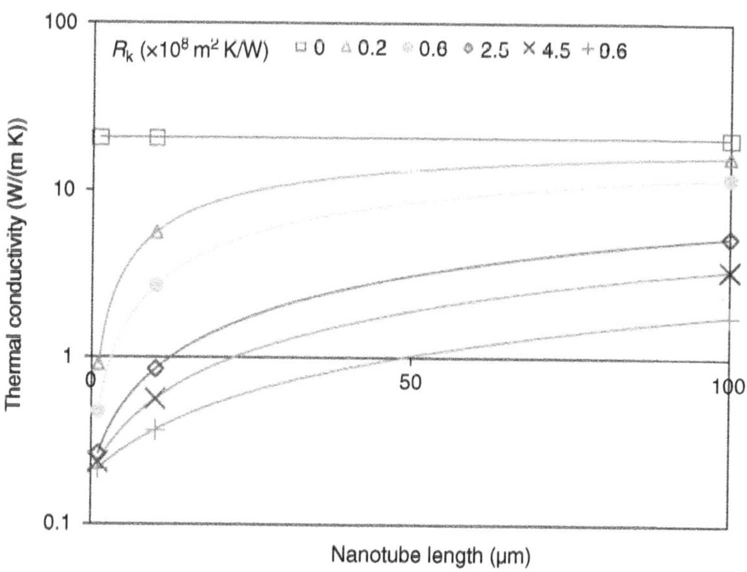

Figure 7.3 Effect of nanotube length on thermal conductivity of a nanotube–polymer composite considering only the effect of Kapitza resistance. Copyright Elsevier Ltd. Reproduced with permission from Ref. 10.

The increase in thermal conductivity as a function of temperature at temperatures below T_g of a nanotube-filled composite should be much larger than that for the pure polymer, which is a result of the temperature dependence of the interfacial resistance.[65] In some cases, a stronger temperature dependence has been found;[16,57,66] however, in other cases the temperature dependence essentially matches that of the pure polymer.[25,63,67] In one case, a broad maximum in thermal conductivity appears at about 320 K,[16] the temperature at which the maximum of nanotube thermal conductivity occurs. Finally, in one unique case, the temperature dependence of the conductivity from room temperature to the melting point of high-density polyethylene was extremely large, and almost a factor of 2 decrease in the thermal conductivity was measured.[68] In a study with polypropylene, the thermal conductivity of the composite had a stronger temperature dependence than the polymer, although this was only clear at at 10 wt% and higher nanotube fractions. Further, at 15% tubes, melting of the polymer did not cause the normal drop in thermal conductivity.[69] These data are reproduced in Figure 7.4.

Using a second filler to adjust macroscopic nanotube dispersion in order to increase the thermal conductivity has had some minor effectiveness. Graphene nanoplatelets are themselves thermally conducting and, not surprisingly, generally yield composites with higher thermal conductivities than nanotube composites because of the fewer number of filler-filler/filler-polymer contacts required for transport. When nanotubes were added to a filled graphene epoxy, there was a

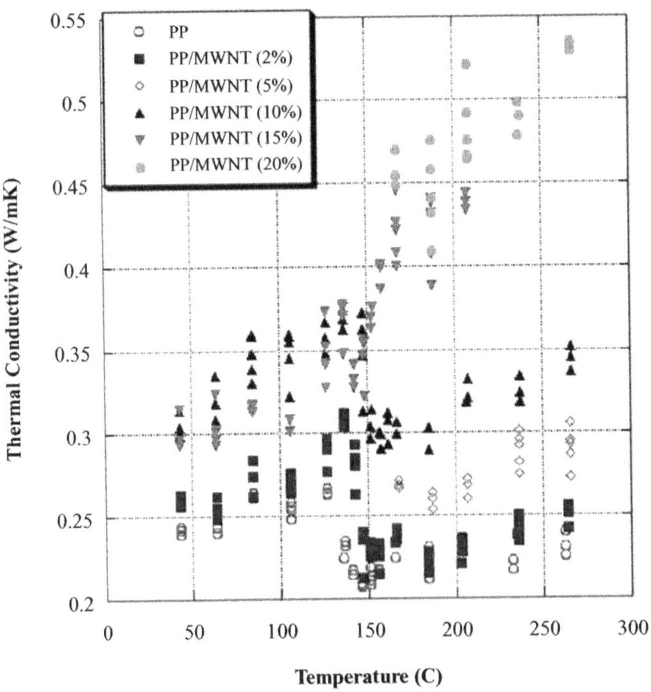

Figure 7.4 Increase in thermal conductivity with temperature and nanotube addition in polypropylene. The discontinuity at around 150°C represents the melting point of the polymer. Copyright Elsevier Ltd. Reproduced with permission from Ref. 69.

synergistic effect; that is, the maximum thermal conductivity at a constant filler volume percent occurred at an intermediate graphene:nanotube ratio. The percentage increase over the pure graphene material was on the order of 20%. However, this enhancement occurred only at filler volume fractions between about 10 and 20 vol%, and at higher volume fractions, the composite filled with pure graphene had the highest conductivity.[70] Nanotubes can also be added to a resin that in turn is added to continuous carbon fiber composites, as detailed in Section 3.7.4. In one case, the thermal conductivity increased from 250 to 393 W/(m K) in the direction perpendicular to the carbon fiber alignment.[71] Such a high value of thermal conductivity perpendicular to the alignment direction is difficult to understand, and the author of this book is not entirely sure what direction was measured in that study.

7.5 CHALLENGES

Clearly reaching higher thermal conductivities is the challenge. For many circumstances, thermal conductivities between 10 and 100 W/(m K) would be more than sufficient to justify new commercial products filled with nanotubes. The single study that achieved a conductivity of 250 W/(m K)[26,27] gives hope that perhaps with the right interfacial modification, much higher thermal conductivities might be possible for cases where the nanotubes do not span one part of the object to another. Hence, even though it is probably not realistic for thermal conductivities to behave according to a mixing law, a factor of 10 reduction in thermal conductivity because of interfacial resistance would still yield a very useful high thermally conductive material. The biggest hope in this area lies in the production of longer nanotubes, possibly in combination with graphene or some other high thermally conductive filler.

REFERENCES

1. Kim, P., Shi, L., Majumdar, A., McEuen, P. L. (2001). Thermal transport measurements of individual multiwalled nanotubes, *Phys. Rev. Lett.*, *87* (21), 215502.
2. Fujii, M., Zhang, X., Xie, H. Q., Ago, H., Takahashi, K., Ikuta, T., Abe, H., Shimizu, T. (2005). Measuring the thermal conductivity of a single carbon nanotube, *Phys. Rev. Lett.*, *95*(6), 065502.
3. Choi, T. Y., Poulikakos, D., Tharian, J., Sennhauser, U. (2005). Measurement of thermal conductivity of individual multiwalled carbon nanotubes by the 3-omega method, *Appl. Phys. Lett.*, *87*, 013108.
4. Pop, E., Mann, D., Wang, Q., Goodson, K., Dai, H. J. (2006). Thermal conductance of an individual single-wall carbon nanotube above room temperature, *Nano Lett.*, *6*, 96–100.
5. Wang, Z. L., Tang, D. W., Li, X. B., Zheng, X. H., Zhang, W. G., Zheng, L. X., Zhu, Y. T. T., Jin, A. Z., Yang, H. F., Gu, C. Z. (2007). Length-dependent thermal conductivity of an individual single-wall carbon nanotube, *Appl. Phys. Lett.*, *91*, 123119.
6. Swartz, E. T., Pohl, R. O. (1989). Thermal-boundary resistance, *Rev. Mod. Phys.*, *61*, 605–668.
7. Cahill, D. G., Ford, W. K., Goodson, K. E., Mahan, G. D., Majumdar, A., Maris, H. J., Merlin, R., Phillpot, S. R. (2003). Nanoscale thermal transport, *J. Appl. Phys.*, *93*, 793–818.
8. Shenogin, S., Xue, L. P., Ozisik, R., Keblinski, P., Cahill, D. G. (2004). Role of thermal boundary resistance on the heat flow in carbon-nanotube composites, *J. Appl. Phys.*, *95*, 8136–8144.
9. Nan, C. W., Liu, G., Lin, Y. H., Li, M. (2004). Interface effect on thermal conductivity of carbon nanotube composites, *Appl. Phys. Lett.*, *85*, 3549–3551.
10. Clancy, T. C., Gates, T. S. (2006). Modeling of interfacial modification effects on thermal conductivity of carbon nanotube composites, *Polymer*, *47*, 5990–5996.

11. Bryning, M. B., Milkie, D. E., Islam, M. F., Kikkawa, J. M., Yodh, A. G. (2005). Thermal conductivity and interfacial resistance in single-wall carbon nanotube epoxy composites, *Appl. Phys. Lett.*, *87*, 161909.
12. Huxtable, S. T., Cahill, D. G., Shenogin, S., Xue, L. P., Ozisik, R., Barone, P., Usrey, M., Strano, M. S., Siddons, G., Shim, M., Keblinski, P. (2003). Interfacial heat flow in carbon nanotube suspensions, *Nat. Mater.*, *2*, 731–734.
13. Haggenmueller, R., Guthy, C., Lukes, J. R., Fischer, J. E., Winey, K. I. (2007). Single wall carbon nanotube/polyethylene nanocomposites: thermal and electrical conductivity, *Macromolecules*, *40*, 2417–2421.
14. Gaxiola, D. L., Keith, J. M., King, J. A., Johnson, B. A. (2009). Nielsen thermal conductivity model for single filler carbon/polypropylene composites, *J. Appl. Polym. Sci.*, *114*, 3261–3267.
15. Guthy, C., Du, F., Brand, S., Winey, K. I., Fischer, J. E. (2007). Thermal conductivity of single-walled carbon nanotube/PMMA nanocomposites, *J. Heat Transfer*, *129*, 1096–1099.
16. Xu, Y. S., Ray, G., Abdel-Magid, B. (2006). Thermal behavior of single-walled carbon nanotube polymer–matrix composites, *Compos. Part A*, *37*, 114–121.
17. Gao, L., Zhou, X. F., Ding, Y. L. (2007). Effective thermal and electrical conductivity of carbon nanotube composites, *Chem. Phys. Lett.*, *434*, 297–300.
18. Bagchi, A., Nomura, S. (2006). On the effective thermal conductivity of carbon nanotube reinforced polymer composites, *Compos. Sci. Technol.*, *66*, 1703–1712.
19. Zhong, C. L., Yang, Q. Y., Wang, W. C. (2001). Correlation and prediction of the thermal conductivity of amorphous polymers, *Fluid Phase Equilib.*, *181*, 195–202.
20. Sombatsompop, N., Wood, A. K. (1997). Measurement of thermal conductivity of polymers using an improved Lee's disc apparatus, *Polym. Test.*, *16*, 203–223.
21. Zhang, X., Hendro, W., Fujii, M., Tomimura, T., Imaishi, N. (2002). Measurements of the thermal conductivity and thermal diffusivity of polymer melts with the short-hot-wire method, *Int. J. Thermophys.*, *23*, 1077–1090.
22. Zhang, R., Fujii, M. (2003). Measurements of the thermal conductivity and thermal diffusivity of polymers, *Polym. Eng. Sci.*, *43*, 1755–1764.
23. Fuller, T. R., Fricke, A. L. (1971). Thermal conductivity of polymer melts, *J. Appl. Polym. Sci.*, *15*, 1729–1736.
24. Lobo, H., Cohen, C. (1990). Measurement of thermal conductivity of polymer melts by the line-source method, *Polym. Eng. Sci.*, *30*, 65–70.
25. Peters, J. E., Papavassiliou, D. V., Grady, B. P. (2008). Unique thermal conductivity behavior of single-walled carbon nanotube–polystyrene composites, *Macromolecules*, *41*, 7274–7277.
26. Sihn, S., Ganguli, S., Roy, A. K., Qu, L. T., Dai, L. M. (2008). Enhancement of through-thickness thermal conductivity in adhesively bonded joints using aligned carbon nanotubes, *Compos. Sci. Technol.*, *68*, 658–665.
27. Ganguli, S., Sihn, S., Roy, A. K., Dai, L. M., Qu, L. (2009). Metalized nanotube tips improve through thickness thermal conductivity in adhesive joints, *J. Nanosci. Nanotechnol.*, *9*, 1727–1733.
28. Huang, H., Liu, C. H., Wu, Y., Fan, S. S. (2005). Aligned carbon nanotube composite films for thermal management, *Adv. Mater.*, *17*, 1652–1656.
29. Borca-Tasciuc, T., Mazumder, M., Son, Y., Pal, S. K., Schadler, L. S., Ajayan, P. M. (2007). Anisotropic thermal diffusivity characterization of aligned carbon nanotube–polymer composites, *J. Nanosci. Nanotechnol.*, *7*, 1581–1588.
30. Ivanov, I., Puretzky, A., Eres, G., Wang, H., Pan, Z. W., Cui, H. T., Jin, R. Y., Howe, J., Geohegan, D. B. (2006). Fast and highly anisotropic thermal transport through vertically aligned carbon nanotube arrays, *Appl. Phys. Lett.*, *89*, 223110.
31. Zhang, J. M., Tanaka, M., Matsumoto, T. (2004). A simplified approach for heat conduction analysis of CNT-based nano-composites, *Comput. Methods Appl. Mech. Eng.*, *193*, 5597–5609.
32. Zhang, J. M., Tanaka, M., Matsumoto, T., Guzik, A. (2004). Heat conduction analysis in bodies containing thin-walled structures by means of hybrid BNM with an application to CNT-based composites, *JSME Int. J. Ser. A*, *47*, 181–188.
33. Yao, Z. H., Xu, J. D., Wang, H. T., Zheng, X. P. (2009). Simulation of CNT composites using fast multipole BEM, *J. Mater. Sci. Technol.*, *17*, 194–202.

34. Duong, H. M., Papavassiliou, D. V., Mullen, K. J., Maruyama, S. (2008). Computational modeling of the thermal conductivity of single-walled carbon nanotube–polymer composites, *Nanotechnology, 19*, 065702.
35. Duong, H. M., Yamamoto, N., Papavassiliou, D. V., Maruyama, S., Wardle, B. L. (2009). Inter-carbon nanotube contact in thermal transport of controlled-morphology polymer nanocomposites, *Nanotechnology, 20*, 155702.
36. Singh, I. V., Tanaka, M., Zhang, J., Endo, M. (2007). Evaluation of effective thermal conductivity of CNT-based nano-composites by element free Galerkin method, *Int. J. Numer. Methods Heat Fluid Flow, 17*, 757–769.
37. Singh, I. V., Tanaka, M., Endo, M. (2007). Meshless method for nonlinear heat conduction analysis of nano-composites, *Heat Mass Transf., 43*, 1097–1106.
38. Jeong, T. T., Song, Y. S. In: Alexandrov, V. N., Van Albada, G. D., Sloot, P. M. A., Dongarra, J. Eds., *Computational Science—Proceedings of ICCS 2006, Part 3*, Vol. 3993, Springer, Berlin, 2006, pp. 105–112.
39. Song, Y. S., Youn, J. R. (2006). Evaluation of effective thermal conductivity for carbon nanotube/polymer composites using control volume finite element method, *Carbon, 44*, 710–717.
40. Seidel, G. D., Lagoudas, D. C. (2008). A micromechanics model for the thermal conductivity of nanotube–polymer nanocomposites, *J. Appl. Mech., 75*, 041025.
41. Shenogin, S., Bodapati, A., Xue, L., Ozisik, R., Keblinski, P. (2004). Effect of chemical functionalization on thermal transport of carbon nanotube composites, *Appl. Phys. Lett., 85*, 2229–2231.
42. Wang, S. R., Liang, R., Wang, B., Zhang, C. (2009). Dispersion and thermal conductivity of carbon nanotube composites, *Carbon, 47*, 53–57.
43. Yu, A. P., Itkis, M. E., Bekyarova, E., Haddon, R. C. (2006). Effect of single-walled carbon nanotube purity on the thermal conductivity of carbon nanotube-based composites, *Appl. Phys. Lett., 89*, 133102.
44. Yang, K., Gu, M. Y., Guo, Y. P., Pan, X. F., Mu, G. H. (2009). Effects of carbon nanotube functionalization on the mechanical and thermal properties of epoxy composites, *Carbon, 47*, 1723–1737.
45. Yang, K., Gu, M. Y. (2009). The effects of triethylenetetramine grafting of multi-walled carbon nanotubes on its dispersion, filler–matrix interfacial interaction and the thermal properties of epoxy nanocomposites, *Polym. Eng. Sci., 49*, 2158–2167.
46. Gojny, F. H., Wichmann, M. H. G., Fiedler, B., Kinloch, I. A., Bauhofer, W., Windle, A. H., Schulte, K. (2006). Evaluation and identification of electrical and thermal conduction mechanisms in carbon nanotube/epoxy composites, *Polymer, 47*, 2036–2045.
47. Kim, H. S., Chae, Y. S., Park, B. H., Yoon, J. S., Kang, M., Jin, H. J. (2008). Thermal and electrical conductivity of poly(L-lactide)/multiwalled carbon nanotube nanocomposites, *Curr. Appl. Phys., 8*, 803–806.
48. Hone, J., Llaguno, M. C., Nemes, N. M., Johnson, A. T., Fischer, J. E., Walters, D. A., Casavant, M. J., Schmidt, J., Smalley, R. E. (2000). Electrical and thermal transport properties of magnetically aligned single wall carbon nanotube films, *Appl. Phys. Lett., 77*, 666–668.
49. Itkis, M. E., Borondics, F., Yu, A. P., Haddon, R. C. (2007). Thermal conductivity measurements of semitransparent single-walled carbon nanotube films by a bolometric technique, *Nano Lett., 7*, 900–904.
50. Prasher, R. S., Hu, X. J., Chalopin, Y., Mingo, N., Lofgreen, K., Volz, S., Cleri, F., Keblinski, P. (2009). Turning carbon nanotubes from exceptional heat conductors into insulators, *Phys. Rev. Lett., 102*, 105901.
51. Xu, Z. P., Buehler, M. J. (2009). Nanoengineering heat transfer performance at carbon nanotube interfaces, *ACS Nano, 3*, 2767–2775.
52. Shenogina, N., Shenogin, S., Xue, L., Keblinski, P. (2005). On the lack of thermal percolation in carbon nanotube composites, *Appl. Phys. Lett., 87*, 133106.
53. Abbasi, S., Carreau, P. J., Derdouri, A., Moan, M. (2009). Rheological properties and percolation in suspensions of multiwalled carbon nanotubes in polycarbonate, *Rheol. Acta, 48*, 943–959.
54. Bonnet, P., Sireude, D., Garnier, B., Chauvet, O. (2007). Thermal properties and percolation in carbon nanotube–polymer composites, *Appl. Phys. Lett., 91*, 201910.
55. Yang, Y., Grulke, E. A., Zhang, Z. G., Wu, G. F. (2006). Thermal and rheological properties of carbon nanotube-in-oil dispersions, *J. Appl. Phys., 99*, 114307.

56. Diez-Pascual, A. M., Naffakh, M., Gomez, M. A., Marco, C., Ellis, G., Gonzalez-Dominguez, J. M., Anson, A., Martinez, M. T., Martinez-Rubi, Y., Simard, B., Ashrafi, B. (2009). The influence of a compatibilizer on the thermal and dynamic mechanical properties of PEEK/carbon nanotube composites, *Nanotechnology, 20*, 315707.
57. Kim, S. W., Kim, J. K., Lee, S. H., Park, S. J., Kang, K. H. (2006). Thermophysical properties of multi-walled carbon nanotube-reinforced polypropylene composites, *Int. J. Thermophys., 27*, 152–160.
58. Song, Y. S., Youn, J. R. (2005). Influence of dispersion states of carbon nanotubes on physical properties of epoxy nanocomposites, *Carbon, 43*, 1378–1385.
59. Savin, A. V., Hu, B. B., Kivshar, Y. S. (2009). Thermal conductivity of single-walled carbon nanotubes, *Phys. Rev. B, 80*, 195423.
60. Cao, B. Y., Hou, Q. W. (2008). Thermal conductivity of carbon nanotubes embedded in solids, *Chin. Phys. Lett., 25*, 1392–1395.
61. Prasher, R. (2007). Thermal conductance of single-walled carbon nanotube embedded in an elastic half-space, *Appl. Phys. Lett., 90*, 143110.
62. Ghose, S., Watson, K. A., Working, D. C., Connell, J. W., Smith, J. G., Sun, Y. P. (2008). Thermal conductivity of ethylene vinyl acetate copolymer/nanofiller blends, *Compos. Sci. Technol., 68*, 1843–1853.
63. Choi, E. S., Brooks, J. S., Eaton, D. L., Al-Haik, M. S., Hussaini, M. Y., Garmestani, H., Li, D., Dahmen, K. (2003). Enhancement of thermal and electrical properties of carbon nanotube polymer composites by magnetic field processing, *J. Appl. Phys., 94*, 6034–6039.
64. Jin, S. H., Kang, I. H., Kim, Y. S., Park, C. Y., Lee, D. S. (2008). Thermal and electrical properties of nanocomposites based on acrylic copolymers and multiwalled carbon nanotube, *J. Nanosci. Nanotechnol., 8*, 5076–5079.
65. Stoner, R. J., Maris, H. J. (1993). Kapitza conductance and heat-flow between solids at temperatures from 50 to 300 K, *Phys. Rev. B, 48*, 16373–16387.
66. Biercuk, M. J., Llaguno, M. C., Radosavljevic, M., Hyun, J. K., Johnson, A. T., Fischer, J. E. (2002). Carbon nanotube composites for thermal management, *Appl. Phys. Lett., 80*, 2767–2769.
67. Evseeva, L. E., Tanaeva, S. A. (2008). Thermal conductivity of micro- and nanostructural epoxy composites at low temperatures, *Mech. Compos. Mater., 44*, 87–92.
68. Wu, F., He, X., Zeng, Y., Cheng, H. M. (2006). Thermal transport enhancement of multi-walled carbon nanotubes/high-density polyethylene composites, *Appl. Phys. A, 85*, 25–28.
69. Kashiwagi, T., Grulke, E., Hilding, J., Groth, K., Harris, R., Butler, K., Shields, J., Kharchenko, S., Douglas, J. (2004). Thermal and flammability properties of polypropylene/carbon nanotube nanocomposites, *Polymer, 45*, 4227–4239.
70. Yu, A. P., Ramesh, P., Sun, X. B., Bekyarova, E., Itkis, M. E., Haddon, R. C. (2008). Enhanced thermal conductivity in a hybrid graphite nanoplatelet–carbon nanotube filler for epoxy composites, *Adv. Mater., 20*, 4740–4744.
71. Kim, Y. A., Kamio, S., Tajiri, T., Hayashi, T., Song, S. M., Endo, M., Terrones, M., Dresselhaus, M. S. (2007). Enhanced thermal conductivity of carbon fiber/phenolic resin composites by the introduction of carbon nanotubes, *Appl. Phys. Lett., 90*, 093125.
72. Yang, Y., Gupta, M. C., Zalameda, J. N., Winfree, W. P. (2008). Dispersion behaviour, thermal and electrical conductivities of carbon nanotube–polystyrene nanocomposites, *Micro Nano Lett., 3*, 35–40.
73. Ji, L. J., Stevens, M. M., Zhu, Y. F., Gong, Q. M., Wu, J. J., Liang, J. (2009). Preparation and properties of multi-walled carbon nanotube/carbon/polystyrene composites, *Carbon, 47*, 2733–2741.
74. Yuen, S. M., Ma, C. C. M., Chiang, C. L., Chang, J. A., Huang, S. W., Chen, S. C., Chuang, C. Y., Yang, C. C., Wei, M. H. (2007). Silane-modified MWCNT/PMMA composites: preparation, electrical resistivity, thermal conductivity and thermal stability, *Compos. Part A, 38*, 2527–2535.
75. Hong, W. T., Tai, N. H. (2008). Investigations on the thermal conductivity of composites reinforced with carbon nanotubes, *Diam. Relat. Mater., 17*, 1577–1581.
76. Du, F. M., Guthy, C., Kashiwagi, T., Fischer, J. E., Winey, K. I. (2006). An infiltration method for preparing single-wall nanotube/epoxy composites with improved thermal conductivity, *J. Polym. Sci. Polym. Phys., 44*, 1513–1519.
77. Kim, J., Hong, S. M., Kwak, S., Seo, Y. (2009). Physical properties of nanocomposites prepared by *in situ* polymerization of high-density polyethylene on multiwalled carbon nanotubes, *Phys. Chem. Chem. Phys., 11*, 10851–10859.

78. Chen, X. G., He, G. H., Du, J. H., Pei, S. F., Guo, J. F. (2009). Investigation on the thermal conductivity of HDPE/MWCNT composites by laser pulse method, *Sci. China Ser. E*, *52*, 2767–2772.
79. George, J. J., Bhowmick, A. K. (2009). Influence of matrix polarity on the properties of ethylene vinyl acetate–carbon nanofiller nanocomposites, *Nanoscale Res. Lett.*, *4*, 655–664.
80. Shen, Z. Q., Bateman, S., Wu, D. Y., McMahon, P., Dell'Olio, M., Gotama, J. (2009). The effects of carbon nanotubes on mechanical and thermal properties of woven glass fibre reinforced polyamide-6 nanocomposites, *Compos. Sci. Technol.*, *69*, 239–244.
81. Li, C. C., Lu, C. L., Lin, Y. T., Wei, B. Y., Hsu, W. K. (2009). Creation of interfacial phonons by carbon nanotube–polymer coupling, *Phys. Chem. Chem. Phys.*, *11*, 6034–6037.
82. Grunlan, J. C., Kim, Y. S., Ziaee, S., Wei, X., Abdel-Magid, B., Tao, K. (2006). Thermal and mechanical behavior of carbon-nanotube-filled latex, *Macromol. Mater. Eng.*, *291*, 1035–1043.
83. Samuel, J., Dikshit, A., DeVor, R. E., Kapoor, S. G., Hsia, K. J. (2009). Effect of carbon nanotube (CNT) loading on the thermomechanical properties and the machinability of CNT-reinforced polymer composites, *J. Manuf. Sci. Eng.*, *131*, 031008.
84. Yu, C., Kim, Y. S., Kim, D., Grunlan, J. C. (2008). Thermoelectric behavior of segregated-network polymer nanocomposites, *Nano Lett.*, *8*, 4428–4432.
85. Bangarusampath, D. S., Ruckdaschel, H., Altstadt, V., Sandler, J. K. W., Garray, D., Shaffer, M. S. P. (2009). Rheology and properties of melt-processed poly(ether ether ketone)/multi-wall carbon nanotube composites, *Polymer*, *50*, 5803–5811.
86. Abdullah, S. A., Iqbal, A., Frormann, L. (2008). Melt mixing of carbon fibers and carbon nanotubes incorporated polyurethanes, *J. Appl. Polym. Sci.*, *110*, 196–202.
87. Cai, D., Song, M. (2007). Water-based polyurethane filled with multi-walled carbon nanotubes prepared by a colloidal-physics method, *Macromol. Chem. Phys.*, *208*, 1183–1189.
88. Cai, D. Y., Song, M. (2008). Latex technology as a simple route to improve the thermal conductivity of a carbon nanotube/polymer composite, *Carbon*, *46*, 2107–2112.
89. Hong, S. M., Hwang, S. S. (2008). Physical properties of thin PVDF/MWNT (multi-walled carbon nanotube) composite films by melt blending, *J. Nanosci. Nanotechnol.*, *8*, 1860–4863.
90. Mamunya, Y., Boudenne, A., Lebovka, N., Ibos, L., Candau, Y., Lisunova, M. (2008). Electrical and thermophysical behaviour of PVC–MWCNT nanocomposites, *Compos. Sci. Technol.*, *68*, 1981–1988.
91. Zeng, J. L., Liu, Y. Y., Cao, Z. X., Zhang, J., Zhang, Z. H., Sun, L. X., Xu, F. (2008). Thermal conductivity enhancement of MWNTs on the PANI/tetradecanol form-stable PCM, *J. Therm. Anal. Calorim.*, *91*, 443–446.
92. Moisala, A., Li, Q., Kinloch, I. A., Windle, A. H. (2006). Thermal and electrical conductivity of single- and multi-walled carbon nanotube–epoxy composites, *Compos. Sci. Technol.*, *66*, 1285–1288.
93. Lazarenko, A., Vovchenko, L., Matsui, D., Prylutskyy, Y., Matzuy, L., Ritter, U., Scharff, P. (2008). Electrical and thermal conductivity of polymer–nanocarbon composites, *Mol. Cryst. Liq. Cryst.*, *497*, 397–407.
94. Assael, M. J., Antoniadis, K. D., Metaxa, I. N. (2009). Measurements on the enhancement of the thermal conductivity of an epoxy resin when reinforced with glass fiber and carbon multiwalled nanotubes, *J. Chem. Eng. Data*, *54*, 2365–2370.
95. Assael, M. J., Antoniadis, K. D., Tzetzis, D. (2008). The use of the transient hot-wire technique for measurement of the thermal conductivity of an epoxy-resin reinforced with glass fibres and/or carbon multi-walled nanotubes, *Compos. Sci. Technol.*, *68*, 3178–3183.
96. Lin, W., Moon, K. S., Wong, C. P. (2009). A combined process of *in situ* functionalization and microwave treatment to achieve ultrasmall thermal expansion of aligned carbon nanotube–polymer nanocomposites: toward applications as thermal interface materials, *Adv. Mater.*, *21*, 2421–2424.
97. Sanada, K., Tada, Y., Shindo, Y. (2009). Thermal conductivity of polymer composites with close-packed structure of nano and micro fillers, *Compos. Part A*, *40*, 724–730.
98. Liu, C. H., Fan, S. S. (2005). Effects of chemical modifications on the thermal conductivity of carbon nanotube composites, *Appl. Phys. Lett.*, *86*, 123106.
99. Park, K. C., Mahiko, T., Morimoto, S., Takeuchi, K., Endo, M. (2008). Hydrophobicity-induced selective covering of carbon nanotubes with sol–gel sheaths achieved by ultrasound assistance, *Appl. Surf. Sci.*, *254*, 7438–7445.

100. Gau, C., Chen, S. Y., Tsai, H. L., Jenq, S. T., Lee, C. C., Chen, Y. D., Chien, T. H. (2009). Synthesis of functionalized carbon nanotubes/phenolic nanocomposites and its electrical and thermal conductivity measurements, *Jpn. J. Appl. Phys.*, *48*, 06ff10.
101. Fujigaya, T., Fukumaru, T., Nakashima, N. (2009). Evaluation of dispersion state and thermal conductivity measurement of carbon nanotubes/UV-curable resin nanocomposites, *Synth. Met.*, *159*, 827–830.
102. Das, A., Stockelhuber, K. W., Jurk, R., Saphiannikova, M., Fritzsche, J., Lorenz, H., Kluppel, M., Heinrich, G. (2008). Modified and unmodified multiwalled carbon nanotubes in high performance solution–styrene–butadiene and butadiene rubber blends, *Polymer*, *49*, 5276–5283.
103. Tonpheng, B., Yu, J. C., Andersson, O. (2009). Thermal conductivity, heat capacity, and cross-linking of polyisoprene/single-wall carbon nanotube composites under high pressure, *Macromolecules*, *42*, 9295–9301.
104. Tonpheng, B., Yu, J. C., Andersson, O. (2008). Polyisoprene single-wall carbon nanotube composites synthesized under high pressure, *High Press. Res.*, *28*, 587–590.

CHAPTER 8

APPLICATIONS OF POLYMER–NANOTUBE COMPOSITES

8.1 OVERVIEW

This chapter gives a brief overview of the applications for combinations of nanotubes and polymers that have been experimentally tested, at least to some extent. However, as you read this, it is almost certain that new applications not discussed here have been developed. Approximately once a month an article appears highlighting a possible new application for carbon nanotubes, with many of these applications involving polymers.

As stated in Chapter 1, interactions of polymers and nanotubes specific to the field of biology, including applications, are not within the scope of this tome. In particular, sensors for biological macromolecules or sensors made with nanotubes and polymers for biological sensing are not part of this tome. Other diagnostic uses and the use of nanotubes for therapeutics (e.g., the use of electromagnetic radiation to cause localized heating in biological systems) are not part of this discussion either.

Each section will briefly describe first the application and then the experimental work in the area. In particular, how nanotube–polymer materials may offer some advantage over currently available materials will be highlighted. This chapter is organized by properties: what is the particular property that lends itself for a given application?

8.2 ELECTRICAL CONDUCTIVITY: EMI SHIELDING, ESD, AND TRANSPARENT ELECTRODES

Nanotube composites (nor in fact any polymeric material, including conducting polymers) do not have high enough conductivity to replace metal wiring. However, a number of applications exist where the conductivity requirement is not as severe. Two of those applications are electromagnetic interference (EMI) shielding and electrostatic dissipation (ESD). The difference between EMI shielding and ESD is the conductivity required; the former requires significantly higher conductivities than the latter. Specifically, a particular material is classified as conductive, static

Carbon Nanotube–Polymer Composites: Manufacture, Properties, and Applications, First Edition.
Brian P. Grady.

conductive, or static dissipative. The surface resistivity regimes for each are approximately less than 10^4, 10^4–10^6, and 10^6–10^{12} Ω/sq, respectively. These surface resistivities correspond roughly to bulk conductivity ranges of >0.1, 10^{-3} to 10^{-1}, and 10^{-3} to 10^{-9} S/m (assuming the fillers are dispersed approximately uniformly). Materials classified as conductive are useful for EMI shielding. Metal cages are used where the shielding requirements are severe. For lesser requirements, carbon black or metal powders are currently the major competitors to nanotubes as fillers in conductive composites. These fillers are inferior to carbon nanotube-filled materials in terms of the reduction in flexibility due to the much higher amounts of filler required. Graphene and smaller diameter vapor-grown carbon fibers are emerging competitors to carbon nanotubes.

The third application that takes advantage of nanotube conductivity is due to an inherent property of all metals: metals are not optically transparent. Flat panel displays require transparent conducting materials; other devices requiring transparent conductors include solar cells, touch panels, organic light-emitting diodes (OLEDs), and electroluminescent lighting. Specifically, the requirement is for transparent electrodes, for example, the connection between a current source and an active component or between an active component and a current sink. Very thin films of nanotubes, for example, buckypaper, with high conductivities can be made and polymers can assist in this process. Flexible electronics, where the nanotubes are deposited on a polymeric substrate, are probably the most common use of nanotubes as transparent electrodes. Transparent electrodes will be dealt with in Section 8.2.3. Finally, there are a number of other applications that utilize the conductivity of nanotubes, which, at this time, have not been as fully explored as these three. These other applications will be discussed in Section 8.2.4.

8.2.1 Electromagnetic Shielding

The purpose of electromagnetic shielding is to reduce or deny exit or entry of electromagnetic radiation from or into a given space. Applications where such shielding is used include microwave doors (shielding is used to prevent microwaves from escaping and heating up something outside the door!), shielded cables (shielding is necessary to prevent interference from the outside from affecting the signal carried by the cable), and various electronic devices such as mobile phones, computers, televisions, and so on, where the purpose can be to prevent both leakage and interference. The EMI shielding effectiveness (SE) in decibel is defined as $\log(P_I/P_T)$, where P_T is the power of the transmitted electromagnetic wave and P_I is the power of the incident wave. In other words, a value of 20 dB (sometimes this will be written as −20 dB) means that 1% of the wave is transmitted, while 30 dB means that 0.1% of the wave is transmitted. In general, the lower the frequency (or the longer the wavelength), the more effective a given shield in reducing the amount of transmitted waves. However, at gigahertz frequencies, which correspond to microwave radiation, an increase in shielding effectiveness with an increase in frequency is often observed.

In a general sense, shielding effectiveness scales with the DC electrical conductivity of the material; that is, higher conductivity means higher shielding effectiveness. For nanotube composites, there are three contributions to the shielding effectiveness. The first two are absorption and single reflection (reflection from the surface), and the empirical Simon formula has been used to describe shielding effectiveness in this situation:

$$\mathrm{SW(dB)} = 50 + 10\log\frac{\sigma}{f} + 1.7t\sqrt{f\sigma} \tag{8.1}$$

where σ is the conductivity of the material (S/cm), t is the thickness of the sample (cm), and f is the frequency (MHz) of the electromagnetic radiation. The first two terms are reflective terms, while the latter is an absorptive term. Some studies have found qualitative and quantitative correspondence between conductivity and SE. For example, percolation measurements using both shielding effectiveness and normal conduction measurements as the ordinate were very similar in a poly(methyl methacrylate) filled with nanotubes.[1,2] Other resins show an increase in shielding effectiveness with increased conductivity.[3–5] In fact, a plot of shielding effectiveness at a measurement frequency of 10 GHz versus log(conductivity) in the plateau region found a slope of 10 just as the second term of the Simon formula predicts.[6] Also, conducting polymer–nanotube composites yield shielding effectiveness values much higher than typical values for insulating polymer–nanotube composites in gigahertz frequency range: 27–40 dB versus typically 15–30 dB.[7] Shielding effectiveness has been found to scale with sample thickness, in agreement with the Simon formalism.[8] Coating with a metal[9] or orientation of nanotubes[10] also improves shielding effectiveness. A final example showing that shielding effectiveness roughly scales with conductivity is that the best single-walled nanotube type for conductivity (e.g., longer nanotubes) also yields the highest shielding effectiveness.[11–13]

However, in not all cases does the shielding effectiveness scale precisely with conductivity. The efficiency of EMI shielding was compared between dissolution–dispersion–precipitation and dispersion–reaction preparation schemes for maleic anhydride-functionalized tubes in poly(methyl methacrylate).[14] The percolation curves were almost identical, but the efficiency of EMI shielding was superior in the dispersion–reaction scheme. The authors attributed the difference in SE to a difference in dispersion. If true, then not only is conductivity unclear in terms of whether dispersion is better or worse, but different dispersions could also yield the same conductivity. In another case comparing nanotube composites made with polyethylene and poly(phenylene sulfide), the conductivity was higher for the former but the shielding effectiveness was higher for the latter.[15] Also, non-nanotube structures with very specific geometric shapes showed much better shielding efficiency than percolation/conduction arguments would suggest, showing that the shape of agglomerated tubes may have an important effect on EMI behavior.[2] Finally, the addition of a small amount of nanotubes to a carbon nanofiber composite caused an increase in shielding efficiency much larger than the change in conductivity would have predicted.[16]

The dependence of shielding effectiveness on dispersion is due to a third mechanism for shielding: multiple reflections by interfaces contained within a material. In noninfusion processes, that is, more or less randomly mixed systems, one study claims that shielding effectiveness in the gigahertz frequency range is dominated by absorption while multiple reflections play a minor role[17]; however, a different paper comes to the exactly opposite conclusion.[18] As expected, as the

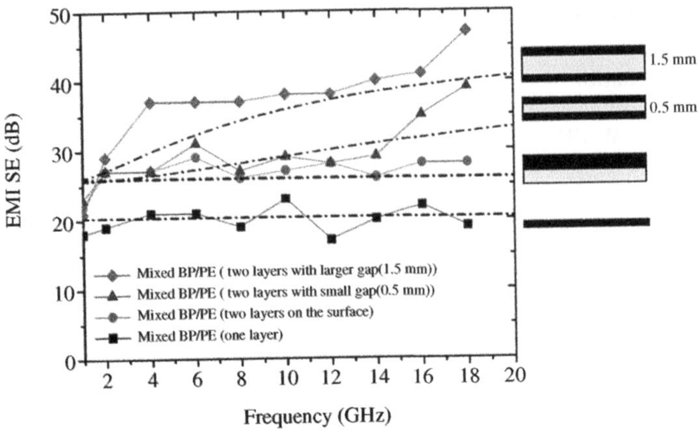

Figure 8.1 Shielding effectiveness of mixed polymer–buckypaper layers with different geometries and configurations. Dashed lines represent a calculation based on a measured buckypaper layer of 2000 S/m (after infusion of low-density polyethylene), while data represent the results for a buckypaper layer (black) of 25 μm and polyethylene layer (gray) with thicknesses as indicated. Copyright IOPscience. Reproduced with permission from Ref. 19.

nanotube content increases, absorption becomes more important relative to reflection.[18] However, reflection has been shown conclusively to dominate the response in appropriately structured systems, as was illustrated quite well by a paper from Park et al.[19] Typical SE values for nonstructured nanotube composites with nonconducting polymers are no higher than 20–30 dB in the gigahertz frequency range (except at extremely high nanotube fractions, for example, 50% nanotubes[20]); however, values as high as 100 dB are possible at low overall tube fractions using buckypapers and alternating arrangements of insulating and conducting layers. As shown in Figure 8.1, the shielding effectiveness increases dramatically as the insulating layer (polyethylene in this case) increases in thickness. Even more remarkable, by not infusing the buckypaper layer with polymer and using higher quality nanotubes, the shielding effectiveness rose from 70 to approximately 100 dB over the same frequency range with three buckypaper layers and two insulating layers. Another interesting processing technique that led to extremely high shielding effectiveness due to multiple reflections was the use of supercritical CO_2 to foam a MWCNT/polycaprolactone composite. In this case, the shielding effectiveness was between 60 and 80 dB over the frequency range of 25–40 GHz.[21] Foaming with a non-supercritical fluid by a different group led to shielding effectiveness of 20 dB, that is, not that high.[22] Similar results showing the effectiveness of laminated composites were also found for poly(methyl methacrylate),[23,24] polycarbonate,[25] and epoxy[26] composites.

8.2.2 Electrostatic Dissipation

For a material to be electrostatically dissipative, static charge must not build up on the surface. Static charge is a significant safety hazard since a discharge can lead to fire in certain situations. For example, static discharge means that one should not reenter his/her motor vehicle when filling the gas tank because car fabric can lead to a static charge and, if discharged too close to the flowing gasoline, an explosion can ensue. Another very common safety issue (although certainly not with the possible dire consequences of an explosion!) is that computer circuit boards are sensitive to the very high voltages (thousands of volts) of a static discharge and one static discharge can ruin a board. Hence, a very common use of polymers filled with conductive particles is in the microelectronics area; all polymers that are used to transport devices (bags, trays, etc.) in microelectronics are filled or surface treated to achieve sufficient surface conductivity.

As with EMI shielding, higher surface conductivities yield better dissipative behavior; however, in this case there is direct correspondence. However, at a certain point, the improvement becomes irrelevant, that is, higher performance does not improve the dissipative behavior, and this point roughly occurs at the conductivity corresponding to the transition between static conductive and conductive behavior. Hence, for most applications, a material that is classified as conductive is perfectly acceptable from an electrical performance perspective (although cost may be higher and flexibility lower). One very important exception is materials designed to withstand lightning strikes (i.e., airframes) where high conductivities are critical. In one case, buckypaper added to the outside of an epoxy–carbon composite significantly improved the composites ability to resist mechanical property degradation due to a lightning strike.[27]

An important point for all dissipative materials is that surface conductivity, not bulk conductivity, controls performance. Hence, metallized coatings are perfectly fine; the problem with metallized coatings compared to nanotube-filled materials is that the coating often falls off easily. As the next section will describe, coating polymers with nanotubes is quite common for transparent electrodes. The cost and difficulty of coating have limited applications for ESD coatings of polymer-coated nanotubes; however, a method was described to coat with a thin film of nanotubes followed by infusion of resin and reaction to build up a static dissipative coating.[28] Almost all ESD uses of nanotubes involve mixing the nanotubes with the polymer.

Since performance requirements are not severe, most dissipative applications use low-cost materials; that is, nanotubes would not be the material of choice. One exception is where flexibility is required; in this case, nanotubes can be competitive because of the lower volume fractions required as opposed to carbon black. A few applications require a relatively precise value of the conductivity in the static dissipative regime, which for most fillers falls into the percolation region. For example, in some cases, bulk conductivity must be avoided. Designing a system with conductivity at a certain value is impractical when the conductivity lies in the percolation region because the steep rise in conductivity means poor conductivity control. For such conductivities, a filler should be used that yields a composite conductivity in the plateau regime with the desired value. Currently, composite conductivity can be lowered by coating the conducting filler with an electrically conducting polymer. A simpler process to control final conductivity should be to reduce the conductivity of nanotubes via functionalization and thereby precisely control the plateau conductivity of polymer-nanotube composites. The author is not aware of any studies with that specific goal, however.

8.2.3 Transparent Electrodes

Carbon nanotube films, for example, buckypaper, can be made conducting at very low thicknesses leading to clear conducting films. In many cases, carbon nanotubes are not mixed with a polymer; however, the most important application of nanotubes as transparent electrodes will be on polymeric substrates. In other cases, the nanotubes are mixed with a polymer to improve dispersion or mixed with an electrically conducting polymer ideally to improve conductivity. The current workhorse of the field of transparent electrodes, indium–tin oxide (ITO), has a number of drawbacks especially when applied to flexible, for example, polymeric, substrates. ITO is brittle and tends to crack when flexed. The deposition temperature of ITO is typically 600°C, which is higher than most polymers can withstand (although there are procedures to deposit films at room temperature[29]). ITO is sensitive to acid environments. Indium is expensive and rare, and sources are not infinite (unlike carbon!). Carbon nanotubes coated on plastic can provide a conductive film that can be bent/stretched with little or no effect on conductivity. By far, the most common polymer substrate is poly(ethylene terephthalate) (PET), although polycarbonate[30] has also been tested. Irrespective of flexibility, nanotube films do have one important performance advantage over ITO: above a radiation wavelength of 1200 nm (i.e., in the infrared region), the transmission of ITO drops off dramatically. Carbon nanotubes, on the other hand, have a much wider transparency range in the infrared region.[31,32] ITO does have one significant performance advantage over nanotube films: ITO films have approximately 10 times lower sheet resistance at comparable transmissions in the visible regime, even for tubes that have been doped after deposition (typically nitric acid or $SOCl_2$ are used as dopants). Commercial ITO films have a transmittance of 90–95% at 550 nm at a sheet resistance of around 20 Ω/sq, while nanotubes typically have transmittance smaller than this value at roughly a factor of 10 higher sheet resistances.

As films on polymers, buckypapers can be made using the filtration process described in Section 3.7.2. To improve conductivity, the dispersing agent (i.e., small-molecule surfactant) is removed via copious rinsing with solvent, usually water. The transfer of the films to the plastic substrate is the key; methods that have been used include dissolving the substrate,[31] flowing water through the filter to force separation of the film,[33] and contact printing.[34–36] Another way to produce nanotube films on flexible substrates is to spray a solution of nanotubes directly on the substrate followed by evaporation.[37] Other methods include direct collection of the nanotubes from a gas-phase reaction,[38] spin coating,[39–41] dip coating (e.g., layer-by-layer coating),[42,43] screen printing,[44] electrophoretic deposition,[45] inkjet printing,[46,47] and Mayer rod coating.[48] A detailed description of the advantages and drawbacks of each method is beyond the scope of this tome; however, dispersion in solution and evaporation rate are extremely important, as noted in Chapter 3 for dispersions in polymers. In addition, the surface energy of the substrate plays a role; adhesion of the nanotube film to the flexible substrate can be an issue. A polyelectrolyte[49] and poly (vinyl alcohol)[50] have been used to precoat PET film to improve adhesion. A nanotube film below an ITO film can improve the ability of the ITO to withstand flexing on a PET substrate.[29]

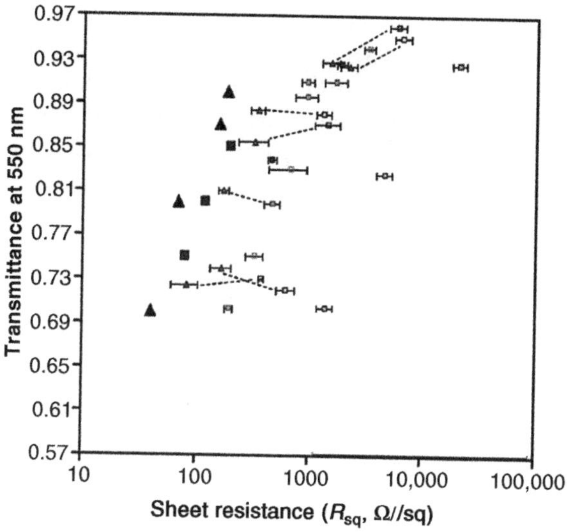

Figure 8.2 Relationship between transmittance and sheet resistance for SWCNT films. Symbols with no lines represent the best performance for various other deposition methods that were taken from the literature by the authors of Ref. 52. The leftmost data that are connected to a dashed line represent performance after doping with $SOCl_2$, while the data on the right represent performance before doping. Further details can be found in Ref. 52. Copyright 2009 American Chemical Society. Reprinted in part with permission from Ref. 52.

Since the limiting performance criterion for transparent electrodes is conductivity at a given transmission, the addition of insulating polymers to the buckypaper is not usually advantageous. However, since dispersion is important with respect to performance, a trade-off may be possible. Adsorbed polymers can be used to promote the dispersion of nanotubes when otherwise pure nanotube films are being made for transparent electrodes.[51] In a paper by Bao and coworkers, it was shown that the addition of a nonconducting polymer that assists in dispersion can yield films with performance comparable to that of the best films made with other methods, as shown in Figure 8.2. The authors found that doping not only improved nanotube conductivity but also seemed to degrade the polymer.[52]

The addition of conducting polymers may reduce the sheet resistance of nanotube films by reducing the tube–tube resistance to charge transfer. The obvious choice is poly(3,4-ethylenedioxythiophene) (PEDOT) doped with poly(styrene sulfonate) (PSS) because this polymer pair has also been investigated extensively as a transparent electrode. PEDOT–PSS has the same problems as nanotube films; the conductivity of the polymer (i.e., without nanotubes) is more than an order of magnitude higher than that of ITO at lower transmittance. In a study, the conductivity of PEDOT–PSS was lowered from around 1000 to about 225 Ω/sq through the use of a 75/25 combination of PEDOT–PSS and nanotubes using an inkjet application process on PET.[46] Another study found no substantial improvement in PEDOT–PSS conductivity with the addition of nanotubes.[53] Nanotube films also have roughness issues, which can lead to shorting (i.e., a nanotube goes across whatever is in the middle to the other electrode) especially if tubes are extending out perpendicular to the surface. In a recent paper, a proprietary PEDOT–PSS mixture was shown to significantly reduce the roughness of the nanotube film, making it more desirable for use as a transparent electrode;[54] a similar argument for planarization was made for the same polymer via multiple spin castings.[55] The issue of planarization of nanotube films should see a great deal of work in the future.

8.2.4 Other Applications Based on Nanotube Conductivity on Polymeric Substrates

The number of applications that have used carbon nanotubes without a polymer backing or not mixed with a polymer is far larger than detailed in this section; however, these applications are outside the scope of this book. Only those applications that involve polymers will be described in this section. This section attempts to describe each of them briefly, although the possibility of missing something cannot be ruled out.

Organic light-emitting diodes are a rapidly growing and emerging field; display panels made from these materials could be brighter and clearer than those currently made with inorganic LEDs or liquid crystals. Carbon nanotube films have applications as transparent electrodes in these systems; however, they have also been shown to improve the hole injection layer in PEDOT–PSS anodes.[56,57] Partially oxidized nanotubes have been shown as good replacements for PEDOT–PSS as hole-collecting electrode in organic solar cells[58,59] as well as improving the performance of ITO.[60] Nanotubes have been shown to be part of supercapacitors when filled with an electrolyte between two nanotube films on an insulating polymer.[61] Nanotubes have been sprayed using ink-jet technology into patterns for thin-film transistors on a polyimide substrate.[62] Nanotubes sprayed on elastomers have served as electrodes so as to squeeze the elastomer between the electrodes when a high-voltage field is applied.[63] These types of systems could possibly have better tolerance to short-circuiting than other approaches, for example, metal electrodes.[64] Finally, nanotube films have found to be suitable replacements for metals in electrowetting devices in microfluidics.[65]

8.3 THERMAL PROPERTIES: FLAME RETARDANCY

As stated in Chapter 7, significant commercial volume for carbon nanotubes in polymers exists if high thermal conductivities can be achieved. Without a doubt, infused systems offer some promise and perhaps in the future nanotubes will be used as layers and coatings designed to dissipate heat. Another application where the high thermal conductivity of nanotubes may play a role in improved performance (but likely a minor one relative to the geometry of nanotubes) is flame retardancy. Polymers have extremely poor flammability properties compared to metals or ceramics; polymers will burn easily and often release toxic fumes. The process by which a polymer burns is actually quite involved. In short, above a certain temperature a polymer will decompose liberating gaseous by-products; these by-products react with the oxygen in the air, in other words burn. Air is needed to burn; however, it is not necessarily needed for the decomposition to occur, although there is some disagreement on that issue.[66]

The action of flame retardants can be divided broadly into two types. The first type actually consists of two separate categories: the retardant can form a protective layer or a cooling layer on the surface. Nanotubes belong to the first category; they form a protective layer on the surface inhibiting the mass transfer gases. As an example of the second category, certain hydroxides will release water around a temperature of 300°C (the typical temperature at which polymers will begin to degrade), which, because the release is endothermic, lowers the temperature of the surface. In the second type, a chemical in the material releases free radicals that will react strongly with decomposition products; halogenated flame retardants are of this type. Halogenated flame retardants work on the basis of releasing a Br or Cl free radical upon heating, which reacts with the free radicals released by the polymer. In fact, much of the current research into polymer flammability is motivated by the fact that halogenated compounds are being phased out because of environmental concerns. Some retardants or mixtures of retardants use both principles to reduce flammability; for example, phosphorous-based flame retardants react with free radicals and form a protective layer.

Since the mechanism of flame retardancy for nanotubes is as a protective layer, it is worth considering this mechanism in more detail. A protective surface layer reduces flammability by reducing the diffusion of oxygen to the surface (assuming that oxygen plays a role in the degradation process), reducing the diffusion of gaseous degradation products to the air, and finally acting as a thermal insulating layer reducing the temperature of the polymer underneath. The increase in melt viscosity caused by nanotubes could also contribute to the reduction in flammability. An increase in thermal conductivity could also play a role if the coating is sufficiently conductive. Nanotubes themselves are flammable at high temperature; because polymers are less thermally stable than nanotubes, nanotubes are effective flame retardants for polymers. In tests that involve heat release, nanotubes tend to improve performance. However, time to ignition is quicker with nanotube addition because of heat localization due to the high thermal conductivity and low specific heat.[67,68] Nanotube-filled materials have also been reported to have problems with mass loss on ignition tests.[69–71] Nanoclays have also been investigated extensively as barrier-type flame retardants; generally nanotubes have been found to be more effective retardants with respect to heat release at similar loading levels.[72,73]

As might be expected based on the mechanism, forming a good protective layer necessitates well-dispersed nanotubes, as illustrated in a study involving PMMA and represented by Figure 8.3.[74] The top part shows the heat release rate as a function of time; a lower rate shows that flammability is reduced, although the total heat release is the same because eventually all the polymer will combust (the conditions of the test do not cause combustion of the nanotubes). The bottom part shows the result of the same test as well as another flammability test; the material with good dispersion (as measured via optical microscopy) on the left formed a good uniform char layer while that on the right did not. Other examples of the importance of dispersion are detailed in the literature. For example, a flame retardant phosphate was grafted to MWCNTs, and the

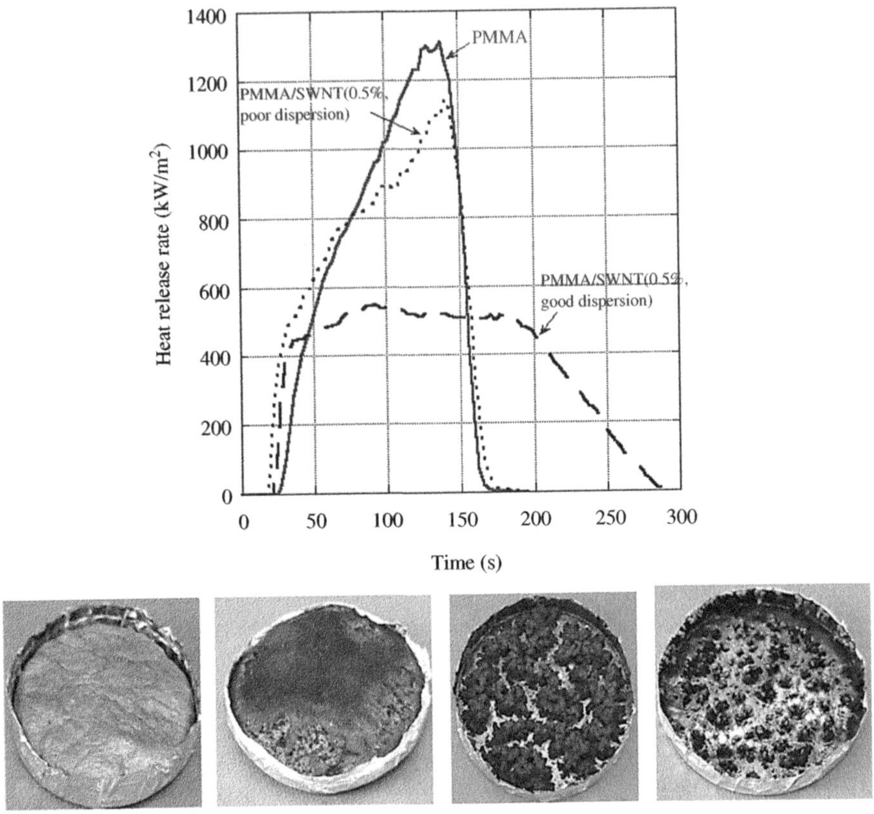

Figure 8.3 *Top*: Heat release rate at an input energy of 50 kW/m^2 in air. *Bottom*: Samples after heat flammability testing. The pictures on the left show the resulting uniform char layer after the heat release test and a gasification test, while those on the right show a poorly formed char layer after the same tests. Copyright Elsevier Ltd. Reproduced with permission from Ref. 74.

acrylonitrile–butadiene–styrene terpolymer nanotube composite showed significantly improved retardancy when grafted tubes were used, which was attributed primarily to improved dispersion.[75] A similar result was found for functionalized nanotubes in polystyrene.[70] A similar reason was given when polyethylene-coated nanotubes were added to ethylene–vinyl acetate as opposed to uncoated tubes.[76]

Nanotube level also affects behavior, but the effect is complicated. In a study on PMMA, there was virtually no difference in heat release rate improvement between 0.5% and 1% tubes.[74] A thermal oxidative study of nanotubes in linear low-density polyethylene showed a significant improvement with the addition of nanotubes versus pure polymer, but no improvement between a nanotube level of 0.5% and 3%.[77] With polypropylene, an optimal level of 1% was found for maximum improvement in heat release rate; higher levels led to poorer performance.[78] Polycarbonate also showed optimal behavior with respect to heat release rate at intermediate nanotube loading. Mass loss and ignition time data with nanotube-containing materials had significantly worse performance than the unfilled polymer. This study discussed factors involved with the various tests and showed the difficultly of making blanket statements about the flame retardancy of nanotube-filled systems due to the complexity of the tests and phenomena that underlay the responses to each of the tests.[79]

The physical nature of the nanotubes is also important. Stabilization, as measured by the time to ignition, improved with mechanically crushed nanotubes versus uncrushed tubes in ethylene–vinyl acetate.[80] The authors attributed this to radical trapping by the crushed nanotubes; a different paper speculated on a catalytic mechanism helping to reduce volatile compounds.[81] Decorating nanotubes with fullerenes also improved flammability retardancy in polypropylene, also attributed to radical trapping.[82] A very interesting study showed that in an immiscible polystyrene/poly(methyl methacrylate) blend, shorter nanotubes were able to diffuse during heating. With the addition of clays to the fast diffusing nanotubes, a very effective barrier could be formed, much better than any single filler.[67] Note that this result conflicts with the results of a previous study on polystyrene with only nanotubes, where longer tubes promoted flame retardancy.[83]

Especially given the poor performance of nanotubes with respect to mass loss on ignition and time required to ignite, one obvious step would be to combine nanotubes will materials that improve resistance to these aspects of retardancy and check for synergisms. In this context, synergism is defined as performance that is better than either individual component at the same volume or mass fraction (whether to use volume or mass fraction depends on the other component). At constant wt% of filler, the use of a 50/50 mixture of nanoclay and nanotubes was slightly better than pure nanotubes and significantly better than pure nanoclay in ethylene–vinyl acetate[84] and acrylonitrile–butadiene–styrene copolymer.[85] A slight improvement in the performance of nickel oxide[86] and a much more substantial one in magnesium hydroxide[87] were found when a small fraction of the inorganic was replaced by nanotubes. However, in the author's opinion, not enough work has taken place to reduce ignition problems that seem to be inherent with carbon nanotubes and hence more work in this area is expected in the future.

8.4 ELECTROMECHANICAL PROPERTIES: STRAIN SENSING AND ACTUATORS

Nanotubes have the property that a change in their mechanical state can induce a change in their ability to conduct electricity, and vice versa. A piezoelectric material is one in which charge accumulates in response to a mechanical strain as well as changes shape in response to an electrical field; an electrostrictive material is one that does the latter but not the former. In Section 8.2, applications depending on nanotube conductivity were highlighted. More specifically, applications where current (electrons) flowed from one part of the nanotube to another, or between nanotubes, were discussed. In this section, electronic properties are of interest, but in this case the purpose of the conductivity is not to transport electrons. In electromechanical actuation, nanotubes or nanotube-containing materials are connected to a voltage

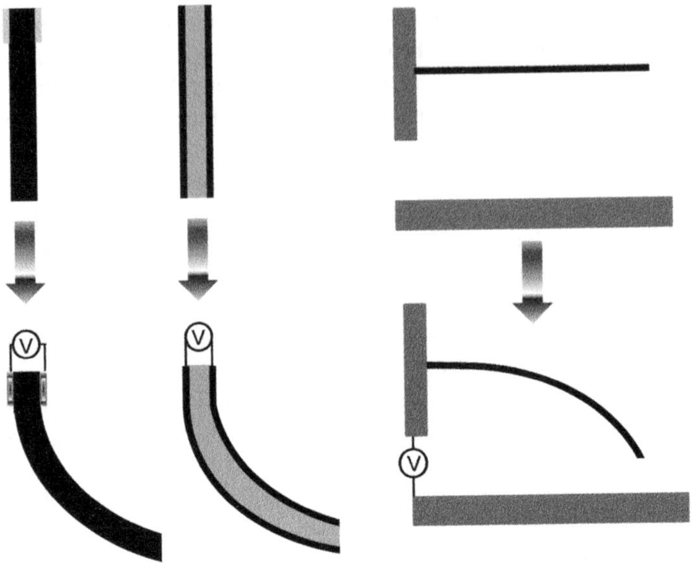

Figure 8.4 Schematic of electromechanical actuation for a cantilever type of effect. When a voltage is applied across buckypaper or polymer-filled nanotubes (black) separated by an insulating layer (gray) and the device is in an electrolytic solution, one side of the nanotube will shorten and the other will lengthen because of charge buildup at the surface.

source only at one end. Since no connection exists, current is not able to flow, instead there is a buildup of charge. Simple schematics of various setups are shown in Figure 8.4. This buildup of charge can cause a change in the shape of the object, with the change depending on the original object shape, voltage applied, and so on. In strain sensors, a mechanical strain imposed on the nanotubes alters the conductivity, either in an individual sense with respect to individual nanotubes or in a global sense with respect to a nanotube network. Strain sensors work on the principle of measuring a change in electronic properties and converting that change to a strain (or stress, assuming the relationship between stress and strain is known). A number of studies have examined pristine nanotube films as strain sensors[88,89] including nanotubes directly deposited on a polymer substrate.[90] In fact, freestanding nanotube films do not generally make good strain sensors because of slippage between nanotubes.[91] The focus of this section will be on strain sensing with polymer–nanotube composites. Chemical sensing, which is a change in conductivity caused by a change in chemical environment, will be discussed in Section 8.5.

8.4.1 Electromechanical Actuation

From an individual nanotube perspective, charge buildup and hence electromechanical actuation occurs because of the ends of nanotubes; without ends charge buildup would not occur. As a corollary to this statement, shorter tubes have higher charge distribution at the ends. The strain that is possible is quite small, less than 1%, but the forces that can build up are quite large because of the very high modulus of

Figure 8.5 Picture showing the dramatic change in shape with an applied voltage (10 V) for a few minutes for a bilayer sample with one layer as buckypaper and the other layer an epoxy: (a) before applying voltage; (b) after applying voltage. Copyright IOPscience. Reproduced with permission from Ref. 94.

nanotubes.[92] With polymers, it is possible to build a system, similar to transparent electrodes, with nanotubes deposited on a plastic substrate that in turn can be electromechanically actuated to cause a change in the shape of the substrate. In fact, the first demonstration of electromechanical actuation occurred in such a setup as shown on the left-hand side of Figure 8.4.[93] A two-layer analogue, with one layer of buckypaper and the other epoxy, using the schematic setup shown on the right-hand side of Figure 8.4, has shown remarkable actuation as shown in Figure 8.5.[94] A similar approach was used for Nafion®, which swells in water and either expels or further swells with charge and hence creates a mechanical response. By coating Nafion® with nanotubes, it was possible to achieve an electromechanical response without the use of electrolyte[95] (normally Nafion® must be in an electrolyte to show electromechanical actuation). Further, light can actuate a nanotube-coated Nafion® with the actuation at a given wavelength being proportional to the absorption by nanotubes at that wavelength.[96] Electroactive paper, that is, paper that has incorporated fillers that allow the paper to be actuated with electrical charge, has been coated with a combination of polyaniline and nanotubes to improve performance.[97,98]

Rather than applying the nanotubes as a coating, another approach is to mix polymer with nanotubes. Two types of polymers can be considered with this type of approach: those that are piezoelectric and those that are not. Nafion® is an example of the first type, and the addition of nanotubes intimately mixed with the polymer has been shown to improve the response of the material.[99,100] The electromechanical response of other swellable ionic polymers in water has been shown to improve with the inclusion of nanotubes.[101,102] In dry polymers, improvement in electromechanical actuation was found in polyaniline[103] and a copolymer containing poly(vinylidene fluoride)[104,105] with the addition of nanotubes. Hysteresis in the response of polyaniline–carbon nanotube fibers has been investigated.[106] In certain cases, the purpose of the carbon nanotubes is simply to strengthen the fiber and does not seem to contribute to the response.[107]

Nanotubes can also be added to a polymer that is inert electromechanically to produce electromechanical actuators; such an approach has been shown to be effective for polyimides.[108] The mechanism of the response in a nonpiezoelectric material is not just a result of the nanotube piezoelectric properties, rather interfacial polarization and donor–acceptor interactions between the nanotubes and the polymer may also play a role. A study on polyimide–nanotube composites suggested that interfacial polarization was primarily responsible for the increase in piezoelectric behavior.[109]

Piezoelectric or electrostrictive behavior essentially scales with conductivity; in fact, electrostrictive behavior has percolation behavior as shown for a polyimide.[108] Dispersion also plays a role. In a composite with Nafion®, a maximum in electromechanical behavior occurred at 1% nanotube content; the reduction at higher nanotube contents was attributed to an inhomogeneous nanotube dispersion above 1% loading.[110] Nanotube orientation can significantly improve performance as well. A rather involved procedure using liquid crystalline elastomers where the elastomer was stretched during the curing process showed an extremely high electromechanical actuation that was only attributable to the nanotubes and was a result of the high orientation of the tubes.[111]

8.4.2 Strain Sensing

With nanotube networks, either in a polymer or as a thin film, sensitivity to strain occurs as a result of two possible mechanisms: a change in nanotube conductivity due to strain on the individual nanotubes or a change in the number or quality of network contacts (where network contacts in a polymer include tunneling contacts). These two mechanisms, individual nanotube strain and network changes, have been examined in detail in epoxies and the authors could not determine which mechanism was dominant.[112] In general, however, the author is of the opinion that the second is more important in most composites and is certainly most important in unsupported buckypaper.

One important possible application of nanotubes as strain sensors is in damage identification, in particular in continuous glass-reinforced thermoset composites. Nanotubes have been shown to be effective damage sensors in glass–epoxy networks and laminated composites; at failure, the conductivity decreases dramatically and the area where conductivity decreases can be identified.[112–114] In fact, in a cyclic loading study, a quantitative measure of damage was developed from the resistance profile with strain in a glass–fiber epoxy composite.[115] Interestingly, nanotubes were able to be used as sensors of matrix stiffness in unloaded epoxy composites, since the molded-in stresses will change with matrix stiffness.[112] In the latter case, the change in nanotube conductivity with stress, and not any change in network configuration, likely causes the difference in response.

The optimal strain sensitivity can be defined in one of two ways. In terms of relative change in resistance (resistance after strain/resistance before strain), which is the more common way to define strain sensitivity, the maximum sensitivity will occur in the percolation regionm[116,117] and has been attributed to the importance of tunneling current in this region.[116] In fact, in the percolation region, a study showed that the relative sensitivity of nanotube-filled polyimide was far superior to that of aluminum-filled polyimide, as well as a factor of ~5 larger than the relative sensitivity at nanotube concentrations above the percolation region.[118] In terms of absolute change in the resistance, the region just above the end of the percolation region will offer the largest change in resistance. In the work done with polyethylene and polypropylene in this concentration regime, carbon black-filled materials were significantly more sensitive versus the nanotube-filled analogues, which was attributed to the geometry of the nanotubes relative to that of the carbon black.[119] Carbon black was also found to be superior in terms of strain sensitivity than MWCNTs in an epoxy composite.[120] The difference in these two regimes can likely be attributed to the relative importance of tunneling current; within the percolation region, changes in tunneling current due to changes in gaps between nanotubes will be relatively more important than outside the percolation region. The sample-to-sample reproducibility would also be expected to be much higher for a sample just outside the percolation region, although reproducibility was also found to be an issue outside the percolation region.[119]

The two regions, in the percolation region or above the percolation region (where exactly in both was not given), were compared in a study of nanotubes mixed with poly(ethylene oxide). The authors found linear and nonlinear regions of resistance versus strain for both nanotube concentrations, but the onset of the nonlinear region, which was attributed to where changes in tunneling resistance became important, occurred at substantially higher strains for the higher nanotube concentration. Surprisingly, the relative change in resistance was also much higher for the higher concentration material. Also, sample-to-sample variability in this study was quite high for both concentrations.[121] Another study with polyurethane and MWCNTs looked at a number of different concentrations within the percolation region, including a concentration region that could best be described as being at the beginning of the plateau region, and found a concentration-independent logarithmic dependence of the strain up from 5% to 100%. The dependence was assigned to a change in tunneling resistance; at less than 5% strain, the authors attributed the change to deformation of the nanotube network.[122]

A very interesting application uses the effect of a thermal gradient during processing of a low-viscosity thermoset to change carbon network characteristics that in turn changes the qualitative nature of the response of the material. Given the resistance values measured (10^{-5} S/m), it is likely that the concentration was in the percolation region. A 3 or 6 mm thick sample was heated from the bottom during curing, causing an order of magnitude higher conductivity at the top of the sample than at the bottom. As shown in Figure 8.6, when the higher conductivity side was under compression in a three-point bend test, that is, nearer the single pushing nose, the response was a decrease in resistivity (increase in conductivity); while when the higher conductivity side was under tension, that is, nearer the two supports, an increase in conductivity with strain occurred.[123] Other than mechanical strain, a strain-sensitive response has been found in that expansion of a polymer because of a molecule that will swell the polymer has been shown in nanotube composites made with polycarbonate,[124] poly(methyl methacrylate),[125] polystyrene,[126,127] polyisoprene,[128] and poly(*p*-phenylene).[129] A similar principle has been used for liquid sensors based on polylactic acid.[130,131] In some cases, nanotubes were deposited on the surface of the polymer that was swollen, including polyamide 6[132] and polydimethylsiloxane.[133]

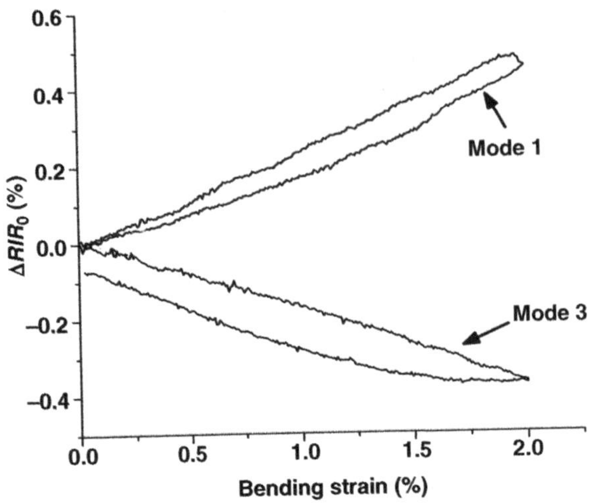

Figure 8.6 Change in resistivity with strain for epoxy–MWCNT made with a conductivity gradient in an epoxy matrix. Mode 1 is the case where the higher conductivity side is facing the single pushing nose, while mode 3 is the case where the higher conductivity side is facing the two supports. Hysteresis was attributed to the local stress fields in the vicinity of the nose. Copyright IOPscience. Reproduced with permission from Ref. 123.

An individual carbon nanotube will change its Raman spectra when strained. If perfect stress transfer is assumed, then the strain on the sample can be ascertained using the shift in maximum absorption intensity of the G′ band[134] or the intensity of the RBM.[135] A very unique application of this concept was performed in single-fiber fragmentation tests of glass in epoxy; this research showed that the model normally applied for the stress along the fiber length was in fact accurate as the Raman spectra were recorded using a microscope along the fiber length.[136] Single-fiber fragmentation tests are very important tests with respect to determination of the interfacial adhesion, so the confirmation that the stress distribution is modeled correctly is an important result. A similar Raman method was used to assess the interfacial strength in polypropylene–glass fibers with different sizings on the glass[137] and the stress distribution in tension around a hole.[138]

8.5 OTHER APPLICATIONS

The use of carbon nanotubes as part of a sensor is a fairly significant application, mostly focused on the biological arena. Regardless of whether the sensor is for a biological molecule or some other molecule, normally the nanotube (unless it is functionalized) is not affected by the molecule; however, some other component changes in response to the stimulus. Nanotubes are used because this change in the other ingredient effects some change upon the nanotubes that in turn can be monitored. Normally, conductivity is used to monitor the change, ideally by a radio-frequency powered system, so that no batteries are required. A number of papers have examined nanotube systems, in conjunction with other materials, as chemical sensors. For example, a layer-by-layer deposition of MWCNTs and a conducting polymer were used to create flexible nitrous oxide sensors on a poly(ethylene terephthalate) substrate.[139] Ethanol[140] and carbon monoxide[141] sensors were created by combining polypyrrole and nanotubes. Triethylamine sensors have been constructed using poly(styrene sulfonate)-grafted nanotubes and polyaniline.[142] Various gases have been sensed using nanotubes mixed with poly (3,3′-dialkylquarterthiophene).[143] Both polypyrrole[144] and polyaniline[145] have been used with MWCNTs to fabricate ammonia sensors. The sensitivity of ammonia sensors based on polyaniline as well as the resistance to humidity was improved via the addition of MWCNTs.[146] Polypyrrole and polyaniline alone,[147] as well as in combination,[147,148] have been combined with nanotubes to form pH sensors. A layer-by-layer technique was used to create both a pH and a strain sensor.[149,150] Rather than simple mixing, grafting of a polymer sensitive to the gas in question onto a nanotube has also been shown to yield a sensor.[151,152] This section does not provide a complete listing of all the sensors built using carbon nanotubes, and biological sensors that are by far the largest class of sensors that contain carbon nanotubes are outside the scope of this text. Further, many other sensors have likely not been reported in the open literature. The issue with sensors made with carbon nanotubes is selectivity, since only electrical response is usually monitored and a variety of chemicals can cause a change in the signal.

Because nanotubes adsorb electromagnetic radiation and convert that energy into heat, nanotubes can serve as photothermal actuators; that is, nanotubes heat up with light causing an increase in temperature and hence, if encapsulated in a polymer, a significant increase in volume due to the high thermal expansion coefficient of a polymer.[153] Sensitivity to light has allowed for nanotubes either as thin films on a polymer or as a composite to be examined as UV detectors[154] and IR detectors.[155] A related topic to electromechanical actuation is photomechanical actuation, for example, a change in dimensions caused by exposure to light. Under strong enough light, nanotubes will change their dimensions and this effect is not simply due to thermal expansion. A number of papers have examined this effect in detail with nanotubes encapsulated in an elastomeric polymer.[156–159] There are a couple of extremely interesting phenomena that have been found in these systems. First, depending on alignment, the change can be expansive or contractive for nanotube films;[160] presumably the same would be found for a polymer. Second, in a polymer, the response can be either expansive or contractive, depending on the prestrain of the polymer. Also, the response did not depend on whether the nanotubes were intimately mixed or whether a laminate type of composite was considered.[161] Of course, the absorbance of light by the polymer as well as the heating caused by the light must be considered.

Carbon nanotube-filled polymers have been tested as vibrational dampers, a property utilized for noise reduction. In this application, poor bonding between the nanotubes and the matrix enhances damping; in fact, nanocomposites in general are being studied for this application, as discussed in a recent review paper.[162] A good vibration dampener has a large loss modulus over a wide range of frequencies, although sometimes temperature is used instead of frequency. One interesting result was found in an elastomeric epoxy, where a factor of 2 increase in loss modulus was found going from 2 to 3 wt% nanotubes, without any appreciable change in the storage modulus.[163] Other polymers that have been examined in this fashion include poly(ethylene oxide),[164] polycarbonate,[164–166] and a rigid epoxy.[167] In the latter, multiwalled carbon nanotubes were found to be better than single-walled tubes, although dispersion differences could have affected the results. A study on polycarbonate-filled materials indicates that prestrain can increase the loss modulus which was attributed to the critical stress for tube–matrix interfacial slip to be reached at lower strain amplitudes.[168] Also, better dispersion caused an increase in loss modulus in polycarbonate.[164] Absorption at sound frequencies was directly measured in a polyurethane foam and was found to increase significantly with only fraction of a percent of added nanotubes.[169]

8.6 CHALLENGES

Most applications that do not involve polymers for carbon nanotubes would be immeasurably improved if nanotubes of only one chirality or at least of one type (semiconducting versus metallic) could be produced. Although producing all metallic tubes would likely improve conductivity that would in turn improve many of the applications described in this chapter, many polymer applications are not that dependent on nanotube type or even nanotube purity (although certainly residual catalyst is a problem!). The most important challenge is the one raised in the first section, namely, cost. Most applications of polymers, at least by volume, are extremely cost-driven, which also explains why roughly three-fourths of all polymer production uses olefin monomers. The important challenge here is to reduce cost, which in turn means increasing yield.

As with cost, the other issues that provide barriers to nanotube applications have been mentioned throughout this text. Beyond cost, control of dispersion is probably the second most important requirement, in order to improve both performance and reproducibility. Improving dispersion (which may in fact not mean increasing dispersion!) is always desired. Control of nanotube quality, that is, nanotubes must have the same characteristics 6 months in the future when purchased from the same manufacturer, is also critical for applications.

REFERENCES

1. Das, N. C., Liu, Y. Y., Yang, K. K., Peng, W. Q., Maiti, S., Wang, H. (2009). Single-walled carbon nanotube/poly(methyl methacrylate) composites for electromagnetic interference shielding, *Polym. Eng. Sci.*, *49*, 1627–1634.
2. Mazov, I., Kuznetsov, V., Moseenkov, S., Usoltseva, A., Romanenko, A., Anikeeva, O., Buryakov, T., Kuzhir, P., Maksimenko, S., Bychanok, D., Macutkevic, J., Seliuta, D., Valusis, G., Banys, J., Lambin, P. (2009). Electromagnetic shielding properties of MWCNT/PMMA composites in Ka-band, *Phys. Status Solidi B*, *246*, 2662–2666.
3. Park, S. H., Thielemann, P., Asbeck, P., Bandaru, P. R. (2009). Enhanced dielectric constants and shielding effectiveness of, uniformly dispersed, functionalized carbon nanotube composites, *Appl. Phys. Lett.*, *94*, 243111.

4. Li, Y., Chen, C. X., Zhang, S., Ni, Y. W., Huang, J. (2008). Electrical conductivity and electromagnetic interference shielding characteristics of multiwalled carbon nanotube filled polyacrylate composite films, *Appl. Surf. Sci.*, *254*, 5766–5771.
5. Yang, Y. L., Gupta, M. C., Dudley, K. L., Lawrence, R. W. (2005). A comparative study of EMI shielding properties of carbon nanofiber and multi-walled carbon nanotube filled polymer composites, *J. Nanosci. Nanotechnol.*, *5*, 927–931.
6. Mathur, R. B., Pande, S., Singh, B. P., Dhami, T. L. (2008). Electrical and mechanical properties of multi-walled carbon nanotubes reinforced PMMA and PS composites, *Polym. Compos.*, *29*, 717–727.
7. Saini, P., Choudhary, V., Singh, B. P., Mathur, R. B., Dhawan, S. K. (2009). Polyaniline–MWCNT nanocomposites for microwave absorption and EMI shielding, *Mater. Chem. Phys.*, *113*, 919–926.
8. Das, N. C., Maiti, S. (2008). Electromagnetic interference shielding of carbon nanotube/ethylene vinyl acetate composites, *J. Mater. Sci.*, *43*, 1920–1925.
9. Ma, C. C. M., Huang, Y. L., Kuan, H. C., Chiu, Y. S. (2005). Preparation and electromagnetic interference shielding characteristics of novel carbon-nanotube/siloxane/poly(urea urethane) nanocomposites, *J. Polym. Sci. Polym. Phys.*, *43*, 345–358.
10. Jou, W. S., Cheng, H. Z., Hsu, C. F. (2006). A carbon nanotube polymer-based composite with high electromagnetic shielding, *J. Electron. Mater.*, *35*, 462–470.
11. Huang, Y., Li, N., Ma, Y. F., Feng, D., Li, F. F., He, X. B., Lin, X., Gao, H. J., Chen, Y. S. (2007). The influence of single-walled carbon nanotube structure on the electromagnetic interference shielding efficiency of its epoxy composites, *Carbon*, *45*, 1614–1621.
12. Jou, W. S., Cheng, H. Z., Hsu, C. F. (2007). The electromagnetic shielding effectiveness of carbon nanotubes polymer composites, *J. Alloys Compd.*, *434*, 641–645.
13. Li, N., Huang, Y., Du, F., He, X. B., Lin, X., Gao, H. J., Ma, Y. F., Li, F. F., Chen, Y. S., Eklund, P. C. (2006). Electromagnetic interference (EMI) shielding of single-walled carbon nanotube epoxy composites, *Nano Lett.*, *6*, 1141–1145.
14. Huang, Y. L., Yuen, S. M., Ma, C. C. M., Chuang, C. Y., Yu, K. C., Teng, C. C., Tien, H. W., Chiu, Y. C., Wu, S. Y., Liao, S. H., Weng, F. B. (2009). Morphological, electrical, electromagnetic interference (EMI) shielding, and tribological properties of functionalized multi-walled carbon nanotube/polymethyl methacrylate (PMMA) composites, *Compos. Sci. Technol.*, *69*, 1991–1996.
15. Han, M. S., Lee, Y. K., Lee, H. S., Yun, C. H., Kim, W. N. (2009). Electrical, morphological and rheological properties of carbon nanotube composites with polyethylene and poly(phenylene sulfide) by melt mixing, *Chem. Eng. Sci.*, *64*, 4649–4656.
16. Yang, Y., Gupta, M. C., Dudley, K. L. (2007). Towards cost-efficient EMI shielding materials using carbon nanostructure-based nanocomposites, *Nanotechnology*, *18*, 345701.
17. Al-Saleh, M. H., Sundararaj, U. (2009). Electromagnetic interference shielding mechanisms of CNT/polymer composites, *Carbon*, *47*, 1738–1746.
18. Liu, Z. F., Bai, G., Huang, Y., Ma, Y. F., Du, F., Li, F. F., Guo, T. Y., Chen, Y. S. (2007). Reflection and absorption contributions to the electromagnetic interference shielding of single-walled carbon nanotube/polyurethane composites, *Carbon*, *45*, 821–827.
19. Park, J. G., Louis, J., Cheng, Q. F., Bao, J. W., Smithyman, J., Liang, R., Wang, B., Zhang, C., Brooks, J. S., Kramer, L., Fanchasis, P., Dorough, D. (2009). Electromagnetic interference shielding properties of carbon nanotube buckypaper composites, *Nanotechnology*, *20*, 415702.
20. Chang, C. M., Chiu, J. C., Jou, W. S., Wu, T. L., Cheng, W. H. (2006). New package scheme of a 2.5-Gb/s plastic, transceiver module employing multiwall nanotubes for low electromagnetic interference, *IEEE J. Sel. Top. Quantum Electron.*, *12*, 1025–1031.
21. Thomassin, J. M., Pagnoulle, C., Bednarz, L., Huynen, I., Jerome, R., Detrembleur, C. (2008). Foams of polycaprolactone/MWNT nanocomposites for efficient EMI reduction, *J. Mater. Chem.*, *18*, 792–796.
22. Yang, Y. L., Gupta, M. C. (2005). Novel carbon nanotube–polystyrene foam composites for electromagnetic interference shielding, *Nano Lett.*, *5*, 2131–2134.
23. Yuen, S. M., Ma, C. C. M., Chuang, C. Y., Yu, K. C., Wu, S. Y., Yang, C. C., Wei, M. H. (2008). Effect of processing method on the shielding effectiveness of electromagnetic interference of MWCNT/PMMA composites, *Compos. Sci. Technol.*, *68*, 963–968.
24. Pande, S., Singh, B. P., Mathur, R. B., Dhami, T. L., Saini, P., Dhawan, S. K. (2009). Improved electromagnetic interference shielding properties of MWCNT–PMMA composites using layered structures, *Nanoscale Res. Lett.*, *4*, 327–334.

25. Hornbostel, B., Leute, U., Pötschke, P., Kotz, J., Kornfeld, D., Chiu, P. W., Roth, S. (2008). Attenuation of electromagnetic waves by carbon nanotube composites, *Physica E*, *40*, 2425–2429.
26. Wang, L. L., Tay, B. K., See, K. Y., Sun, Z., Tan, L. K., Lua, D. (2009). Electromagnetic interference shielding effectiveness of carbon-based materials prepared by screen printing, *Carbon*, *47*, 1905–1910.
27. Mall, S., Ouper, B. L., Fielding, J. C. (2009). Compression strength degradation of nanocomposites after lightning strike, *J. Compos. Mater.*, *43*, 2987–3001.
28. Valentini, L., Bon, S. B., Kenny, J. M. (2007). Electrodeposited carbon nanotubes as template for the preparation of semi-transparent conductive thin films by *in situ* polymerization of methyl methacrylate, *Carbon*, *45*, 2685–2691.
29. Li, J. F., Hu, L. B., Liu, J., Wang, L., Marks, T. J., Gruner, G. (2008). Indium tin oxide modified transparent nanotube thin films as effective anodes for flexible organic light-emitting diodes, *Appl. Phys. Lett.*, *93*, 083306.
30. Castro, M. R. S., Al-Dahoudi, N., Oliveira, P. W., Schmidt, H. K. (2009). Multi-walled carbon nanotube-based transparent conductive layers deposited on polycarbonate substrate, *J. Nanopart. Res.*, *11*, 801–806.
31. Wu, Z. C., Chen, Z. H., Du, X., Logan, J. M., Sippel, J., Nikolou, M., Kamaras, K., Reynolds, J. R., Tanner, D. B., Hebard, A. F., Rinzler, A. G. (2004). Transparent, conductive carbon nanotube films, *Science*, *305*, 1273–1276.
32. Hu, L. B., Hecht, D. S., Gruner, G. (2009). Infrared transparent carbon nanotube thin films, *Appl. Phys. Lett.*, *94*, 081103.
33. Shin, J. H., Shin, D. W., Patole, S. P., Lee, J. H., Park, S. M., Yoo, J. B. (2009). Smooth, transparent, conducting and flexible SWCNT films by filtration–wet transfer processes, *J. Phys. D*, *42*, 045305.
34. Hu, L. B., Gruner, G., Li, D., Kaner, R. B., Cech, J. (2007). Patternable transparent carbon nanotube films for electrochromic devices, *J. Appl. Phys.*, *101*, 016102.
35. Zhou, Y. X., Hu, L. B., Gruner, G. (2006). A method of printing carbon nanotube thin films, *Appl. Phys. Lett.*, *88*, 123109.
36. Zhang, D. H., Ryu, K., Liu, X. L., Polikarpov, E., Ly, J., Tompson, M. E., Zhou, C. W. (2006). Transparent, conductive, and flexible carbon nanotube films and their application in organic light-emitting diodes, *Nano Lett.*, *6*, 1880–1886.
37. Paul, S., Kim, D. W. (2009). Preparation and characterization of highly conductive transparent films with single-walled carbon nanotubes for flexible display applications, *Carbon*, *47*, 2436–2441.
38. Heras, A., Colina, A., Lopez-Palacios, J., Kaskela, A., Nasibulin, A. G., Ruiz, V., Kauppinen, E. I. (2009). Flexible optically transparent single-walled carbon nanotube electrodes for UV–Vis absorption spectroelectrochemistry, *Electrochem. Commun.*, *11*, 442–445.
39. Zou, J. H., Tran, B., Huo, Q., Zhai, L. (2009). Transparent carbon nanotube/poly(3,4 ethylenedioxythiophene) composite electrical conductors, *Soft Mater.*, *7*, 355–365.
40. Yim, J. H., Kim, Y. S., Koh, K. H., Lee, S. (2008). Fabrication of transparent single wall carbon nanotube films with low sheet resistance, *J. Vac. Sci. Technol. B*, *26*, 851–855.
41. Tung, V. C., Chen, L. M., Allen, M. J., Wassei, J. K., Nelson, K., Kaner, R. B., Yang, Y. (2009). Low-temperature solution processing of graphene–carbon nanotube hybrid materials for high-performance transparent conductors, *Nano Lett.*, *9*, 1949–1955.
42. Song, Y. I., Yang, C. M., Kim, D. Y., Kanoh, H., Kaneko, K. (2008). Flexible transparent conducting single-wall carbon nanotube film with network bridging method, *J. Colloid Interface Sci.*, *318*, 365–371.
43. Lee, S. W., Kim, B. S., Chen, S., Shao-Horn, Y., Hammond, P. T. (2009). Layer-by-layer assembly of all carbon nanotube ultrathin films for electrochemical applications, *J. Am. Chem. Soc.*, *131*, 671–679.
44. Choi, J. H., Park, J. H., Moon, J. S., Nam, J. W., Yoo, J. B., Park, C. Y., Lee, C. G., Choe, D. H. (2006). Fabrication of carbon nanotube emitter on the flexible substrate, *Diam. Relat. Mater.*, *15*, 44–48.
45. Valentini, L., Bagnis, D., Cagnoli, R., Meloni, F., Mucci, A., Schenetti, L., Kenny, J. (2008). Electrodeposition of carbon nanotube semi-transparent thin films: a facile route for preparing photoactive polymeric hybrid materials, *Diam. Relat. Mater.*, *17*, 1573–1576.
46. Denneulin, A., Bras, J., Blayo, A., Khelifi, B., Roussel-Dherbey, F., Neuman, C. (2009). The influence of carbon nanotubes in inkjet printing of conductive polymer suspensions, *Nanotechnology*, *20*, 385701.

47. Song, J. W., Kim, J., Yoon, Y. H., Choi, B. S., Kim, J. H., Han, C. S. (2008). Inkjet printing of single-walled carbon nanotubes and electrical characterization of the line pattern, *Nanotechnology, 19*, 095702.
48. Dan, B., Irvin, G. C., Pasquali, M. (2009). Continuous and scalable fabrication of transparent conducting carbon nanotube films, *ACS Nano, 3*, 835–843.
49. Zhao, X., Chu, B. T. T., Ballesteros, B., Wang, W. L., Johnston, C., Sykes, J. M., Grant, P. S. (2009). Spray deposition of steam treated and functionalized single-walled and multi-walled carbon nanotube films for supercapacitors, *Nanotechnology, 20*, 065605.
50. Kim, M. J., Yoo, J. B. (2008). Electrical properties of transparent conductive CNT composite films, *Electron. Mater. Lett., 4*, 57–61.
51. Bocharova, V., Kiriy, A., Oertel, U., Stamm, M., Stoffelbach, F., Jerome, R., Detrembleur, C. (2006). Ultrathin transparent conductive films of polymer-modified multiwalled carbon nanotubes, *J. Phys. Chem. B, 110*, 14640–14644.
52. Hellstrom, S. L., Lee, H. W., Bao, Z. N. (2009). Polymer-assisted direct deposition of uniform carbon nanotube bundle networks for high performance transparent electrodes, *ACS Nano, 3*, 1423–1430.
53. Mustonen, T., Kordas, K., Saukko, S., Toth, G., Penttilla, J. S., Helisto, P., Seppa, H., Jantunen, H. (2007). Inkjet printing of transparent and conductive patterns of single-walled carbon nanotubes and PEDOT–PSS composites, *Phys. Status Solidi B, 244*, 4336–4340.
54. Ou, E. C. W., Hu, L. B., Raymond, G. C. R., Soo, O. K., Pan, J. S., Zheng, Z., Park, Y., Hecht, D., Irvin, G., Drzaic, P., Gruner, G. (2009). Surface-modified nanotube anodes for high performance organic light-emitting diode, *ACS Nano, 3*, 2258–2264.
55. Williams, C. D., Robles, R. O., Zhang, M., Li, S., Baughman, R. H., Zakhidov, A. A. (2008). Multiwalled carbon nanotube sheets as transparent electrodes in high brightness organic light-emitting diodes, *Appl. Phys. Lett., 93*, 183506.
56. Wang, G. F., Tao, X. M., Wang, R. X. (2008). Fabrication and characterization of OLEDs using PEDOT:PSS and MWCNT nanocomposites, *Compos. Sci. Technol., 68*, 2837–2841.
57. Wang, G. F., Tao, X. M., Chen, W., Wang, R. X., Yang, A. (2007). Improvement in performance of organic light-emitting devices by inclusion of multi-wall carbon nanotubes, *J. Lumin., 126*, 602–606.
58. Hatton, R. A., Blanchard, N. P., Tan, L. W., Latini, G., Cacialli, F., Silva, S. R. P. (2009). Oxidised carbon nanotubes as solution processable, high work function hole-extraction layers for organic solar cells, *Org. Electron., 10*, 388–395.
59. Ulbricht, R., Jiang, X., Lee, S., Inoue, K., Zhang, M., Fang, S., Baughman, R., Zakhidov, A. (2006). Polymeric solar cells with oriented and strong transparent carbon nanotube anode, *Phys. Status Solidi B, 243*, 3528–3532.
60. Ulbricht, R., Lee, S. B., Jiang, X. M., Inoue, K., Zhang, M., Fang, S. L., Baughman, R. H., Zakhidov, A. A. (2007). Transparent carbon nanotube sheets as 3-D charge collectors in organic solar cells, *Sol. Energ. Mater. Sol. Cells, 91*, 416–419.
61. Kaempgen, M., Chan, C. K., Ma, J., Cui, Y., Gruner, G. (2009). Printable thin film supercapacitors using single-walled carbon nanotubes, *Nano Lett., 9*, 1872–1876.
62. Vaillancourt, J., Zhang, H. Y., Vasinajindakaw, P., Xia, H. T., Lu, X. J., Han, X. L., Janzen, D. C., Shih, W. S., Jones, C. S., Stroder, M., Chen, M. Y. H., Subbaraman, H., Chen, R. T., Berger, U., Renn, M. (2008). All ink-jet-printed carbon nanotube thin-film transistor on a polyimide substrate with an ultrahigh operating frequency of over 5 GHz, *Appl. Phys. Lett., 93*, 243301.
63. Yuan, W., Brochu, P., Ha, S. M., Pei, Q. B. (2009). Dielectric oil coated single-walled carbon nanotube electrodes for stable, large-strain actuation with dielectric elastomers, *Sens. Actuators A, 155*, 278–284.
64. Yuan, W., Hu, L. B., Yu, Z. B., Lam, T. L., Biggs, J., Ha, S. M., Xi, D. J., Chen, B., Senesky, M. K., Gruner, G., Pei, Q. B. (2008). Fault-tolerant dielectric elastomer actuators using single-walled carbon nanotube electrodes, *Adv. Mater., 20*, 621–625.
65. Hu, L. B., Gruner, G., Gong, J., Kim, C. J., Hornbostel, B. (2007). Electrowetting devices with transparent single-walled carbon nanotube electrodes, *Appl. Phys. Lett., 90*, 093124.
66. Laoutid, F., Bonnaud, L., Alexandre, M., Lopez-Cuesta, J. M., Dubois, P. (2009). New prospects in flame retardant polymer materials: from fundamentals to nanocomposites, *Mater. Sci. Eng. R, 63*, 100–125.

67. Pack, S., Kashiwagi, T., Stemp, D., Koo, J., Si, M., Sokolov, J. C., Rafailovich, M. H. (2009). Segregation of carbon nanotubes/organoclays rendering polymer blends self-extinguishing, *Macromolecules, 42*, 6698–6709.
68. Tong, L. F., Ma, H. Y., Fang, Z. P. (2008). Thermal decomposition and flammability of acrylonitrile–butadiene–styrene/multi-walled carbon nanotubes composites, *Chin. J. Polym. Sci., 26*, 331–339.
69. Bourbigot, S., Duquesne, S., Fontaine, G., Bellayer, S., Turf, T., Samyn, F. (2008). Characterization and reaction to fire of polymer nanocomposites with and without conventional flame retardants, *Mol. Cryst. Liq. Cryst., 486*, 325–339.
70. Bourbigot, S., Duquesne, S., Jama, C. (2006). Polymer nanocomposites: how to reach low flammability? *Macromol. Symp., 233*, 180–190.
71. Schartel, B., Pötschke, P., Knoll, U., Abdel-Goad, M. (2005). Fire behaviour of polyamide 6/multiwall carbon nanotube nanocomposites, *Eur. Polym. J., 41*, 1061–1070.
72. Kashiwagi, T., Mu, M. F., Winey, K., Cipriano, B., Raghavan, S. R., Pack, S., Rafailovich, M., Yang, Y., Grulke, E., Shields, J., Harris, R., Douglas, J. (2008). Relation between the viscoelastic and flammability properties of polymer nanocomposites, *Polymer, 49*, 4358–4368.
73. Costache, M. C., Heidecker, M. J., Manias, E., Camino, G., Frache, A., Beyer, G., Gupta, R. K., Wilkie, C. A. (2007). The influence of carbon nanotubes, organically modified montmorillonites and layered double hydroxides on the thermal degradation and fire retardancy of polyethylene, ethylene–vinyl acetate copolymer and polystyrene, *Polymer, 48*, 6532–6545.
74. Kashiwagi, T., Du, F. M., Winey, K. I., Groth, K. A., Shields, J. R., Bellayer, S. P., Kim, H., Douglas, J. F. (2005). Flammability properties of polymer nanocomposites with single-walled carbon nanotubes: effects of nanotube dispersion and concentration, *Polymer, 46*, 471–481.
75. Ma, H. Y., Tong, L. F., Xu, Z. B., Fang, Z. P. (2008). Functionalizing carbon nanotubes by grafting on intumescent flame retardant: nanocomposite synthesis, morphology, rheology, and flammability, *Adv. Funct. Mater., 18*, 414–421.
76. Peeterbroeck, S., Laoutid, F., Taulemesse, J. M., Monteverde, T., Lopez-Cuesta, J. M., Nagy, J. B., Alexandre, M., Dubois, P. (2007). Mechanical properties and flame-retardant behavior of ethylene vinyl acetate/high-density polyethylene coated carbon nanotube nanocomposites, *Adv. Funct. Mater., 17*, 2787–2791.
77. Bocchini, S., Frache, A., Camino, G., Claes, M. (2007). Polyethylene thermal oxidative stabilisation in carbon nanotubes based nanocomposites, *Eur. Polym. J., 43*, 3222–3235.
78. Kashiwagi, T., Grulke, E., Hilding, J., Groth, K., Harris, R., Butler, K., Shields, J., Kharchenko, S., Douglas, J. (2004). Thermal and flammability properties of polypropylene/carbon nanotube nanocomposites, *Polymer, 45*, 4227–4239.
79. Schartel, B., Braun, U., Knoll, U., Bartholmai, M., Goering, H., Neubert, D., Pötschke, P. (2008). Mechanical, thermal, and fire behavior of bisphenol A polycarbonate/multiwall carbon nanotube nanocomposites, *Polym. Eng. Sci., 48*, 149–158.
80. Peeterbroeck, S., Laoutid, F., Swoboda, B., Lopez-Cuesta, J. M., Moreau, N., Nagy, J. B., Alexandre, M., Dubois, P. (2007). How carbon nanotube crushing can improve flame retardant behaviour in polymer nanocomposites? *Macromol. Rapid Commun., 28*, 260–264.
81. Fina, A., Bocchini, S., Camino, G. (2008). Catalytic fire retardant nanocomposites, *Polym. Degrad. Stab., 93*, 1647–1655.
82. Song, P. A., Shen, Y., Du, B. X., Guo, Z. H., Fang, Z. P. (2009). Fabrication of fullerene-decorated carbon nanotubes and their application in flame-retarding polypropylene, *Nanoscale, 1*, 118–121.
83. Cipiriano, B. H., Kashiwagi, T., Raghavan, S. R., Yang, Y., Grulke, E. A., Yamamoto, K., Shields, J. R., Douglas, J. F. (2007). Effects of aspect ratio of MWNT on the flammability properties of polymer nanocomposites, *Polymer, 48*, 6086–6096.
84. Beyer, G. (2002). Short communication: carbon nanotubes as flame retardants for polymers, *Fire Mater., 26*, 291–293.
85. Ma, H. Y., Tong, L. F., Xu, Z. B., Fang, Z. P. (2007). Synergistic effect of carbon nanotube and clay for improving the flame retardancy of ABS resin, *Nanotechnology, 18*, 375602.
86. Yu, H., Liu, J., Wang, Z., Jiang, Z. W., Tang, T. (2009). Combination of carbon nanotubes with Ni_2O_3 for simultaneously improving the flame retardancy and mechanical properties of polyethylene, *J. Phys. Chem. C, 113*, 13092–13097.

87. Ye, L., Wu, Q. H., Qu, B. J. (2009). Synergistic effects and mechanism of multiwalled carbon nanotubes with magnesium hydroxide in halogen-free flame retardant EVA/MH/MWNT nanocomposites, *Polym. Degrad. Stab.*, *94*, 751–756.
88. Song, X. H., Liu, S., Gan, Z. Y., Lv, Q., Cao, H., Yan, H. (2009). Controllable fabrication of carbon nanotube–polymer hybrid thin film for strain sensing, *Microelectron. Eng.*, *86*, 2330–2333.
89. Vemuru, S. M., Wahi, R., Nagarajaiah, S., Ajayan, P. M. (2009). Strain sensing using a multiwalled carbon nanotube film, *J. Strain Anal. Eng. Des.*, *44*, 555–562.
90. Chen, C. L., Lopez, E., Jung, Y. J., Muftu, S., Selvarasah, S., Dokmeci, M. R. (2008). Mechanical and electrical evaluation of parylene-C encapsulated carbon nanotube networks on a flexible substrate, *Appl. Phys. Lett.*, *93*, 093109.
91. Kang, I. P., Schulz, M. J., Kim, J. H., Shanov, V., Shi, D. L. (2006). A carbon nanotube strain sensor for structural health monitoring, *Smart Mater. Struct.*, *15*, 737–748.
92. Baughman, R. H., Zakhidov, A. A., de Heer, W. A. (2002). Carbon nanotubes: the route toward applications, *Science*, *297*, 787–792.
93. Baughman, R. H., Cui, C. X., Zakhidov, A. A., Iqbal, Z., Barisci, J. N., Spinks, G. M., Wallace, G. G., Mazzoldi, A., De Rossi, D., Rinzler, A. G., Jaschinski, O., Roth, S., Kertesz, M. (1999). Carbon nanotube actuators, *Science*, *284*, 1340–1344.
94. Yun, Y. H., Shanov, V., Schulz, M. J., Narasimhadevara, S., Subramaniam, S., Hurd, D., Boerio, F. J. (2005). Development of novel single-wall carbon nanotube–epoxy composite ply actuators, *Smart Mater. Struct.*, *14*, 1526–1532.
95. Levitsky, I. A., Kanelos, P., Euler, W. B. (2004). Electromechanical actuation of composite material from carbon nanotubes and ionomeric polymer, *J. Chem. Phys.*, *121*, 1058–1065.
96. Levitsky, I. A., Kanelos, P. T., Woodbury, D. S., Euler, W. B. (2006). Photoactuation from a carbon nanotube–Nafion bilayer composite, *J. Phys. Chem. B*, *110*, 9421–9425.
97. Yun, S., Kim, J. (2006). Multiwalled-carbon nanotubes and polyaniline coating on electro-active paper for bending actuator, *J. Phys. D*, *39*, 2580–2586.
98. Yun, S. Y., Kim, J., Ounaies, Z. (2006). Single-walled carbon nanotube/polyaniline coated cellulose based electro-active paper (EAPap) as hybrid actuator, *Smart Mater. Struct.*, *15*, N61–N65.
99. Landi, B. J., Raffaelle, R. P., Heben, M. J., Alleman, J. L., VanDerveer, W., Gennett, T. (2002). Single wall carbon nanotube–Nafion composite actuators, *Nano Lett.*, *2*, 1329–1332.
100. Lee, D. Y., Kim, K. J., Heo, S., Lee, M. H., Kim, B. Y (2006). In: Nakamura, T., Yamashita, K., Neo, M.Eds., Bioceramics 18, Parts 1 and 2, Vols. *309–311*, Trans Tech Publications Ltd., Zurich-Uetikon, pp. 593–596.
101. Dai, C. A., Hsiao, C. C., Weng, S. C., Kao, A. C., Liu, C. P., Tsai, W. B., Chen, W. S., Liu, W. M., Shih, W. P., Ma, C. C. (2009). A membrane actuator based on an ionic polymer network and carbon nanotubes: the synergy of ionic transport and mechanical properties, *Smart Mater. Struct.*, *18*, 085016.
102. Du, F. P., Tang, C. Y., Xie, X. L., Zhou, X. P., Tan, L. (2009). Carbon nanotube enhanced gripping in polymer-based actuators, *J. Phys. Chem. C*, *113*, 7223–7226.
103. Tahhan, M., Truong, V. T., Spinks, G. M., Wallace, G. G. (2003). Carbon nanotube and polyaniline composite actuators, *Smart Mater. Struct.*, *12*, 626–632.
104. Zhang, S. H., Zhang, N. Y., Huang, C., Ren, K. L., Zhang, Q. M. (2005). Microstructure and electromechanical properties of carbon nanotube/poly(vinylidene fluoride–trifluoroethylene–chlorofluoroethylene) composites, *Adv. Mater.*, *17*, 1897–1901.
105. Laxminarayana, K., Jalili, N. (2005). Functional nanotube-based textiles: pathway to next generation fabrics with enhanced sensing capabilities, *Text. Res. J.*, *75*, 670–680.
106. Sohn, K., Shin, S. R., Park, S. J., Kim, S. J., Yi, B. J., Han, S. Y., Kim, S. I. (2007). Hysteresis in a carbon nanotube based electroactive polymer microfiber actuator: numerical modeling, *J. Nanosci. Nanotechnol.*, *7*, 3974–3979.
107. Spinks, G. A., Shin, S. R., Wallace, G. G., Whitten, P. G., Kim, I. Y., Kim, S. I., Kim, S. J. (2007). A novel "dual mode" actuation in chitosan/polyaniline/carbon nanotube fibers, *Sens. Actuators B*, *121*, 616–621.
108. Deshmukh, S., Ounaies, Z. (2009). Single walled carbon nanotube (SWNT)–polyimide nanocomposites as electrostrictive materials, *Sens. Actuators A*, *155*, 246–252.

109. Kang, J. H., Park, C., Gaik, S. J., Lowther, S. E., Harrison, J. S. (2006). The effect of single wall carbon nanotubes on the dipole orientation and piezoelectric properties of polymeric nanocomposites, *Nano*, *1*, 77–85.
110. Lee, D. Y., Park, I. S., Lee, M. H., Kim, K. J., Heo, S. (2007). Ionic polymer–metal composite bending actuator loaded with multi-walled carbon nanotubes, *Sens. Actuators A*, *133*, 117–127.
111. Courty, S., Mine, J., Tajbakhsh, A. R., Terentjev, E. M. (2003). Nematic elastomers with aligned carbon nanotubes: new electromechanical actuators, *Europhys. Lett.*, *64*, 654–660.
112. Fernberg, P., Nilsson, G., Joffe, R. (2009). Piezoresistive performance of long-fiber composites with carbon nanotube doped matrix, *J. Intell. Mater. Syst. Struct.*, *20*, 1017–1023.
113. Nofar, M., Hoa, S. V., Pugh, M. D. (2009). Failure detection and monitoring in polymer matrix composites subjected to static and dynamic loads using carbon nanotube networks, *Compos. Sci. Technol.*, *69*, 1599–1606.
114. Gao, L. M., Thostenson, E. T., Zhang, Z., Chou, T. W. (2009). Sensing of damage mechanisms in fiber-reinforced composites under cyclic loading using carbon nanotubes, *Adv. Funct. Mater.*, *19*, 123–130.
115. Thostenson, E. T., Chou, T. W. (2008). Real-time *in situ* sensing of damage evolution in advanced fiber composites using carbon nanotube networks, *Nanotechnology*, *19*, 215713.
116. Hu, N., Karube, Y., Yan, C., Masuda, Z., Fukunaga, H. (2008). Tunneling effect in a polymer/carbon nanotube nanocomposite strain sensor, *Acta Mater.*, *56*, 2929–2936.
117. Pham, G. T., Park, Y. B., Liang, Z., Zhang, C., Wang, B. (2008). Processing and modeling of conductive thermoplastic/carbon nanotube films for strain sensing, *Compos. Part B*, *39*, 209–216.
118. Kang, J. H., Park, C., Scholl, J. A., Brazin, A. H., Holloway, N. M., High, J. W., Lowther, S. E., Harrison, J. S. (2009). Piezoresistive characteristics of single wall carbon nanotube/polyimide nanocomposites, *J. Polym. Sci. Polym. Phys.*, *47*, 994–1003.
119. Hatami, K., Grady, B. P., Ulmer, M. C. (2009). Sensor-enabled geosynthetics: use of conducting carbon networks as geosynthetic sensors, *J. Geotech. Geoenviron. Eng.*, *135*, 863–874.
120. Boeger, L., Wichmann, M. H. G., Meyer, L. O., Schulte, K. (2008). Load and health monitoring in glass fibre reinforced composites with an electrically conductive nanocomposite epoxy matrix, *Compos. Sci. Technol.*, *68*, 1886–1894.
121. Park, M., Kim, H., Youngblood, J. P. (2008). Strain-dependent electrical resistance of multi-walled carbon nanotube/polymer composite films, *Nanotechnology*, *19*, 055705.
122. Zhang, R., Baxendale, M., Peijs, T. (2007). Universal resistivity–strain dependence of carbon nanotube/polymer composites, *Phys. Rev. B*, *76*, 195433.
123. Wichmann, M. H. G., Buschhorn, S. T., Boger, L., Adelung, R., Schulte, K. (2008). Direction sensitive bending sensors based on multi-wall carbon nanotube/epoxy nanocomposites, *Nanotechnology*, *19*, 475503.
124. Lu, J. B., Kumar, B., Castro, M., Feller, J. F. (2009). Vapour sensing with conductive polymer nanocomposites (CPC): polycarbonate–carbon nanotubes transducers with hierarchical structure processed by spray layer by layer, *Sens. Actuators B*, *140*, 451–460.
125. Shang, S. M., Li, L., Yang, X. M., Wei, Y. Y. (2009). Polymethylmethacrylate–carbon nanotubes composites prepared by microemulsion polymerization for gas sensor, *Compos. Sci. Technol.*, *69*, 1156–1159.
126. Li, Q., Li, L., Yan, G. P., Jiang, D. B., Cheng, Z. Y. (2009). Polystyrene–carbon nanotubes composites prepared by microemulsion polymerization for vapour sensors, *Polym. Polym. Compos.*, *17*, 557–561.
127. Zhang, B., Dong, X. M., Fu, R. M., Zhao, B., Zhang, M. Q. (2008). The sensibility of the composites fabricated from polystyrene filling multi-walled carbon nanotubes for mixed vapors, *Compos. Sci. Technol.*, *68*, 1357–1362.
128. Knite, M., Ozols, K., Zavickis, J., Tupureina, V., Klemenoks, I., Orlovs, R. (2009). Elastomer–carbon nanotube composites as prospective multifunctional sensing materials, *J. Nanosci. Nanotechnol.*, *9*, 3587–3592.
129. Ilcham, A., Srisurichan, A., Soottitantawat, A., Charinpanitkul, T. (2009). Dispersion of multi-walled carbon nanotubes in poly(*p*-phenylene) thin films and their electrical characteristics, *Particuology*, *7*, 403–407.

130. Kobashi, K., Villmow, T., Andres, T., Haussler, L., Poetschke, P. (2009). Investigation of liquid sensing mechanism of poly(lactic acid)/multi-walled carbon nanotube composite films, *Smart Mater. Struct.*, *18*, 035008.
131. Kobashi, K., Villmow, T., Andres, T., Pötschke, P. (2008). Liquid sensing of melt-processed poly(lactic acid)/multi-walled carbon nanotube composite films, *Sens. Actuators B*, *134*, 787–795.
132. Lala, N. L., Thavasi, V., Ramakrishna, S. (2009). Preparation of surface adsorbed and impregnated multi-walled carbon nanotube/nylon-6 nanofiber composites and investigation of their gas sensing ability, *Sensors*, *9*, 86–101.
133. Lim, C., Choi, W. I., Choi, E. S., Choi, K., Ahn, S., Lee, J., Choi, S. J., Park, B., Lee, S. B. (2008). Toxic-gas-sensing characteristics of flexible carbon-nanotube-network thin-film devices fabricated on elastomer substrates, *J. Korean Phys. Soc.*, *53*, 2039–2044.
134. Lachman, N., Bartholome, C., Miaudet, P., Maugey, M., Poulin, P., Wagner, H. D. (2009). Raman response of carbon nanotube/PVA fibers under strain, *J. Phys. Chem. C*, *113*, 4751–4754.
135. Lucas, M., Young, R. J. (2007). Unique identification of single-walled carbon nanotubes in composites, *Compos. Sci. Technol.*, *67*, 2135–2149.
136. Sureeptanapas, P., Young, R. J. (2009). SWNT composite coatings as a strain sensor on glass fibres in model epoxy composites, *Compos. Sci. Technol.*, *69*, 1547–1552.
137. Barber, A. H., Zhao, Q., Wagner, H. D., Baillie, C. A. (2004). Characterization of E-glass–polypropylene interfaces using carbon nanotubes as strain sensors, *Compos. Sci. Technol.*, *64*, 1915–1919.
138. Zhao, Q., Frogley, M. D., Wagner, H. D. (2002). Direction-sensitive strain-mapping with carbon nanotube sensors, *Compos. Sci. Technol.*, *62*, 147–150.
139. Su, P. G., Lee, C. T., Chou, C. Y., Cheng, K. H., Chuang, Y. S. (2009). Fabrication of flexible NO_2 sensors by layer-by-layer self-assembly of multi-walled carbon nanotubes and their gas sensing properties, *Sens. Actuators B*, *139*, 488–493.
140. Wu, R. J., Huang, Y. C., Yu, M. R., Lin, T. H., Hung, S. L. (2008). Application of *m*-CNTs/$NaClO_4$/ Ppy to a fast response, room working temperature ethanol sensor, *Sens. Actuators B*, *134*, 213–218.
141. Wanna, Y., Srisukhumbowornchai, N., Tuantranont, A., Wisitsoraat, A., Thavarungkul, N., Singjai, P. (2006). The effect of carbon nanotube dispersion on CO gas sensing characteristics of polyaniline gas sensor, *J. Nanosci. Nanotechnol.*, *6*, 3893–3896.
142. Li, Y., Wang, H. C., Cao, X. H., Yuan, M. Y., Yang, M. J. (2008). A composite of polyelectrolyte-grafted multi-walled carbon nanotubes and *in situ* polymerized polyaniline for the detection of low concentration triethylamine vapor, *Nanotechnology*, *19*, 015503.
143. Vieira, S. M. C., Beecher, P., Haneef, I., Udrea, F., Milne, W. I., Namboothiry, M. A. G., Carroll, D. L., Park, J., Maeng, S. (2007). Use of nanocomposites to increase electrical "gain" in chemical sensors, *Appl. Phys. Lett.*, *91*, 203111.
144. Van Hieu, N., Dung, N. Q., Tam, P. D., Trung, T., Chien, N. D. (2009). Thin film polypyrrole/ SWCNTs nanocomposites-based NH_3 sensor operated at room temperature, *Sens. Actuators B*, *140*, 500–507.
145. He, L. F., Jia, Y., Meng, F. L., Li, M. Q., Liu, J. H. (2009). Gas sensors for ammonia detection based on polyaniline-coated multi-wall carbon nanotubes, *Mater. Sci. Eng. B*, *163*, 76–81.
146. Yoo, K. P., Kwon, K. H., Min, N. K., Lee, M. J., Lee, C. J. (2009). Effects of O_2 plasma treatment on NH_3 sensing characteristics of multiwall carbon nanotube/polyaniline composite films, *Sens. Actuators B*, *143*, 333–340.
147. Ferrer-Anglada, N., Kaempgen, M., Roth, S. (2006). Transparent and flexible carbon nanotube/ polypyrrole and carbon nanotube/polyaniline pH sensors, *Phys. Status Solidi B*, *243*, 3519–3523.
148. Kaempgen, M., Roth, S. (2006). Transparent and flexible carbon nanotube/polyaniline pH sensors, *J. Electroanal. Chem.*, *586*, 72–76.
149. Loh, K. J., Lynch, J. P., Kotov, N. A. (2008). Passive wireless sensing using SWNT-based multifunctional thin film patches, *Int. J. Appl. Electromagn. Mech.*, *28*, 87–94.
150. Loh, K. J., Kim, J., Lynch, J. P., Kam, N. W. S., Kotov, N. A. (2007). Multifunctional layer-by-layer carbon nanotube–polyelectrolyte thin films for strain and corrosion sensing, *Smart Mater. Struct.*, *16*, 429–438.
151. Niu, L., Luo, Y. L., Li, Z. Q. (2007). A highly selective chemical gas sensor based on functionalization of multi-walled carbon nanotubes with poly(ethylene glycol), *Sens. Actuators B*, *126*, 361–367.

152. Zhang, T., Mubeen, S., Bekyarova, E., Yoo, B. Y., Haddon, R. C., Myung, N. V., Deshusses, M. A. (2007). Poly(*m*-aminobenzene sulfonic acid) functionalized single-walled carbon nanotubes based gas sensor, *Nanotechnology*, *18*, 165504.
153. Yang, L. Q., Setyowati, K., Li, A., Gong, S. Q., Chen, J. (2008). Reversible infrared actuation of carbon nanotube–liquid crystalline elastomer nanocomposites, *Adv. Mater.*, *20*, 2271–2275.
154. Najeeb, C. K., Lee, J. H., Chang, J., Kang, W. S., Kim, J. H. (2009). Ultra fast UV-photo detector based on single-walled carbon nanotube/PEDOT–PSS composites, *J. Nanosci. Nanotechnol.*, *9*, 6928–6933.
155. Aliev, A. E. (2008). Bolometric detector on the basis of single-wall carbon nanotube/polymer composite, *Infrared Phys. Technol.*, *51*, 541–545.
156. Ahir, S. V., Terentjev, E. M. (2005). Photomechanical actuation in polymer–nanotube composites, *Nat. Mater.*, *4*, 491–495.
157. Ahir, S. V., Terentjev, E. M. (2006). Fast relaxation of carbon nanotubes in polymer composite actuators, *Phys. Rev. Lett.*, *96*, 133902.
158. Ahir, S. V., Squires, A. M., Tajbakhsh, A. R., Terentjev, E. M. (2006). Infrared actuation in aligned polymer–nanotube composites, *Phys. Rev. B*, *73*, 085420.
159. Vaia, R. (2005). Nanocomposites: remote-controlled actuators, *Nat. Mater.*, *4*, 429–430.
160. Lu, S. X., Ahir, S. V., Terentjev, E. M., Panchapakesan, B. (2007). Alignment dependent mechanical responses of carbon nanotubes to light, *Appl. Phys. Lett.*, *91*, 103106.
161. Lu, S. X., Panchapakesan, B. (2007). Photomechanical responses of carbon nanotube/polymer actuators, *Nanotechnology*, *18*, 305502.
162. Sun, L. Y., Gibson, R. F., Gordaninejad, F., Suhr, J. (2009). Energy absorption capability of nanocomposites: a review, *Compos. Sci. Technol.*, *69*, 2392–2409.
163. Auad, M. L., Mosiewicki, M. A., Uzunpinar, C., Williams, R. J. J. (2009). Single-wall carbon nanotubes/epoxy elastomers exhibiting high damping capacity in an extended temperature range, *Compos. Sci. Technol.*, *69*, 1088–1092.
164. Suhr, J., Koratkar, N. A. (2008). Energy dissipation in carbon nanotube composites: a review, *J. Mater. Sci.*, *43*, 4370–4382.
165. Koratkar, N. A., Suhr, J., Joshi, A., Kane, R. S., Schadler, L. S., Ajayan, P. M., Bartolucci, S. (2005). Characterizing energy dissipation in single-walled carbon nanotube polycarbonate composites, *Appl. Phys. Lett.*, *87*, 063102.
166. Ajayan, P. M., Suhr, J., Koratkar, N. (2006). Utilizing interfaces in carbon nanotube reinforced polymer composites for structural damping, *J. Mater. Sci.*, *41*, 7824–7829.
167. Rajoria, H., Jalili, N. (2005). Passive vibration damping enhancement using carbon nanotube–epoxy reinforced composites, *Compos. Sci. Technol.*, *65*, 2079–2093.
168. Suhr, J., Koratkar, N. (2006). Effect of pre-strain on interfacial friction damping in carbon nanotube polymer composites, *J. Nanosci. Nanotechnol.*, *6*, 483–486.
169. Verdejo, R., Stampfli, R., Alvarez-Lainez, M., Mourad, S., Rodriguez-Perez, M. A., Bruhwiler, P. A., Shaffer, M. (2009). Enhanced acoustic damping in flexible polyurethane foams filled with carbon nanotubes, *Compos. Sci. Technol.*, *69*, 1564–1569.

GLOSSARY

Addition polymerization: A type of polymerization where only monomers can add to a growing chain, as opposed to step-growth polymerization where monomers and oligomers can add to a growing chain.

Amorphous: A term that describes a substance where the arrangement of atoms has no long-range order in any direction. All liquids and gases are amorphous, and synthetic polymers are either totally amorphous or partially amorphous.

Armchair tubes: The case where $n = 0$ (see Equation 2.1 and 2.2 and associated discussion) for single-walled carbon nanotubes; the orientation of the carbon hexagons is 30° different from the orientation of the long axes but hexagons on opposite sides of the tube are parallel to one another.

Chirality: A term that indicates that a particular structure has a counterpart that differs only in the fact that the two are mirror images of one another. In nanotubes, all nanotubes except armchair and zigzag tubes are chiral.

Crystalline material: A substance where the arrangement of atoms has long range order. If the position of a single atom is given for a perfect crystal, then the position of every other atom in the material is known.

Complex Viscosity: In an oscillatory experiment done in shear, the complex viscosity is defined as the square root of the sum of the squares of the in-phase and out-of-phase viscosities, i.e. $(\eta^*)^2 = (\eta')^2 + (\eta'')^2$. In terms of the storage and loss modulus, $\eta' = G''/\omega$ and $\eta'' = G'/\omega$.

Conductivity (electrical): The ability of a material to carry electrical current. Bulk conductivity has SI units of Siemens/meter. The typical electrical conductivity of a metal is on the order of 10^6 S/m while that of an electrically insulative polymer is on the order of 10^{-16} S/m. Surface conductivity measures the ability of a surface to carry charge, and is normally expressed in terms of surface resistance, having units of Ohms/square.

Conductivity (thermal): The ability of a material to carry heat. Thermal conductivity has SI units of W/m K (or W/m °C). The typical thermal conductivity of a metal is on the order of 10^2 W/m K while that of an electrically insulative polymer is on the order of 10^{-1} W/m K.

D-Band: A movement of the atoms in pure carbon materials due to sp^3 hybridization. Important in Raman spectroscopy.

Diffusion Constant: A parameter that is a measure of how fast a molecule is able to move when the driving force is movement from a region of high concentration to region of low concentration.

Electrostrictive Material: See Piezoelectric material.

Extruder: A pump/mixer that works by turning a screw inside a barrel. There are two common types, a single screw extruder where only one screw is used and a twin-screw

Carbon Nanotube–Polymer Composites: Manufacture, Properties, and Applications, First Edition.
Brian P. Grady.

extruder where two intermeshing screws are used. The latter is superior for dispersing solids.

Enantiomers: When an atom has four different chemical fragments bonded to it, there are two different possible compounds depending on the way in which the fragments are bonded; the two structures are mirror images of one another. In polymers having the general structure $-[CH_2CHX]-$ where X is something other than hydrogen, each C connected to the X is enantiomeric. This characteristic leads to the tacticity classifications of isotactic, syndiotactic and atactic.

(Critical) Entanglement length: A critical length for rods or strings that signifies when removing one string or rod from a bundle with large size becomes very difficult because of multiple entanglements. In general, the persistence length (stiffness) scales with the entanglement length although other factors do contribute to the entanglement length. The viscosity changes its dependence on length at the critical entanglement length.

Equilibrium Melting Temperature: In polymers, the melting temperature depends on the crystal thickness, with higher melting temperatures for thicker crystals. The equilibrium melting temperature is the temperature at which a crystal of infinite thickness will melt.

Forests: Nanotubes that are grown with the long axis perpendicular to a flat substrate. See Figure 2.4 in the text.

Fractional Crystallinity: In a semicrystalline polymer, it is the fraction of the polymer that is crystalline. Typical values for high-density polyethylene are 60–80%, while typical values for low-density polyethylene are 25–45%.

Functionalized: A tube is functionalized if a moiety is chemically reacted with a carbon atom belonging to a carbon nanotube.

Fullerene: A pure carbon material that has an atomic arrangement based on planar carbon (sp^2 hybridization) and assumes a non-planar shape.

G-Band: A movement of the atoms in pure carbon materials due to atoms moving 180° relative to one another, in carbon nanotubes this occurs tangential to the tube. Important in Raman spectroscopy.

Glass Transition: A material that is amorphous will change from a hard glass to a molten liquid over a finite temperature range in a process termed the glass transition. The glass transition temperature is the midpoint (or some other consistent definition) of this temperature range. For a semi-crystalline material, the material usually changes from a brittle to a flexible material.

"Grafting from": A nanotube functionalized with a polymer chain that forms by a single monomer first reacting with a carbon nanotube, and then subsequent monomers adding to that monomer that is grafted to the tube.

"Grafting to": An already-formed polymer chain that is reacted with a carbon nanotube to form a grafted nanotube.

Graphite: A material that consists of graphene sheets that are layered so that each plane is parallel with one another. The distance between planes has been measured as 0.335 nm.

Graphene sheet: A plane of carbon atoms. The carbon atoms are located on the vertices of a hexagon, and the hexagons are connected together in a repeating pattern. See Figure 2.1 in the text.

Interfacial Energy or Interfacial tension: The interfacial energy is the energy necessary to create a unit area of interface between two substances with the starting condition being two separate bulk phases. The surface energy is simply the interfacial energy where the other material is a vacuum (which by definition has a surface energy of zero). As with surface tension, the term interfacial tension is generally not used for two solids. The units of interfacial energy are energy/area while interfacial tension units are usually given in force/length; of course the two are equivalent.

Interfacial Fracture Energy: A parameter that is determined in the same manner as the interfacial shear stress, but requires a model to determine its value. The units of this measure are energy/area, while that of interfacial shear stress are force/area.

Interfacial Shear Stress (or Interfacial Shear Strength): The maximum force required to pull out a filler from a matrix divided by the contact area between the filler and the matrix. This is calculated usually for fibers, by imbedding a short length of fiber into a matrix and measuring the force required to pull the fiber out. Alternatively, one can imbed a long fiber into a matrix, and then strain the matrix and measure the distribution of fiber lengths.

Lamallae: The unit cell is the fundamental building block of any crystalline material. In polymers, the number of unit cells that repeat in one direction is 10–100 times lower than that in two other directions, forming long, wide and thin objects called lamellae. The average direction of the chain axis in a unit cell is always perpendicular to the lamellar thickness. Figure 4.6 illustrates this structure.

Latex: A dispersion of polymeric spheres in water.

Linear Viscoelastic Region: In an oscillatory test, the oscillating strain is small enough so that there is no change in the material with the number of oscillating cycles. A simple way to test for this is to change the strain, if the results do not change then the system must be in the linear viscoelastic region.

Liquid crystal material: A substance that is in between crystal and amorphous materials; the atoms have translational symmetry in either one or two dimensions, but not three.

Loss Modulus: In an oscillatory experiment, the loss modulus is the portion of the response that out-of-phase with the applied strain. In tension, the symbol given is E'', in shear the symbol given is G''. This definition assumes that the strain is such so that the total stress can be described by in-phase and out-of-phase components, i.e. the sample behaves in a linear viscoelastic manner.

Melt Index: A measurement that is related to the viscosity of a polymer, but taken in such a way that a direct correspondence to a viscosity and shear rate is not possible.

Melting temperature: In a semicrystalline polymer, the temperature at which the crystals become amorphous. The material changes from a flexible material to a melt. Polymers melt over a temperature range, the melting temperature is typically defined as either the temperature at which the maximal amount of material makes this transition, or the temperature where melting begins. The latter is standard in most materials, but with polymers the difficulty in determining this temperature usually leads to the former definition.

Metallic Tubes: Tubes that do not have a threshold voltage required for electrical conduction, e.g. the current vs. voltage plot is linear to a voltage of zero. Tubes with $n\text{-}m = 3^{*}\text{integer}$ are metallic.

Modulus: The stress divided by the strain in the limit as the strain approaches zero.

Overlap Concentration: In a polymer solution, the change in viscosity increases in an approximately linear fashion on a plot of viscosity vs. log (concentration), at the overlap concentration there is a discontinuous change in slope. Molecularly, this concentration is a result where chains begin to overlap with one another, causing chain-chain entanglements.

Percolation and Percolation Threshold: Percolation is a geometric state where a path through a given phase in a two phase material can be traced from one side to the other. The percolation threshold for a given phase is the lowest concentration where such a path can be traced.

Persistence length: A parameter that describes the stiffness of a rod-like/string-like object; a higher persistence length means a more stiff rod/string.

Piezoelectric Material: A material that when strained accumulates electrical charge within the material; alternatively a material that when external charge is applied will change its shape. A piezoelectric material differs from an electrostrictive material in that the latter will not accumulate charge upon strain, but will change its strain in response to an electric field.

Polymorphism: A molecule that crystallizes into more than one type of crystal structure. The crystal structure that forms depends on the conditions used to crystallize the molecule.

Primary crystallization: For polymers, crystallization that occurs where spherulites do not impinge upon one another.

Radius of Gyration (R_g): A parameter that is used to describe the size of an individual polymer molecule. Equation 4.1 gives a mathematical definition of this parameter.

Raman spectroscopy: The intensity of scattered light is measured as a function of energy difference between the incoming and scattered light. This technique is used to characterize tube purity and quality.

RBM (Radial Breathing Mode): A movement of a nanotube due to expansion and contraction of the radius. Important in Raman spectroscopy.

Relative Crystallinity: In the case where a semicrystalline material is crystallizing with time, the relative crystallinity is the fractional crystallinity at a certain time divided by the fractional crystallinity at infinite time.

Secondary Crystallization: Crystallization that occurs at a much slower rate than primary crystallization and is due to crystallinity that occurs between two relatively closely spaced lamellae, i.e. between lamaller arms of a spherulite.

Semiconducting Tubes: Tubes that require a threshold voltage to begin conduction.

Shish-kebab: A type of morphology that has extended chains as the shish, and crystalline lamellae as the kebabs. In nanotube composites, nanotubes can serve as the shish. The nucleation density of crystallization along the nanotube axis must be much less than the lamellar thickness or a transcrystalline morphology will result.

Step-growth polymerization: A type of polymerization where non-monomers (e.g. dimers, trimers, tetramers. . .) can add to a growing chain, as opposed to addition polymerization where only monomers can add to a growing chain.

Storage Modulus: In an oscillatory experiment, the storage modulus is the portion of the response that is in-phase with the applied strain. In tension, the symbol given is E', in shear the symbol given is G'. This definition assumes that the strain is such so that the total stress can be described by in-phase and out-of-phase components, i.e. the sample behaves in a linear viscoelastic manner.

Strain: (Change in a dimension value)/(dimension value). In tension, the dimension value is length; for other geometries the expression is more complicated. There are two types of strain, true strain and engineering strain, in the former the strain is defined on an instantaneous basis (so the definition is a derivative) and in the latter the strain is defined on a starting basis (the denominator and the subtracted value in the numerator becomes the dimension value at the time when the experiment begins). As an example, in tensile tests the engineering strain is the change in (change in length)/(initial length). Note that throughout this book, only engineering properties are used unless otherwise noted.

Stress: (Force/Cross-sectional area). As with strain, there are both true stress and engineering stress, with the former being based on the instantaneous area and the latter being based on the area prior to starting the experiment. As an example, the engineering stress in a tensile test is the force at any time divided by initial area. Note that throughout this book, only engineering properties are used unless otherwise noted.

Surface Energy or Surface Tension: The energy at the surface must be greater than the energy in the bulk [or else the material would completely sublimate (if a solid) or boil (if a liquid!)]; this excess energy per unit area is the surface energy. The term "surface tension" and "surface energy" are essentially interchangeable. The classic definition of the former which involves stretching a surface is only applicable to liquids, while the latter is applicable to both liquids and solids. The units of interfacial energy are energy/area while interfacial tension units are usually given in force/length; of course the two are equivalent.

Tacticity: A polymer that has the general form -[CH_2CHX]- where X is something other than hydrogen, can be either isotactic, syndiotactic or atactic. Isotactic materials have X all on the same side of the chain, syndiotactic materials have X on alternating sides of the chain, and atactic materials have X randomly arranged.

Transcrystallinity: A type of morphology where a cylindrical filler causes the nucleation of crystallinity so that the growth direction is perpendicular to the fiber axis and the polymer chain axis is parallel. The nucleation density must be high enough so that individual lamellae are not distinguishable; if distinguishable then a shish-kebab morphology results.

Unit Cell: The geometric structure which has a box-like shape that contains atoms within. Unit cells stacked in three-dimensional space describe the bulk arrangement of atoms of the crystal.

Zigzag Tubes: The case where $n = 0$ (see Equation 2.1 and 2.2 and associated discussion); the orientation of the carbon hexagons is along the long axes of the nanotube.

INDEX

Carbon Nanotube–Polymer Composites: Manufacture, Properties, and Applications, First Edition. Brian P. Grady.

Printed and bound by CPI Group (UK) Ltd, Croydon, CR0 4YY

16/09/2026

14963456-0004